# *Across This Land*

*CREATING THE*
*NORTH AMERICAN LANDSCAPE*

Gregory Conniff
Edward K. Muller
David Schuyler
*Consulting Editors*

George F. Thompson
*Series Founder and Director*

Published in cooperation with
the Center for American Places,
Santa Fe, New Mexico,
and Harrisonburg, Virginia

# *Across This Land*

## A Regional Geography of the United States and Canada

*JOHN C. HUDSON*

*The Johns Hopkins University Press*
BALTIMORE AND LONDON

Printed in the United States of America on acid-free paper
9 8 7 6 5 4 3 2

The Johns Hopkins University Press
2715 North Charles Street
Baltimore, Maryland 21218-4363
www.press.jhu.edu

Library of Congress Cataloging-in-Publication Data

Hudson, John C.
Across this land : a regional geography of the United States and Canada / John C. Hudson.
p. cm. — (Creating the North American landscape)
Includes bibliographical references and index.
ISBN 0-8018-6567-0 (pbk.: alk. paper)
1. North America—Geography. 2. Regionalism—North America. 3. United States—Geography. 4. Regionalism—United States. 5. Canada—Geography. 6. Regionalism—Canada. I. Title. II. Series.
E40.5 .H83 2001
917—dc21

00-011515

A catalog record for this book is available from the British Library.

*To William and Clare*

# Contents

# Maps

# Preface

When I joined the Northwestern University faculty in the early 1970s, I spent some time looking through the geography catalog to find a course that might be fun to teach. Although I had never taught regional geography, I was intrigued by Geography C13: North America. The course had no title or description apart from the name of the continent it was supposed to cover. With plenty of room to experiment, I offered the course as a seminar, then introduced some lectures the next year. Within a few more years it had become a lecture course and was moved to a larger classroom. I began illustrating the lectures with slides and then moved to an auditorium. I have taught "North America" every spring quarter since 1973. This book is the product of teaching that course.

Regional geography begins with the premise that it is possible to gain the sense of a place by reading about it. It is more a teaching field than a research subject, and most regional geography books—including this one—have come from efforts to teach students. It is also a subject that tends to be long on facts. Although the emphasis on factual information sometimes invites comparisons with a trivia contest, regional geography is designed to be precisely the opposite of geographical trivia. Unconnected facts are simply that and lend themselves to little other than memorization. The purpose of regional geography is to offer a framework for those facts and an interpretation of their relevance, so that they may take on significance for the reader.

It was long believed that the relevant facts about a place were given and that any well-trained geographer would know how to go about acquiring them. The facts were commonly cataloged in a sequence, beginning with bedrock geology and proceeding through climate, soils, and vegetation. This was followed by a description of all of the forms of economic activity, beginning with the primary ones (agriculture, forestry, mining, and fishing) and then moving to the industries that fashioned those products into manufactured goods (the secondary activities). The catalog continued through the service industries, then turned to demographics and urbanization and typically culminated in a discussion of a region's problems.

The strength of such an approach is that it offers a framework in which the topics can be organized for any region. Its most obvious weakness is that it is difficult to weave through such a framework a story line that will make the facts come alive. Some authors take a thematic approach, surveying a whole continent, topic by topic, but the essence of regional geography is lost in the process.

Another strategy is to adopt a historical perspective, but this makes it necessary to repeat the entire catalog for every time period.

In recent years a more vexing problem has emerged regarding the facts of geography. Increased separation between the physical and human dimensions of geographical research and the continued evolution of new methodologies in the social sciences and humanities have rendered null and void those comfortable, old agreements about the relevant facts. The eclipse of regional geography has been one casualty of this evolution, even as the public decries its loss in calls for reinstating geography in the school curriculum. Much of the "new geography" that has appeared since the 1960s has not yet found its place in a regional approach.

These problems notwithstanding, I believe that the basic premise of regional geography remains sound and that the subject will be as useful in the 21st century as it was in the 20th. New data, new theories, new terminology, and new perspectives from geography and its related disciplines are waiting to be included. My own biases about the best way to write regional geography include a belief that books on the subject have suffered not from being too place-specific, but rather from being insufficiently so. If the geographical approach to assimilating a worldview is to be successful, then I believe the greatest care must be given to organizing the myriad details comprising the end product.

Regions are admirably suited to this pedagogical task because they are capable of division and subdivision down to almost any scale. The definition of a region involves a (usually implicit) thesis that proposes what is important about a place. All of the chapters in *Across This Land* are organized around regions. The chapters aggregate into larger regions, and they also are subdivided into smaller ones as indicated by the subheadings. The sequence in most cases follows a geographical order that is adopted in order to maximize the value of a regional approach.

Regionalizations of North America generally begin with geology. There are eight major subdivisions that form a rough symmetry around the core of ancient rocks, the Canadian Shield, which is exposed at the surface in the center of the continent (map P.1). Surrounding this continental core, or craton, is a stable platform, undeformed by mountain-building forces, known as the Central Lowland on the east and the Great Plains on the west. The stable platform, in turn, is fringed by two mountainous belts, the Appalachians on the east and the Rocky Mountains on the west. The Coastal Plain is a margin of more recent sedimentary rocks, derived from both land and sea, bordering the Appalachians on the east. The two major divisions of the Far West are the Basin and Range section of low, widely separated mountain ranges west of the Rockies and the Sierra Nevada–Cascade region of higher mountains that forms the Pacific border.

Canada and the United States differ in terms of the relative areas covered by these eight regions (map P.2). The Shield forms much of Canada's interior but only a small part of that of the United States. The Central Lowland, the Coastal Plain, and the Basin and Range lie primarily within the United States. Numerous further subdivisions, based on geologic structures, surface materials, or to-

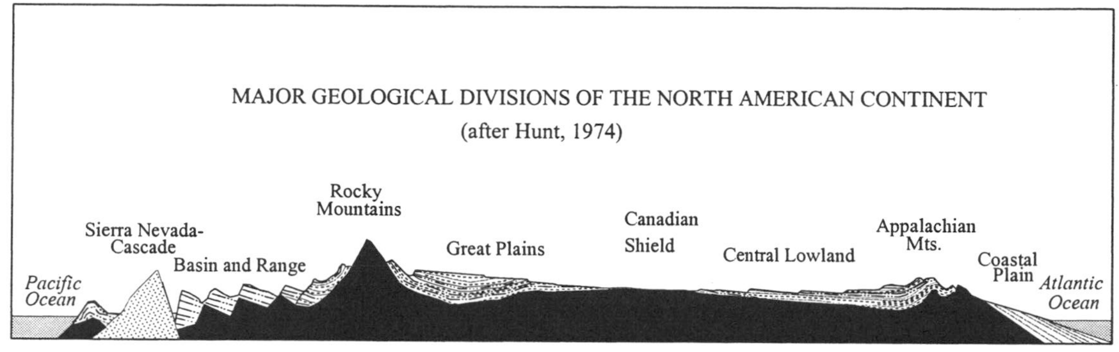


MAP P.1

pography, are possible as well. The Hudson Bay Lowland, the Colorado Plateau, and the Ozark-Ouachita regions are examples of smaller subdivisions; each has a unique character. Because the record of human occupancy is nearly everywhere influenced by the physical environment, cultural regions—especially those based on a long presence by a given cultural group—tend to follow natural boundaries to some extent, although they are by no means determined by them.

Political boundaries constitute a second fundamental regionalization. Politics is considered essential in the writing of history; yet it rarely plays a central role in regional geographies. Geographers emphasize natural regions but rarely focus on political regions (countries, states, provinces) as primary divisions of

MAP P.2

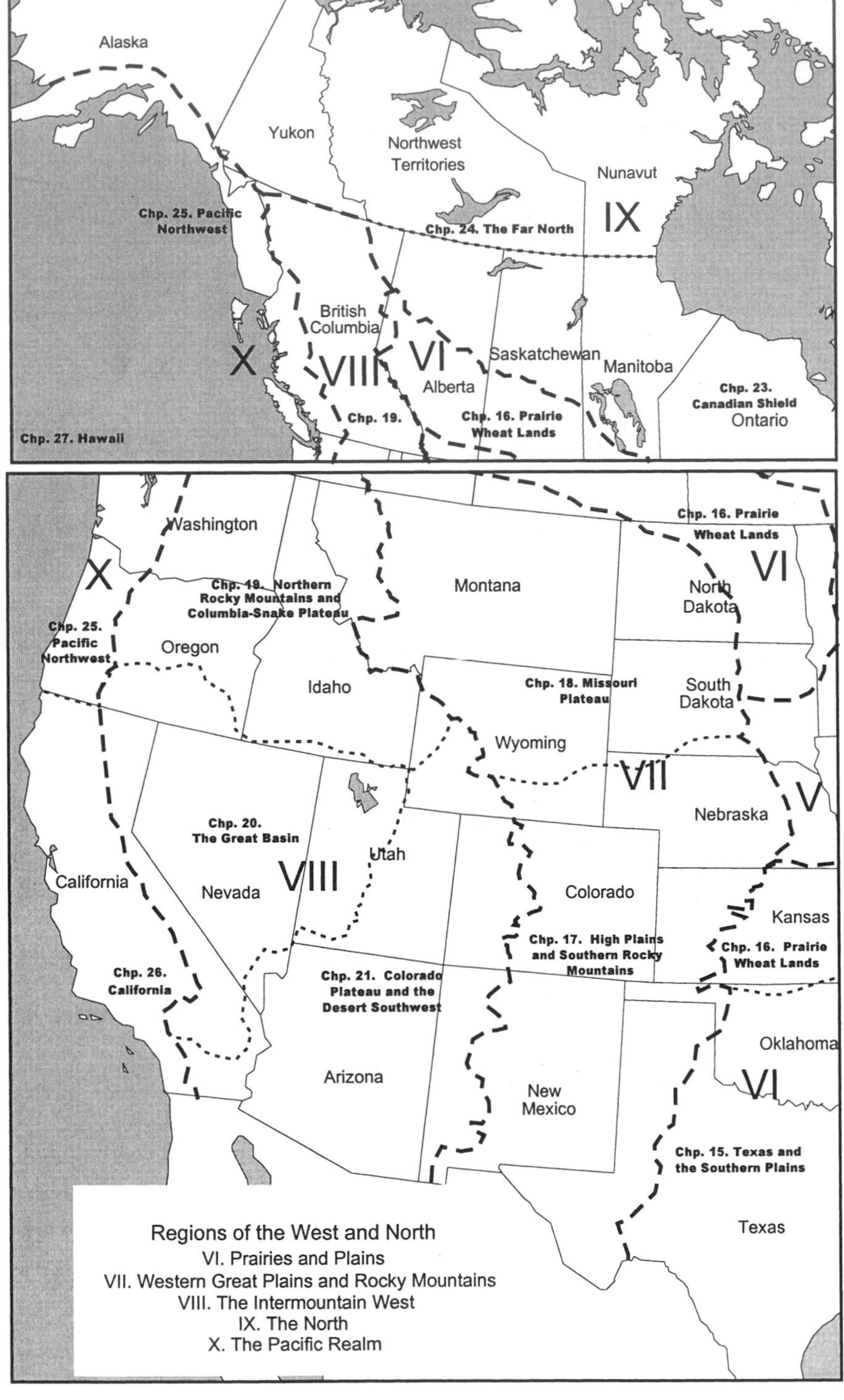
Alaska
Yukon
Northwest Territories
Nunavut
IX
Chp. 25. Pacific Northwest
Chp. 24. The Far North
British Columbia
X
VIII
VI
Saskatchewan
Manitoba
Alberta
Chp. 23. Canadian Shield
Ontario
Chp. 19.
Chp. 16. Prairie Wheat Lands
Chp. 27. Hawaii
Washington
Chp. 16. Prairie Wheat Lands
VI
X
Chp. 19. Northern Rocky Mountains and Columbia-Snake Plateau
Montana
North Dakota
Chp. 25. Pacific Northwest
Oregon
Chp. 18. Missouri Plateau
South Dakota
Idaho
Wyoming
VII
Nebraska
V
Chp. 20. The Great Basin
Utah
VIII
California
Nevada
Colorado
Kansas
Chp. 17. High Plains and Southern Rocky Mountains
Chp. 16. Prairie Wheat Lands
Chp. 26. California
Chp. 21. Colorado Plateau and the Desert Southwest
Oklahoma
Arizona
VI
New Mexico
Chp. 15. Texas and the Southern Plains
Texas
Regions of the West and North
VI. Prairies and Plains
VII. Western Great Plains and Rocky Mountains
VIII. The Intermountain West
IX. The North
X. The Pacific Realm

MAP P.3

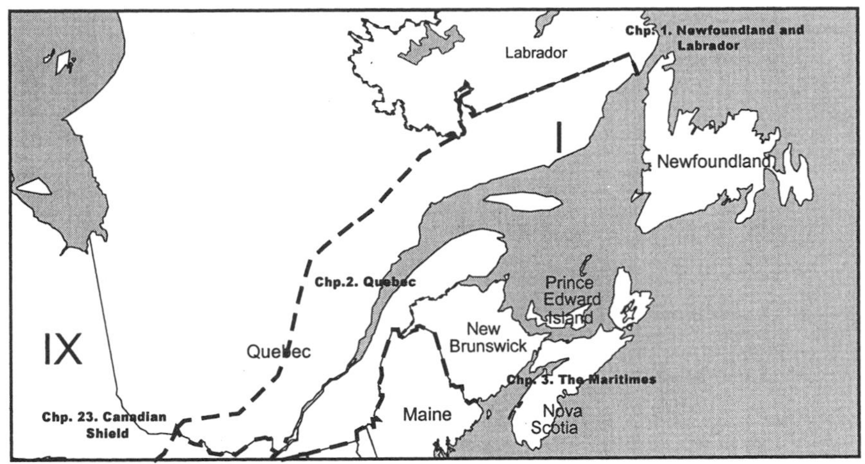
Chp. 1. Newfoundland and Labrador
Labrador
I
Newfoundland
Chp.2. Quebec
Prince Edward Island
New Brunswick
IX
Quebec
Chp. 3. The Maritimes
Chp. 23. Canadian Shield
Maine
Nova Scotia

IX
Quebec
Chp. 23. Canadian Shield
Maine
Chp. 22. Upper GreatLakes
Vt.
N. H.
Ontario
Minnesota
Chp. 5. New York and Ontario
Chp. 4. New England
Mass.
Wisconsin
Michigan
New York
II
Conn.
R. I.
Chp. 14. Lower Great Lakes
Iowa
Pennsylvania
N. J.
V
Chp. 6. Middle Atlantic
Chp. 13. The Corn Belt
Ohio
Md.
Illinois
Indiana
Del.
West Virginia
Chp. 7. Southern Appalachians
Virginia
VI
Missouri
Kentucky
Chp. 8. Interior Low Plateaus
North Carolina
III
Chp. 9. Ozarks and Ouachitas
Tennessee
Chp. 10. Southeastern Piedmont and Coastal Plain
IV
Arkansas
South Carolina
Mississippi
Georgia
Alabama
Chp. 12. Gulf Coastal Plain and Alluvial Mississippi Valley
Louisiana
Chp. 11. The Florida Peninsula
Florida
Regions of the East and South
I. Atlantic Canada and Quebec
II. The Northeast
III. The Upland South
IV. The Lowland South
V. The Middle West

territory. Politics plays a larger role in *Across This Land* because I believe it is not the mere presence of a given physical environment, various natural resources, or certain cultural groups that define regional character; rather, it is actions undertaken for a purpose, by groups and individuals, that have shaped the distinct landscapes that each generation leaves in modified form to the next. This book focuses on places and how they have come to acquire their present geographies. Millions of years of geological evolution, ten thousand years of human presence, and five hundred years of European influence telescope into the two centuries of intensive development that surrounds us.

What is the measure of significance in regional geography? The answer can only be suggestive. Area or territory is one measure, but it is not an absolute one; if it was, much of this book would focus on the uninhabited North, which is the largest subdivision of the continent in terms of area. Neither is population the sole measure of significance: that would mean that most of this book would be about large cities and that the North would scarcely be mentioned. Those who expect scenic or picturesque areas to be emphasized at the expense of the less attractive will be disappointed, although I hope readers will find that their interest in all places increases as a result of learning more about them.

*Across This Land* covers the geography of the United States and Canada—separately in places, jointly in others—but it barely mentions Mexico, North America's third major country. The restriction is based both on limitations of the author's experience and on a traditional agreement among geographers about the division of the continent. The United States and Canada formerly were grouped together as Anglo-America. It takes nothing away from the important contribution Anglos have made to both countries to admit that such a term is culturally biased in the extreme and to note with approval that it has fallen into almost total disuse. Mexico belonged in the other major realm, Latin America, a term that, for better or worse, remains in common usage. In our NAFTA-conscious age, Mexico probably ought to be considered jointly with the United States and Canada, but I must respectfully yield to others, who know Mexico far better than I do, to take that step.

The twenty-seven chapters that follow are grouped by region (map P.3), beginning with six on the Northeast that are delimited mainly in terms of political boundaries. The next six chapters, on the South, follow its conventional regionalization into Upland and Lowland segments. Two chapters on the Middle West make a distinction between the agriculturally oriented Corn Belt and the more industrial, Lower Great Lakes portions of the region. The eastern Great Plains states and the Prairie Provinces are covered in the next two chapters.

I chose to discuss the western Great Plains together with the Rocky Mountains for both geological and historical reasons, tying both the High Plains and the Missouri Plateau to the Rockies. The mountains are thus split along the continental divide, with the northern Rockies assigned to the chapter on the Columbia-Snake Plateau and the southern Rockies west of the divide to two chapters, on the Great Basin and the Colorado Plateau–Desert Southwest. The book's last six chapters consist of three on the North, beginning with the Upper

Great Lakes region, then following north to Alaska; and three on the Pacific realm, beginning in coastal Alaska, tracing south to California, and concluding with a short chapter on Hawaii.



References are grouped at the end of each chapter for the purpose of convenient access. The individual chapter lists are by no means exhaustive, and the titles selected include mainly books rather than articles and book chapters. A very useful general guide to the literature, which has the virtue of regional organization itself, is the bibliography by Conzen, Rumney, and Wynn (1993). More general works, treating the entire continent or entire countries, are listed at the end of this preface.

Of the many people who have helped shape the content of *Across This Land*, special mention must be made of the late Professor R. Barry Farrell. A Canadian and a political scientist, Barry directed a successful Canadian Studies program at Northwestern and urged me to apply for a travel-study grant from the Canadian Embassy in Washington. Two such grants supported study in parts of Canada that would have been difficult for me to reach otherwise. I am grateful to the Canadian government for its financial assistance and to Barry Farrell for rekindling my interest in Canada.

I owe a great debt to a long list of geographers whose regional geographic studies form much of the basis for this book. I have acknowledged that debt in part through citing their works at the end of individual chapters. Two geographers have influenced my thinking to such an extent that chapter references alone will not suffice. I am greatly indebted to John Fraser Hart, who encouraged me to go forward on this project, contributed substance to nearly every chapter, and lent his editorial skills to the final manuscript; one could not possibly ask for more from a friend and colleague. I am also greatly indebted to D. W. Meinig, whose insights into the development of American regions are without parallel. As I wrote my way back and forth "across this land" I came to appreciate Don Meinig's innovative geographical thinking ever more deeply. I recommend Professor Meinig's books to all who would delve into the study of North American geography.

The maps and photographs in this volume are the author's own. The photographs, taken over a roughly twenty-year period, were chosen to illustrate the recurring themes of environment, settlement, and landscape that are central to every chapter. The maps, all of which were produced digitally, are intended to lend a different perspective to those same themes. I would like to thank my students, especially Ryan Baxter, Jeffrey Gray, and Brian Schretzmann, for the patience they showed teaching their professor how to make maps on the computer.

I would like to thank Kimberly Johnson, Lee Sioles, and Martha Farlow of Johns Hopkins University Press for their help. My copy editor, Lois Crum, made many useful suggestions for improving the text. George F. Thompson, president of the Center for American Places, provided critical support for the project when it was needed and expertly guided the manuscript through the publication process. To Anne, Jane, and Debby go my love and my sincerest appreciation for their encouragement.

## *References*

### REGIONAL GEOGRAPHY

Birdsall, Stephen S., John W. Florin, and Margo L. Price. *Regional Landscapes of the United States and Canada.* 5th ed. New York: John Wiley, 1999.

Hart, John Fraser. "The Highest Form of the Geographer's Art." *Annals of the Association of American Geographers* 72 (1982): 1–29.

McKnight, Tom L. *Regional Geography of the United States and Canada.* 3d ed. Englewood Cliffs, N.J.: Prentice-Hall, 2001.

Warkentin, John. *Canada: A Regional Geography.* Scarborough, Ont.: Prentice-Hall, 1997.

White, C. Langdon, and Edwin J. Foscue. *Regional Geography of Anglo-America.* 2d ed. Englewood Cliffs, N.J.: Prentice-Hall, 1954.

### PHYSICAL GEOGRAPHY

Bally, Albert W., and Allison R. Palmer, eds. *The Geology of North America: An Overview.* Boulder: Geological Society of America, 1989.

Bryson, Reid A., and F. Kenneth Hare, eds. *Climates of North America.* New York: Elsevier Scientific, 1974.

Fenneman, Nevin M. *Physiography of Western United States.* New York: McGraw-Hill, 1931.

———. *Physiography of Eastern United States.* New York: McGraw-Hill, 1938.

Foth, Henry D., and John W. Schaefer. *Soils Geography and Land Use.* New York: John Wiley, 1980.

Graf, William L. *Geomorphic Systems of North America.* Boulder: Geological Society of America, 1987.

Hunt, Charles B. *Natural Regions of the United States and Canada.* San Francisco: W. H. Freeman, 1974.

Porter, Stephen C., ed. *Late-Quaternary Environments of the United States.* Vol. 1, *The Late Pleistocene.* Minneapolis: University of Minnesota Press, 1983.

Ruddiman, W. F., and H. E. Wright, Jr. *North America and Adjacent Oceans during the Last Deglaciation.* Boulder: Geological Society of America, 1987.

Speed, R. C., ed. *Phanerozoic Evolution of North American Continent-Ocean Transitions.* Boulder: Geological Society of America, 1994.

Thornbury, Willam D. *Regional Geomorphology of the United States.* New York: John Wiley, 1965.

Wright, H. E., Jr., ed. *Late-Quaternary Environments of the United States.* Vol. 2, *The Holocene.* Minneapolis: University of Minnesota Press, 1983.

### HUMAN GEOGRAPHY

Brown, Ralph H. *Historical Geography of the United States.* New York: Harcourt, Brace, 1948.

Carver, Craig M. *American Regional Dialects: A Word Geography.* Ann Arbor: University of Michigan Press, 1987.

Conzen, Michael P., ed. *The Making of the American Landscape.* Boston: Unwin Hyman, 1990.

Conzen, Michael P., Thomas A. Rumney, and Graeme Wynn, eds. *A Scholar's Guide to Geographical Writing on the American Past.* Geography Research Paper no. 235. Chicago: University of Chicago Press, 1993.

Gentilcore, R. Louis, ed. *Historical Atlas of Canada.* Vol. 2, *The Land Transformed, 1800–1891.* Toronto: University of Toronto Press, 1993.

Glassie, Henry. *Pattern in the Material Folk Culture of the Eastern United States.* Philadelphia: University of Pennsylvania Press, 1968.

Harris, R. Cole, ed. *Historical Atlas of Canada.* Vol. 1, *From the Beginning to 1800.* Toronto: University of Toronto Press, 1987.

Harris, R. Cole, and John Warkentin. *Canada before Confederation: A Study in Historical Geography.* New York: Oxford University Press, 1974.

Hart, John Fraser. *The Rural Landscape.* Baltimore: Johns Hopkins University Press, 1998.

Kerr, Donald, and Deryck W. Holdsworth, eds. *Historical Atlas of Canada.* Vol. 3, *Addressing the Twentieth Century.* Toronto: University of Toronto Press, 1990.

McCann, L. D., ed. *Heartland and Hinterland: A Geography of Canada.* 2d ed. Scarborough, Ont.: Prentice-Hall Canada, 1987.

Meinig, Donald W. *The Shaping of America: A Geographical Perspective on 500 Years of History.* Vol. 1, *Atlantic America, 1492–1800.* New Haven: Yale University Press, 1993.

———. *The Shaping of America.* Vol. 2, *Continental America, 1800–1867.* New Haven: Yale University Press, 1993.

———. *The Shaping of America.* Vol. 3, *Transcontinental America, 1850–1915.* New Haven: Yale University Press, 1998.

Reps, John W. *The Making of Urban America: A History of City Planning in the United States.* Princeton, N.J.: Princeton University Press, 1965.

———. *Cities of the American West: A History of Frontier Urban Planning.* Princeton, N.J.: Princeton University Press, 1979.

Vance, James E., Jr. *The North American Railroad: Its Origin, Evolution, and Geography.* Baltimore: Johns Hopkins University Press, 1995.

Ward, David. *Cities and Immigrants.* New York: Oxford University Press, 1971.

Wright, Louis B. *The Atlantic Frontier: Colonial American Civilization, 1607–1763.* New York: Alfred A. Knopf, 1947.

Zelinsky, Wilbur. *The Cultural Geography of the United States.* Englewood Cliffs, N.J.: Prentice-Hall, 1973.

## MAPS

*Geologic Map of the United States.* 1:2,500,000. Washington, D.C.: U.S. Geological Survey, 1974.

Hammond, Edwin H. *Classes of Land-Surface Form in the Forty-Eight States.* 1:5,000,000. Washington, D.C.: Association of American Geographers, 1964.

Küchler, A. W. *Potential Natural Vegetation of the United States.* 1:3,168,000. New York: American Geographical Society, 1964.

*The National Atlas of Canada.* 4th ed. Ottawa: Department of Energy, Mines, and Resources, 1973.

Raisz, Erwin. *Landforms of the United States.* 6th ed. Privately printed, 1957.

Thelin, Gail P., and Richard J. Pike. *Landforms of the Conterminous United States: A Digital Shaded Relief Portrayal.* Map I-2206. 1:3,500,000. Reston, Va.: U.S. Geological Survey, 1991.

*PART I*

# Atlantic Canada and Quebec

CHAPTER 1

# Newfoundland and Labrador

Newfoundland is the site of the earliest European habitation that has thus far been discovered in North America. During the 1960s archaeologists unearthed the remains of an outpost built by seafaring Norse people about A.D. 1000 at L'Anse-aux-Meadows, on Belle Isle Strait at the northern tip of Newfoundland's Northern Peninsula (map 1.1). The site is thought by some to have been the Vinland of Leif Eirikksson as described in the *Greenlander's Saga.* Trees cut at L'Anse-aux-Meadows provided timber that was shipped to the Greenland settlements. The Norse foothold in North America was weak and lasted for only a short time, although Norse pastoral settlements survived in southern Greenland until A.D. 1500. The climate had begun to grow colder by that time (because of the onset of the "Little Ice Age"), and marginal areas like Greenland were abandoned. A once-advancing zone of Norse settlement in the cold lands bordering the North Atlantic turned to one of retreat.

Norse people had looked west and discovered Iceland, Greenland, and the coast of the North American continent—all by accident. Other Europeans, seeking a route to the East, would make more than a dozen voyages in the 16th and early 17th centuries searching for an assumed Northwest Passage around this continent. The modern European discoverer of North America, John Cabot (Giovanni Caboto), therefore was something of a latecomer. His voyage of exploration from Bristol, England, in 1497 is associated more importantly with the beginnings of cod fishing, an industry that dominated Newfoundland's economy for nearly five hundred years thereafter. Cabot's party reported that the waters swarmed with fish, especially the fat, succulent Atlantic cod, by far the most prized of species. Cod was in great demand at European ports, where it often brought a higher price than beef. Oil pressed from cod livers came into use as a vitamin-rich tonic.

French, Portuguese, and Basque fishermen were the first to spread their activities to the waters off Newfoundland. They also established scattered seasonal habitations along the island's long, rockbound coastline, a presence that still survives in place names such as Burgeo and Fogo Island (Portuguese); Bonavista, Catalina, and Trepassey (French); and Burin and Placentia (Basque). By the 1570s the English also were taking part in the fishery, especially along the Avalon Peninsula. Closer to Europe than any other part of North America, Newfoundland was the temporary home for many who sought a more permanent life elsewhere and for an assortment of others, including a number of notori-

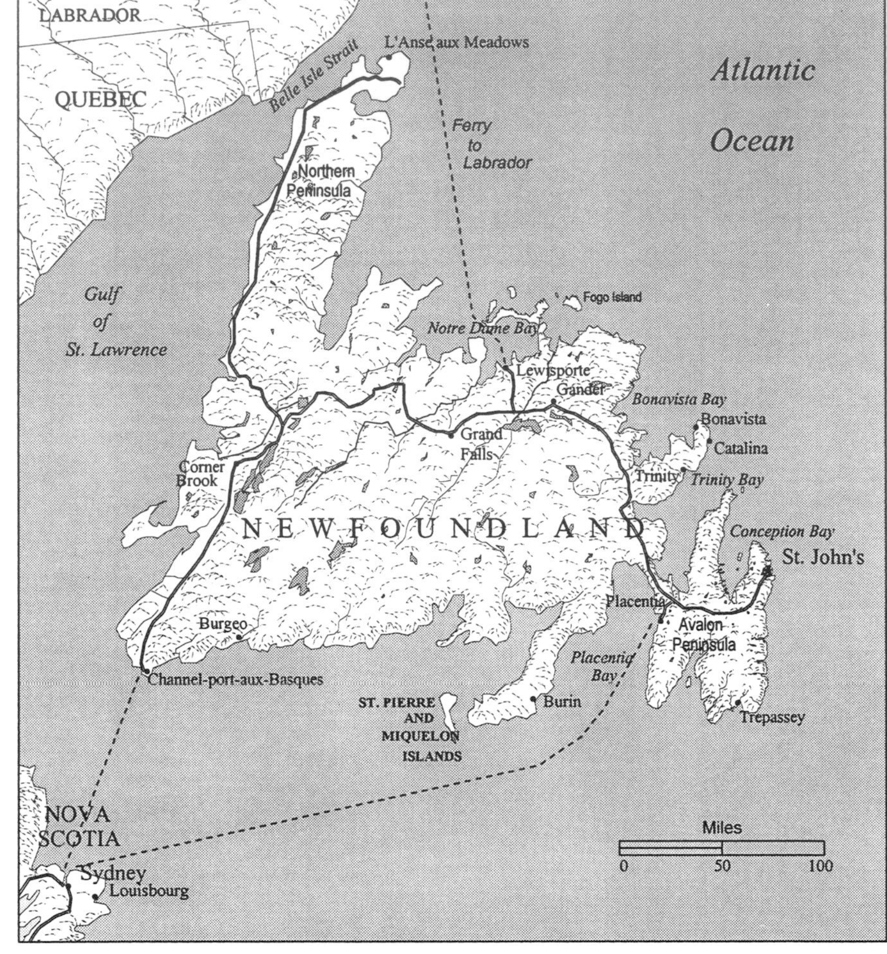


MAP 1.1

ous pirates who took shelter in its many bays. Newfoundland's history through the early 18th century is filled with struggles to maintain authority by the various nations or groups involved.

Newfoundland's coast, like its interior, is rocky and occasionally appears mountainous (fig. 1.1). The irregular terrain is the result of the Avalonian orogeny, a mountain-building episode that began more than 600 million years ago. A proto–Atlantic Ocean existed at that time, which separated North America from Europe and Africa. As this former Atlantic Ocean closed because of convergent movements on the underlying plates, the continents drew closer together. Belts of volcanic mountains appeared near the shrinking ocean's margins, and new mountain ridges were thrust upward on land. Eastern Newfoundland, including all of the Avalon Peninsula, now rests on a portion of what was the Iberian Peninsula of Europe but which broke apart and was welded to the North America continent before the present Atlantic Ocean opened.

The mountains were gradually worn down until today they form the island's

FIG. 1.1. *John Cabot's landfall in North America probably took place on Newfoundland's Bonavista Peninsula, a jutting rock platform tilted by the force of continental collision.*

low hills and peninsulas. Newfoundland's indented coastline offered numerous snug harbors and sheltered coves, where eventually more than one thousand outports ("out" from St. John's) were built (fig. 1.2). Outports are the typical Newfoundland coastal communities, where life always has been centered around fishing.

## Newfoundland Settlement

For most of the 17th and 18th centuries the Newfoundland cod fishery operated as a seasonal, overseas business that was based on numerous western European ports. Fishermen from the West Country of England predominated after Great Britain obtained sovereignty over Newfoundland by the Treaty of Utrecht in 1713. France retained some fishing rights in Newfoundland, and even today the

FIG. 1.2. *Trinity, one of Newfoundland's many outports, was an important center of fishing activity and trading by 1800. The flattish-roofed buildings in the foreground are one version of the Newfoundland folk house.*

islands of St. Pierre and Miquelon, a few miles off the Newfoundland coast, remain a Collectivitée Territoriale of France. In the 18th century fishing vessels came to Newfoundland by the hundreds each year. The summer's catch was lightly salted, dried on outdoor racks known as *flakes,* and then taken to Europe by the returning fishermen. Spain and Portugal were the largest consumers of Newfoundland salt cod.

In 1750 Newfoundland had only about two thousand permanent residents; yet many times that number of people participated in the seasonal fishery. By 1800, however, seasonal migrants were outnumbered by larger numbers of permanent residents, who settled in St. John's and in the outports. A dozen counties in southwestern England and southeastern Ireland were the most important sources of migration to Newfoundland. Descendants of these migrants form the greatest share of the island's population today. The Avalon Peninsula, including St. John's, was, and remains, dominantly Irish Catholic, whereas the outport communities more distant from St. John's were most often settled from England and are dominantly Protestant.

St. John's became Newfoundland's largest city, the seat of its government, and the focus of most of its economic activity (fig. 1.3). The outports were more often connected to one another and to St. John's by sea than by land. Mercantile houses once established as monopolies in isolated locations around Newfoundland's coast eventually moved to the principal city. The trading companies that handled Newfoundland's fish exports and goods imports were attracted to the busy wharves lining the sheltered harbor of St. John's.

By the late 19th century, the St. John's merchants developed a credit system: they employed fishermen to catch fish that the merchants then sold in foreign markets. The merchants did not pay wages but rather outfitted each fisherman with provisions and gear necessary for a year's operation and later settled the balance according to the quantity and quality of the salt cod that was brought in. The outports where the fishing families lived were so chronically short of

FIG. 1.3. *St. John's older residential neighborhoods climb the hills bordering the city's inner harbor.*

cash that inhabitants supplemented their subsistence by raising food crops on little patches of arable land and by raising animals for their own consumption. The Newfoundland fisherman was thus also a farmer and a jack-of-all-trades.

## The Forest Industry



Interior Newfoundland is only sparsely settled, because of the glaciated terrain, in which swamps, bogs, and barrens have formed. Large areas are covered with an open, lichen woodland, dotted with stunted spruce trees. Commercially exploitable forests are confined to a swath approximately one hundred miles wide across the central portion of the island. It was the forest that inspired the first landward turn of development. In the late 19th century, the Newfoundland government borrowed money from foreign banks to build an expensive but poorly constructed railroad that wound 567 miles through the better forestlands between St. John's and Port-aux-Basques on the Gulf of St. Lawrence. Until the Trans-Canada Highway was completed in the 1960s, the railroad was the only land route across Newfoundland.

Newsprint mills were built along the line by British and American interests at Grand Falls and Corner Brook in the early 20th century. Black spruce and balsam fir account for roughly equal amounts of Newfoundland's forest, and both are cut for pulp even though the island's trees are small compared with forests farther south. More than five hundred thousand tons of Newfoundland newsprint are now shipped annually to newspapers around the world. Just two paper companies control two-thirds of Newfoundland's productive forests, a legacy of the government's generosity with forest land in earlier times. Operations at Grand Falls were the work of English newspaper magnates Alfred and Harold Harmsworth (Viscounts Northcliffe and Rothermere), who desired a continuous supply of newsprint for their *London Daily Mail* and other presses. As in many remote or bypassed places that have desired economic growth, the benefits to Newfoundlanders have been less than anticipated, whereas the assets given away to attract capital have helped earn good profits for outsiders.

## Political Change

Great Britain granted responsible government to Newfoundland in 1855. Although Newfoundland was part of British North America, it did not join Canada when the Canadian Confederation was proclaimed in 1867. The island's economy was more closely tied to the North Atlantic trade—especially with Great Britain and the United States—than it was to Canada. Newfoundland sought to turn its independent status to advantage by brokering its own relationship with the United States but was rebuffed by both the British and Canadian governments. Not really Canadian, not British, and certainly not American, Newfoundland remained peripheral to all three of the larger economies and nations with which it conducted most of its business.

Newfoundland's economic problems worsened in the 1920s as it slowly sank under the foreign debt that had accumulated, and when the 1930s depression came, its economy collapsed. In 1933 Newfoundland took the drastic step of forfeiting its responsible government and accepted the direct authority of Great Britain once again. A Commission of Government was imposed, which, although it included elected Newfoundlanders, took its orders from London. Life under the Commission was made easier by the economic recovery stimulated by military activity in Newfoundland during World War II. With its forward position in the Atlantic, Newfoundland was a strategic location for Canadian and American air and naval bases. The international airport constructed at Gander became a refueling point for transatlantic flights and continued in this role for the duration of the piston-aircraft era. Newfoundlanders took construction jobs at the naval bases and air fields during World War II and came to know Canadians and Americans better than they ever had before.

In 1949 Newfoundland cast off its Commission of Government and finally voted to join Canada. Victory by the proconfederation forces, led by Joseph Smallwood (who served as Newfoundland's premier into the 1970s), was accomplished in part on the promise that Canadian social programs would benefit Newfoundlanders. "Never again," mothers were told, "would there be a hungry child in Newfoundland." Fishermen voted for confederation believing that better fishery management would come under the Canadian government. That the promise of welfare support could win so many votes suggests the desperate conditions of life that so many Newfoundlanders knew. Even at present, the operation of the Newfoundland fishery revolves largely around the issues of 1949, those of federal versus provincial authority and of Canada's role in providing additional economic support for the tens of thousands of Newfoundlanders who derive their living from the sea.

## The North Atlantic Fishery

North America's Atlantic fishery depends on the dispersal of many species of ocean fish from concentrated areas, known as banks, in the relatively shallow waters of the continental shelf off Labrador, Newfoundland, Nova Scotia, and New England (map 1.2). The banks, which are submerged in only 250 feet of water, support abundant phytoplankton life. Of the dozen banks within two hundred miles of Canada's coastline, the Grand Bank off Newfoundland is the largest and traditionally the most important spawning ground of cod. Still shallower, inshore waters of the Maritime Provinces and Newfoundland yield lobster, mussels, and snow crab. Capelin and lumpfish are caught primarily for their roe, which commands a high price in Japanese markets. Turbot, redfish (marketed as ocean perch), flounder, and herring have added to the large catches made here, one of the world's most productive fisheries. About two-thirds of Newfoundland's fish products are consumed in the United States.

The economics of fishing and the lives of those who make their living from it have been anything but stable in the past half century. Once Newfoundland joined Canada, the provincial government embarked on a series of development initiatives to strengthen the island's economy. Most Newfoundland fish-

MAP 1.2



ermen, then and now, operate small vessels as part of the seasonal inshore fishery, defined as the zone within twelve miles of shore. The cod they catch have already migrated away from the spawning grounds on the Grand Bank. Inshore fishing is labor intensive and the output per fisherman is comparatively small. It is a business ideally suited to the Newfoundland outports.

Believing that a small population scattered over many outports was an inefficient basis upon which to modernize the fishery, Newfoundland's government began to induce migration out of some isolated outports during the 1950s. The role of outports that produced only salted, dried cod declined as the demand for frozen fish fillets increased. Residents of those outports for which extinction had been planned were assisted financially if they relocated to communities that had been marked for growth as frozen-fish processing centers. Growing communities, it was argued, could provide services unavailable in the outports. Newfoundlanders would get jobs in the fish plants and on large fishing trawlers that would supply enough fish to keep the processing plants in operation.

Fishing might have become more productive with the introduction of larger vessels, but the size of the catch dropped each year into the 1960s as a result of overfishing by foreign vessels. In 1977 Canada extended its fishery jurisdiction to two hundred miles offshore in an attempt to regain control, a limit that leaves a portion of the Grand Bank (the "nose" and "tail") outside Canadian waters. With a new emphasis on the offshore fishery, larger fishing trawlers were favored in order to extend the fishery's reach. The Newfoundland government subsidized the construction of a fleet of trawlers that could drag the Grand Bank itself. Most of the fleet was purchased by the large, frozen-fish processing firms, thereby making thousands of inshore fishermen, with their small craft, the economic rivals of companies that would modernize the industry. By the 1980s Newfoundland, Quebec, and the Maritime Provinces had seven hundred fish-processing plants in operation, roughly half of which concentrated on the groundfish species—cod, haddock, and flounder.

Problems of overcapacity in fish processing, decreased demand for fish products, reduced catches, and seasonal income supplements necessary to support inshore fishermen continued into the 1990s. Worst of all, it became clear that the cod were disappearing because their spawning grounds had been massively disrupted by the bottom-dragging trawlers. A moratorium on cod fishing was declared in 1992 and has remained in effect for some coastal waters indefinitely. The centuries-old cod fishery—the basis for settlement in Newfoundland ever since John Cabot's discovery—was destroyed in only a few decades of overfishing and habitat disruption. Newfoundland has taken the position that its waters should not be open to fishing even by other Canadian provinces, and the federal government has taken action to drive non-Canadian fishing vessels from the nose and the tail of the Grand Bank.

The extension of all-weather roads into Newfoundland's major peninsulas revitalized the outports that once were thought to be an outmoded system of settlement. Many outports have grown since 1960 as a result of improved access, and this has led to more diversified local economies. Newfoundlanders once were identified with the bays around which they lived, but today the population is organized more in terms of the peninsulas separating those bays, a reflection of the shift from water to road transportation. The standard of living in Newfoundland has improved markedly in the past half century, in part because of transfer payments from the federal government. St. John's harbor receives two or three ocean-going container ships each week. They are brimming with food and consumer goods that support a Newfoundland lifestyle now much more like that of the Maritime Provinces or New England. But the future of the Newfoundland fishery remains in doubt (fig. 1.4).

## Labrador

Just as Newfoundland began as a seasonal, overseas fishery of Europeans, fishing and sealing off Labrador have long been the preoccupation of many Newfoundlanders. World opinion against the killing of seals has driven that business into near extinction, but for more than a century Newfoundlanders took

FIG. 1.4. *The moratorium on cod fishing idled many Newfoundlanders and created new economic problems in the outports.*

part in the annual winter seal hunt in the sea ice off Labrador. Ancestors of the Innu and the Inuit who live along the Labrador coast today moved down from the eastern Arctic less than five hundred years ago. With their kayaks and floating harpoons, they were efficient hunters, and they replaced an earlier, land-based Inuit culture. Labrador's Inuit once depended heavily on the whales and ringed seals they took, but now they live in westernized villages under the protection of the Canadian government.

Labrador lies in roughly the same latitude as Great Britain, yet the two have vastly different environments. Western Europe is warmed by waters of the Gulf Stream, whereas the cold, south-flowing Labrador Current, which originates in Baffin Bay between Greenland and Canada, moves a stream of icebergs down the coast of Labrador most of the year (fig. 1.5). The coast is cold and rainy, whereas the interior is warmer in summer and colder in winter. Except for its

FIG. 1.5. *Icebergs, drifting with the Labrador Current, pass Notre Dame Bay on Newfoundland's north coast during midsummer.*

northern tip, Labrador is forested, although the trees are even smaller than they are on the island of Newfoundland. No scheme to exploit Labrador's forests has succeeded as a commercial venture despite several attempts.

With its severe environment and a population of perhaps thirty-five thousand, Labrador seems an unlikely focus of conflicting claims of political jurisdiction, but its borders have yet to be established to the satisfaction of all parties concerned. When France lost its remaining control of Canada to Great Britain in the Treaty of Paris (1763), Labrador was placed under the authority of the British-appointed governor of Quebec and the Labrador fishery was awarded to Newfoundland. Canada and Newfoundland—which, in effect, remained separate countries until 1949—engaged in several disputes over Labrador. In 1927 Great Britain intervened to settle the matter and awarded all of what is now Labrador to Newfoundland. This was far from the end of the matter, however. Quebec claimed that Canada did not confirm its boundary with Labrador when Newfoundland entered Canada in 1949. Much of the Labrador-Quebec boundary is defined in terms of the watersheds of rivers that flow either to the Atlantic Ocean or to Hudson Bay.

Large bodies of iron ore that were discovered along the Labrador-Quebec boundary in 1896 made the issue of Quebec-versus-Newfoundland sovereignty all the more important. Mines were developed near Schefferville, Quebec, in the 1950s by the Iron Ore Company of Canada, a creation of several steel companies in the United States. An ore-carrying railroad was built to the deepwater port of Sept Isles, Quebec, where ores were shipped to steel mills near Philadelphia. When the switch was thrown to load the first ore boats at Sept Isles in 1954, premiers Maurice Duplessis of Quebec and Joseph Smallwood of Newfoundland both had their hands on the controls. By then a large deposit of lower-grade ore had been discovered at Wabush, Labrador; it was sold to American steelmakers. This was followed by another discovery near Gagnon, Quebec. Canadian steel mills at Sydney, Nova Scotia, and Hamilton, Ontario, also purchased ore from the Labrador and Quebec mines.

All settlements based on nonrenewable resources have a predictably limited life span. Before iron ore was mined in Quebec and Labrador, the major iron mine in Atlantic Canada was at Bell Island (Wabana), Newfoundland, near St. John's. Germany considered it so important during World War II that U-boats attacked it twice and sank four ore carriers. But the Bell Island ore was costly to extract, and the mine was closed in 1966. Operations at Schefferville, where the ore was of lower quality, were phased out in the 1980s. Mines at Labrador City–Wabush and at Gagnon continue to produce several million tons of ore annually, which is shipped to steel mills via Sept Isles, Pointe Noire, and Port-Cartier on the St. Lawrence. Import substitution of German and Japanese products for North American steel has led to mill closures on the East Coast, however, and the Quebec-Labrador mines now produce less ore than in former times.

Newfoundland's inherent claim to Labrador is to some extent compromised because the former is an island, and Labrador's land boundary is with Quebec. These facts were made more real to Newfoundlanders in the 1960s when the province developed what was then the largest hydroelectric project in the Western Hemisphere at Churchill Falls, Labrador. Although the island of Newfoundland could have used some of this power that would be generated within its own provincial boundaries, the prospect of laying high-voltage transmission cables in the waters of Belle Isle Strait was just as costly as the new hydroelectric project's principal beneficiary was obvious.



Quebec's power authority, Hydro Québec, was a substantial investor in the Churchill Falls project. Newfoundland receives payments from the operating profits of the Churchill Falls facility, but Hydro Québec gets most of the electricity, and at a cost that is lower than that of power generated in its other major installations. Quebec's large sales of electricity to the United States have been made possible, in part, by this arrangement, which continues as a major issue between the two provinces. And the contest continues. After years of isolation from the North American road network, Labrador now is reachable by highway from Baie-Comeau, Quebec, although it remains linked to Newfoundland only by ferry. The question of whose government controls Labrador seems no easier to answer today than in the past.

## Offshore Oil and Gas

Hydrocarbon resources have long been known to exist in buried sediments of the continental shelf off the Atlantic coasts of the United States and Canada. Interest in their extraction was heightened when oil production began in the North Sea oil field of Europe during the 1960s. Oil companies in the United States began acquiring drilling rights off the coasts of Nova Scotia, Newfoundland, and Labrador during the 1970s when worldwide oil prices were rising and market supplies appeared to be declining. More than six dozen exploration wells were drilled on Nova Scotia's continental shelf. A major oil discovery was made at the site known as Hibernia, 175 miles southeast of St. John's at the edge of the Grand Bank.

Hibernia began producing on a small scale in the late 1990s, although it is a high-risk location for offshore petroleum production. Hibernia is located at the end of "iceberg alley," where the Labrador Current carries icebergs south even during summer months, posing an economic and environmental risk to offshore drilling platforms and shipping. Natural gas trapped in geologic structures associated with the oil reserves also is abundant; but its extraction remains uneconomical, because it is located so far away from markets. Because the Atlantic offshore fields have oil and gas reserves exceeded only by those of Alaska's North Slope and the Beaufort Sea among North American producing regions, their eventual full exploitation seems certain.

## *References*

Alexander, David G. *Atlantic Canada and Confederation: Essays in Canadian Political Economy.* Toronto: University of Toronto Press, 1983.

———. "Newfoundland's Traditional Economy and Development to 1934." In *Atlantic Canada after Confederation,* ed. P. A. Buckner and David Frank, 11–33. Fredericton, N.B.: Acadiensis Press, 1988.

Brox, Ottar. *Newfoundland Fishermen in the Age of Industry: A Sociology of Economic Dualism.* St. John's: Memorial University of Newfoundland, Institute of Social and Economic Research, 1972.

Economic Council of Canada. *Newfoundland: From Dependency to Self-Reliance.* Ottawa: Economic Council of Canada, 1980.

Horwood, Harold. *Newfoundland.* New York: St. Martin's Press, 1969.

Mannion, John. *The Peopling of Newfoundland: Essays in Historical Geography.* St. John's: Institute of Social and Economic Research, Memorial University of Newfoundland, 1977.

Neary, Peter. *The Political Economy of Newfoundland, 1929–1972.* Vancouver: Copp Clark, 1973.

———. *Newfoundland in the North Atlantic World, 1929–1949.* Kingston, Ont.: McGill-Queen's University Press, 1988.

Weeks, Ernie P., and Leigh Mazany. *The Future of the Atlantic Fisheries.* Montreal: Institute for Research in Public Policy, 1983.

Wood, Clifford, and Gary E. McManus. *Atlas of Newfoundland and Labrador.* St. John's, Newf.: Breakwater, 1991.

## CHAPTER 2

# Quebec

Half a century before England made its first claim to Newfoundland, France sponsored several transatlantic voyages by Jacques Cartier in search of a short route to Asia—the presumed Northwest Passage. In 1534 Cartier explored the Atlantic coastline and the next year sailed up the St. Lawrence, which he discovered to be a river rather than the arm of an ocean. Cartier's party described two prosperous Iroquoian villages along the St. Lawrence. The men wintered at Stadacona, the future site of Quebec City, and visited a larger agricultural village, Hochelaga, on the island that Cartier named Montreal. To go beyond Hochelaga required navigating what was later named LaChine Rapids, a reference to the impossibility of a water route between Montreal and China. Unable to go farther upriver, Cartier traded a few furs with the natives and made an unsuccessful attempt to found a colony. Fur trading posts were begun by others in subsequent years, but permanent European settlement was not established until Samuel de Champlain founded the city of Quebec in 1608.

Where Cartier had reported large native villages, decades later Champlain found a countryside that had been largely abandoned. The St. Lawrence Iroquoians disappeared some time after Cartier's visit, perhaps because of warfare between native groups who vied with one another for access to the new trade in European goods. When Champlain first ascended the St. Lawrence in 1603, he was greeted by several Algonquian groups who were celebrating a successful raid they had made against a weakened Iroquois population. Champlain himself was persuaded to join the battles brought by the Algonquians and by a more western Iroquoian group, the Hurons (sometimes called "good Iroquois") against the Iroquois Five Nations (Seneca, Cayuga, Onondaga, Oneida, and Mohawk).

The French thus became allied militarily with Indians of what are now Quebec and Ontario against the Iroquois south of Lake Ontario, who became allies of first the Dutch and then the English. Given the geographical trend of these alliances, the Montreal-based fur trade expanded westward, up the tributaries of the St. Lawrence. It did so at a pace so rapid that by 1635 the French had entered the eastern Great Lakes. Numerous alliances already established by the Hurons opened a large territory to the north that supplied more beaver peltry for the growing European market.

In the middle of the 17th century, however, the balance of power shifted against the French and their native allies. Smallpox broke out among the Algonquians in the St. Lawrence and Ottawa River Valleys, and the dreaded disease then spread west to afflict the Hurons. The Senecas, part of the Iroquois Five Nations, attacked the weakened Huron settlements and reduced the population by two-thirds. French trading posts were placed under siege wherever the Iroquois found them. After the Hurons' demise, the role of middleman was assumed by independent French traders, known as coureurs de bois, who adopted the natives' birchbark canoes and woodland ways. French fur trading expanded still farther to the west, and by the 1750s traders had built posts on the Saskatchewan River. A French presence was relayed westward through these linked outposts of the fur trade.

## The St. Lawrence Lowland

Even though their trading empire was far flung, the area of present-day Quebec that was actually settled by the early French included little more than the St. Lawrence Lowland between Montreal and Quebec City (map 2.1). The Lowland itself is a gap between the Canadian Shield to the north and the Appalachian Mountains to the south. The Gaspé Peninsula, which is part of the Appalachians, was thrust northward, toward the Shield, during the Appalachian orogeny. Quebec is an Algonquian word meaning strait or narrow passage, a reference to the narrow St. Lawrence channel at the point where the Lowland terminates and the river enters the widening gap between the Appalachians and the Shield. The city is named for the narrows; the province of Quebec, in turn, takes its name from the city.

The St. Lawrence Lowland is a roughly triangular region that is more than one hundred miles wide at the Quebec–New York border. In late glacial times an extensive flooded area known as the Champlain Sea transgressed this low-lying land now drained by the St. Lawrence River. The glacial ice had melted, but the land surface had not yet rebounded as the weight of the Laurentide ice sheet was slowly removed. The lowland soils that emerged above sea level were fertile, especially near Montreal, but a short growing season and long, cold winters severely limit agricultural activity. Despite Quebec's location in the same latitude as France, both Montreal and Quebec City have January temperatures 25°F colder than those of Paris. Quebec has a continental climate because it lies too far inland to benefit from the same warm ocean currents that moderate western Europe's winters (fig. 2.1).

The entire claim of France in North America was called New France, and the St. Lawrence settlement was known simply as Canada, there being no more need to qualify it as "French Canada" than there would have been to speak of an "English Virginia" in the 17th century. Canada was French in population, language, law, and customs, a direct extension of ancien regime France. Beginning in 1663, its affairs were directed by King Louis XIV. Canada was a feudal society that embraced a landed nobility, an influential clergy, a bourgeoisie, and a peasant class. All were enmeshed in another French institution exported to Canada, the seigneurial system of land tenure in which were specified the obligations and rights of the seigneur, or landowner, who had received the land directly or indirectly

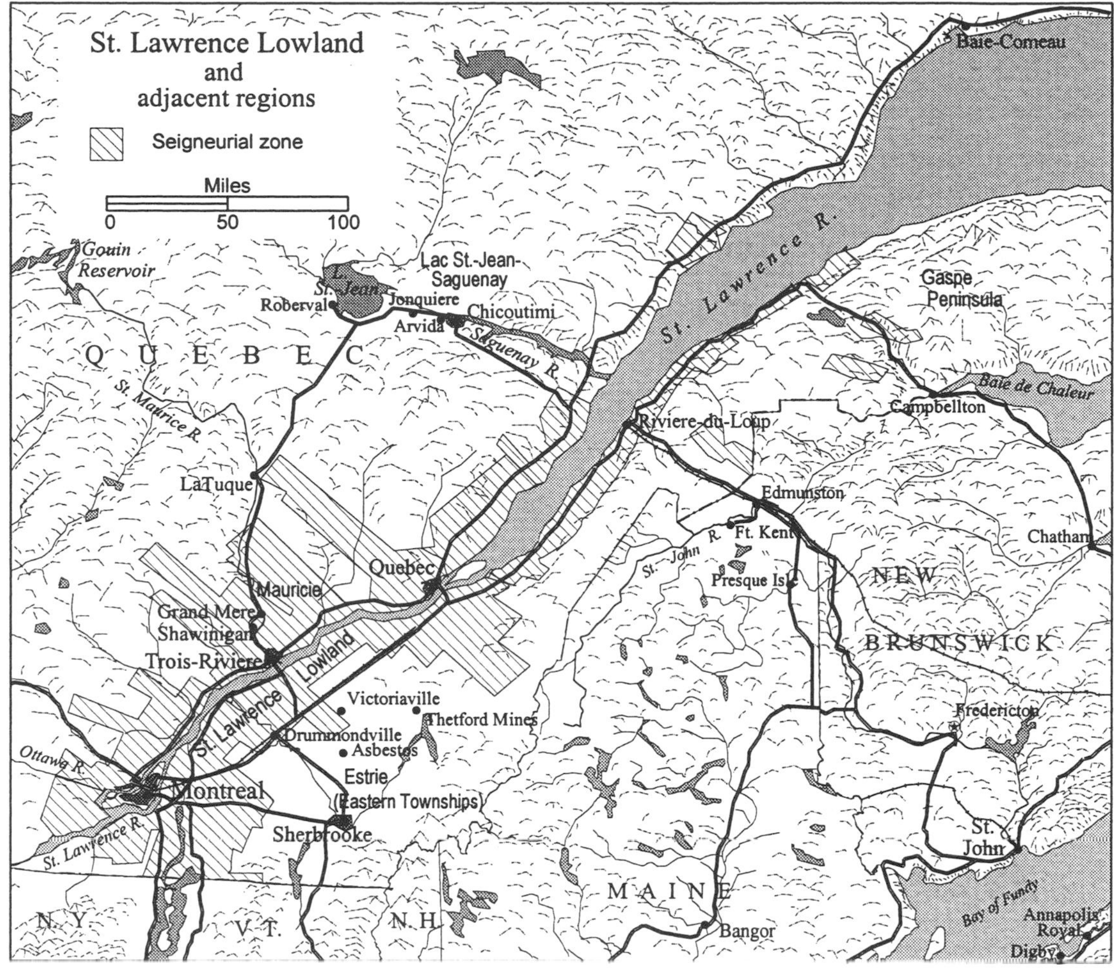


MAP 2.1

from the Crown; and the *cenistaire,* who held a concession of land for which he paid an annual rent (*cens*) to the seigneur. About two-thirds of the seigneurs were members of the nobility.

Seigneuries usually were trapezoidal in shape with one side formed by the river's bank. The entire St. Lawrence Lowland was divided into several dozen seigneuries in this manner, and each seigneury was, in turn, subdivided into *rotures,* which were the parcels of land farmed by the cenistaires. The rotures were laid out in long strips, about one-tenth as wide as they were long (fig. 2.2). When the land along the St. Lawrence River had been made into farms, the procedure was repeated, in an interior *rang* facing a road parallel to the river. The rates of forest clearing and land improvement were slow in many seigneuries, and thus the establishment of new rangs of settlement, away from the river, took years to complete.

One benefit of the long-lot system was the simplicity by which rotures could

FIG. 2.1. *Riviere Bécancour is one of many small tributary streams flowing through the St. Lawrence Lowland. The latitude of this place (46°N) is the same as central France, but Quebec has a much harsher climate.*

be surveyed. For the *habitant* (farmer) it meant that farmhouses were less dispersed than in the typical rectangular system of survey later used in North America; neighbors therefore lived closer to one another. Perhaps more visible than significant, the distinctive long-lot geometry of the seigneuries has nonetheless endured as a marker of early French settlement in many parts of Canada and the United States.

Although France lost its claim to Quebec well over two hundred years ago, it is the French origins of Quebec (and therefore of Canada) that remain as perhaps the most important political issue faced not only by Quebecois but by all other Canadians at the present time. Quebec has unsuccessfully sought recognition from the other provinces as a "distinct society" as part of the failed 1987 Meech Lake Accord that some saw as a solution to Canada's problem of providing a federal constitution that all the provinces could embrace. Even the no-

FIG. 2.2. *St. Lawrence Lowland near Bic, Quebec. The distinctive pattern of long-lot fields and fences has long outlived the seigneurial land-tenure system of which it was a part.*

tion of a bilingual Canada—or an ethnic French Canada that presumably exists wherever French speakers are found—is considered irrelevant by the many who seek establishment of a sovereign Quebec state rather than to continue as a Francophone province within the Canadian state.

Contemporary Quebec-Canada tensions originate in the long memories of a conquered people. In North America the fall of France and the ascendancy of Great Britain took place between 1758 and 1763 as a result of the Seven Years' War between the two countries. Ever since the 1670s, when the Hudson's Bay Company emerged as a trading force in the north, there had been aggressive competition over rights to a monopoly trade with various native North American groups. South of the Great Lakes, French traders were in the vanguard, but they were also in constant conflict with the British (usually Americans) who were making the rounds of native villages in the Ohio Valley by the 1720s. The French went into retreat.



To Britain's surprise, however, the French reemerged and regained control over the empire of stockaded trading posts that made up the fur trade. The French formed new alliances with crucial native groups that lived in the Great Lakes and Upper Mississippi Valley region, where furs were most plentiful. In 1758 the British retaliated by attacking Canada. They captured the French fortress at Louisbourg, Nova Scotia, and then moved on to Quebec. Having already set fire to the St. Lawrence habitants' dwellings and barns and laid waste their crops, British troops under the command of Major-General James Wolfe decided to move against the capital. On September 13, 1759, some forty-five hundred British troops landed at Quebec City, scaled the cliffs, and engaged a like number of French troops, under the command of General Louis-Joseph Marquis de Montcalm, in a pitched battle. Before French replacements could arrive, Wolfe's army broke through the French lines and surrounded the walls of the city. The French surrendered five days later.

Beaten still farther up the St. Lawrence by British troops, the French finally made an unconditional surrender at Montreal one year later. Under terms of the Treaty of Paris, signed in 1763, France ceded to Great Britain all her North American possessions except for the tiny islands of St. Pierre and Miquelon off the southern Newfoundland coast. There would be no new French settlements made after that date in the territory that would become the United States, but the vibrant French society of Quebec survived. In fact it was not until the early 19th century that French-speakers finally were outnumbered by English-speakers in Canada as a whole.

At the time of the conquest in 1760, more than 70,000 people of French descent lived in Quebec. Of this total, some 8,000 lived at Quebec City, 5,200 at Montreal, and 550 at Trois-Rivières. The remaining majority, some 56,000 people, lived in seigneuries bordering the St. Lawrence River. The habitants practiced a largely subsistence agriculture and supplemented their livelihood by working seasonally in the fur trade. Their lives were centered on the seigneury and the parish in which they lived, an existence not all that different from the European past.

FIG. 2.3. *Chateau Frontenac at Quebec City is a latter-day re-creation of a French castle, but it remains as a well-known symbol of Quebec's heritage.*

By 1760 Quebec City had evolved in a fashion reminiscent of European cities (fig. 2.3). Champlain's habitation of 1608 was at the river's edge; it became a commercial enclave lined with wharves, known as the lower town. Quebec City's upper town was the center of government and religious authority, a walled city that nonetheless had fallen to the British. Trois-Rivières, which would later become the center of an important industrial region of Quebec, was a small market town when the British arrived.

Quebec City has remained the capital, but it has never been a close rival of Montreal in terms of economic control. Montreal began as a religious mission in 1642. The original, walled town was enlarged and redesigned by the religious order of Sulpicians, who became seigneurs of Montreal Island in the 18th century. Montreal's location, near a rapids that required portaging and at the confluence of the Ottawa and St. Lawrence Rivers, helped make it the center of the fur trade, a role that continued into the 19th century. The fertile soils of the surrounding Montreal Plain also enabled the city to become a minor center of agricultural production. Because of its location and its well-developed trade, Anglophone businesspeople moved to Montreal soon after it became British territory. Quebec City, outside the mainstream of North American economic transactions, remained French to the core. Montreal did not, in part because it grew as a commercial and industrial center that was open to new capital and new entrepreneurship in the decades following the conquest.

The terms under which Quebec residents would coexist with the new British regime were put forth in the Quebec Act of 1774, a rather remarkable edict in terms of the tolerance it showed. The act guaranteed the future existence of some important Quebec institutions. Roman Catholicism was granted a degree of official recognition, the seigneurial system was guaranteed, and French civil law was maintained. The British also enlarged Quebec's territory by attaching to it most of the Great Lakes and Upper Ohio Valley, simultaneously challenging the westward push of settlement from the American seaboard colonies and giving new territory to the Montreal traders, who thereby were situated at the apex of British North American commerce (briefly, as it turned out, until the American Revolution). American troops in the command of Richard Montgomery and Benedict Arnold marched against British-controlled Quebec in 1775 but were soundly defeated the next year. The present southern boundary of Quebec lies approximately where Great Britain and the United States drew the line in 1783.



With the Great Lakes–St. Lawrence water route established as the only feasible artery for shipment into the Canadian interior, Montreal grew faster than any of its rivals. It surpassed Quebec City in size by 1830 and became Canada's largest industrial center. The Lachine Canal, completed in 1825, allowed ships from the interior to bypass the rapids that had stopped Jacques Cartier centuries earlier. Mills, foundries, and refineries built along the canal, on the southern edge of Montreal Island, stimulated new directions of urban growth. Over the 19th century Montreal became Canada's undisputed center of banking and finance. Steel mills, oil refineries, flour mills, sugar refineries, and food-processing plants gave it a diversified industrial base. Canada's major railroad companies made Montreal their headquarters and built extensive shops and yards that employed thousands of workers in and around the city.

In many respects Montreal—which was Canada's largest city until the mid-20th century—was atypical of Quebec as a whole. Anglo capital investment was largely responsible for building Montreal's industries, a circumstance that also attracted a large Anglophone workforce to the city. From 1830 until the 1870s, Francophone Montrealers were outnumbered by non-Francophone European immigrants, who, as a share of Quebec's population, were concentrated in Montreal. The new immigrants swelled the ranks of laborers and accounted for increasing numbers within the city's business elite. Although Montreal slowly regained a Francophone majority as a result of internal migration, there was no overwhelming move to Canada's metropolis from most parts of Quebec, where the population remained agricultural and overwhelmingly rural.

Ruralization has a long history in Quebec. In the 18th century land was so plentiful in the seigneuries along the St. Lawrence—compared with Quebec's comparatively small population and slow rate of growth—that the ratio of country dwellers to urban dwellers increased for a century after the conquest. Anglos became the largest purchasers of seigneuries after 1800, and they introduced to their lands a more intensive agriculture than the typical habitant had practiced. Wheat was Quebec's staple crop in the French period, and it continued in that role after the conquest. Wheat was once grown as far north as Rivière-du-Loup, more than seventy miles downstream from Quebec City, but a series of cold years in the first decades of the 19th century hastened a southward retreat of wheat production.

More important in the crop's decline was the emergence of wheat exports from Upper Canada (Ontario). The shipment of Ontario wheat overseas caused Montreal's grain trade with interior Canada to grow, but Ontario's crop marked an end to large-scale wheat production in Quebec. Oats, hay, and dairy products became the principal outputs of Quebec agriculture thereafter, a shift that reflected some adjustment to climatic realities and also the abandonment of agricultural specialties that became more profitable elsewhere as population expanded to the west. Quebec cheese and butter found markets in the United States.

The seigneurial system itself fell victim to the changes in Quebec agriculture and society; it was abolished in 1854. The influence of Quebec's landed nobility had declined, diluted in part by the presence of a merchant class of seigneurs, both Anglo and French, and much more by a shift in wealth toward the cities.

Among the institutions of ancien regime France that remained part of the fabric of Quebec society, the ultramontanist Roman Catholic church was by far the most important. The church's influence increased in the late 19th century and on into the 20th. It gradually took responsibility for most of Quebec's schools, its hospitals, and its programs of social security and welfare. Intellectual life was enmeshed with the church's teachings, which emphasized authority, obedience, and family. The church was an active agent promoting an ethnic and religious Quebec nationalism: in Canada, what was French was Quebec; and what was Quebec was overwhelmingly Catholic. Beyond that, the church emphasized the superiority of rural life—the "nobility of agriculture"—where, in the open air, the Christian life could best be realized.

Rural colonization schemes promoted by the church during the late 19th century had the dual purpose of relieving population pressure in the St. Lawrence Lowland (where decades of natural increase had put agricultural land in short supply) and establishing new Francophone colonies, away from settled territory, in the spirit of cultural expansion. Quebec faced an increasing gap between rural population growth and a roughly constant agricultural base. The general trend toward smaller family sizes that swept late-19th-century western Europe, the United States, and most of Canada was not operative in rural Quebec.

The church's efforts to spread out the population were timely but also comparatively unimportant when compared with the number of Quebecois who left Canada. Far more attractive to the thousands of families struggling to keep body and soul together on a Quebec farm was the lure of jobs in the cotton textile mills of New England. Quebec labor—men, women, and children—provided nearly half the workforce in some Merrimack Valley textile mills north of Boston. The outpouring was strongest in the 1880s, although the social contacts between homeland and destination that such migrations produce led to even more migration in later years. Eventually whole Quebec neighborhoods appeared in the industrial cities of New England.

## Estrie (the Eastern Townships)

European settlement was sparse outside of the seigneurial zone until the 19th century. As a result of Great Britain's loss of the American colonies under the 1783 Treaty of Paris—only twenty years after the first Treaty of Paris gave Britain authority over Quebec—there was a natural tendency for those American colonists who were loyal to the British Crown to seek new homes north of the border. Loyalist migrations created new colonies of English-speakers in scattered areas from Ontario to New Brunswick, including Quebec. But the Loyalists avoided the St. Lawrence seigneuries. Whether they are regarded as having been more British or more American, the large majority of Loyalists were Anglophone Protestants. They settled in an irregular-shaped zone more than one hundred miles long and within fifty miles of the U.S. border that was given the informal name Cantons de l'Est (Eastern Townships) and later called Estrie. Estrie is overwhelmingly Francophone today, but its distinctiveness derives from its Anglophone past.



The character of the Eastern Townships was stamped even more boldly with the arrival of thousands of British immigrants in the early decades of the 19th century. English law, freehold tenure, and a township land-survey system added further contrasts with the French traditions of Quebec. For a time land companies held large tracts that were sold primarily to Anglophone settlers, a practice that temporarily isolated the Eastern Townships from the rest of Quebec. Sherbrooke became the region's principal city and developed a manufacturing economy based on direct railroad linkages with the year-round ocean ports of St. John, New Brunswick, and Portland, Maine. Textile and leather industries similar to those in New England appeared in the Eastern Townships by the mid–19th century. Under Anglo influence, the area developed more as an outlier of New England than of Quebec.

Growth in manufacturing industries created a new demand for labor, which was met in part by immigration from the St. Lawrence seigneurial zone after 1850. Two-thirds of the Eastern Townships' population was of British descent at that time, but by the early 1870s the area had a Francophone majority. Continued growth based on industrial expansion produced a six-to-one ratio of Francophone to Anglophone ancestry by 1931. Quebec's textile and leather industries, like New England's, declined in the face of foreign competition beginning in the 1960s. High-tech industries and transportation equipment manufacturing are the growth industries in the Eastern Townships at present.

Quebec's asbestos mines, located slightly east of the old Cantons de l'Est, were an important part of the Canadian economy from the 1880s until recent times. Asbestos is a filament-like mineral with extremely low heat conductivity that makes an excellent, chemical-resistant insulation. Heat and pressure applied to a deep vein of peridotite rock during the Appalachian orogeny transformed it into asbestos. Approximately 85 percent of the world's asbestos was blasted out of a fifty-mile vein that runs between the communities of Asbestos and Thetford Mines, Quebec. The mines were subsidiaries of British or American firms until the Quebec government nationalized part of the industry in 1977. By then Brazil, Cyprus, and British Columbia also produced quantities of asbestos, and Quebec's share of the world market was declining. But that was a minor problem compared with what followed.

Accumulating evidence of deaths caused by asbestos-related carcinogens led the U.S. Environmental Protection Agency to propose an eventually total ban on its use in building construction. Although only about 240 Quebec mine workers were receiving workmen's compensation payments for lung ailments in 1982, by that year the Johns-Manville Corporation, a leading American fabricator of Quebec-mined asbestos, sought bankruptcy protection after it was served with nearly 20,000 health-related lawsuits. Asbestos production plummeted, causing massive unemployment and population loss in the mining area. New products, such as asbestos-cement pipe, have been successful in some markets. But the long-term prospects for the industry are bleak even though the "asbestos scare" of the 1980s has since been revealed as something of an overreaction by the United States.

## Gaspé

When the northern Appalachian Mountains were formed more than 400 million years ago, ocean sediments at the edge of the continent were compacted by convergence of the land masses. The Gaspé Peninsula was formed as a range of mountains thrust upward between the colliding continents. Geologically speaking, Gaspé is more closely related to the northern Appalachians than it is to the old, worn-down mountains of New England to the south. Except for several peaks over 3,000 feet in elevation near the St. Lawrence River, however, most of Gaspé consists of gently rolling uplands that are creased by the deep valleys of fast-flowing streams. The coastal lowland is only a few miles wide along the St. Lawrence, where villages are crowded into a narrow strip of arable land between river and mountain. About a dozen seigneuries were scattered along the somewhat broader lowlands on the eastern and southern sides of Gaspé, but the land proved to be little more than marginal for agriculture.

Gaspesians traditionally looked to the sea more than to the land. Some native groups lived relatively undisturbed in the peninsula's interior well into the 20th century. A railway built across the peninsula to reach the Maritimes in 1876 was the first land route between Gaspé and Quebec City. Paved roads did not reach around the peninsula until the 1960s. Gaspé's remote fishing villages date from the end of the French period, when British and American settlers took over the seigneuries and established a commercial fishery in the Gulf of St. Lawrence. A credit system was imposed, similar to that in Newfoundland, whereby merchants equipped fishermen with provisions in exchange for their annual catch of cod. Gaspé's merchants sold their product in European markets, creating stronger economic ties with London than with the rest of Quebec.

Francophone migration from the St. Lawrence Lowland to Gaspé reversed the population trend that had begun with British and American settlement, and by the 1930s Gaspé was overwhelmingly French once more (fig. 2.4). Attention turned inland, toward the spruce and fir forests, which were cut for wood-pulp manufacture. Although Gaspé's forest industry is small by Quebec standards, the combination of logging and subsistence farming in cleared woodlands supports thousands of families. During the 1930s the Quebec government and the church created new, inland communities on Gaspé's farming-forestry frontier,

FIG. 2.4. *The traditional Quebec folk house with its bell-cast roof.*

but most were short-lived, and the population thereby retained was small in comparison to the flow of migrants out of the region.

Overfishing in the Gulf of St. Lawrence and economic problems in the fishing industry eventually led to population loss in coastal areas as still more people left the region in search of work. The peninsula's long, rugged coastline, dotted with fishing villages and small farming communities, no longer sees the kind of economic production that once characterized Gaspé. But tourism based on those very attractions has become an important growth industry.

## The Saguenay River–Lac St.-Jean

Most of Quebec is either too cold or its soils are too poor to allow agricultural development. Of these two limits, climate is actually of lesser importance. Even though the growing season may be too brief for grain crops to mature, summer days are long and are warm enough to produce good hay fodder for dairy cows. But if the soil is too thin, too rocky, or too acid, then even hay crops are not possible. Such soils are typical of the Canadian Shield, the massive core of ancient granitic rocks that forms the heart of the North American continent and which is exposed at the surface over most of Quebec. The Shield meets the St. Lawrence River near Quebec City and rises abruptly at the edge of the St. Lawrence Lowland just north of Montreal. Within the Shield only relatively small, confined areas have a cover of glacial sediments (often, glacial lake margins) thick enough for the development of agricultural soils.

One such area is the margin of Lac St.-Jean and the adjacent broad plain fringing the upper Saguenay River. Although Lac St.-Jean lies one hundred miles north of the edge of settled territory at Quebec City, agricultural settlers began moving up the Saguenay Valley in the 1860s, making farms out of forest clearings in an area where soils were known to be relatively fertile. The hard labor required to clear forested land was provided by the large families typical of rural Quebec. New farms were made for sons who married and moved on north in a continuing cycle of settlement expansion. Farming communities on the forest fringe were promoted by the church as part of its program of rural colonization and French cultural expansion within Quebec.

The lower Saguenay River is a fjord, a former river valley overdeepened by the seaward movement of glacial ice, which allows ocean vessels to pass inland more than sixty miles from the St. Lawrence River. British-owned companies sent shiploads of Saguenay Valley timber to Great Britain during the 19th century. As the local population grew, lumber interests built sawmills that, in turn, became the nuclei of some of the region's first urban centers. One example is Chicoutimi, now the focus of a metropolitan area whose population exceeds 150,000. Wood-pulp production for paper manufacturing began in the 1890s. Five mills with a total capacity of more than 1 million tons of pulp and paper annually now dominate the local manufacturing economy.

A large generation of hydroelectric power takes place at waterfalls on the Saguenay River as it emerges from Lac St.-Jean. Electricity is a vital source of energy in the manufacture of pulp and paper, and it is used in even larger quantities in the production of aluminum. No commercial deposits of bauxite, the ore from which aluminum is extracted, lie within a thousand miles of Quebec. But ocean-going ships laden with bauxite from Jamaica or Guyana can deliver the raw material to within a few miles of the Saguenay region's hydroelectric plants. The Pittsburgh-based Aluminum Company of American (Alcoa) constructed an aluminum smelter and refinery at Arvida, Quebec, in 1924. Arvida, a new town named for Alcoa's president, Arthur Vining Davis, became one of the most important aluminum manufacturing centers in North America at a time when the demand for the product was soaring. Like the rolls of newsprint produced by the Lac St.-Jean and Saguenay mills, most of the aluminum is shipped to the United States.

## Mauricie

Although the Saguenay Valley and the fringes of Lac St.-Jean were settled slowly during the 19th century, the region became an urban-industrial enclave, many miles from the manufacturing heartland of North America, by the 1930s. Similar developments took place in Mauricie—the valley of the St. Maurice River north of Trois-Rivières—during roughly the same span of time. The St. Maurice and its tributaries tap a region of black spruce and balsam fir that extends more than two hundred miles north into the Canadian Shield. Rural colonization schemes were promoted here by the church and by the Quebec government, but agriculture did not develop in Mauricie to the extent that it did in other areas.

The typical sequence of rural expansion in Quebec began with small-scale forest clearing and farming, which gave way to sawmilling industries and urban growth, followed by a shift to wood-pulp production and larger cities. In Mauricie the first two phases were brief; the transition to wood-pulp production had begun by the late 1870s. About 60 percent of Quebec's commercially exploitable softwood forest consists of spruce varieties, which are considered the best for pulping; in Mauricie the percentage of spruce is even greater. "Improvements" were made to the St. Maurice River in order to expedite log driving from the upstream forests to pulp mills located at downstream rapids, where logs are ponded and power is generated for the mills.

The adoption of the Fourdrinier method of continuous paper formation after 1900 increased efficiency and led to a marked increase in paper production. Eighty percent of Quebec's newsprint was sold to newspapers in the United States, a market that also was served by Ontario. In the early years of pulp and paper manufacture, however, American firms preferred to purchase Canadian pulp but manufacture their own newsprint in the United States. Seeking to establish greater newsprint production in Canada, between 1900 and 1910 the governments of Ontario and Quebec placed an embargo on exports of pulp made from trees cut on Crown (government) lands. The United States countered with a high tariff on Quebec and Ontario paper, but pressure from American newspaper publishers led to the removal of the tariff in 1913. One effect of the trade skirmish was to quintuple the size of Quebec's newsprint industry. From the 1920s onward, Quebec has exported more than 90 percent of its paper production to the United States.



Another of Quebec's industrial developments began in 1898 when a hydroelectric power dam was constructed on a 135-foot falls on the St. Maurice River. Shawinigan Falls became the focus of an industrial complex that included pulp and paper mills, chemical plants, and an aluminum smelter. Between 1900 and 1920, the Quebec government built hydroelectric dams that would flood new reservoirs for (usually American-owned) pulp mills. The Canadian government financed a new railway in the St. Maurice Valley to expand the zone of development. All of these efforts were part of a deliberate policy to attract foreign investment. Montreal's financial community held a major interest in Quebec's pulp and paper mills, but a wave of mergers and consolidation of smaller companies led to an increase in American control after 1920.

Shawinigan, Grand-Mère, and La Tuque—Mauricie's smaller cities—all were the work of the paper and power companies. Trois-Rivières was a regional trade and service center whose industries consisted of a few sawmills before the boom in wood-pulp production got under way after 1900. Today the Trois-Rivières metropolitan area contributes the largest volume of pulp and newsprint of any Canadian city, leading the industry that accounts for the largest share of Canada's exports to the United States. British Columbia now produces nearly as much pulp and paper as does Quebec, and both have tens of billions of cubic feet of softwood forests in reserve, but Quebec's location remains advantageous in reaching its traditional market, the publishing industry of the eastern United States.

## The Quiet Revolution and Its Aftermath

The role played by foreign capital in the pulp and paper industry illustrates the pronounced shift in Quebec's political economy that took place in the first half of the 20th century. Foreign influence, as represented by British authority, had been unwelcome historically and was perceived by Quebec's Francophone majority as a threat to their cultural survival. But American direct investment was invited, because it carried no taint of the past and, more importantly, was regarded as the means by which Quebec could acquire a modern industrial economy. Natural resources were given outright to American corporations that would develop them, an arrangement that became increasingly difficult to justify in later years. In 1949 a five-month strike by Johns-Manville miners at Asbestos, Quebec, in which provincial police intervened on the company's behalf, showed how closely the Quebec government was linked to American business. More strikes fed a growing concern over the state's role in promoting and protecting foreign investment.

The 1960s period in Quebec history, known as the Quiet Revolution, was characterized by a rise in popular dissatisfaction with established institutions such as foreign corporations. An even greater resentment developed toward the conservative authority of the church. Long-stifled dissent had already emerged in the writings of Catholic intellectuals like Pierre Trudeau, who attacked the church's privileged role and argued for a revitalization of public life. The church's influence declined in the 1960s as the provincial government expanded its programs of education, health, and social welfare for all of Quebec's people—activities that the church had controlled for many years. The high birthrates once typical of rural Quebec declined as family planning became the norm.

Another aspect of *la révolution tranquille* was an increased resistance to the role played by the Anglophone business elite of Montreal. Quebec's classical colleges prepared their graduates for the professions but not for careers in engineering or business. English was the language of the business community, and neither by training nor inclination were many Francophones attracted to it. Educated Quebecois spoke English, but their Anglophone counterparts typically did not speak French. Young Francophones from outlying areas of Quebec who under other circumstances would have been expected to gravitate toward the metropolis were effectively shut out of Montreal's business world unless they also spoke English. A durable slogan in Quebec politics, to be *maîtres chez nous* (masters of our house), suggests the nature of the problem and the underlying desire to "retake Montreal."

Two popular movements emerged from the Quiet Revolution; both reflected a growing concern for the politics of language. The federalist approach, typified by the leadership of Pierre Trudeau, who became prime minister of Canada in 1968, was to impose upon Canada as a whole greater recognition for Quebec and acceptance of French as a language. The Official Languages Act of 1969 made Canada bilingual (French and English) by fiat, although native languages were thereby ignored and no provisions were made for implementing the act beyond the level of federal agencies. Ontario and New Brunswick, both with substantial Francophone minorities, were to be bilingual provinces. Ontario refused to recognize the French language as official, which left only one province, New Brunswick, as bilingual in practice. Quebec responded in 1974 by passing a law that declared it to be French—and unilingual.

FIG. 2.5. *Law 101 made French mandatory on outdoor signs in Quebec. The Canadian Pacific Railway, which was Canada's largest corporation for many years, was a major Anglo presence in Montreal's business community. In 1997 the company moved its headquarters to Calgary.*

A second popular thrust that drew stimulus from the Quiet Revolution was the Mouvement souveraineté-Association, led by Rene Levesque. Quebeckers who sought outright independence joined forces with the more moderate faction that favored Quebec sovereignty but also sought an economic alliance with what would remain of Canada. Levesque became premier of Quebec in 1976 when his Partí Québecois won the provincial election. Among the ambitious programs that Quebec developed was the expansion of Hydro Québec, the provincial power authority, which constructed the gigantic power dams that allowed the province to sell electricity to the United States. The dams and generating stations built in the north incidentally demonstrated both French and Quebec engineering prowess and drew attention to the province's growing industrial might.

The Levesque government enacted a new and much tougher language law (fig. 2.5). Although the language issue served as a proxy for other disagreements, there were immediate and practical reasons for making only French the language of Quebec. It had been observed for more than a century that immigrants who spoke neither French nor English were assimilating into the Anglophone community in Montreal. In 1977 the new language act (Law 101) mandated French-language education for the children of new immigrants to the province. More and more, French became the language of the workplace, but language laws also were cited as a provocation by some Montreal companies that chose to relocate their corporate headquarters to English-speaking Ontario.

Despite such unfavorable reactions, Quebec's language laws gradually produced their intended effect, that of making French the language of public discourse in Quebec. In the 1990s the language laws were relaxed to allow some choice for schooling children. Outdoor commercial signs, which formerly had to appear only in French, now could include other languages in smaller print. Even the commission that monitored compliance with the laws, the "language police," was abolished.

Francophones now dominate Quebec's business sector, and they are gaining control of its manufacturing economy as well. In 1980 the Levesque government put the sovereignty-association question before the people in a referendum. It was defeated, by roughly a three-to-two margin, an outcome that was welcomed by Canadian federalists and which met with approval in New York's financial community. Subsequent referendums on the issue have moved closer to an evenly divided electorate.

The failure to achieve a clear mandate in favor of separation has drawn attention away from the movement at times. Opinion remains so closely divided that no drastic actions are deemed advisable, yet Quebec has worked so long renegotiating its ties with the rest of Canada that in some respects it has become a separate state already, even while remaining part of Canada. Canada's provinces have broad powers within the Canadian federation that American states lack; thus, the maneuvering probably will continue, with frequent polls testing public opinion on the issue. In 1998 Canada's Supreme Court ruled that Quebec does not have the right to declare independence from Canada unilaterally, under either the Canadian Constitution or international law. The 1998 provincial elections in Quebec returned the Partí Québecois to power but without a mandate to proceed toward another referendum on separation.

## *References*

Bernier, Gerald, and Daniel Salee, *The Shaping of Quebec Politics and Society.* Washington, D.C.: Crane Russak, 1992.

Bouchard, Gerard, and Isabelle de Pourbaix. "Individual and Family Life Courses in the Saguenay Region, Quebec, 1842–1911." *Journal of Family History* 12 (1987): 225–42.

Fraser, Graham. *Rene Levesque and the Partí Québecois in Power.* Toronto: Macmillan of Canada, 1984.

Freeman, Alan. *Dividing the House: Planning for a Canada without Québec.* Toronto: Harper-Collins, 1995.

Grenier, Fernand, ed. *Quebec: Studies in Canadian Geography.* Toronto: University of Toronto Press, 1972.

Guindon, Hubert. *Québec Society: Tradition, Modernity, and Nationhood.* Toronto: University of Toronto Press, 1988.

Harris, Richard Colebrook. *The Seigneurial System in Early Canada.* Madison: University of Wisconsin Press, 1968.

Linteau, Paul André, Rene Durocher, and Jean-Claude Robert. *Québec: A History, 1867–1929.* Toronto: James Lorimer & Co., 1983.

Louder, Dean, ed. *The Heart of French Canada.* New Brunswick, N.J.: Rutgers University Press, 1992.

Ouellet, Fernand. *Economy, Class, and Nation in Québec.* Toronto: Copp Clark Pitman, 1991.

Thomas, David. *Whistling Past the Graveyard: Constitutional Abeyances, Québec, and the Future of Canada.* New York: Oxford University Press, 1997.

Young, Brian, and John A. Dickinson. *A Short History of Québec: A Socio-Economic Perspective.* Toronto: Copp Clark Pitman, 1988.

CHAPTER 3

# The Maritimes

A third major division of eastern Canada is comprised of the three Maritime Provinces: Nova Scotia, New Brunswick, and Prince Edward Island. Newfoundland did not join Canada until the middle of the 20th century, and Quebec has acted as though it wants out of Canada altogether, but the Maritime provinces have long sought a closer integration with the country whose mainland periphery they occupy. Handicapped by a location hundreds of miles east of Canada's metropolitan industrial centers, Maritimers support national economic policies to reduce the regional isolation they face. In the early years of European settlement, the Maritimes were a distant and little-known land to the French and the English, a forested domain of hard rocks, cold winters, and few apparent resources, yet one over which the two powers contested frequently and sometimes bitterly.

## Acadia

French settlement in the Bay of Fundy region began with a few families who lived at Port-Royal (Annapolis Royal), Nova Scotia, in the 1640s (map 3.1). Other settlements or outposts founded by the French were scattered around the Bay of Fundy (Baye Françoise), from the mouth of the Penobscot River in Maine north into New Brunswick and south along the entire western coast of Nova Scotia. Though the term Acadia (Acadie) is properly applied to all French settlements in what are now the Maritime provinces, the heart of the region in the 17th and 18th centuries was the Nova Scotia side of the Bay of Fundy. The early French settlement of Acadia was roughly contemporaneous with that of Quebec, although Acadia was much the smaller of the two, with a population only about one-tenth of Quebec's.

Approximately fourteen hundred Acadians were living around the Annapolis River and in the Minas and Cumberland Basins when the British captured the French garrison at Port-Royal in 1710. British control imposed no immediate threat to the Acadians for the next four decades, however, and life in the settlements, which revolved around farming, fishing, and some trading of furs, continued as the population grew.

Most Acadian settlers avoided the rugged uplands, which were largely unfit for agriculture. They favored the much less extensive (but level) tidewater marshlands instead. Dikes were constructed across the low-lying salt-water marshes

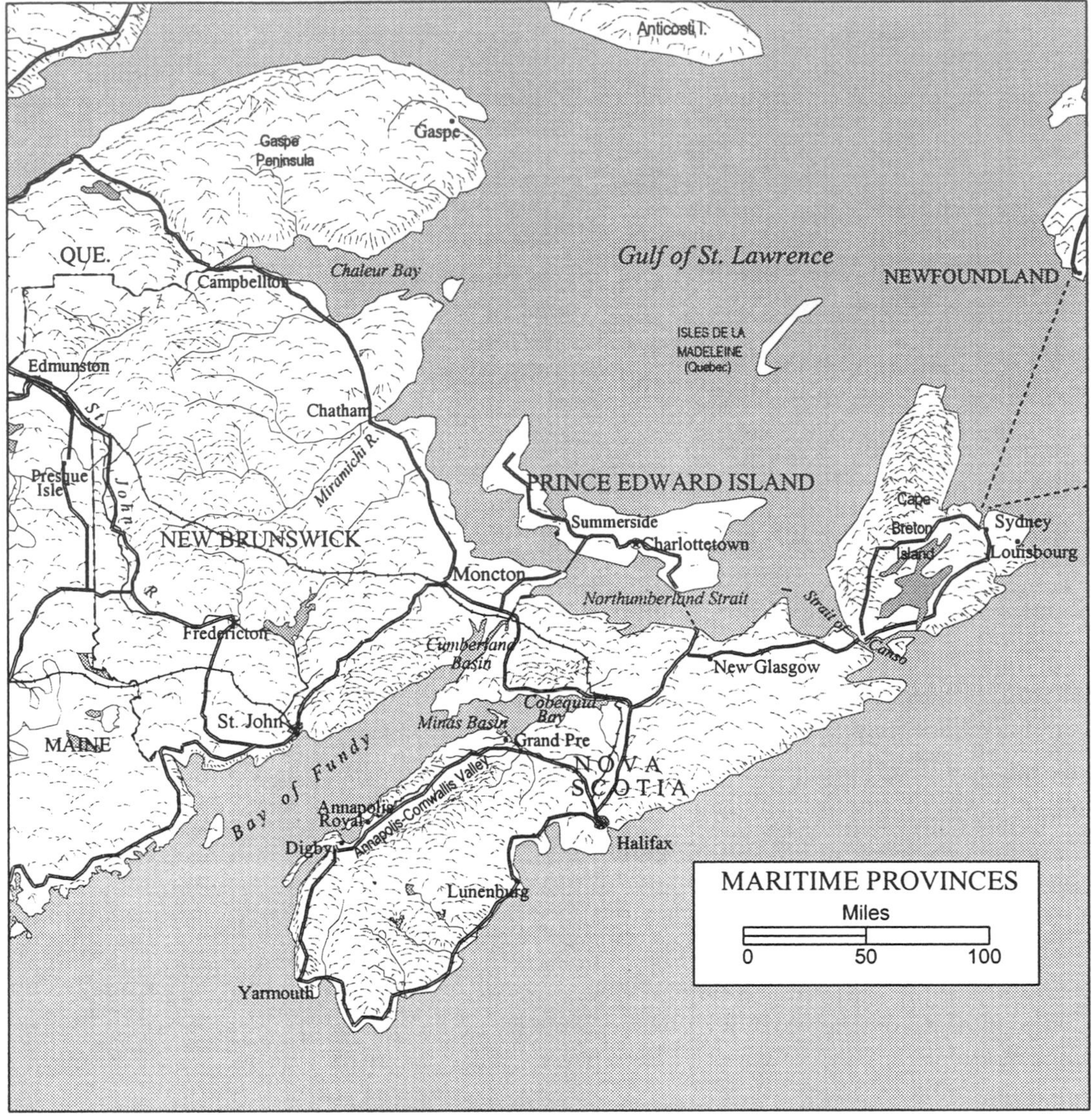


MAP 3.1

in the Annapolis Valley. Sluices built into the dikes allowed freshwater streams to drain to the sea, but the passages closed tightly when the flood tide rose against the dikes. New farmland was reclaimed from the sea in this manner. It is a practice that still continues in the broad, marshy valleys that open out to the Bay of Fundy (fig. 3.1).

Between 1710 and the early 1750s, the official posture of Acadians living under British rule was one of neutrality; but the times were not to last, as the level of conflict between France and Great Britain escalated. The French constructed their fortress of Louisbourg on Cape Breton Island, and the British countered with a rival fortification at Halifax. British officials came to doubt claims of neutrality in the Acadian population, which numbered nearly thirteen thousand by 1750. Acadians' lands were seized and their homes were destroyed.

Deportations began in 1755. More than half of the Acadians were sent to British colonies (American or West Indian) to the south; many others were deported to France, a country that, despite their French heritage, few Acadians had seen. Henry Wadsworth Longfellow later portrayed the tragic deportations from Nova Scotia to Louisiana in his poem *Evangeline: A Tale of Acadie* (1847). Although the origin of Acadian (Cajun) settlement in Louisiana is associated with the Acadian diaspora from Nova Scotia, most of Louisiana's Acadians went there in a second migration, after they had first been deported to France or the West Indies (see chapter 12, fig. 12.5). Their culture acquired new influences in the West Indies that were not traceable to the life they had known in Acadian Nova Scotia.



By 1800 there were only half as many Acadians in the Maritimes as had lived there fifty years earlier, but this low point did not signify the end of Acadia as much as it marked a new phase in its growth. A slow return of displaced people eventually led to both a larger and a more geographically dispersed Francophone population in the Maritimes. Migration from Quebec to lumbering and milling centers in northern New Brunswick in the late 19th and early 20th centuries added to the Francophone population. Acadians were less inclined to leave their region of birth than were their Quebecois neighbors. New Acadian population centers appeared where there had been none before.

An increased level of Acadian nationalism and a rebirth of cultural consciousness emerged with time. By the 1950s Acadia was no longer just the vanished "Land of Evangeline" used in tourist promotions of this scenic land. Acadian political and cultural movements emerged in the 1960s, especially in the railway and manufacturing city of Moncton, New Brunswick, which had acquired a large and growing Acadian population. These popular movements were comparatively small in comparison with those of Quebec, but they led to

FIG. 3.1. *Land reclamation in the Annapolis-Cornwallis Valley continues a tradition of creating arable tracts out of the tidal marshes dating back to the 18th-century Acadian settlers.*

the achievement of official bilingualism in New Brunswick in 1969 and to a much greater degree of de facto bilingualism in succeeding decades.

## The Annapolis-Cornwallis Valley



Acadian colonists had come to the Annapolis Valley in the 17th century in part because the flat, tidal marshes offered potential farming land. Annapolis Valley drainage is separated from that of a second river, the Cornwallis, only by a low divide, thus creating an almost continuous lowland in western Nova Scotia. The Annapolis-Cornwallis lowland, in turn, is walled in by a nearly unbroken ridge of volcanic rock, North Mountain, to the northwest and by the broader, upland massif of South Mountain, which trends in parallel fashion a few miles away. The valley, which varies between three and seven miles in width, is bordered by the prominent North and South Mountain uplands for more than seventy-five miles from Grand Pré to Digby.

Unaffected by fog and sheltered from the winds that batter the Fundy coast, Annapolis-Cornwallis Valley farmland was attractive to many groups, including thousands of immigrants from New England who took over the Acadians' Nova Scotia lands, opportunistically, after the deportations. The new immigrants dominated the population of Nova Scotia by 1775. Annapolis Valley farmers grew fruits, vegetables, and grains for local consumption and for the export market. Agriculture became more intensive after a railway was completed linking the valley to Halifax in 1869.

Apples became a major crop for Valley farmers, who produced millions of bushels annually by the early 20th century. Most of the crop was exported, especially to Great Britain, where the apples enjoyed great popularity. Eventual problems with the quality of the fruit, greater apple production in Great Britain, and the inability to capture markets elsewhere in North America caused the industry to decline by the 1950s. Faced with the burden of supporting too many farmers with too many apple trees, the government paid some farmers to cut down the orchards that had once been a hallmark of the region. Agriculture has declined generally in the Annapolis-Cornwallis Valley during the past forty years, part of a general trend observed in many areas of the Maritimes and New England that once had farming economies. The thousands of tourists who visit the region each year represent an increasingly important part of the local economy.

The Acadian farmers who diked and drained the Annapolis-Cornwallis marshlands more than three centuries ago were taking advantage of one of the region's most unique and renewable of resources—the tides. The Bay of Fundy has a funnel-shaped cross section that accentuates the differences between low and high tides and creates a noticeable rush of water inland (the "tidal bore") when waters rise. The tidal range averages about 28 vertical feet at the Digby Gut, where the Annapolis River enters the Bay of Fundy, and exceeds 50 feet—one of the largest tidal ranges in the world—on Cobequid Bay at the apex of the Minas Basin. A massive, turbulent mixing of salt and fresh water and bottom sediments pushes inland on the flood tide.

Plans to harness the ebb and flow of the tides to create electrical power were realized, at least on a small scale, when the Annapolis Tidal Generating Station was completed in 1984. The plant generates 30 million kilowatt hours per year, enough electricity to supply forty-five hundred homes. Plans have been laid to construct much larger tidal generating stations in the Cumberland and Minas Basins, although concerns about the environmental impacts of such facilities, which would require dams more than six miles wide, probably will prevent their construction.

## Eastern Nova Scotia

Eighteenth-century life in the Maritimes was dominated by a series of pitched battles waged by the French and the British for control of the Atlantic Coast. The victors one year often were the vanquished a few years later. In 1713 the Treaty of Utrecht confirmed British control over most of the Maritimes but left Île-Royale (Cape Breton Island) to the French, who soon began building the massive fortress at Louisbourg. Louisbourg fell in a lengthy siege brought by four thousand New Englanders in 1745 but was returned to French control in 1749. It fell for the second and last time to a much larger British force in 1758, a prelude to the British invasion of Quebec. The British then destroyed the fort, which remained in ruins for two hundred years until it was reconstructed by Parks Canada as a tourist attraction in the late 1960s.

What Louisbourg was to the 18th-century French, Halifax was to the British. A fortified city was begun along Halifax Harbor in 1749 to offset the return of Louisbourg to France. Halifax became one of Great Britain's most fortified imperial cities in North America, a base of naval operations, the colonial capital, and a growing mercantile center that traded with the American colonies and with the West Indies. With the French removed as a threat, British strategic concerns became focused on the United States. Early-19th-century Halifax continued to be, above all else, a military town.

Nova Scotia's rugged upland forests offered few incentives to European-derived farming people, but the better soils in the broader river valleys did attract immigrants. The land was cleared of its heavy timber, and farms were established. Immigration to Nova Scotia was brisk at times. Coastal fishing settlements at Yarmouth, Shelburne, and Lunenburg attracted more European immigrants and became, in addition to Halifax and Digby, Nova Scotia's most important fishing ports (fig. 3.2). Nova Scotia's banks fishery was based on the use of fast-sailing schooners and cod trawl lines in the Georges and Sable Island Banks, where American fishermen competed for the annual catch (see chapter 1, map 1.2). These fisheries, like those elsewhere in Atlantic Canada and New England, are in jeopardy today due to overfishing, estuarine pollution, and seabed disruption caused by bottom trawling by large fishing vessels.

Cape Breton Island became home for thousands of Scottish Highlanders and islanders in the early decades of the 19th century. Patches of arable land around Cape Breton's margins supported small-scale agriculture that was supplemented by an inshore fishery similar to that which developed in Newfound-

FIG. 3.2. *Peggy's Cove, near Halifax, was once a fishing village, but now it bears a heavy influx of tourists, who come to the Maritimes in search of scenes such as this.*

land. Near Sydney and Glace Bay, exposed seams of coal led to the development of the Cape Breton coal mining industry in the 1830s; nearly the entire output of the mines was exported, however, mainly to Great Britain. A steel industry, using Cape Breton coal and Newfoundland iron ore, was established at Sydney and New Glasgow in the late 19th century.

Industrial growth in the Maritimes was favored by local resources, but it was inherently restricted by isolation from markets. Beginning in 1879, the Canadian government pursued its National Policy of economic growth for the nation as a whole. A system of protective tariffs was imposed to stimulate industrial growth. Government support of railway construction was designed to link all of the Canadian provinces with the national market. Railway freight rates were adjusted to make it relatively cheaper for industrial products originating in the Maritimes to reach central Canada. Although national economic growth was the primary objective, policies relating to railway rate reductions also attempted to overcome regional inequalities—to draw the Maritimes closer to the Canadian economy.

The National Policy was successful in sheltering and developing Canadian industries that might otherwise have been overwhelmed by products from competitors in the United States. Textile and woolen industries, shipbuilding, and machinery works were some of the industries that thrived in Halifax, St. John, and smaller Maritime cities at the beginning of the 20th century. But the National Policy could not prevent changes in regional economic conditions within Canada. The coal-mining and steel manufacturing and fabricating industries near Sydney and New Glasgow constituted one of Canada's most important industrial complexes. Originally based on local capital, control was successively lost, first to British interests and then to investors in Montreal. Pressure from central and western Canada led to the removal of the railway rate structure that favored Maritime industries. Steel production in Nova Scotia declined. Halifax, once an important banking center in its own right, saw its principal financial, wholesale, and retail institutions remove their headquarters to Montreal or

Toronto by the 1930s. Economically speaking, the Maritimes became increasingly peripheral to Canada by the middle of the 20th century.

More recent attempts to reindustrialize Nova Scotia have met with partial success, especially in the growth of its automobile industry. Nova Scotia's steel industry has been unable to compete successfully with mills in the Great Lakes region, however. What remains of the Maritimes' strategic locational value to Canada is a circumstance of geography. Halifax and St. John are Canada's major year-round Atlantic ports. The shipping season at Montreal is foreshortened by ice in the St. Lawrence River, a condition that led railway interests to link Montreal with the Maritime ports beginning in the 1870s. With its important role as a naval and shipbuilding center already established, for more than a century Halifax expanded its role as a grain-shipping and general cargo port and became the largest seaport in Atlantic Canada. The advent of time-sensitive containerized shipping in international trade has reemphasized the value of Halifax's forward position in the North Atlantic. An importer in Chicago can obtain shipments from Europe several days earlier if the container is routed via Halifax and then carried by rail to Chicago, rather than taking the longer and slower ocean haul and using the port of New York.



## Prince Edward Island

Generalizations about the harsh physical environment of eastern Canada ("the land God gave to Cain," mused Jacques Cartier) cannot be applied uniformly throughout the region, especially not to Prince Edward Island. Île Saint-Jean, as the French named this small island (about two thousand square miles, roughly the size of Delaware), is an arc-shaped plateau of gently dipping, red sandstone that rises only a few hundred feet above the waters of the Northumberland Strait, which separates it from Nova Scotia. Glaciation provided a deep mantle of soil materials derived from the sandstone. Although the reddish-hued Prince Edward Island soils are not especially fertile, they are better for agriculture than most soils elsewhere in eastern Canada. No point on the Island is more than about seven miles from the sea. The cool, maritime climate is comparatively mild and offers a three- to four-month growing season. Potatoes, grain, and hay crops do well in this environment.

Great Britain took possession of Île St.-Jean under terms of the Treaty of Paris of 1763, at the same time that it gained sovereignty over Quebec and Nova Scotia. In 1799 the island was renamed Prince Edward. As an experiment in replicating the stratified social structure of rural Great Britain, Prince Edward Island was subdivided into more than five dozen townships (called "lots") of twenty thousand acres each. The lots were awarded to an assortment of court favorites and land speculators on the condition that they were to populate their lots with taxpaying, Protestant settlers who would farm the land. Acadians who returned to Prince Edward Island after 1800, like the new immigrant Scots, found it difficult to obtain title to land under Prince Edward Island's proprietorship system.

Despite problems of land tenure in the early years, thousands of acres of for-

FIG. 3.3. *A Prince Edward Island landscape at Hunter River, west of Charlottetown.*

est land were cleared—far more in proportion to Prince Edward Island's small size than were cleared for agriculture elsewhere in the Maritimes. Fields of potatoes, turnips, wheat, and other small grains, fringed by the native forest, constitute an open, settled landscape of farms that contrasts sharply with the endless stretches of dark, coniferous forest that are more typical elsewhere in the Atlantic region (fig. 3.3).

Charlottetown has been Prince Edward Island's principal city since colonial times, the seat of its government, and its most important harbor. In 1864 delegates from New Brunswick, Nova Scotia, Ontario, and Quebec met here to discuss union, a conference that led to the creation of the Dominion of Canada in 1867. Prince Edward Island, then part of Nova Scotia, became Canada's fifth province in 1873. The island's population of roughly 95,000 at the time it achieved provincial status grew slowly to reach a total of perhaps 100,000 for the first time in the early 1950s. Agriculture, based on potato production, and some fur (fox) farming were the principal means of support for rural people, although 10 percent of the island's residents also derived at least a partial living from the lobster, herring, and cod fisheries.

Prince Edward Island's economy began to modernize in the 1950s as it entered a period of tourism-based growth, a trend that has become even more pronounced at present. Agriculture declined, but at a pace slower than observed elsewhere in the Maritimes. Prince Edward Island's potatoes, many of which are grown for seed, remain an important crop, even though disease has sometimes led to embargoes preventing the crop's shipment elsewhere. Water-pollution problems that once threatened the nearshore shell fisheries have been brought under control in recent years. The demand for Atlantic shellfish is so strong that experiments in aquaculture (already used to grow mussels and other shellfish) probably will appear in more of the island's sheltered bays.

The quiet, rural appearance of Prince Edward Island's landscape of farms and fishing villages remains, however, to the delight of tourists, who now provide the largest share of the island's earned income. The bucolic charms of Canada's "Garden Province" are apparent to anyone who visits, but they became best known to the world through the eyes of a fictional orphan girl, Anne Shirley, whose love for her new home illuminated the pages of Lucy Maud Montgomery's children's classic, *Anne of Green Gables* (1908). Tourist parks, hotels, golf courses, and restaurants began to grow in number during the 1960s as more tourists flocked to see the island where the fictional Anne had lived. Farmers were advised to enter the bed-and-breakfast trade, one result of which was to create an unrealistically large number of rural inns and cabins along the island's winding highways.

After many years of planning, the Confederation Bridge across the Northumberland Strait, linking Prince Edward Island to New Brunswick, opened for traffic in 1997. Automobiles and tour buses once had to wait their turn getting to the island on one of the two ferry connections, but now they can reach it quickly. Prince Edward Island has no freeways or highway bypasses—not even in the capital city of Charlottetown, where traffic across the island still passes through the city center. The bridge link will bring additional traffic, but whether tourism will reach levels to make the island lose its charms remains a matter of debate.

## New Brunswick

The aftermath of the American Revolution brought changes to Canada that were as important as those wrought in the former American colonies. As in any revolution, the losing side sought some accommodation with the new order. Loyalists—Americans who remained loyal to Great Britain throughout the Revolution—began to desert the United States in large numbers beginning in the early 1780s, as the new American nation was forming. Nearly 40,000 of them came to the British colony of Nova Scotia, a number that included 3,000 black colonists, most of them ex-slaves from Tidewater plantations who, like their former masters, had remained loyal to Great Britain. Most Loyalists demanded grants of land from the British Crown as a price for their allegiance and sacrifice. Cape Breton Island was set aside as a Loyalist reserve for a time, although better lands in the Annapolis Valley as well as opportunities in existing cities like Sydney and Halifax were most attractive to the incoming Americans.

New Brunswick was established as a direct result of the Loyalist migrations. In 1783 British ships sailing from New York landed approximately 14,000 Loyalists at the mouth of the St. John River. Many of the new arrivals settled the St. John Valley; others founded the city of St. John at the river's entrance to the Bay of Fundy. Their numbers were sufficient to secure a separation of New Brunswick from the larger colony of Nova Scotia, of which it was originally a part. Although St. John's growth depended on changing British mercantile policies, the city became the focus of New Brunswick's commercial developments in the 19th century. St. John became an important shipbuilding center and a major port for

FIG. 3.4. *A log "herringbone" cut in a New Brunswick industrial forest of the St. John River Valley.*

importing trade goods and exporting timber. By 1850 St. John and Halifax, with about 25,000 people each, were the major urban centers of the Maritimes.

Good agricultural land was even harder to find in New Brunswick than it was in Nova Scotia. The forest was New Brunswick's major resource from the early decades of the 19th century onward. Hardwoods, including maple, birch, and aspen, dominate the St. John Valley and the margins of the Bay of Fundy. Softwoods, especially spruce and fir, are most abundant in the interior. Two major rivers and their tributaries—the St. John, which flows from western New Brunswick to the Bay of Fundy, and the Miramichi, which has a large drainage basin in the eastern interior and flows to the Gulf of St. Lawrence—made it comparatively easy to gain access to all varieties of timber. Logging continues today in large portions of interior New Brunswick that never have had permanent settlement (fig. 3.4).

Historically the industry depended on thousands of lumberjacks, who took to the woods toward the end of each winter, when the ground remained frozen and logs could be skidded out more easily. Once the spring breakup of ice on the rivers had taken place, river drivers did their part by guiding booms of logs downstream toward the sawmills. New Brunswick's timber was sawn into ships' masts, barrel staves, deals (large planks), and boards, which were then shipped mostly to Great Britain. Sawmills were the basis for many of New Brunswick's cities and towns, both as the source of lumber from which the entire town was constructed and as the means of livelihood for those who lived there.

In the early 20th century pulp and paper manufacturing began to replace sawmilling as the major consumer of New Brunswick's softwoods. Pulp and paper mills constructed at the head of Chaleur Bay and near the mouth of the Miramichi depended on the interior spruce and fir stands even more heavily than before. New Brunswick's approximately one dozen pulp and associated paper mills remain today as the province's largest industry. The ownership of New Brunswick's forest lands at present reflects the dominance of these interests. It is about evenly divided between Crown land and privately held tracts; many of the latter are controlled by a few major paper producers. The introduction of recycled paper has reduced the rate of cutting for pulpwood, although not the total cut. Larger trees that once would have gone into paper now leave the woods trimmed for lumber mills.



## The St. John Valley

The push of settlement up the St. John Valley, begun by the Loyalists, continued in later years with the arrival of new immigrants from Great Britain. Beyond Reversing Falls (a low waterfall in St. John's harbor that flows inland on the flood tide), the river was navigable as far inland as Fredericton, where New Brunswick's provincial capital was founded. The St. John Valley contained most of the potential farmland in New Brunswick, and the inland penetration of new settlement followed its course, first westward and then northward along the international boundary. Portions of the St. John Valley that would later lie within the borders of Quebec, New Brunswick, or Maine were in dispute. The Aroostook country of Maine, bounded by the St. John River's great northward arc, did not become part of the United States until the Webster-Ashburton Treaty was negotiated between the United States and Great Britain in 1842.

Those who settled the upper St. John Valley comprised both an international and a multiethnic mixture of native, Quebec-born, Acadian, British Canadian, and British-American peoples. Typical of many border zones, its population retains an international flavor because the same cultural groups live on both sides. Here are brought together the two major subpopulations of French Canada, Acadians and Quebecois. Because northern Maine is part of the United States, it is neither officially bilingual (like New Brunswick) nor unilingual French (like Quebec). But French surnames are common in the northern Maine population, the region is heavily Roman Catholic, and many families trace their roots either to Acadia or to the St. Lawrence Lowland. Both Canadian and American investment capital is behind the local pulp and paper mills. Like the hydroelectric dams across the St. John that supply the mills with power and which are anchored in both American and Canadian soil, the pattern of investment is binational and shows little regard for the border.

## *References*

Acheson, T. W. *Saint John: The Making of a Colonial Urban Community.* Toronto: University of Toronto Press, 1985.

Buckner, P. A., and David Frank, eds. *Atlantic Canada after Confederation.* Fredericton, N.B.: Acadiensis Press, 1988.

Clark, Andrew Hill. *Three Centuries and the Island: A Historical Geography of Settlement and Agriculture in Prince Edward Island, Canada.* Toronto: University of Toronto Press, 1959.

———. *Acadia: The Geography of Early Nova Scotia to 1760.* Madison: University of Wisconsin Press, 1968.

Day, Douglas, ed. *Geographical Perspectives on the Maritime Provinces.* Halifax, N.S.: Department of Geography, St. Mary's University, 1988.

Forbes, E. R., and D. A. Muise, eds. *The Atlantic Provinces in Confederation.* Toronto: University of Toronto Press, 1993.

Harris, R. Cole, and John Warkentin. *Canada before Confederation.* New York: Oxford University Press, 1974.

Hornsby, Stephen J. *Nineteenth Century Cape Breton: A Historical Geography.* Montreal: McGill-Queen's University Press, 1992.

Smitheram, Verner, David Milne, and Stadal Dasgupta. *The Garden Transformed: Prince Edward Island, 1945–1980.* Charlottetown, P.E.I.: Ragweed Press, 1982.

Wynn, Graeme. *Timber Colony: A Historical Geography of Early Nineteenth Century New Brunswick.* Toronto: University of Toronto Press, 1981.

*PART II*

# The Northeast

CHAPTER 4

# New England

The northern New England states of Maine, New Hampshire, and Vermont are transitional between the rugged uplands of the Maritimes and Quebec and the industrial lowlands of southern New England (map 4.1). Bypassed by the main stream of population expansion during colonial times, northern New England was a slowly developed frontier that depended on the willingness of settlers to move north—even east—when the main direction of American expansion lay clearly to the west. The same circumstances that made "down East" New England a backwater in terms of settlement expansion and economic growth also insulated it from some of the negative impacts of that growth in later years.

Southern New England was more favorably endowed in terms of the resources that mattered to European colonists. It was warmer, had a less broken topography, had better soils, and possessed a mixed hardwood-coniferous forest cover that supplied nearly every use of wood at the time. The highly indented coastline of Long Island Sound provided numerous sheltered harbors. Although the prospects for agriculture were only fair in Massachusetts, they were good in parts of Connecticut and Rhode Island. Southern New England was the colonial hearth from which settlers spread into the northern New England states, just as they moved from southern New England westward, across the Hudson River Valley and into upstate New York, in later years. Despite their small size and many miles of shared borders, all six New England states developed in distinctive ways that were revealed even in colonial times.

## Colonial Lands and Politics

Popular usage of the term *New England* began with the publication of Captain John Smith's *Description of New England* (1616). Smith's presence in the New World is most often associated with Virginia, but he also promoted the idea of fishing-based settlement along what is now the coast of Maine. Efforts to raise expeditions that would colonize New England centered in the West Country port of Plymouth, England. There Smith became associated with Sir Ferdinando Gorges, the principal organizer of the Virginia Company of Plymouth, which had obtained a charter to colonize that part of "Virginia" lying between the 38th and 45th parallels of latitude, roughly from Chesapeake Bay to Maine.

The English settlers who made the first permanent habitation in this zone were, of course, the Puritans. The sea-weary band who crossed the Atlantic on

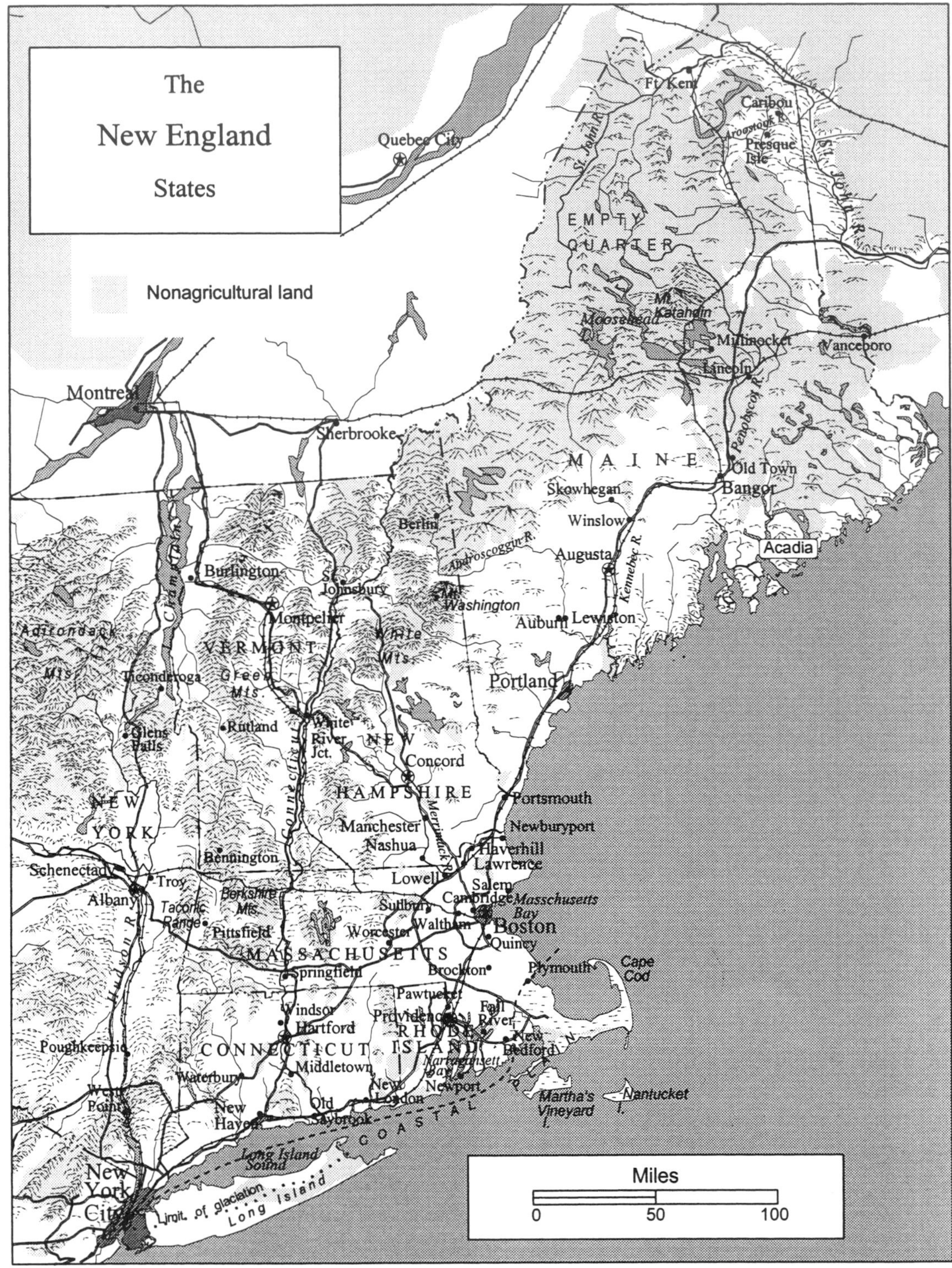
The
New England
States
Nonagricultural land
Quebec City
Montreal
Sherbrooke
Ft Kent
Caribou
Presque Isle
St. John R.
Aroostook R.
EMPTY QUARTER
Mt. Katahdin
Moosehead L.
Millinocket
Vanceboro
Lincoln
Penobscot R.
MAINE
Old Town
Bangor
Skowhegan
Winslow
Acadia
Augusta
Kennebec R.
Androscoggin R.
Berlin
Mt. Washington
Auburn
Lewiston
Portland
White Mts.
Burlington
St. Johnsbury
Montpelier
VERMONT
Adirondack Mts.
Ticonderoga
Green Mts.
Rutland
Glens Falls
White River Jct.
NEW HAMPSHIRE
Concord
Portsmouth
Newburyport
NEW YORK
Manchester
Nashua
Merrimack R.
Haverhill
Lawrence
Bennington
Schenectady
Troy
Albany
Lowell
Salem
Cambridge
Masschusetts Bay
Berkshire Mts.
Taconic Range
Pittsfield
Sudbury
Waltham
Boston
Worcester
Quincy
MASSACHUSETTS
Hudson R.
Connecticut R.
Springfield
Brockton
Plymouth
Cape Cod
Pawtucket
Windsor
Hartford
Providence
Fall River
RHODE ISLAND
New Bedford
Poughkeepsie
CONNECTICUT
Middletown
Narragansett Bay
Newport
Waterbury
New London
Martha's Vineyard I.
Nantucket I.
West Point
New Haven
Old Saybrook
COASTAL
Long Island Sound
Long Island
Limit of glaciation
New York City
Miles
0
50
100

MAP 4.1

the *Mayflower* and landed at Plymouth, Massachusetts, in 1620 decided not to travel south to where they had purchased land. They remained at Plymouth and drew up the Mayflower Compact as an instrument to guide their affairs, knowing that they had no legal right to the land they were settling. The first ten years of settlement in Massachusetts were marginal for the Puritan colonizers. In 1630 the tempo of activity increased, and its focus shifted from Plymouth to Massachusetts Bay, which had a better harbor and was surrounded by lands more suitable for settlement. In 1630 John Winthrop and a group of his fellow Puritans brought nearly two thousand new colonists to Massachusetts Bay, and they became the nucleus for the new city of Boston.

During the 1630s more than twenty thousand additional immigrants joined Winthrop's group and spread out over the scattered patches of arable land fringing the many harbors of Massachusetts Bay. Although the new arrivals were of varied backgrounds, the majority of them were devout Puritans. Once the colony's Puritan leaders decreed that only church members had a right to vote, the affairs of government increasingly came into the hands of a small ruling class centered in Boston. During the period of Puritan rule in England (1649–60), Massachusetts became a holy commonwealth, governed by the iron laws of the local Puritan rulers and supported by a friendly government in the mother country. The restoration of the English monarchy brought a relaxation of these laws, a shift toward a more mercantile basis for social and economic life in the colonies, and a less privileged role for Massachusetts.

The many outlying towns surrounding Boston, including Salem, Newtown (Cambridge), Watertown, Roxbury, and Sudbury, were organized land-settlement schemes where the common forest and pasture lands belonging to each village were administered by a town government. Individually owned parcels of cropland surrounded each village. Although the New England town-meeting form of local, democratic government arose from this arrangement, the tradition of an agricultural village, inhabited by farmers and a few artisans, with strips of cultivated fields radiating from the village, was an import from the European past. In Quebec a landlord (*seigneur*) was part of this system, but in Massachusetts farmers owned their own croplands and cooperated with one another in the use and protection of common woodlands and pasture. The town—a village and all the lands belonging to it—became the basic unit of New England settlement.

Rhode Island began as a refuge for those who sought to escape the authority of Puritan rule in Massachusetts. Roger Williams, an English-born clergyman, secured a separate charter for a new settlement that he founded on Narragansett Bay in 1644. Providence Plantation, which expanded to become the state of Rhode Island, had a less restrictive society than Massachusetts and became a haven for nonconforming religious groups. The lands that became Rhode Island already were well used by their native Algonquian inhabitants when the first Europeans settled there. The first written description, by Giovanni da Verrazano, who happened upon Rhode Island while on a voyage of discovery in 1524, referred to the many islands and peninsulas of Narragansett Bay as a tended landscape of fields and woodlands. Rhode Island had a native population of several thousand during colonial times.

Population growth in Massachusetts was so rapid during the 1630s that it stimulated a movement in search of new lands to the west as more immigrants arrived from England. The best agricultural land in New England was in the Connecticut River Valley, where migrants from Massachusetts settled as early as 1636. The next year the towns of Windsor, Wethersfield, and Hartford formed the self-governing colony of Connecticut. A separate colony was founded at New Haven in 1638. From these origins the general course of Connecticut's development would follow as a clustering of settlements along the Connecticut River and a secondary scattering along the coast. Like Massachusetts, which supplied many of Connecticut's early settlers, most Connecticut towns had a Puritan majority in the early years.

Maine, New Hampshire, and Vermont each had different origins. New Hampshire was one of the original thirteen colonies; Vermont was a self-proclaimed republic for fourteen years before it became a state; and Maine was under the jurisdiction of Massachusetts until it, too, became a state in 1820. Having watched the Massachusetts settlement begin without his involvement, Ferdinando Gorges persuaded the council of King Charles I to grant him the land between the Merrimack and Kennebec Rivers. In 1629 Gorges divided the grant with Captain John Mason. Gorges took the northern part and became proprietor of the province of Maine, and Mason acquired the territory that became New Hampshire.

When the Puritans ruled England, Massachusetts extended its borders northward to include New Hampshire and southern Maine. The restoration of the monarchy brought attempts to curb Massachusetts's appetite for territory, one result of which was the Crown's creation of the separate province of New Hampshire in 1679. Another haven from the excesses of Massachusetts Puritanism, New Hampshire was further enlarged at Massachusetts's expense in 1740. A large tract on both sides of the Connecticut River came under New Hampshire's jurisdiction. New Hampshire's colonial governor, Benning Wentworth, made numerous grants of land, including generous awards to himself and to his father-in-law. Settlers moved into New Hampshire from the older coastal communities to the east and south, where land had become scarce, and from the longer-settled parts of the Connecticut River Valley.

By 1764 Wentworth had granted 138 towns totaling more than 3 million acres of land. The upper Connecticut Valley and the area that is now Vermont were known as the "Hampshire grants," so extensive was the area that had been granted. Unfortunately for those who expected to receive titles to their land, however, New York had been granting land in the same area. In 1664 King Charles II had awarded all land west of the Connecticut River to his brother, the Duke of York (later, King James II), and thus it became part of New York. A century later, in 1764, when New York's settled area still was confined largely to the Hudson River Valley, the Connecticut River was again confirmed as the New York–New Hampshire border. What is now Vermont was thus as much a part of New York as it was an extension of New Hampshire.

New York authorities demanded that settlers in the Hampshire grants pay fees in order to legitimize their land claims and sent surveyors to resurvey the land. Before long the Green Mountain Boys, under the leadership of Ethan Allen, became an organized force, resisting the surveyors and meting out punishment of their own choosing wherever they found a New York surveyor. The colonial governor of New York offered a reward for Allen's capture. In 1775 Allen's band captured Fort Ticonderoga from the British. Heman Allen, Ethan's younger brother, tried to persuade the Continental Congress to strike down New York's authority, but in vain. An independent Vermont republic was formed later that year.

Statehood was Vermont's goal, but New York's objection stood in its way. Great Britain attempted to exploit the dispute by attracting Vermont to its side during the Revolutionary War. In 1778 sixteen towns on the New Hampshire side of the Connecticut River petitioned to join Vermont and were accepted, but they were expelled by Vermont the next year. One proposal for the future envisioned splitting troublesome Vermont along the crest of the Green Mountains and attaching roughly half of it each to New York and New Hampshire. In 1781 Vermont took back the sixteen New Hampshire towns, annexed another eighteen, and claimed all territory west to the Hudson River. Partly because of these actions, those who had underestimated Vermont's seriousness acquired greater respect for the tiny republic's political nerve. The struggles finally ended in 1791 when Vermont paid a small sum to New York and became the first state to join the United States of America.

Maine was doubly leveraged by colonial status, being subjected not only to English authority but also to that of Massachusetts, of which it was a part. The Plymouth colony, which was trading with Maine's native people by 1629, received a royal grant of 3 million acres along the Kennebec River. Other grants of Maine land along the Androscoggin and Penobscot Rivers were made mainly to Bostonians. These absentee landlords, or "proprietors," collected rents from settlers. Acquiring large tracts of Maine land through purchase became an established practice among Boston investors even after the era of colonial land grants had passed. When Maine achieved statehood in 1820, direct political control from Massachusetts ended, but economic control by outsiders continued.

## Northern Maine

By 1790, when most of Vermont and New Hampshire had been settled, the population frontier lay no more than fifty miles from the seacoast in Maine. Northern Maine was a latter-day frontier, an eastward extension of settled New England. Geographically, northern Maine is the mirror image of western New Brunswick. It consists of two land-use regions that are divided unequally: a vast domain of forest and a narrow zone of agricultural land along the St. John River.

Today Maine contains some 17.6 million acres of forested land, nearly nine-tenths of the land in the state. This makes Maine the most heavily forested of the fifty states—including Alaska, whose millions of acres in tundra decrease the proportion in tree cover. Maine's forests are roughly two-thirds softwoods (balsam fir, spruce) and one-third hardwoods (maple, aspen, birch). All of these trees have commercial uses. Manufacturing is Maine's largest economic sector, and the paper industry accounts for the largest share of its manufacturing income. A torrent of paper, lumber, wood paneling, particle board, shingles, containers, toothpicks, and furniture pours out of Maine's factories each year.

Forests of this magnitude often are associated with extensive acreages of public land, especially national forests, in the western states, but only a small fraction of the Maine woods is government land. Most of the productive forest is privately owned, a legacy of Maine's early history, when extensive tracts were sold to investors at nominal prices. Some large land areas still are called *plantations,* a term often used for remote settlements in early New England. Today's plantations are focused on the production of spruce trees for local pulp mills.

Unique among the states is Maine's northwestern region—the Empty Quarter—whose population has now disappeared entirely. It consists of more than 5 million privately owned acres, all of it forested, which has a resident population of zero. Even though no one makes a permanent home in the Empty Quarter today, the region is anything but empty when measured in terms of economic activity. It is an industrial forest and an active part of the lumber, pulp, and paper industry of the Northeast. The unpopulated area includes nearly all land in the state of Maine north and west of Millinocket.

The term *clear-cut* is avoided, but the method of logging is to remove all commercially useful trees from a parcel before moving on to the next area to be cut. Seasonal bonded laborers, imported from Canada, once were used extensively for work in the northern Maine woods, but they have been replaced by freelance cutting crews who live in communities around the fringes of the Empty Quarter.

One negative result of this method of cutting is the creation of a series of even-aged stands of trees rather than a forest composed of variously aged trees, as would be produced under natural conditions. The spruce budworm, which attacks both balsam and spruce trees, has a life cycle that is favored when the forest consists of even-aged stands. Spruce budworm damage was extensive in the 1970s, but the annual menace began to recede in the 1980s. The forests of the Empty Quarter, like those cut for pulpwood elsewhere, have a compressed cycle of human-caused regeneration and are far from natural, despite the apparent wilderness condition.

Because the Empty Quarter has no population, it has no organized local governments and hence lacks public infrastructure. The only highway access is provided by the loggers' haul roads, such as the Golden Road, which are owned by the logging and paper companies that built them (fig. 4.1). The Maine woods ends abruptly at the Quebec border. What for Americans is a little-known industrial forest in a remote corner of Maine is, by Quebec standards, reasonable land for farming. Sawmills on the Quebec side of the border process some logs cut in northern Maine that are taken west via the haul roads. Dairy farms on the Quebec side end in a solid line of forest at the international boundary. Few places offer so vivid a contrast in land uses between the two countries.

FIG. 4.1. *Log truckers pause at a roadhouse near the end of one of northern Maine's privately owned haul roads.*

Maine's northern border is roughly defined by the great arc of the St. John River as it flows first north, then east, and then south into New Brunswick. In contrast to the acidic soils of the Empty Quarter, home to the undemanding spruce and balsam, the St. John Valley has more fertile soils where forests of maple and other hardwoods occupy smooth upland surfaces. In this international zone, an agricultural island has developed within a sea of forest. The soils are stony because of the glacial materials on which they developed, but they are well drained, have deep profiles, and are excellent for producing the potato crops with which the region has long been identified. Aroostook County, Maine, is the most important producer of potatoes in the eastern United States. New Brunswick's portion of the St. John Valley is one of Canada's largest suppliers of the crop as well.

A railway built north from Bangor to the Aroostook region in the 1880s enabled farmers to turn away from the traditional wheat culture they had practiced to concentrate on raising potatoes for outside markets. Like Annapolis–Cornwallis Valley apples, potatoes from the St. John Valley once enjoyed great popularity in the British market. Problems with the quality of the potatoes and the emergence of other sources of supply eventually restricted the sale of the crop largely to domestic markets. Occasional "potato wars" develop across the Maine–New Brunswick border as a result of contrasting tariff and pricing policies on potatoes made by the U.S. and Canadian governments, sometimes in retaliatory fashion.

Potatoes are a difficult crop to grow. Susceptible to disease, blights, and freezing, potatoes also are intolerant of both drought and deluge. Northern Maine's one-hundred-day growing season is barely long enough to bracket the ninety days potatoes require to mature. Commercially salable potatoes typically are produced with substantial amounts of chemical fertilizers, herbicides, and pesticides. Because they are so sensitive to disease, great efforts must be made to produce uncontaminated seed potatoes, whose eyes constitute the stock from which the next year's crop is grown. Maine's seed potatoes are sold widely, although its production of table stock has not increased nearly as rapidly as that of other states in recent years. Maine now produces only about one-fourth as many potatoes as Idaho, the leading potato state.

Potato country is not picturesque. Great stretches of rolling, bare ground are punctuated by farmsteads and the necessary potato sheds, which are low barns surrounded by an insulating blanket of earth as a protection against freezing temperatures. But as one of the few agriculturally productive regions remaining in either New England or Atlantic Canada, the middle St. John Valley provides an interesting contrast with the expanses of spruce and balsam forest that surround it.

Northern Maine's other rivers—the Penobscot, the Kennebec, and the Androscoggin—rise in the rugged uplands and flow southward across the rock outcroppings of New England's bedrock underpinning. Numerous ponds and water powers along the rivers became the sites of mill towns that specialized in the manufacture of paper. Mills at Berlin, New Hampshire, on the Androscoggin; at Winslow, Augusta, and Skowhegan, Maine, on the Kennebec: and at Millinocket, Lincoln, and Old Town, Maine, on the Penobscot or its tributaries —plus an even larger complex of paper mills in and around Portland—constitute the largest paper-manufacturing region in the northeastern United States. New England's manufacturing fortunes, declining in some sectors, have grown stronger in the paper industry during the past four decades.

## The New England Uplands

Outside of northern Maine, coniferous forests of the sort attractive to paper manufacturers are confined mainly to the higher ground of the White and Green Mountains. Lowland areas typically have a vegetation cover of hardwood forest. The White and Green Mountains are not geologically distinct units but rather are regional terms, applying to the uplands of New Hampshire and Vermont, respectively. They are part of a single mountain complex that extends from western Connecticut across northern New England into Quebec. The Taconic Range and the Berkshire Mountains of Massachusetts and Connecticut are part of the same system. All of the ranges consist of parallel, folded ridges separated by parallel valleys and have a core of granitic rocks. The Green Mountains are northern New England's counterpart of the Blue Ridge section of the Appalachians.

New Hampshire's White Mountains are more massive and have higher peaks than the Green Mountains. White Mountain topography is rugged enough to isolate much of the northern portion of New Hampshire, which is a principal watershed for the Connecticut River. Mount Katahdin (elevation 5,200 feet) in Maine is an outlier (a monadnock) of the White Mountains. Impressive though the peaks of New England's mountains are, however, Mount Katahdin and even Mount Washington (6,300 feet) have summits that were overtopped by the Pleistocene ice sheets. Ice moving from the north trended along the existing

FIG. 4.2. *Although the glaciers that left them disappeared thousands of years ago, a substantial crop of boulders heaves to the surface in Vermont's fields every spring. The state's miles of boulder fences bear testimony to the hard labor required to farm such land.*

north-south bedrock valleys, planing off mountain peaks and filling the valleys with glacial debris.

Northern New England's glacial till (unsorted glacial material) is particularly stony, a condition reflected in the miles of boulder fences along field boundaries (fig. 4.2). The fences have been built from the boulders heaved to the surface in each year's spring thaw. Generations of New England farmers have picked boulders from their fields and piled them atop the fence lines. Attempts to farm the rugged uplands have been continuous since colonial times. During the 1830s, Vermonters specialized in sheep grazing, but, as with all agricultural specialties the state has tried, there were always better lands to the west where production costs would be lower. Agriculture has long been in a state of decline in New England as a result. Off-farm migration produced population decline in upland portions of New Hampshire and Vermont even by the early decades of the 19th century.

Glaciation also left the land surface dotted with undrained depressions. "Pond" is among the commonest of topographic names in New England. Thin, rocky soils here contrast sharply with much of the Middle West, which was glaciated at the same time. The Middle West's cover of loess (fine-grained, windblown dust derived from former glacial lakes and streambeds) is absent in New England, a circumstance that makes the latter's soils much less suited for agriculture.

Northern New England still is classified as a dairy region, although poultry (broilers and eggs) provides the largest share of farm income in some parts of the region. Except for the Aroostook County potato-specialty area, agriculture now is a more sporadic than continuous feature of the rural landscape. Vermont is New England's leading dairy producer, although it has fewer than 175,000 milk cows, only about one-fourth as many as the state could count in 1959. Vermont's dairy industry is one-tenth the size of Wisconsin's today, despite proximity to urban markets in the Northeast. The dairy farms of New England were too small to be competitive with larger operations elsewhere. The loading of milk cans, once a daily ritual at dozens of country railroad stations in New England as the daily "down train" to Boston arrived, is only a memory.

Rural land use in the White and Green Mountains has shifted away from farming and now is dominated by a variety of exurban activities, especially tourism. The "soft" look of Vermont's smooth hillsides and rounded mountaintops, attributable to glaciation, represents an important asset for the state's ski industry. Good ski slopes abound. With the completion of interstate highways in the 1960s, skiers throughout the Northeast were able to gain easy weekend access to the growing number of ski resorts. Vermont had more than four dozen mechanized ski lifts in operation by 1975. In that year it was calculated that the state's entire population could be lifted to the tops of its ski slopes in a single afternoon.

## The Champlain Lowland

West of the Green Mountains lies the Champlain Lowland, a continuation of the Hudson River Valley, which separates New York from Vermont (fig. 4.3). Both the Hudson and the Champlain Lowlands represent a northward extension of the Great Valley of the Appalachians. Most of the lowland is occupied by Lake Champlain, a 125-mile-long body of water, averaging 10 miles in width, which extends north into Quebec. Lake Champlain drains to the north, via its outlet into the Richelieu River, a tributary of the St. Lawrence. The entire lowland was occupied by the Champlain Sea at the end of the Pleistocene epoch, when glaciers melted and drainage was blocked by ice sheets to the north but before the earth's crust had rebounded from removal of the ice sheet's weight. Sediments laid down by the ephemeral Champlain Sea became the basis of fertile soils. Good agricultural land is found around the lake's margins and on the larger islands within the lake. Much of Vermont's dairy production is concentrated in the Champlain Lowland.

The direction of regional slope reverses south of Glens Falls, New York, where the lowland carries the south-flowing Hudson River and its tributaries. The entire lowland is a corridor of easy access connecting New York City to the St. Lawrence Valley near Montreal. The level-profile route separating the Green Mountains on the east and the Catskills and Adirondacks on the west was long used as a route of trade, migration, and warfare by native people. It was a route of invasion for both British and American troops during the Revolutionary War. Vermont is unique in having two easy corridors of access, the Connecticut Valley on its eastern border and the Champlain Lowland on the west. Burlington, Vermont's largest city, is in the Champlain Lowland.

## Boston

Boston's early rapid growth made it the political and economic capital of Massachusetts. Almost as quickly, the city extended its influence outward to become

FIG. 4.3. *The Champlain Lowland is a northward continuation of the Great Valley of the Appalachians. It is drained by tributaries of the Hudson River on the south and by the Richelieu River, a tributary of the St. Lawrence, on the north.*

the economic capital of all of New England. Boston is the only major American city to have evolved so early (by the end of the 17th century) anything like the economic role it would later play. The city's growth was dispersed around its many natural harbors. In its early years, Boston lacked a grid-pattern town design such as were provided for New York and Philadelphia. Only after Boston expanded beyond its early, harborside origins did it begin to follow the typical American urban grid form (fig. 4.4).

Colonial Boston was an integral part of England's overseas trade. Its first successful merchants engaged in long-distance trading and shipbuilding. Boston's early focus on trade often is explained in terms of what Massachusetts lacked—because local agriculture was limited and natural resources were few, entrepreneurs turned to trade. Though these limitations were a fact, Bostoni-

FIG. 4.4. *Boston Common. The Massachusetts State House is the domed building at the upper left corner of the Common. The land at the bottom left of the photo was created by filling in the Back Bay along the Charles River.*

ans' ambitions in the overseas trade were grander than what expediency alone would have dictated. Boston merchants were involved in the rum and molasses trade with the West Indies and hence also with the African slave trade. It was Boston traders who made the first American contacts with China and developed the first long-distance economic transactions with the Pacific Coast of North America, thus anticipating the routes of commerce that one day would bind the continent together.

Profits made in overseas trading were invested in new industries, especially in cotton textiles. This set the pattern for Boston's economic growth for years to come: the development of new economic ventures funded by the proceeds of earlier ones. Although Boston became important as a manufacturing city, its influence extended throughout New England and eventually throughout the United States, by innovating production methods in new industries as the technology became available. The shift from overseas trading to the development of textile industries was followed, in the middle and late decades of the 19th century, by a period of investment in American railroads by Boston capitalists.

The pattern continued when high-technology industries were developed following World War II. Boston became the focus of one of two regions of the United States (Silicon Valley, near San Jose, California, is the other) that concentrated on the production of computer technology. The Route 128 Corridor surrounding Boston produced high-technology products that were made possible, in part, by the stream of science-and-engineering-based innovations emerging from its local academic institutions of excellence, such as Harvard University and the Massachusetts Institute of Technology.

## Urban-Industrial Patterns in New England

With more than 4 million inhabitants, Boston's metropolitan area remains by far the largest in New England, ahead of both Providence and Hartford (1.1 million each). The overall zone of metropolitan areas is almost continuous across southern New England, beginning with Auburn-Lewiston and Portland, Maine; extending southward across the Merrimack Valley (Manchester, New Hampshire; Lawrence and Lowell, Massachusetts); and including more than a dozen other sizable cities (map 4.2). Although the industrial fortunes of many of New England's cities have long been in decline, manufacturing industries were responsible for their growth into metropolitan centers.

New England was the first important manufacturing region in the United States. Many theories have been advanced to explain the location of industrial production in general, and, among them, the problem of explaining New England's industrial strength has attracted particular attention. New England's lack of opportunities in agriculture and the scarcity of its mineral resources no doubt played a role by diverting attention to forms of economic activity that could be based on trade and local labor availability. "Yankee ingenuity"—the tendency for New Englanders to invent and innovate—also played a role. During colonial times invention was necessary to overcome the handicaps of isola-

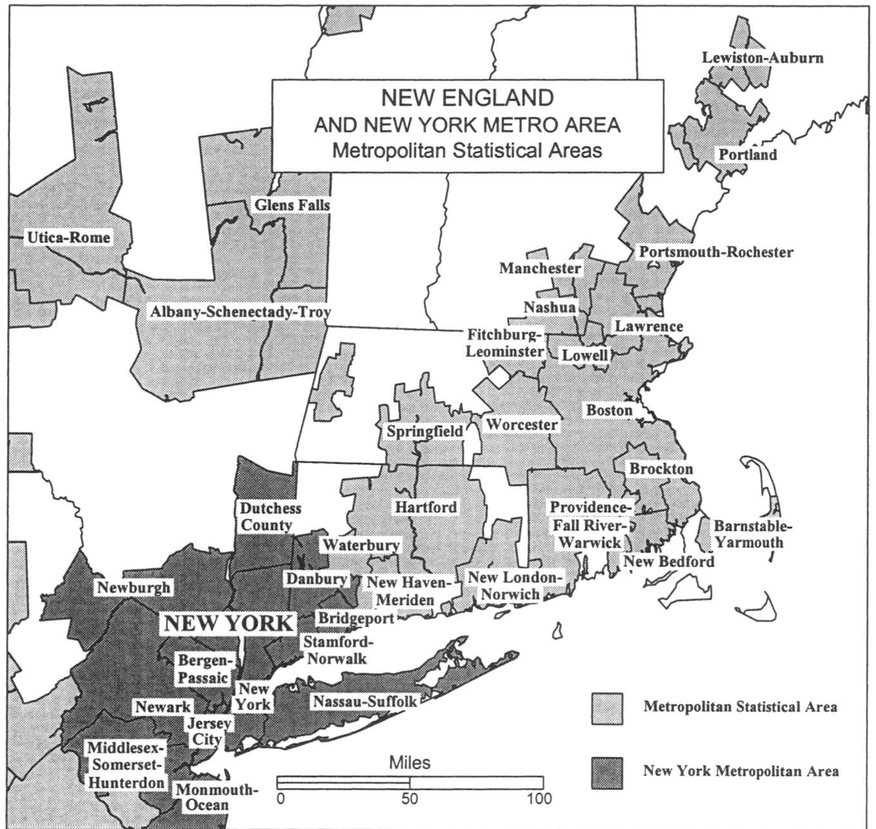


MAP 4.2

tion and lack of capital. As the United States became more industrialized in later years, New England entrepreneurs were among the most skilled at supplying technology to make industrial production more efficient.

Although the technology was employed to greatest advantage elsewhere, New England inventors created the first self-scouring steel plow and the first machinery to plant, husk, and shell corn. Eli Whitney invented the cotton gin in Connecticut, in 1792; Charles Goodyear, also of Connecticut, first vulcanized rubber in 1839. New England's first textile mill was built, following British designs, by Samuel Slater at Pawtucket, Rhode Island, in the 1790s. Francis Cabot Lowell perfected the first power loom for the textile industry at Waltham, Massachusetts, in 1814. Guns, pistols, clocks, machine tools, scales, saws, electric motors, and paddle-wheel steamboats were other inventions of New Englanders that were adopted widely elsewhere.

New England became the center of manufacturing for industries that supplied these products. One result was the creation of a large, skilled labor force. New England's cities had a division between the "hard goods" industries, such as clocks, munitions, and machine tools, which were produced in largest quantity in the Connecticut River Valley; and "soft goods," such as textiles and shoes, which were the manufacturing specialty of cities in eastern Massachusetts and Rhode Island.

The New England mill town was one prototype for the American manufacturing city of the 19th century. It was almost entirely a by-product of the manufacturing system employed in the textile industry. Samuel Slater's mill at Pawtucket, near Providence, continued a labor practice that had become common in the new industrial towns of England at that time. His first labor force consisted of eight children and one adult supervisor. Child labor eventually was replaced by the practice of hiring women, usually in their late teens or early twenties, from the surrounding rural towns as a labor force for the mills.

Because New England's textile mills originated in the era just before widespread use of the steam engine, they used water power as their source of energy. New England lacks large waterfalls, but it has numerous small water-power sites at minor falls and rapids, usually some dozens of miles inland from the coast, along the many streams that enter the Atlantic (fig. 4.5). Reliance on a local labor force and the lack of any single site with great water-power potential dispersed the mills over numerous sites, and thus industrial growth became widespread across southern New England. Some of the later mill towns, such as Lowell, Massachusetts, on the Merrimack River, were founded for a single purpose, manufacturing cotton thread; others specialized in textiles or woolen goods.

The Amoskeag Corporation, which was the economic foundation for the city of Manchester, New Hampshire, for many years was perhaps the largest textile manufacturing complex. Boot and shoe factories, which used many of the same technologies and had similar labor demands, also were centered in the Merrimack Valley. As the demands of the factories grew beyond the local supply of labor, workers migrated to the mill towns from more remote parts of New England and from Quebec.

FIG. 4.5. *Mills along the Souhegan River, a Merrimack River tributary, at Milford, New Hampshire.*

Waves of immigrants from Ireland, Italy, Portugal, and Poland found work in southern New England mill towns and added a new ethnic dimension to the region. Industrial cities, sharply divided between neighborhoods of the working class versus those of the owners and managers of the factories, were one legacy of this system. But whatever the inequities, wages in New England's mill towns eventually increased until the cost of labor was higher than in other regions, such as the southeastern Piedmont. New England's soft-goods industries began to decline in the 1920s. Such industries were only a memory in many of New England's mill towns by the time massive industrial restructuring took place in the American economy during the 1970s.

## Coastal New England

The maritime economy of Canada's Atlantic Provinces was matched by an early emphasis on fishing industries along the coast of New England, although New Englanders depended less on fishing because of the wider range of economic opportunities available to them. The modern New England fishing industry is based on exploitation of the same fish stocks—and, in many cases, the same fishing grounds—on which Newfoundlanders and Nova Scotians also depend. Shellfish, such as the Maine lobster, are taken in the shallow, inshore fishery, whereas the ocean-fishing industry depends on long-distance exploitation of the outer fringes of the continental shelf.

New England has three distinct types of coastal environments. North of Boston the coastline is generally rocky, with numerous wave-cut features and a scatter of small offshore islands. Sandy beaches are uncommon. The cause of this type of topography—which offers many miles of coastal inlets, bays, and snug harbors for fishing—is the underlying geology of New England's bedrock. North of Boston the eroded remnants of New England's ancient mountains form the coastline itself. The many long indentations of Maine's seacoast are drowned river valleys. Numerous sea-level changes have produced the offshore islands, which were peninsulas in a previous era when sea levels were lower.

South of Boston the coastline is part of the Coastal Plain, a geologically much more recent series of formations, which are sandy, easily eroded, and have high-infiltration surfaces. Sandy soils such as these are generally infertile for agriculture, although they are ideal to produce crops such as cranberries and blueberries, both of which are grown on the small acreage of Coastal Plain land in southeastern Massachusetts. The large supply of sand is evident in such coastal landforms as the Cape Cod peninsula and Martha's Vineyard and Nantucket islands. Their low, dune-covered surfaces are the result of a reworking of beach deposits by wind and wave action. Cape Cod is a spit—a recurved, sandy feature produced by the constant sculpting of the ocean currents. Martha's Vineyard and Nantucket are fringed by sandy spits of the same type.

The third type of New England coastline begins just west of Cape Cod and continues along the northern shores of Long Island Sound. Here the zone of contact between New England's bedrock and the inner edge of the Coastal Plain is submerged under the waters of the sound. The New England side of Long Island Sound has a coastline somewhat resembling that of Maine, where river mouths are drowned and former valley margins extend seaward as peninsulas. Narrangansett Bay is a very large feature of this type, and it is surrounded by surface materials that produce fertile soils. The mouths of the Thames, Connecticut, and Housatonic Rivers in Connecticut are drowned by the current sea level; thus, they appear as greatly broadened indentations in the coastline, extending ten to fifteen miles inland and providing excellent harbor conditions.

## The Connecticut River Valley

Level land is rare in New England. Land that is both level enough to farm and also covered with a mantle of fertile soils is even more uncommon. The largest such area is the valley of the Connecticut River south of the Vermont–New Hampshire border. The lower Connecticut Valley flows through a structural valley produced by a downfaulting of Triassic-age rocks. The rock formations, which erode easily, have formed a smooth lowland that is interrupted only by low, discontinuous ridges of volcanic trap rock. The Connecticut River meanders southward across the flat valley floor until it reaches the vicinity of Middletown. There glacial deposits have blocked its southward flow through its original valley, and the river has been diverted to a new course that angles eastward to enter Long Island Sound near Saybrook.

Wheat was an important early crop of the Connecticut Valley; migrants from there took the crop to western New York State in the early decades of the 19th century. Tobacco was the most important of the agricultural specialties that replaced wheat in the Connecticut Valley. A thick-leaf variety, especially suitable for cigar manufacturing, was grown intensively near Hartford until the middle decades of the 20th century, when urban sprawl took over many of the small patches of ground that had been devoted to tobacco culture.

New Haven, Waterbury, Meriden, Hartford, and New Britain, Connecticut—like Springfield and Chicopee, Massachusetts, to the north—formed the heart of New England's durable-goods manufacturing region centered in and around the Connecticut Valley. By the 1840s Hartford also had become an important financial center. Eventually, more than fifty insurance companies chose the city for their headquarters, and insurance became the largest employer in the city.

Southwestern New England's role as a high value-added manufacturing region was overshadowed in the 1960s by the ever-growing suburban overspill from New York City. The western half of Connecticut lies within the commuting zone of New York. As a response to various factors operating at the local, national, and even international level, many firms headquartered in New York relocated their offices to the suburbs in the 1960s. This made southwestern Connecticut not only a zone of commuter "bedroom communities" but also the primary place of employment for thousands of white-collar workers who had once commuted to the city.

Flexible work schedules, further dispersion of workplaces, and a preference for exurban environments were among the centripetal forces that led to a continuation of these trends throughout the remainder of the 20th century. The area west of the Connecticut River Valley became the home for daily or weekly commuters to New York. Those living in the same area who do not commute often are employed in the retail and service sectors that depend on the pull of rural Connecticut for weekending New Yorkers seeking time in the country.



## *References*

Black, John D. *The Rural Economy of New England: A Regional Study.* Cambridge: Harvard University Press, 1950.

Clark, Charles E. *The Eastern Frontier: The Settlement of Northern New England, 1610–1763.* New York: Alfred A. Knopf, 1970.

Conzen, Michael P., and George K. Lewis. *Boston: A Geographical Portrait.* Cambridge, Mass.: Ballinger, 1976.

Cronon, William. *Changes in the Land: Indians, Colonists, and the Ecology of New England.* New York: Hill and Wang, 1983.

Frederic, Paul B. "Upper Saint John Valley Potatoes: An International Trade Issue." In *Geographical Perspectives on the Maritime Provinces,* ed. Douglas Day, 60–72. Halifax: Department of Geography, St. Mary's University, 1988.

McManis, Douglas R. *Colonial New England: A Historical Geography.* New York: Oxford University Press, 1975.

Osborn, William C. *The Paper Plantation: Ralph Nader's Study Group Report on the Pulp and Paper Industry in Maine.* New York: Grossman, 1974.

Parenteau, Bill. "Bonded Labor: Canadian Woods Workers in the Maine Pulpwood Industry." *Forest & Conservation History* 37 (1993): 108–20.

Smith, David C. *The Maine Agricultural Experiment Station: A Bountiful Alliance of Science and Husbandry.* Orono: Life Sciences and Agriculture Experiment Station, University of Maine, 1980.

Vance, James E., Jr. *This Scene of Man: The Role and Structure of the City in the Geography of Western Civilization.* New York: Harper's, 1975.

Wallach, Bret. "Logging in Maine's Empty Quarter." *Annals of the Association of American Geographers* 70 (1980): 542–52.

———. *At Odds with Progress: Americans and Conservation.* Tucson: University of Arizona Press, 1991, chap. 1.

Wood, Joseph S. *The New England Village.* Baltimore: Johns Hopkins University Press, 1997.

CHAPTER 5

# New York and Ontario

Dutch settlement in New York began when traders built a fort at the site of Albany in 1614, following Henry Hudson's visit five years earlier, which had established Dutch claim to the area. In one of the best-known real estate transactions ever made, Peter Minuit purchased Manhattan Island from its native inhabitants for the reputed sum of twenty-four dollars in 1624. The first Dutch settlement was at the southern tip of Manhattan; there a fortified outpost, New Amsterdam, was established. Other Dutch settlements, including Harlem, Bronck's, and Breuckelen, were founded later on lands near the converging river mouths that form the perimeter of New York harbor. The Dutch term *kill,* given to the meandering channels through this marshy estuary, survives in place names unique to the area (map 5.1).

Dutch claims to the Hudson Valley were challenged by the English, to whom the Dutch surrendered their lands in 1664. Apart from a brief reemergence of Dutch authority in 1672, the area remained under British control until the American Revolution. New York, as New Amsterdam was renamed, grew slowly during the British period and was twice ravaged by fire during the Revolutionary War. When trade resumed in 1789, New York quickly took advantage of its central location among the former American colonies and, given its excellent natural harbors, it became the most important seaport on the Atlantic coast. When the first U.S. Census was taken in 1790, New York, with 33,131 inhabitants, was revealed to be the largest city in the nation, slightly larger than its nearby rival, Philadelphia.

Although metropolitan Los Angeles's population is now more than four-fifths the size of metropolitan New York's, the five boroughs (counties) comprising New York City have grown to a population of 7.3 million, which is more than twice that of the city of Los Angeles. The current New York metropolitan-area population is slightly larger than that of the state of New York as a whole—an apparent paradox until one realizes that the adjoining urbanized areas of New Jersey and Connecticut have a combined population that is larger than that of New York State outside of New York City. In terms of census definitions, the New York metropolitan area contains the city itself, plus another nine urban agglomerations that are metropolitan areas (each containing more than fifty thousand people) in their own right, plus all of the intervening suburbs and urban fringes (see chapter 4, map 4.2). The region stretches more than one hundred miles from Trenton, New Jersey, to Bridgeport, Connecticut. It encom-

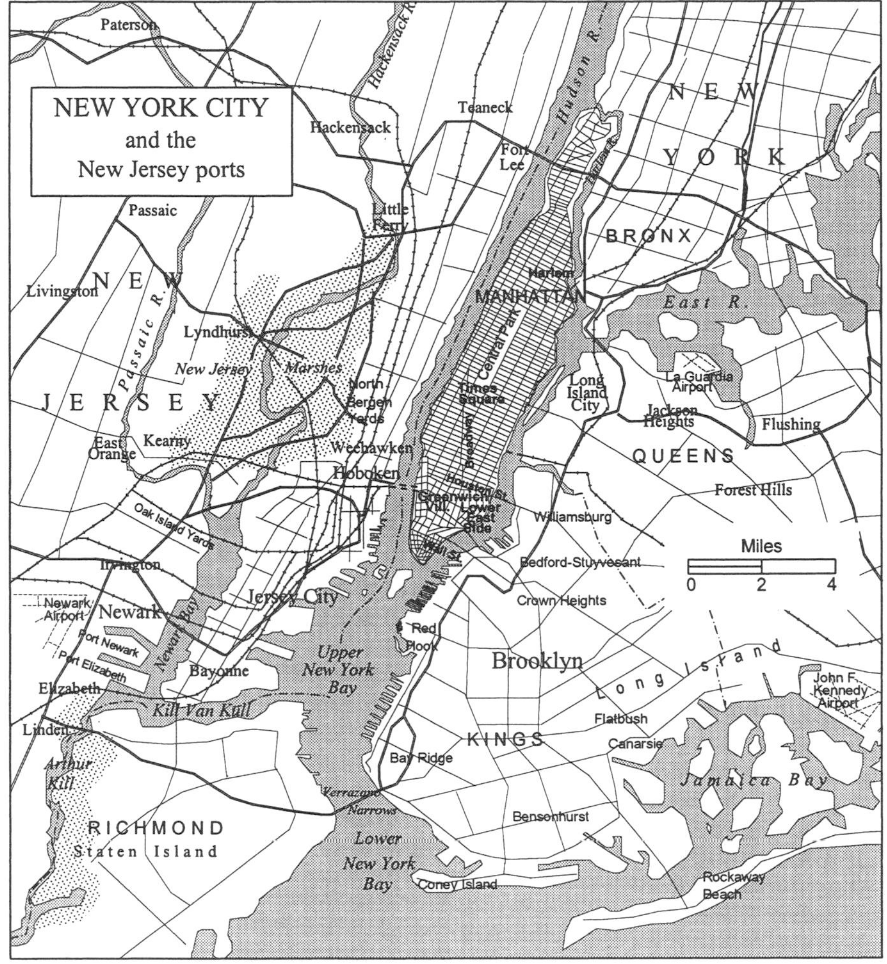


MAP 5.1

passes all of Long Island and even includes one county in extreme eastern Pennsylvania.

## The Geography of New York City

New York's emergent urban pattern was foreshadowed in the layout of its predecessor settlement, New Amsterdam. New York's major north-south street, Broadway, is the northward extension of New Amsterdam's first street. When the English threatened to attack New Amsterdam in 1633, its inhabitants erected a wall along the northern edge of the settlement and built a road following its inside perimeter. The wall was later removed, but the road survived and is now Wall Street, which retains the slightly irregular alignment of the Dutch fortification's wall in 1633 (fig. 5.1).

Lower Manhattan (the southern end of the island) has the densest concentration of tall buildings in the city. The World Trade Center dominated its sky-

FIG. 5.1. *Wall Street's slightly irregular alignment follows the shape of fortifications built around the Dutch settlement of New Amsterdam.*

line until September 11, 2001, when terrorists crashed two jet liners into the twin towers, reducing them to rubble, resulting in thousands of deaths. Even as New York rebuilds, lower Manhattan remains the heart of the city's financial district. More than $200 billion of international financial transactions are cleared through New York's banks each business day, and more than half a million people work in the city's financial sector.

A sharp drop in building height along the northern edge of the financial district marks the transition to Greenwich Village, which, along with its satellite neighborhoods SoHo (south of Houston Street), TriBeCa (the triangle below Canal Street), and Chelsea, are both home and workplace for thousands of artists, writers, designers, and producers engaged in multimedia entertainment and advertising. The linkage between media and advertising is in some respects an outgrowth of the connections between New York City's art world and its financial institutions. The many businesses specializing in these connections have long been an important part of the city's economy.

Contrasting with the low density of buildings in the Greenwich Village neighborhood was the one-time concentration of tenements on Manhattan's Lower East Side. Italians and Chinese people lived west of the "the Bowery" (Third Avenue), and Jewish immigrants lived to the east, in one of the most densely settled urban neighborhoods anywhere in the United States. Between 1880 and 1914, 2 million Jews migrated to New York City, many of them to the Lower East Side. As the population grew and became more affluent, out-migration took place north to the Bronx, east into Queens, and eventually into all parts of the metropolitan area. New York's Jewish population grew to become the largest in the Western Hemisphere. The tenements of the Lower East Side were razed to make room for some of the first multistory public housing projects built in the United States.

The somewhat irregular street pattern of Lower Manhattan, dating from

New York's colonial origins, was replaced by a regular grid of city blocks north of Houston Street when lands were resurveyed after the American Revolution. The 1782 Act of Confiscation declared as forfeit all lands that had belonged to the Loyalists, who, by this time, were relocating to Canada. These tracts became the "common lands" of New York City, and by 1796 they had been subdivided into blocks running northward across the remainder of Manhattan Island. North-south streets were broad and few in number, whereas the east-west streets, spaced only two hundred feet apart, were more numerous but much narrower. The arrangement made travel between uptown and downtown relatively easy because of the wide streets, but crosstown traffic has always moved more slowly.

The major exception to Manhattan's grid of streets is Central Park, running from 59th Street north to 110th, between 5th and 8th Avenues. Not part of the city's original design, it was a later creation by Calvert Vaux and Frederic Law Olmsted, who designed its ponds, hills, and walkways in the late 1850s. Although Olmsted later influenced the design of many other urban environments, New York's Central Park was his grandest creation, and it has remained unique among American city parks for its substantial green space in the center of a metropolis.

Midtown Manhattan is the commercial heart of New York City. Here were built the distinctive skyscrapers of the interwar period, including the Chrysler Building and the Empire State Building, plus dozens of other office buildings that, together, provide office space that now totals in the hundreds of millions of square feet. New York's leading role in the worlds of art, music, literature, theater, and dance is especially evident in Midtown, where the Metropolitan Museum of Art, the Museum of Modern Art, the Guggenheim Museum, the American Museum of Natural History, and the Metropolitan Opera are located. Running north from Times Square (Broadway and Forty-second Street) is the city's Theater District and the Rockefeller Center–Radio City Music Hall entertainment and office complex. Madison Avenue, one birthplace of the advertising industry, borders the commercial and entertainment districts on the east.

Seventh Avenue in Midtown Manhattan remains the center of the New York garment industry. Although employment has declined in recent years, more than ninety thousand workers here still design, cut, and sew men's and women's clothing, manufacture buttons and zippers, and coordinate fashion merchandising for much of the nation—all in a densely packed urban neighborhood. Truckers supplying the garment district have no hope of parking and little chance even of moving along its crowded streets during busy times of the day. Garment manufacturing is New York's leading industrial sector, although the city's printing and publishing industry is nearly as large. Many activities associated with book publishing followed the trend and relocated to suburban office parks during the past four decades, but the garment industry has remained in the city.

In addition to New York's preeminence in finance, entertainment, the arts, advertising, and merchandising, it is the largest corporate-headquarters city in the world. The presence of the United Nations, which built its General Assembly and Secretariat buildings along the East River in the late 1940s, helped reinforce New York's role as a center for all types of international transactions. The wealth amassed in such a concentration of activity is reflected by affluent residential neighborhoods on Manhattan's Upper East Side, which extend from Central Park toward the United Nations building.

A city of such size and complexity also is home to many whose labor is poorly paid, and New York is no exception. Harlem, on the northern edge of Manhattan, is the traditional center of the city's African American and Puerto Rican populations. Ethnic enclaves scattered about Manhattan and in the boroughs of Brooklyn, Bronx, and Queens also include some of the city's least affluent inhabitants. By itself, the borough of Brooklyn (Kings County) is the fourth largest city in the nation and contains perhaps the largest ethnic diversity of any U.S. city. Suburbanization, first from Manhattan to Queens and later throughout Long Island, created dozens of bedroom communities that have been linked to the city by rapid transit ever since they were built. The western half of Long Island has been included within the continuous built-up area of New York City since the 1950s, whereas the eastern half remains a zone of suburbs and weekend-home retreats.

Because New York is home both to the advertising business and to the entertainment industry, its neighborhoods and attractions have become famous around the world through the associations they carry. Coney Island, Broadway, Park Avenue, Grand Central Station, Madison Square Garden, and the Bowery are places, but they are also labels or concepts, connoting various qualities, that are known to millions of people who have never seen these locations except, perhaps, in the movies. Only a few other cities in the world are known in such detail that their geographies carry ready-made significance through the medium of popular culture, and New York's status in this respect is rivaled only by that of Los Angeles among American cities. Such a distinction undoubtedly is one measure of what it means to be a "world-class city"; New York fulfills that role perhaps as comfortably as any city that could be named.

## The Port of New York

New York is the world's busiest ocean-shipping port. The easy access offered by its extensive natural harbors was one of the most important attractions that drew the early Dutch and English colonists to the site, which became an integral part of their worldwide systems of trade. The Hudson River is the major feature forming New York Harbor, but the East River (which actually is a neck of the Atlantic Ocean, separating Manhattan from Long Island) and the Hackensack and Passaic Rivers on the New Jersey side also enter New York Bay near the mouth of the Hudson. The converging rivers created a coastline that is several hundred miles in length around New York Harbor. Ocean-shipping access was afforded to the entire perimeters of Manhattan and Staten Island, most of Brooklyn and Queens, and the New Jersey cities of Newark, Jersey City, Elizabeth, and Hoboken.

The Hudson River side of Manhattan once was lined with the long piers where transatlantic passenger ships docked. Freight cargo terminals, which require more space, were forced into less accessible, less expensive locations. Even some transcontinental railroads were unable to afford passenger terminals in Manhattan and reached no closer to it than the New Jersey ferry landings at Weehawken and Hoboken. Brooklyn's busy wharves were the hub of ocean cargo shipping, but numerous other freight terminals and warehouses were located at waterfront sites surrounding the harbor zone. Ferries, tugboats, and car floats plied the waters of New York Harbor, shuttling passengers and freight between the various specialized components of the port.

Transportation revolutions since the 1950s have produced massive changes in this pattern. Except for cruise ships, ocean-passenger liners have disappeared, and the Manhattan termini of the great steamship lines have been replaced by waterfront commercial and residential buildings. International travelers pass mainly through the massive John F. Kennedy (Idlewild) Airport in Queens. Railroads retreated from much of New England during the 1970s and 1980s; partly for this reason, little surface freight traffic now moves through New York City. Long-distance trucking of freight understandably avoids the congestion of the city and moves by the numerous freeway bypasses. The former transportation focus on the city has thus been diffused in various directions.

The intermodal shipping revolution, in which trucks, railroads, and ocean ships all participate in the handling of cargoes, was made possible by the adoption of all-purpose containers, which now carry the freight that once was loaded and reloaded at warehouses each time it was transshipped. Intermodal shipping through New York is concentrated at a series of container terminals, where the nation's shipping "lanes"—the transcontinental railroads and a maze of interstate highways—converge on the New Jersey side of the Hudson. The cities of Elizabeth, Newark, Kearny, North Bergen, and Little Ferry embrace an almost continuous zone of intermodal terminals for nearly twenty miles along Newark Bay and the Hackensack River. Domestic mail and parcel express services located here receive trainload shipments from all points in the United States, transfer the New York–bound containers to trucks for local delivery, and relocate the containers bearing international destinations to the nearby ocean-shipping docks. In many respects, this small section of New Jersey has become the "port of New York," in all of its many functions, both foreign and domestic.

## The Hudson River Valley

New England's mountains extend as a ridge of Precambrian rocks more than 150 miles, across the Hudson River, through northern New Jersey, and into eastern Pennsylvania (map 5.2). Known as the Reading Prong, this lengthy upland is resistant to erosion, and thus it is surprising that the Hudson River has cut through it instead of being diverted around it. The cut comes at the Highlands, near West Point, where the river flows through a gorge between the Taconic Range on the east and the Hudson Palisades on the west. The Palisades are named for their vertical, columnar structure of joints that were produced when

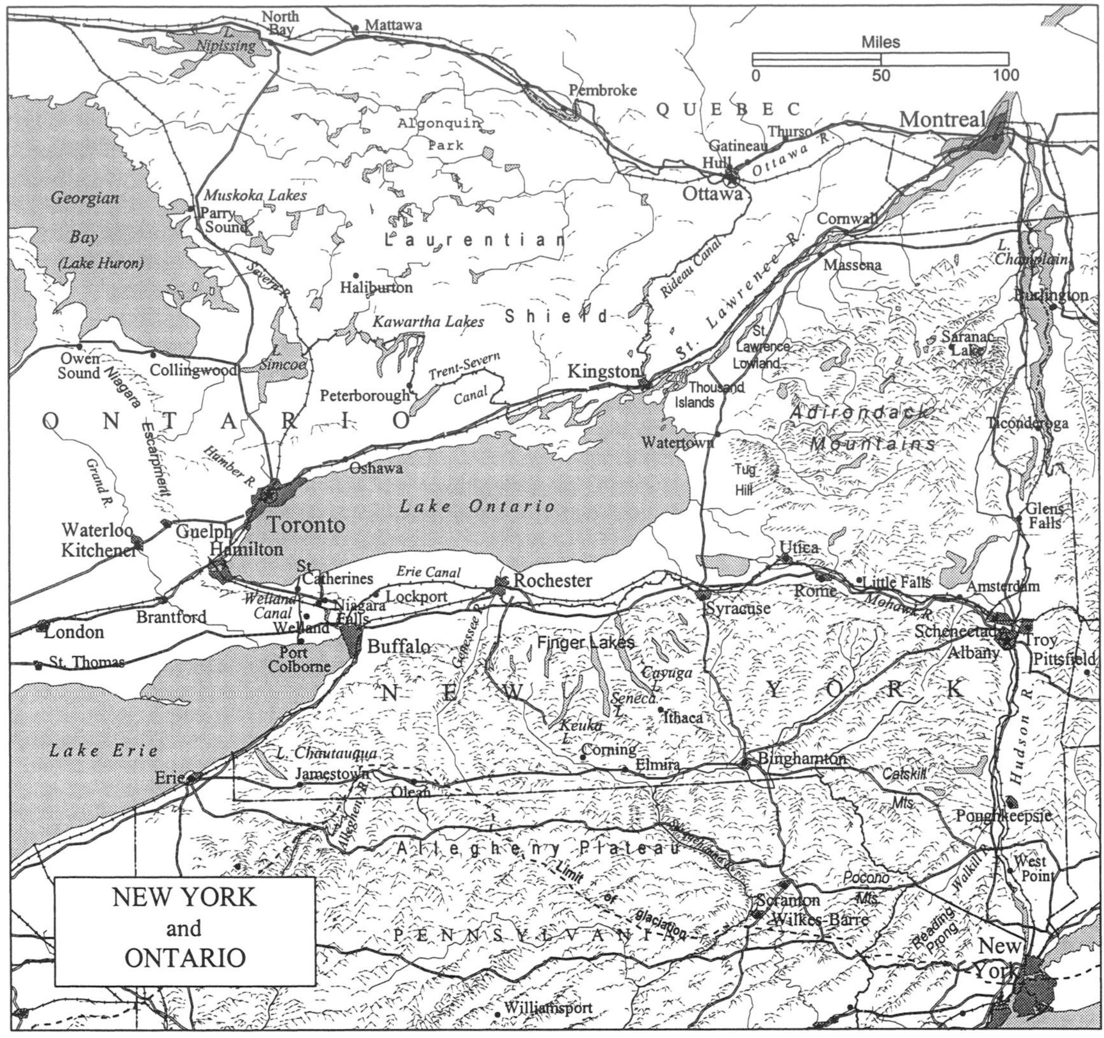


MAP 5.2

molten material intruded into Tertiary sedimentary rocks and cooled into basalt. The Palisades form an abrupt wall south of Storm King Mountain, on the west side of the Hudson, which continues southwestward as the Ramapo Mountains of northern New Jersey.

The river's channel is more than 700 feet deep at West Point. The Hudson extends northward more as an estuary (an inlet of the ocean) than as a true river as far north as Troy. The drowning of the Hudson River's channel by the current sea level has permitted deepwater navigation well inland in New York state and has enabled the Albany-Troy-Rensselaer area to function as an ocean port. Albany is a receiving port for petrochemicals and other bulk materials, and it is a shipping terminal for export grain raised in the Great Lakes region.

The Albany-Troy urban area grew into a major transportation center that joins four important routes. Railroads constructed through the rugged Berkshire Hills in Massachusetts forged a link to Boston and thereby connected New England directly to the Middle West through Albany. To the west of Albany stretches the level-profile Mohawk Valley and, beyond that, a lowland extending west to Buffalo; this became the route of the Erie Canal. Albany's position at the head of navigation on the Hudson River and simultaneously at the southern extremity of the Champlain Lowland placed it at the heart of the north-south commerce between Montreal and New York (fig. 5.2).



Seventeenth-century Dutch interests in the Hudson Valley were focused as much on Albany (Fort Orange) as they were on New Amsterdam. Fort Orange played an active role in the fur trade. The Dutch patroon system of land grants in the Hudson Valley resembled the seigneuries of Quebec. Blocks of territory fringing the river were awarded to landlords who then found rent-paying farmers to settle the land. Except for the large Rensselaer grant south of Fort Orange, however, the patroon system produced little effective settlement. Dutch land tenure in the Hudson Valley harked back to the European manorial system. In that sense it contrasted sharply with New England, which was being settled simultaneously but under the system of freehold land tenure that became the common pattern in the United States.

The Dutch persisted as the dominant Euro-American element in the Hudson Valley's population even after they surrendered their lands in 1664. The manorial system was continued by the British, who made additional grants to landlords and thereby reaffirmed the Hudson Valley's social contrast with New England. Although the Hudson River occupies a broad, smooth-floored valley, the lands along it are not especially fertile for agriculture. Urbanization was retarded by the growing concentration of commercial activity at New York. Growth also was slow because this was a dangerous frontier for European settlers during the long period known as the French and Indian War. Raiding parties from the Iroquois Five Nations held in check any tendency for white settlement to expand westward until the 1763 Treaty of Paris resolved control of the

FIG. 5.2. *Albany, in the Hudson River Valley. The Mohawk Valley corridor (New York Thruway) enters the Hudson Valley at the extreme left. The broad valley stretches to the foothills of the Taconic Range.*

region in favor of the British. But by the time of the American Revolution, New Englanders had begun to look west beyond the Hudson River Valley for new opportunities and better lands.

## The Mohawk Valley, the Erie Canal, and the Lake Ontario Plain

The population growth in New England was sufficient to produce pressure on the land base by the 1750s, which stimulated out-migration, including the planting of satellite New England colonies in Nova Scotia (on lands confiscated from the Acadians), Long Island, and the fringes of the Hudson Valley. Before the American colonies were brought together as a single nation, it had been the practice for colonial governments to claim areas to the west as outlets of expansion for their growing populations, usually on the basis of their colonial charters; in some cases they cited the dubious authority of treaties made with native groups.

Massachusetts, which had the largest population in New England, was especially short of arable land to support expansion. In earlier times Massachusetts had reserved all the lands directly to its west for future use. Although settlement of the Hudson Valley by other groups prevented a literal interpretation of this claim, the lands beyond the Hudson, west of the Albany area, had just begun to receive new settlers in 1785 when the old colonial land claims were being incorporated into the new land system of the United States. Massachusetts insisted on ownership of a swath of land covering most of western New York state. Because New York wanted to expand into the same area, a compromise was reached whereby lands west of the seventy-seventh meridian were reserved for Massachusetts and lands to the east were reserved mainly for New York. These were the background conditions for the establishment of a "new New England" in upstate New York at the close of the 18th century.

Persons to whom tracts of western land had been awarded (often for service in the Revolutionary War) had the right of resale, and thus a market for land was established. Both Massachusetts and New York sold large blocks of their lands at low prices to speculators. Lands were sold and resold in a frenzy of activity that nonetheless had the intended effect of placing actual settlers on the land. Many Connecticut settlers also came to western New York, reinforcing the Yankee character of the region. By 1810 all of western New York, save for the more rugged lands of the Appalachian Plateau, had been settled, very largely by migrants from New England.

The best lands were in the northern two tiers of counties in western New York, where hardwood forests predominate and a smooth layer of glacial materials covers the Paleozoic bedrock. Numerous short streams descend to Lake Ontario from the uplands at the south, and some of the valleys, especially that of the Genessee River, developed strong agricultural economies. Wheat was the Genessee Valley's early specialty; waterfalls on the river at Rochester provided power to turn the local wheat crops into flour. Although wheat was raised successfully in many of the eastern states, the Genesee Valley's wheat crop and the cluster of flour mills at Rochester formed the first of many "breadbaskets" that would emerge in the westward migration of wheat culture in later years.

With natural water-power sites, a fertile lowland for agriculture, and an easy corridor for access to the Hudson Valley by the Mohawk River, both agriculture and manufacturing developed rapidly in western New York. What was needed most was improved transportation access, which in that era meant the construction of a canal. The general route for such a canal was easy to predict. The Mohawk River extends only as far west as Oneida Lake, north of Syracuse, but the Lake Ontario Plain continues as a swath of level land at the foot of the Niagara Escarpment west to the Niagara River and beyond into Canada. Plans for a canal eventually became fixed on the goal of reaching Lake Erie, and the route that was selected followed the smooth lands of the lake plain west nearly to Niagara Falls. A series of five locks built at Lockport raised the canal's level to that of Lake Erie.



The construction of the first Erie Canal was completed in 1825. There was an almost immediate growth in population and business activity along its route and also in New York City, which became a major beneficiary of the commerce brought by the canal. The city of Buffalo had been founded by one of the groups of real estate speculators who had purchased millions of acres of western New York land with the intention of reselling it. Organized as the Holland Land Company, the promoters laid out a new city that they first named New Amsterdam in 1803, but they soon changed the name to Buffalo. Because it was the point of embarkation for travelers who boarded Great Lakes vessels sailing west, Buffalo became a city familiar to many westerners. It grew as the market for products of the Great Lakes region, receiving shiploads of grain, lumber, and other products of the interior that were milled or processed in Buffalo and then forwarded east, first by the Erie Canal and in later years by railroads. Whereas Rochester's early flour mills processed mainly the wheat crop of the Genesee Valley, Buffalo became a flour-milling center for wheat raised in the entire Great Lakes region.

The Erie Canal also stimulated industrial growth in the Mohawk Valley. Schenectady, Amsterdam, Little Falls, Utica, and Rome developed diversified manufacturing economies much like those of the cities of southern New England. In later years the Mohawk Valley specialized in high-value-added consumer durable goods. The General Electric Company, headquartered at Schenectady, added thousands of manufacturing jobs in that city as well as in Syracuse, Utica, and Rome. Rochester's industrial growth focused on the Eastman Kodak Company and later on scientific optical equipment, photocopiers, and telecommunications equipment. The Mohawk Valley–western New York urban corridor manufactured steel, home appliances, shoes, kitchen utensils, and hundreds of other products. A general decline in manufacturing began in this region, as it did in New England, during the 1970s, when companies relocated to cheaper labor markets or to states with tax structures more favorable to business. The corridor of cities across upstate New York no longer specializes in manufacturing, but its professional and service-sector employment continues to grow.

FIG. 5.3. *The Niagara Escarpment is a prominent landscape feature near Guelph, Ontario. The escarpment curves southwestward to form the Niagara Peninsula, separating Lakes Erie and Ontario.*

## The Niagara Peninsula

The Niagara Peninsula is a bedrock feature, an erosion-resistant dolomite ledge separating Lake Ontario from Lake Erie (fig. 5.3). It is crossed by the Niagara River, which flows placidly northward out of Lake Erie at Buffalo, only to cascade abruptly some 180 feet over its famous, namesake falls just twenty miles downstream. All the waters flowing from Lakes Superior, Michigan, Huron, and Erie pass over Niagara Falls. The discharge, which exceeds two hundred thousand cubic feet per second, generates more than 2 million kilowatts of electricity, the largest source of hydroelectric power in the eastern United States.

The total descent of the Niagara River across the narrow peninsula is in excess of 330 feet. It was an impediment to commerce between the lakes in the days when all goods had to be unloaded at the falls and portaged over the Niagara Escarpment. The first of several waterways known as the Welland Canal was built to bypass the falls on the Canadian side of the border in 1829. The first canal overcame the difference in elevation with forty locks, which were reduced to seventeen in the waterway's second reconstruction during the 1850s. The channel was widened and straightened, and the locks were enlarged to accommodate larger lake vessels, most recently in the early 1970s. The present canal raises and lowers vessels with only eight locks. Heavy industries located along the canal in the Ontario cities of Port Colborne, Welland, and St. Catharines and along the Niagara River (which is the international boundary) at Buffalo, Tonawanda, and Niagara Falls.

Like all waterfalls, Niagara's turbulent descent actively undercuts the resistant rock ledge on which it is formed. The gorge of the Niagara River extends 6.5 miles downstream (north) from the falls, which means that the river has cut that distance back into the escarpment during the 8,500 to 9,000 years that the present channel has been active. At that rate, the Niagara River will be shortened to nothing in another 75,000 years, which, among other consequences, would result in the draining of Lake Erie.

The Niagara Escarpment continues as a well-defined ridge west of the falls, bending around the western extremity of Lake Ontario at Hamilton. Steep slopes that permit cold-air drainage and the moderating climatic effect of Lake Ontario are combined here to produce one of Canada's warm microclimates. Orchards and vineyards once covered the slopes, although urban encroachment in recent years has substantially reduced the fruit acreage. The backslope of the Niagara Escarpment (the gentle slope, descending toward Lake Erie) also has a favorable climate and has been Canada's principal area of tobacco production for many years. The proximity of the lakes on either side of the Niagara Peninsula makes this small area one of the most favorable for agriculture in all of Canada.



Major steel industries were founded at Hamilton and Buffalo based on Quebec-Labrador or Upper Great Lakes iron ore. The availability of western Pennsylvania coking coal and a strong market for steel in the Toronto-Hamilton-Buffalo region helped the steel industry grow, and the area became a center for heavy-machinery manufacturing. Niagara Falls, St. Catharines, and Buffalo are leading gateway cities in the international trade between Canada and the United States. Part of the heavy cross-border traffic is not international, however, because the Ontario route is the shortest one for shipments between Detroit and the eastern United States; it is competitive even for eastbound traffic from Chicago to New York or Boston.

## Upper Canada

Upper Canada was the first name given to what is now the province of Ontario. Several thousand Loyalists, many from New York, were living temporarily in Quebec in 1783, awaiting resettlement. Because they were both Anglophone and Protestant, neither the Loyalists nor the authorities in Quebec wanted them to remain there. In 1791 a new colony, Upper Canada, was created for these Loyalists, as New Brunswick had been a few years earlier. Although Upper Canada was to the west and south of Montreal, it was "upper" in the sense that it was farther up the St. Lawrence River. By 1800 Upper Canada had a European-derived population of several thousand, nearly all of whom had gone north of the border seeking refuge. Early Loyalist centers included several on the north side of Lake Ontario, especially near Kingston, near Windsor, and on the Niagara Peninsula.

Americans who believed that the Loyalist–Upper Canadians would become sympathetic toward the United States were proved wrong when the United States declared war on Great Britain in 1812. American forces invaded Canada believing that the Canadians desired liberation from British rule and would not fight. But they did fight, and the war essentially turned into a stalemate. Although the War of 1812 had little impact on the United States, it was an early and important test of Canada's nationhood and demonstrated that Americans could not expect that the territory north of the border eventually would become part of the United States. Americans still "invade" Canada, but nowadays as summer tourists. Those who visit the historic military fortifications on the north side of

Lake Ontario may notice that the gun emplacements face south—the direction from which the last enemy came.

Upper Canada's capital was at York (the name was changed to Toronto in 1834), which was a small village comprised mainly of Loyalists in 1812. American forces captured York during the war and burned its parliament buildings, although the larger settlement of Kingston, farther east along Lake Ontario, emerged from the war unscathed. Toronto's site, on the level plain bordering Lake Ontario, does not appear to be strategic, but it marks a navigation shortcut that was used by native people and fur traders long before the Loyalists arrived. The Humber River, which enters Lake Ontario at Toronto, was followed inland to a low ridge running west from the vicinity of Lake Simcoe. A short portage over this ridge allowed access to streams flowing north into Georgian Bay. The route, known as the Toronto Passage, was well known to the Huron people, who built a thriving agricultural civilization in the zone between Lake Simcoe and Georgian Bay before the arrival of Europeans. The same route was chosen later for railroad lines that linked Toronto with Owen Sound and Georgian Bay. The purpose was to create a shortcut that avoided the more circuitous all-water route via Detroit, Lake Erie, and the Niagara Peninsula for traffic between the eastern and western portions of Ontario.

Lingering questions concerning American threats to Canada's integrity led to the construction of two canals that were built through inland portions of Ontario, away from the vulnerable frontier of the Great Lakes, during the 19th century. The Trent Canal, linking Georgian Bay with Lake Ontario, and the Rideau Canal, between Kingston and Ottawa, were constructed partly because they offered safe passage through landlocked areas, away from possible American interference. Immigration was another strategy for securing Canadian territory. Groups of Scots Highlanders were resettled in the Glengarry colony north of Cornwall, Ontario, in 1815. Thousands more migrants, especially Protestants from Ireland, Scotland, and England, came in the next two decades. They took up lands over a wide area of eastern Ontario and rapidly brought the land into production. When the Dominion of Canada was formed in 1867, Upper Canada (which for a time also had been known as Canada West) became the province of Ontario.

## The Growth of Ontario

Toronto is Canada's largest city, and Ontario is its most populous province (fig. 5.4). The province and the city have surpassed Quebec and Montreal, despite Quebec's longer history, partly because of Ontario's relatively favored physical environment. Although it has no common name, Ontario's southward-protruding "peninsula," protruding into latitudes equal to those of New York, Michigan, and Wisconsin, gave it a strong advantage in terms of agricultural production compared with the rest of eastern Canada. Wheat was Ontario's staple export crop during the mid-19th century. Its production was concentrated on the better lands west of Lake Ontario, and the crop eventually spread west-

FIG. 5.4. *Summer festival celebrations in downtown Toronto.*

ward throughout the southwestern portion of the province. Livestock and dairying replaced the emphasis on wheat in the older producing areas by the 1890s, but wheat continued as a frontier crop as agricultural settlers expanded to the fringes of Lake Huron.

By the 1840s Toronto was the major commercial center for Ontario agriculture. Like Chicago, it became a center for grain milling, meatpacking, and machinery manufacturing based on the agricultural economy of its hinterland. Also like Chicago, Toronto expanded in a grid-pattern city form along the lakeshore, on a level site that offered no impediments to its future growth. The wealth amassed in its agricultural industries was invested in a wide range of manufacturing industries, and Toronto became the leading center of Canadian industry. Toronto grew as a center for wholesaling and railway transportation; it developed Canada's major department-store and catalog-retailing businesses; and by the late 19th century it ranked second to Montreal as Canada's leading banking center. In these respects, too, Toronto's growth resembled Chicago's, and it occurred during the same time period. The two cities' growth paths began to diverge in the 20th century. Chicago found no new roles to play in the U.S. economy after the 1930s, but Toronto continued to emerge as the economic capital of Canada.

Ontario's economy was larger than Quebec's well back into the 19th century, even though Montreal was the largest business and financial center in Canada. The evolution of financial institutions in Canada led to a migration of bank headquarters to the two cities, but increasingly toward Toronto. Manufacturing growth after World War II was concentrated in Toronto and in Kitchener, Waterloo, Guelph, London, and Brantford to its west. The Toronto-Hamilton-London region, which formed Ontario's agricultural heartland during the mid–19th century, became the center of Canada's manufacturing belt in the 20th century.

Although Montreal's business community retained its Anglophone orientation until recent times, Toronto's developed closer ties with businesses throughout Anglophone western Canada. Investment capital flowed more to Toronto, especially in critical sectors such as natural resources and manufacturing. Step by step, Toronto passed Montreal to become Canada's largest and wealthiest city. French-language legislation in Quebec beginning in the 1970s produced some flight of businesses from Montreal to Toronto, but the shift toward Ontario had been under way for decades before that time. Metropolitan Toronto's current population of 3.9 million is about one-fourth larger than Montreal's.

The favorable conditions for agriculture in peninsular southern Ontario have no counterpart east and north of Toronto, where the hard rocks and infertile soils of the Canadian Shield severely limit crop production. Eastern Ontario thus lacks the productive agricultural base of western Ontario, but several eastern Ontario cities, including Oshawa, Kingston, and Cornwall, became important manufacturing centers. Tourism and second-home developments dominate the scenic portions of eastern Ontario away from urban areas. Lake Simcoe and the Muskoka Lakes region are intensively developed, and a still larger area to the north, including Algonquin Provincial Park near the Ottawa River, is similarly oriented to recreational activities.

Ottawa became Canada's capital city when Canada's separate colonies were evolving the government that was established in the confederation of 1867. Montreal, Toronto, and Kingston all sought to be Canada's capital, but Queen Victoria chose Ottawa in 1857. What was then only a small urban center gradually grew into the role for which it had been chosen. Ottawa, as well as the suburbs of Hull, Gatineau, and Thurso on the Ottawa River, was a center of pulp and paper manufacturing and hydroelectric power generation. Ottawa thus had a local industrial base, and for some years industry overshadowed the city's governmental role. Ottawa's urban landscape was redesigned in the early 20th century, befitting its status, and it became one of the world's most attractive national capitals.

Like Washington, D.C., Ottawa occupies the borderland between the two most important cultural regions within the national territory it governs. In Canada, the division involves French-speaking versus English-speaking realms, whereas in the United States the fundamental cleavage was between North and South. The French influence in Ottawa has always been strong, given its location immediately across the Ottawa River from Quebec, and the Ottawa Valley comes as close to being traditionally bilingual as any region of Canada.

## The St. Lawrence Lowland in New York and Ontario

After the end of the last glaciation, the St. Lawrence River emerged as the lowest-level drainage outlet for the Great Lakes. The river found a course through the maze of hard-rock islands along the New York–Ontario border, known as the Thousand Islands, that constrict navigation and limit the depth of the river's channel. The Thousand Islands (many of which are privately held vacation properties) are part of a glacially scraped surface of Precambrian rocks that form a saddle-shaped lowland from the Canadian Shield of Ontario through the Adirondack Mountains of New York. The Thousand Island section still is rebounding following the removal of the mass of the Laurentide ice sheet. Treacherous navigation conditions made it necessary to construct ship canals through the Thousand Islands, most recently in the 1950s, when the St. Lawrence Seaway was built.

Hydroelectric power dams built at Massena, New York, brought the aluminum industry to the St. Lawrence Valley. Other industries have been attracted to the area since the Seaway was constructed, but the benefits have been less than expected. In an era when nearly all shipments (except bulk commodities) demand speedy delivery, the sight of ocean liners slowly winding their way through the Great Lakes to deliver imported goods to the likes of Milwaukee, Toronto, or Chicago has become an anachronism. Traffic in iron ore, crude oil, coal, and grain has been brisk at times, but the prospects for the Seaway's future growth as a general route of commerce, linking Europe with the North American heartland, are bleak.

The St. Lawrence Lowland and the Ottawa River Valley were important areas of lumber production in the 19th century. Lumberjacks, some from Quebec, others from New England and Ireland, worked here in the camps and sawmills during the era of active lumbering before the 1880s. Gradually, often in small groups, they migrated west to the Upper Great Lakes region as the latter's forest industries evolved, and there they found similar work. The St. Lawrence region has a tradition of being an international zone, and through migration it developed strong ties with the upper peninsula of Michigan, northern Wisconsin, and northern Minnesota. St. Lawrence Valley lands cleared of timber eventually were used for dairy farming. Although the region has a short growing season and is remote from markets, it produces substantial quantities of milk. The pulp and paper industry is the largest economic sector in the region at present.

## The Adirondack Mountains

The Adirondacks are a roughly symmetrical dome of granitic rock that was eroded, in part by glaciation, to form mountainous topography. Several peaks in the Adirondacks are more than one mile high, the highest elevations in the state of New York. The Adirondacks once had a covering of younger sedimentary rocks, but the newer rocks were stripped away as the land surface rose and the underlying granite was exposed. Although the Adirondacks are adjacent to the Appalachians, they are not part of the same set of geologic formations. The Adirondacks instead represent a southward extension of the Canadian Shield. Like the Upper Great Lakes region, the Adirondacks are part of the exposed core of Precambrian rocks underlying the North American continent.

Rivers flow in all directions down the flanks of the Adirondack Mountains. The largest stream originating there is the Hudson River, which flows southeastward, away from the high ground and through several deep gorges before it enters the Hudson-Champlain Lowland. The Adirondacks' broken topography and rocky soils made a good hunting ground for the Iroquois, but the area was never home to many native people; they preferred the better lowlands to the south. Nor were these mountains attractive to settlers moving west from New England in later years. The area was skipped over by the westward progress of settlement and was not "discovered" until its natural beauty attracted artists, nature lovers, and sportsmen from New York City in the mid–19th century.

The Adirondacks were one of the first summer resort areas in the eastern United States. Rustic hotels and lodges catered to wealthy, urban-based visitors who began arriving in larger numbers after railroad lines connected them to New York City. Five million acres of the Adirondacks have been set aside as a state park, beginning with the first small park created in 1892. The Adirondacks were protected from overdevelopment early enough that the area retains a landscape perhaps much like the one the Iroquois knew when they used it as a hunting ground hundreds of years ago. Small deposits of magnetite, a rock rich in iron ore, once were mined in the Adirondacks and were important to the steel industry of the Buffalo region, but the remaining deposits are lower in grade and are costly to extract.

Immediately west of the Adirondacks lies the Tug Hill Upland, another sparsely settled area. Most of Tug Hill is a sandstone plateau with inferior soils on steep slopes that are largely unfit for agriculture. Because of its high elevations immediately downwind from Lake Ontario, Tug Hill ranks as the snowiest place in the eastern United States, with annual snowfalls exceeding two hundred inches per year. Buffalo, Rochester, and Syracuse also are downwind from Lake Ontario, but at lower elevations, and thus they receive somewhat less snow. The entire Lake Ontario perimeter of New York state, however, from Buffalo to Watertown, qualifies for the "snowbelt" label.

## The Appalachian Plateau in New York

Fringes of the Appalachian Plateau extend across the "Southern Tier" counties of upstate New York. Farthest north lies a series of roughly parallel north-south valleys, originally cut by streams that were tributaries of the Susquehanna River in preglacial times. They were broadened and deepened by the advance of glacial ice and now contain long, narrow lakes. The lakes were formed when drainage was dammed by materials left by the retreating glacier. These are the Finger Lakes; from west to east, they include Lakes Canandaigua, Keuka, Seneca, and Cayuga. Seneca Lake, the largest, is about thirty-five miles long. Warm microclimates on the steep valley sides are produced by free cold-air drainage. A hardy variety of wine grape produced in the Finger Lakes region is manufactured into champagne and other wines that are sold widely on the national market.

The Allegheny Mountains are the western portion of the Appalachian Plateau in New York, and the Catskill Mountains are the major landform feature of the plateau's eastern end. Both the Alleghenies and the Catskills have rugged topography, but apart from a small area near Olean, nearly all of New

York's surface relief was muted by glaciation. Most of the Southern Tier is a region of moderate relief, where the uplands were smoothed by glacial action and the valleys became filled with glacial debris.

Parts of the area are suited to crop farming, and much of the rest can be used as pasture for dairy cows. New York's dairy industry is concentrated in the St. Lawrence Valley and the Allegheny Plateau. New York ranks third among the states in the size of its dairy herds, although it has less than half as many milk cows as either Wisconsin or California, the two leading dairy states. Most of New York's dairy production is marketed as fluid milk in urban areas of the Northeast; much of the rest is used to produce cheese.

In addition to the generally lower relief resulting from glaciation, New York's portion of the Allegheny Plateau is distinct because the Pennsylvania–New York border happens to roughly coincide with the northern limit of Carboniferous (coal-bearing) rock formations. Pennsylvania is rich in coal, but New York has none. This circumstance impacted the industrial development of the two states differently and made the landlocked areas of New York unsuited to many types of industrial production. New York's Allegheny fringe has deposits of natural gas, however, which have been exploited by the local glass and ceramics industries. Although they were not as favored in terms of transportation access, cities of New York's Southern Tier also developed diversified manufacturing industries. Binghamton and Elmira manufactured many of the same products as did cities of the Mohawk Valley to the north.

The Catskill Mountain section is a small area of rugged, sandstone uplands that lies within one hundred miles of New York City. In addition to the poor resource endowment, which limited local population growth, the Catskills also were a barrier to communication between New York City and upstate New York. They diverted transportation routes east through the Hudson Valley or west through the Delaware Valley and eastern Pennsylvania. The Catskills repelled most types of economic development, but their scenic beauty as well as their proximity to New York made them a natural attraction to vacationers from the city.

New York's observant Jewish population sought summertime retreats away from the city, but people found few resorts that were willing to guarantee the observance of kashruth (kosher dietary laws). By the 1930s resort entrepreneurs were operating hotels in the Catskills that catered to a Jewish clientele, and the region became known informally as the Borscht Belt. Although many of the older resorts have been converted to condominium developments, the Catskills still function primarily as a rural retreat for vacationing New Yorkers.

## *References*

Bourne, L. S., R. D. MacKinnon, and J. W. Simmons, eds. *The Form of Cities in Central Canada.* Toronto: University of Toronto Press, 1973.

Craig, Gerald M. *Upper Canada: The Formative Years, 1784–1841.* Toronto: McClelland & Stewart, 1963.

Ellis, David M., James A. Frost, Harold C. Syrett, and Harry J. Carman. *A History of New York State.* Rev. ed. Ithaca, N.Y.: Cornell University Press, 1967.

Goheen, Peter G. *Victorian Toronto, 1850 to 1900.* Research Paper no. 127. Chicago: Department of Geography, University of Chicago, 1970.

Hall, Max, ed. *Made in New York.* Cambridge: Harvard University Press, 1959.

Hoover, Edgar M., and Raymond Vernon. *Anatomy of a Metropolis.* Cambridge: Harvard University Press, 1959.

Jackson, John N. *Welland and the Welland Canal.* Belleville, Ont.: Mika, 1975.

Knight, David B. *Choosing Canada's Capital: Conflict Resolution in a Parliamentary System.* Ottawa: Carleton University Press, 1991.

McCalla, R. J. "The Seaway: Rationale, Progress, and Prospects." *Operational Geographer* 7, no. 3 (1989): 27–33.

McIlwraith, Thomas F. *Looking for Old Ontario: Two Centuries of Landscape Change.* Toronto: University of Toronto Press, 1997.

Osborne, Brian S., and Donald Swainson. *Kingston: Building on the Past.* Westport, Ont.: Butternut Press, 1988.

Spelt, Jacob. *Toronto.* Canadian Cities Series. Toronto: Collier-Macmillan Canada, 1973.

Thompson, John H., ed. *Geography of New York State.* Rev. ed. Syracuse, N.Y.: Syracuse University Press, 1977.

Yeates, Maurice H. *Main Street: Windsor to Quebec City.* Toronto: Macmillan of Canada, 1975.

CHAPTER 6

# The Middle Atlantic

The Middle Atlantic region includes physical environments ranging from sandy tidewater inlets to rugged upland plateaus. From east to west, the sequence begins with the Coastal Plain, which covers the southern two-thirds of New Jersey, nearly all of Delaware, and roughly the eastern half of Maryland (maps 6.1 and 6.2). Two other lowland belts, the Piedmont and the Triassic Lowland, lie immediately west of the Coastal Plain. Both have a more rolling topography and surface rocks that are older and less erodible than the Coastal Plain.

The Appalachians begin at the western edge of the Triassic Lowland and are marked by two prominent features, the Reading Prong on the north and the Blue Ridge on the south, which are separated by the Schuylkill and Susquehanna Rivers. Immediately west of the Reading Prong–Blue Ridge lies the Great Valley of the Appalachians, which extends beyond this region as the Walkill Valley and the Hudson-Champlain Lowland in New York and southward as the Shenandoah Valley in Virginia. The two largest landform subdivisions are the Ridge and Valley section, which covers central Pennsylvania and most of western Maryland, and the Appalachian Plateau, which extends across western Pennsylvania and into Ohio.

Contrasts between these various environments influenced the rates of land settlement in historic times. Because underlying geological differences affected factors of soil formation and vegetation growth, those variations also are reflected in land-use patterns today. Even the configuration of transportation routes and urban centers within the Middle Atlantic region shows the arrangement of environmental divisions.

## The Delaware Valley

The Delaware River, which rises in the Catskill Mountains of New York, also cross-cuts all of these environmental regions, but from north to south. Unable to erode the more resistant Ridge and Valley formations, it follows Kittatinny Mountain for more than fifty miles before crossing the Great Valley through the Delaware Water Gap. The river then winds through the rugged lands of the Reading Prong and enters the Triassic Lowland (an easy stretch of the river in terms of its bounding topography, where General George Washington made his famous crossing). At Trenton it makes another sharp change in direction to follow more easily the limit of Coastal Plain rocks toward Delaware Bay.

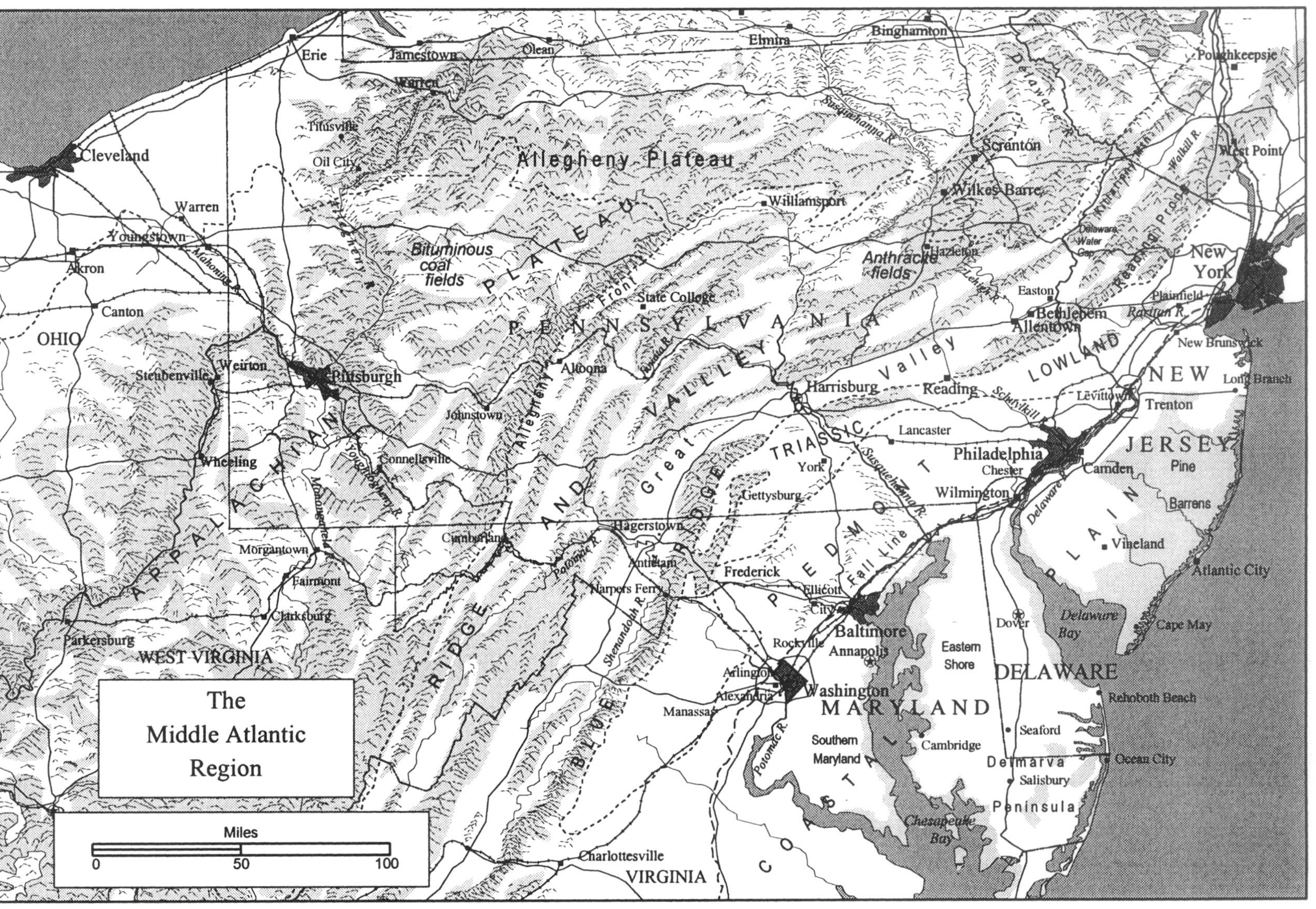
The Middle Atlantic Region
Miles
0
50
100
Cleveland
Erie
Jamestown
Olean
Elmira
Binghamton
Poughkeepsie
West Point
Warren
Titusville
Oil City
Youngstown
Akron
Canton
OHIO
Steubenville
Weirton
Pittsburgh
Wheeling
Johnstown
Connellsville
Morgantown
Fairmont
Clarksburg
Parkersburg
WEST VIRGINIA
Allegheny Plateau
PLATEAU
Bituminous coal fields
PENNSYLVANIA
State College
Altoona
Williamsport
Scranton
Wilkes-Barre
Hazleton
Anthracite fields
Delaware Water Gap
New York
Plainfield
Raritan R.
New Brunswick
Long Branch
NEW
JERSEY
Trenton
Levittown
Easton
Bethlehem
Allentown
LOWLAND
Valley
Reading
Harrisburg
Lancaster
Philadelphia
Chester
Camden
Pine
Barrens
Wilmington
Vineland
Atlantic City
Cape May
Delaware Bay
Dover
DELAWARE
Rehoboth Beach
Seaford
Ocean City
Salisbury
Delmarva
Peninsula
Cambridge
Eastern Shore
MARYLAND
Southern Maryland
Chesapeake Bay
Baltimore
Annapolis
Ellicott City
Washington
Rockville
Arlington
Alexandria
Manassas
Potomac R.
Frederick
Hagerstown
Antietam
Harpers Ferry
Shenandoah R.
Cumberland
Gettysburg
York
TRIASSIC
PIEDMONT
Fall Line
Great
VALLEY
RIDGE AND
APPALACHIAN
BLUE RIDGE
COASTAL PLAIN
Charlottesville
VIRGINIA
Allegheny Front
Susquehanna R.
Delaware R.
Schuylkill
Lehigh R.
Monongahela R.
Youghiogheny R.
Mahoning R.
Juniata R.
Kittatinny Mtn.
Reading Prong
Walkill R.

MAP 6.1

Although Dutch activities were focused on New Amsterdam and Fort Orange, their colonization efforts also extended to the Delaware River Valley. The first nation to claim a portion of the territory, however, was Sweden. Following his dismissal as colonial administrator by the Dutch West India Company, Peter Minuit—the man who had purchased Manhattan Island from the natives—was employed to obtain lands for a Swedish colonization effort. In 1638 Minuit purchased the west bank of the Delaware River, including lands in what are now the states of Pennsylvania and Delaware, to begin the colony of New Sweden. A trading post and a fort built near the mouth of Brandywine Creek later became the nucleus for the city of Wilmington.



Sweden lacked populations of potential colonizers like the Puritans and Quakers, who felt the pressure of religious persecution at home and were willing to move to the New World. The small number of Swedish immigrants to the Delaware Valley was augmented by the arrival of Finnish settlers. Probably it was the Finns—experts in log construction—who introduced the log-building styles that later became a familiar part of frontier settlement in the eastern United States. Sweden's support for the Delaware Valley colony was weak, and in 1655 its efforts to prevent takeover by the Dutch failed. Only nine years later the Dutch, in turn, bowed to the increasing presence of English traders and colonists on all sides and surrendered their possessions to England.

Colonial land grants in the Middle Atlantic region reflected political turmoil in England during the 17th century. The restoration of the British monarchy in 1660 brought Charles II to the throne. His brother and heir apparent, James, duke of York, was made lord of the British Admiralty. The king made ill-defined grants of land along the Atlantic coast, but the duke took possession of New Netherlands (renaming it for himself) and granted lands between the Hudson and Delaware Rivers to two courtiers, John Lord Berkeley and Sir George Carteret, both of whom had received land in the Carolinas already. In 1676 Carteret and Berkeley divided the land. Carteret became the proprietor of East Jersey (roughly east of Trenton) and Berkeley of West Jersey, although Berkeley had already sold his lands to a group of Quakers. Land titles remained in conflict long after the two domains were united as the single colony of New Jersey in 1702.

MAP 6.2

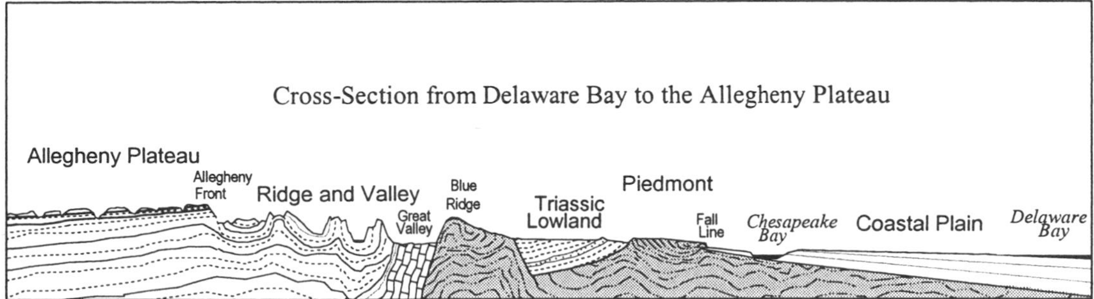

A belt of fertile soils along the landward margin of the Coastal Plain (the Inner Coastal Plain) helped make the lower Delaware Valley productive of grain crops, especially wheat and corn. The reddish-colored soils on shales and sandstones of the Triassic Lowland also are suited for agriculture, and these lands were perceived correctly to be good for settlement. Both the Inner Coastal Plain and the Triassic Lowland (especially the Raritan River Valley) were well occupied by the early 18th century. Although northern New Jersey has been included within the commuter zone of New York City for many years, its landscape retains a rural–urban fringe quality. Like southern New England, Long Island, and eastern Pennsylvania, it remains as one of the more intensive areas of nursery and greenhouse production in the United States.

## Philadelphia and Southeastern Pennsylvania

William Penn was another successful petitioner who received a large grant of land from King Charles II. In 1681 Penn was awarded lands between the Delaware and Susquehanna Rivers to found a colony for England's Quakers. It was one of the last large grants made along the eastern seaboard, but it became one of the most influential because of its effects on later American settlement. The east bank of the Delaware, opposite Penn's grant, was already occupied by Quaker settlers when Penn received his land, and Penn's Philadelphia Quakers were not the first Europeans to settle on the Delaware's west bank, given the earlier presence of the Swedes and the Dutch.

Penn carefully planned the settlement of both Philadelphia and Pennsylvania. Philadelphia was staked out on high ground along the Delaware, with plans that it would grow west to the Schuylkill (fig. 6.1). Penn directed that the city be laid out in a compact grid form, with wide avenues, and with entire city blocks set aside as squares for parks and public buildings. Philadelphia's growth followed Penn's design, although the city inserted new streets running between the

FIG. 6.1. *Society Hill in Philadelphia.*

existing thoroughfares that intensified land use and thereby increased congestion.

Those who bought rural land were given a house lot in the city, an innovation that tied the surrounding territory to Philadelphia. The city grew rapidly in its first half century. It was the largest city in the American colonies in the 1750s and served for a time as the national capital, but Philadelphia did not grow as rapidly as other urban centers after the American Revolution. It lost the distinction of being the nation's largest city to New York by 1790, and it lost the national capital in a series of moves culminating in 1800 when the seat of government was moved permanently to Washington.



Among William Penn's innovations was a procedure for settlement formation then known in some of the frontier towns of Europe but fairly new to North America. Penn designed Philadelphia first, before any settlement occurred. Only after the design was fixed was any property offered for sale. Philadelphia's simple grid pattern made this comparatively easy to accomplish, and it was a procedure easy to imitate: the same sequence was repeated in thousands of other American cities in later years. The significance of Penn's approach was that it created a market for urban land, whereby competition for access to central locations within the city determined a price for urban real estate. Philadelphia soon began growing up as well as out. Penn demonstrated that not only are cities the sites where businesses locate; the creation of cities itself is also a business.

William Penn also was among the first to promote his settlement through advertising. He described Pennsylvania's attractive qualities in articles he wrote for the German-language press in Europe. Groups of German Baptists, Mennonites, and Moravians were influenced by Penn's proselytizing, and they emigrated and settled on the Delaware Valley side of Pennsylvania. The area from Easton through Philadelphia and west across the Piedmont to Lancaster became the first large area of German settlement in the United States. (These Pennsylvania Germans were the forebears of today's Pennsylvania Dutch, a term that comes from a corruption of *deutsch.* They are not ethnic Dutch.)

Scotch-Irish immigrants dominated the movement to Pennsylvania at times, especially during the middle decades of the 18th century. The Scotch-Irish (ethnic Scots, of both lowland and highland origin, who had been living in Ireland) settled within the same areas of Pennsylvania that attracted the Germans. They thrived under the conditions of frontier life. The Scots were aggressive in their dealings with native peoples, acquiring a lasting reputation for being "Indian fighters." Welsh and English immigrants also moved to Pennsylvania in sizable numbers.

Part of Philadelphia's early growth came from the development of agriculture in the Piedmont and the Great Valley. The gently sloping lands had reasonably fertile soils and were easy to convert into productive farms. There were few obstacles to settlement, and the continued growth in population provided a market for crops and livestock. Early Pennsylvanians raised wheat for the export market and supplied local populations through their production of foodstuffs. No other colonial city had a hinterland as rich and agriculturally productive as Philadelphia's, and it became the most densely settled large area of the eastern seaboard by 1800. Despite decades of urban encroachment, southeastern Pennsylvania still is an important agricultural region, one of the few areas of highly productive farmland in the urbanized Northeast.

Southeastern Pennsylvania became a manufacturing region by the early decades of the 19th century and remains so today. Reading, Allentown, Lancaster, and York developed as satellite cities of Philadelphia and later grew into an urban-industrial region surrounding the metropolis. Early iron manufacturing in the Delaware Valley was based on local hardwood forests cut for charcoal manufacture and on small, localized deposits of iron ore. Charcoal iron furnaces became the nuclei for scattered industries where iron implements and utensils were forged. Grist mills for grinding wheat into flour occupied sites along streams flowing down the gentle slopes toward the Delaware River. Most of these industries were independent, home-grown efforts, and in this respect they contrasted with the industrial mills of New England, which operated more as true factories.

The Delaware Valley's steel industry came later; it was based on access to iron ore imported from Quebec-Labrador mines and from South America. Competition from foreign steel suppliers and a shift in industrial activity away from the East Coast in recent decades made the industry vulnerable. The largest integrated iron and steel mills, such as the Fairless works between Philadelphia and Trenton, have reduced operations and now concentrate on specialty products. Heavy machinery, automobiles, trucks, and other transportation equipment continue to be manufactured in the Philadelphia-Wilmington region, however, and southeastern Pennsylvania's cities still supply dozens of products to the national market, ranging from prepared foods to machine tools.

A less tangible contribution made by southeastern Pennsylvania was the blending of cultural influences when diverse European populations encountered one another there in the 18th century. A "Pennsylvania culture region," based on both Scotch-Irish and German influences, emerged in the Susquehanna Valley. It is regarded as the formative area for various patterns of speech and material culture that spread, through the process of frontier expansion, to create the culture region known as the Upland South. In the era when folk practices were transmitted mainly through currents of migration, the implanting of new customs was more easily traceable. Much of the Corn Belt of the Middle West was settled by Euro-Americans whose earliest ancestors on American soil lived in southern and central Pennsylvania. One form of speech, called *Midland,* is common to both the Upland South and lower Middle West, and it emanates from Pennsylvania.

## Baltimore and Washington

The many settlement experiments launched on the eastern seaboard by various European nations suggest that the form colonialism took in one place or another mattered less than the colonists' common ambition to succeed. Pennsylvania was a successful experiment, and William Penn had a benign influence on its growth. Maryland, in contrast, was created as a feudal barony. It originated as a grant of land by King Charles I to George Calvert (Lord Baltimore). Calvert and his son, Cecilius, became proprietors of more than 10 million acres of land between the Potomac and Delaware Rivers with a royal charter allowing them great freedom in levying taxes, collecting rents, and establishing an almost medieval system of land tenure.

Maryland also was the northern limit of plantation agriculture. A staple crop like tobacco could be grown profitably there by employing indentured servants or slave labor. The plantation mode of production was encouraged by a tenure system in which land was owned in large tracts. Maryland's early settlers focused on tobacco just as their contemporaries did in Virginia. Southern Maryland—a regional term denoting the counties south of Annapolis and west of Chesapeake Bay—saw the most intensive tobacco production. The city of Baltimore originated in 1729 around a customs house and a dock built on the Patapsco River for the purpose of shipping tobacco to Europe. Like New York, Baltimore's large natural harbor became one of its most important assets (fig. 6.2).

Baltimore was well situated for growth. The fertile lands of the Maryland Piedmont produced large wheat crops. The level profile across the Piedmont was a natural connection between Baltimore and the Potomac River, and access to the Ohio Valley lay just beyond the Potomac's watershed. In 1811 work began on the National Road, one of the federal government's early public works projects, which was built from Cumberland, Maryland, across the Appalachian Plateau to Wheeling, West Virginia, on the Ohio River. The Baltimore & Ohio Railroad, which was the nation's first important railway line, later followed the same route.

In the era of steamboat commerce on the Ohio and Mississippi Rivers, merchants and wholesalers in Philadelphia and Baltimore dominated trade with the interior, but Baltimore merchants had an advantage over Philadel-

FIG. 6.2. *Baltimore's Inner Harbor once functioned as part of the city's commercial seaport, but it has been transformed into a museum and recreational environment.*

phia because Baltimore was nearly one hundred miles closer to Wheeling on the Ohio River. Baltimore became the obvious port for coal mined in West Virginia and for grain and livestock raised in the Ohio Valley.

The lengthy perimeter of peninsulas and bays within Baltimore's harbor offered sites for industries that processed incoming raw materials. A large steel mill at Sparrows Point, in Baltimore's outer harbor, processes iron ore imported from Venezuela. Sugar refiners, who imported cane from the West Indies, also located along the harbor. Like many port cities and transportation centers, Baltimore reaped profits from these ventures that were invested in further urban growth. Baltimore's metropolitan area now contains 2.4 million inhabitants and accounts for roughly half of Maryland's total population.

Washington, D.C., was not a colonial city. It was a tract of lightly used land when, in 1790, northern and southern factions in Congress compromised on a location for the national capital that would be at the border between the two sections of the country. A location on the Potomac River was fixed, President George Washington chose the land, and arrangements were made for its purchase. In 1791 Major Pierre Charles L'Enfant, a French artist and engineer who had been an officer in the army under Washington, was chosen to prepare a plan for the new city.

L'Enfant drew inspiration from such examples of French architecture and design as the Versailles Palace of King Louis XIV. Thomas Jefferson was one of the first critics of L'Enfant's plan, which imitated the excesses of a European monarchy in creating a capital for the new American democracy. President Washington supported L'Enfant, however, and although quarrels eventually led to L'Enfant's dismissal from the project, the plan he created was followed fairly closely. As at Versailles, long vistas separated prominent buildings. Squares and circles terminating these vistas were the obvious sites upon which to build the monuments and statues commemorating great leaders. Europeans who visited Washington in its early years ridiculed its sprawling extent, noting that the city amounted to little more than scattered buildings linked by trails through the woods.

Washington's design was grander than the city itself throughout most of the 19th century. Its population swelled when Congress was in session, but other times it was more like a small town. As federal agencies were created, one by one they occupied new buildings constructed along Pennsylvania Avenue leading from the White House to Capitol Hill. Monuments eventually occupied the prominent sites reserved for them, and the city's long vistas began to seem less excessive. The federal workforce grew during the Depression of the 1930s and even more during World War II, when the first major expansion took place south of the Potomac with the construction of the Pentagon for the War (Defense) Department.

By the 1960s new buildings occupied by an expanding federal bureaucracy had nearly exhausted the building space in the original city design (fig. 6.3). Lobbyists and interest groups that felt the need for a presence in Washington helped lead a further suburbanization of government-related activities to the

FIG. 6.3. *Washington, D.C., with the Tidal Basin and the Jefferson Memorial in the foreground. Nearly all of the buildings shown house government agencies.*

Capital Beltway. New agencies as well as new quarters for old branches of the government were built in the Virginia and Maryland suburbs and were linked to the central city by rapid transit in the 1980s. Washington remains essentially a single-purpose city, but the sheer volume of its business has grown to require a metropolitan-area population of nearly 4 million.

Just as the Americans invaded Toronto (York) and burned its government buildings, so did the British attempt to capture Washington during the War of 1812. Later, the city's location on the border between North and South drew Confederate military campaigns in its direction during the Civil War. General Robert E. Lee's northward thrusts, which led to the bloody battles at Gettysburg, Pennsylvania, and Antietam, Maryland, were an attempt to encircle Washington. Because Washington is on the northern border of the South, it became an attractive destination for northward migration, especially for African Americans, who encountered fewer barriers to employment in the federal city than they did elsewhere. Blacks outnumber whites more than two to one in the District of Columbia's population today.

## The Fall Line and Megalopolis

Trails and wagon roads that followed the edge of the Coastal Plain during colonial times had been replaced by railroads and highways linking New York, Trenton, Philadelphia, Wilmington, Baltimore, and Washington by the middle of the 19th century. Today no fewer than eight parallel highway and rail routes follow one another within a twenty-mile span across New Jersey, carrying people and goods within this, the busiest of all North American transportation corridors. These are some of the arteries of Megalopolis, a term coined by the French geographer Jean Gottman to describe the urban agglomeration that stretches along the eastern seaboard, not only from Washington to New York, but from there to Boston as well.

Megalopolis is a region of high population densities and urban sprawl. More than three dozen contiguous metropolitan areas form individual peaks of urban density in the region, which includes roughly one-sixth of the total population of the United States. The cities are so numerous and so closely spaced that by 1950 their fringes were growing together into what seemed to be a supercity, a single metropolitan complex that eventually might cover the entire northeastern seaboard. The combination of urban sprawl, freeways, shopping centers, and housing subdivisions—all fairly new developments in the 1950s—later was seen to be less a new form of urbanism than a passing phase of urban growth. In the 1990s attention was focused on other new developments, such as the "edge city" phenomenon, in which homes, shopping centers, and workplaces all cluster around highly accessible nodes in the transportation network. Megalopolis contains numerous examples of all of these types of urban settlement.

The New York–Washington urbanized corridor has other reasons for its existence as well. It lies close to the geological divide between easily eroded rocks of the Coastal Plain and the more erosion-resistant Piedmont or Triassic Lowland formations. Streams crossing this divide have formed waterfalls slightly upstream from the point of contact where the two types of rock meet. Some of the waterfalls, such as the Great Falls of the Potomac, near Washington, are spectacular cascades where waters rush through the hard-rock edges of the Piedmont before tumbling down to the lower, flatter Coastal Plain. From there the rivers meander slowly out to tidewater.

This series of waterfalls is known collectively as the Fall Line. It extends along the entire Atlantic Coast, from Cape Cod to Alabama, although waterfall development is not uniform along the entire edge. In New Jersey, falls or rapids are at Paterson (the Passaic River), New Brunswick (the Raritan River), and Trenton (the Delaware River); in Delaware at Wilmington (Brandywine Creek);

FIG. 6.4. *The falls of the Passaic River at Paterson, New Jersey. Natural water-power sites played an important role in the settlement of the eastern seaboard.*

and in Maryland at Ellicott City (the Patapsco River). During the early era of water transportation, the Fall Line limited navigation to sites no farther inland than these places, and thus locations along the Fall Line were favored for development as transportation centers.

The availability of waterpower at the same sites also led to their early growth as manufacturing centers. Paterson, New Jersey, was created as an industrial new town in 1791 to take advantage of the falls of the Passaic River (fig. 6.4). It became a textile manufacturing center and later specialized in spinning raw silk into fine cloth. Trenton's early start in manufacturing led to its later development of a wide variety of hard- and soft-goods industries. Wilmington's growth was led by the giant Du Pont chemical manufacturing complex; the much smaller Ellicott City became a flour-milling center for wheat raised on the Maryland Piedmont.



## The Coastal Plain: The Pine Barrens, the Eastern Shore, and Delmarva Peninsula

East of the Fall Line the Coastal Plain is flat, sandy, and dotted with numerous ponds and wetlands. Coastal Plain rock formations are largely of marine origin, and they outcrop in bands of progressively younger age approaching the continental margin. Except for the Inner Coastal Plain fringe, the soils are acidic. The land repelled the sorts of colonial agriculture that thrived in the Piedmont.

Although the Coastal Plain begins at Cape Cod, it is a discontinuous region in the north. Long Island is a terminal moraine produced by the last Pleistocene glaciation and resting atop the Coastal Plain formations. The main portion of the Coastal Plain begins only a few miles south of Staten Island (New York) and continues south from there around the entire Atlantic and Gulf margin of the United States.

Eastern New Jersey is known as the Pine Barrens, and although today the vegetation is not noticeably a barrens (referring to large areas of grassy or shrubby plant cover) or pine forest (it is now mainly hardwoods), the Pine Barrens remains a unique region of very low land-use intensity at the fringe of Megalopolis (fig. 6.5). Except for acid-tolerant crops such as blueberries, which are produced commercially in the Pine Barrens, agriculture was slow to develop in the region. It was limited by the sandy soils, which drain rapidly and become droughty despite abundant precipitation during New Jersey's humid summers. Irrigation has extended agricultural production eastward into the fringes of the Pine Barrens in recent years, and although most of the sandy lands still remain uncultivated, the region produces a sizable crop of summer vegetables for the commercial market.

The Pine Barrens' major urban center is Atlantic City, once known for its oceanfront boardwalk and resort amenities; in recent decades it has become a major concentration of gambling casinos. Like Las Vegas, Atlantic City's gaming complex is an intense cluster of human activity within a largely unoccupied territory, and the city must live on the business it draws from a distance. Asso-

FIG. 6.5. *New Jersey's Pine Barrens.*

ciated resort and second-home developments have increased the population of counties along New Jersey's barrier island coast until today no part of the Pine Barrens lies outside the limits of a metropolitan area, despite the generally low population densities.

The term *Eastern Shore* is applied to that portion of Maryland that lies east of Chesapeake Bay. Along with the state of Delaware and two counties of Virginia, it makes up the Delmarva Peninsula, a 150-mile stretch of Coastal Plain separating Delaware and Chesapeake Bays. Euro-American colonists came here early, although the rate of settlement was comparatively slow. Growth was held back by the region's peninsular status until the 1960s, when construction of the Chesapeake Bay bridge-tunnel allowed through traffic between Philadelphia and Norfolk to move via the Delmarva route.

Delmarva currently produces almost half a billion meat-type chickens per year. Although poultry production is even more intensive in parts of Georgia, Alabama, and Arkansas, the modern poultry "factory" was largely a creation of entrepreneurs in the Eastern Shore region of Maryland beginning in the 1930s. Consumer preference for poultry as a healthier alternative to red meat began to emerge in the 1950s. Increased demand led to a new type of operation in which individual farms were large and totally specialized. Poultry-raising was the first of the meat industries in the United States to switch to factory methods of production. Chicken, which once had been considered a delicacy, thereafter became one of the cheapest meats sold in the supermarket. In recent years the larger poultry companies have become diversified food conglomerates that also produce frozen entrees and snack foods.

When irrigated, Delmarva's sandy soils produce quantities of corn and soybeans for chicken feed, although large stocks of grain still must be imported from the Middle West. The southeastern poultry producer is one of the best customers of the Corn Belt cash-grain farmer. Broiler sheds and corn and soybean fields occupy much of the total land area of the Delmarva Peninsula today, and some of the nation's largest poultry processing plants are located in the region's small cities. Over sizable areas, nearly all of the rural landscape is devoted to this single industry.

The most intensive development in Delmarva is "the Shore"—especially its beachfront cities such as Rehoboth Beach, Delaware, and Ocean City, Maryland. Like Atlantic City, their large volume of tourism depends on the proximity of metropolitan centers. Though the sand beaches remain, they are now fringed by continuous rows of high-rise condominium towers. Together these seaside resorts can accommodate hundreds of thousands of visitors at a time.



## The Appalachian Orogeny

The Appalachian Mountains begin at the landward edge of the Piedmont. The term *Appalachians* refers to a collection of mountain types that is divided into distinct north-south-trending subregions. The Appalachians began forming some 500 million years ago when the ancestral Atlantic Ocean, separating the North American and African continents, started to narrow because of the convergence of the underlying plates. Sediments of the ocean floor adjacent to North America were compressed, and new rocks were formed under the intense heat and pressure as the two continental land masses converged (fig. 6.6).

The Blue Ridge Mountains were pushed up in a belt paralleling the colliding land masses. Rocks of what had been the ocean floor plunged under the North American continent, melted, and gave rise to a line of volcanoes (no longer visible). The effects of the collision were felt over a wider area than just the line of contact between the two land masses. Rocks of the Ridge and Valley section, paralleling the Blue Ridge, were compressed into parallel folds from the force. In fact, all ridges of the Appalachians, from Vermont to Alabama, are aligned parallel with the Atlantic coast, a reflection of their origin in the process of continental collision.

FIG. 6.6. *Sideling Hill, a synclinal mountain that was cut through to build Interstate Highway 68 near Hancock, Maryland, typifies the geology of the Ridge and Valley section of the Appalachians.*

A majestic range of mountains graced the eastern edge of North America 225 million years ago. These earlier Appalachians had elevations many times higher than at present, with summits rivaling those of the loftiest peaks in the Andes of South America today. The great mass of the mountains caused the earth's crust to flex downward in partial compensation for the load. This, in turn, produced a deep basin (geosyncline) that filled with sediments eroding from adjacent land areas. The fringing basin was proportionally as deep as the mountains were high, with the greatest basin depth closest to the edge of the mountainous folds. Sediments that collected in the Appalachian Geosyncline eventually solidified into the rock layers of the Appalachian Plateau.

Even though the correct relative positions of the Blue Ridge, the Ridge and Valley, and the Appalachian Plateau can be seen in these glimpses into earth history, the position of the North American continent itself was much different at that time. North America lay far south of its present location, and the Appalachians straddled the equator—a hot and humid location that produced luxuriant forests analogous to those of the Amazon region of Brazil today. The decomposition and burial of the surface vegetation, followed by deep burial and compression of the rock layers, eventually produced layers of coal.

## The Anthracite Region

Bituminous coal and anthracite are associated, respectively, with the Appalachian Plateau and the Ridge and Valley portions of the mountains because of contrasts in the underlying geology. Anthracite—sometimes called "stone coal" for its hardness—is a metamorphic rock that was produced by heat and pressure in the earth's crust during the formation of the Ridge and Valley topography. The extra heat deriving from the compression of rock layers transformed bituminous coal into a product with a higher fixed-carbon content. Anthracite burns at a high temperature, but it produces little smoke; it was a valuable fuel in the steam-power era of American railroads. It was also used for home heating.

Anthracite was the principal natural resource of central Pennsylvania's Ridge and Valley section. The Scranton, Wilkes-Barre, and Hazleton urban areas of the Susquehanna Valley grew for many years because of the demand for anthracite. Although it still is used as a source of industrial carbon, anthracite is no longer used for fuel. The industry has been in decline for many years despite the fact that as much as two-thirds of the total reserve remains unmined. Only about fifteen hundred miners are employed in the anthracite industry today.

## Central Pennsylvania

Central Pennsylvania's topography presents a formidable obstacle for those who would cross it east-west, even today. Although rivers such as the Juniata and the West Branch of the Susquehanna are able to thread their way through the various offset gaps separating anticlinal and synclinal mountains in the Ridge and

Valley section, the gaps are few compared with the miles of ridges in a topography of parallel mountain folds. The Susquehanna's tributaries reach well westward into the Allegheny Plateau, but all streams disappear at the drainage divide running along the highest ground of the Alleghenies.

The state of Pennsylvania had no possibility for a water-level route similar to the Erie Canal across New York. Pennsylvania canal promoters devised various schemes for inclined planes and portage railroads that would carry canal boats or railroad cars over the mountains, but despite some beginnings, an effective transportation system was not available until the Pennsylvania Railroad was completed between Philadelphia and Pittsburgh in 1857. Even this line had its difficulties, because it required an assault on the Allegheny Front—the eastern edge of the Allegheny Plateau—a few miles west of Altoona. There the railroad constructed its famous Horseshoe Curve, a looping of track around three sides of a mountain valley that leads to a tunnel piercing the edge of the plateau. The line has remained in continuous use ever since its construction in the 1850s.



Two of the great streams of westward migration during the late 18th and early 19th centuries started from central and southeastern Pennsylvania. The easier route lay to the south, along the topographic grain of the mountains, especially up the Great Valley separating the Blue Ridge from the Ridge and Valley region. This route was favored by most settlers until the 1790s. The more difficult route was to cross the Allegheny Mountains directly west, by whatever trails, wagon roads, or creeks one could find, to enter the Ohio Valley. Thousands of emigrants took one of those paths out of central Pennsylvania, thereby spreading cultural traits of the Middle Atlantic region to the Upland South and the lower Middle West.

## Western Pennsylvania and the Steel Valleys

West of the Appalachian divide, all rivers drain to a common point, where the Allegheny meets the Monongahela and forms the Ohio River at Pittsburgh. The site of Pittsburgh has been of strategic value for at least hundreds of years and probably longer. The confluence saw the construction of military fortifications during the long struggle for control of the fur trade. The French called this place Fort Duquesne when they were in charge; the British knew it as Fort Pitt. Control of the site was crucial for access to the Ohio Valley. Pittsburgh eventually lost its fortifications, but it became an important trade and transportation center. The Allegheny River was navigable by flatboats downstream from Olean, New York, and thus the forks of the Ohio could be reached easily from the north. From the south came the Monongahela and its principal tributary, the Youghiogheny, flowing north out of West Virginia and Maryland. West of Pittsburgh lay water routes extending two-thirds of the way across the continent and south to the Gulf of Mexico.

The meandering rivers that converge on Pittsburgh are slowly eroding down through Paleozoic rock formations that contain layers of bituminous coal. Unlike the folded deposits of anthracite in central Pennsylvania, western Pennsylvania's bituminous coal is in flat-lying beds that have not been contorted by mountain-building forces. Bituminous coals do not produce as hot a fire as anthracite, but they are clean-burning and produce little ash. Some high-grade bituminous coal in Pennsylvania lies close enough to the surface that it can be strip-mined, although underground mining is necessary in many areas to reach the deeper seams. The coal is used both for heating and for manufacturing coke, which is essential in steel metallurgy.

The availability of coking-quality coal made western Pennsylvania and eastern Ohio attractive locations for building steel mills. It was the first area of the United States where steel (rather than iron) was the main focus of industrial development. The emergence of a steel industry in and around Pittsburgh in the 1870s coincided with the shift away from iron to the much wider variety of products that can be made from steel. Pittsburgh was the nation's leading steelmaking center for many years, and although the industry had expanded westward to the Gary, Indiana, district by 1910, Pittsburgh retained a dominant role. (See chapter 14 for further discussion of the steel industry.)

Competition from foreign steel producers, a slump in domestic demand associated with troubled times in the U.S. automobile industry, and competition from smaller, specialty steel mills created overcapacity conditions in the Pittsburgh-Cleveland steelmaking region by the 1970s. Numerous plant closures in the following decade did not mean steel's total demise, however; a smaller but more vigorous industry emerged in the 1990s after substantial restructuring that involved the sale of some mills to employees and partnership ventures with Japanese firms. Although in this region the fires of many blast furnaces have been extinguished—and, indeed, the city of Pittsburgh itself no longer contains an integrated steel-mill complex—blast furnaces no longer equate with the entire steel industry. Numerous communities near Pittsburgh still produce a variety of types of steel.

A continuous region of steel mills, manufacturing hot rolled and cold rolled steel, steel tubes, galvanized sheets, and a variety of coated steels, stretches from the Monongahela Valley south of Pittsburgh through Wheeling and Weirton, West Virginia, and from the Mahoning Valley (including the cities of Canton, Youngstown, Massillon, and Warren, Ohio) to the Cleveland area. This "steel valley," as it has been called, is the largest cluster of specialty steel mills in the United States, supplying steel to the automobile industry and to all other metal fabricating industries of the Northeast.

## *References*

Borchert, John R. *Megalopolis: Washington, D.C., to Boston.* New Brunswick, N.J.: Rutgers University Press, 1992.

Brush, John E. *The Population of New Jersey.* New Brunswick, N.J.: Rutgers University Press, 1956.

Earle, Carville V. *The Evolution of a Tidewater Settlement System: All Hallow's Parish,*

*Maryland, 1650–1783*. Research Paper no. 170. Chicago: Department of Geography, University of Chicago, 1975.

Gardner, T. W., and W. D. Sevon, eds. *Appalachian Geomorphology.* New York: Elsevier, 1989.

Garreau, Joel. *Edge City: Life on the New Frontier.* New York: Doubleday, 1991.

Gottman, Jean. *Megalopolis: The Urbanized Northeastern Seaboard of the United States.* New York: Twentieth Century Fund, 1961.

Lemon, James T. *The Best Poor Man's Country: A Geographical Study of Early Southeastern Pennsylvania.* Baltimore: Johns Hopkins Press, 1972.

Lewis, Peirce F. "Small Town in Pennsylvania." *Annals of the Association of American Geographers* 62 (1972): 323–51.

Olson, Sherry H. *Baltimore: The Building of an American City.* Baltimore: Johns Hopkins University Press, 1980.

Reps, John W. *Monumental Washington: The Planning and Development of the Capital Center.* Princeton, N.J.: Princeton University Press, 1967.

Wacker, Peter O., and Paul G. E. Clemans. *Land Use in Early New Jersey: A Historical Geography.* Newark: New Jersey Historical Society, 1995.

Zelinsky, Wilbur. "The Pennsylvania Town: An Overdue Geographical Account." *Geographical Review* 67 (1977): 137–47.

*PART III*

# The Upland South

*CHAPTER 7*

# The Southern Appalachians

The Southern Appalachian Mountain, Interior Low Plateau, and Ozark-Ouachita regions commonly are grouped together under the regional label Upland South. Less precisely, and with a slightly different purpose, the three regions also are described as the Upper South. The reason for making both distinctions is to separate the two major divisions of the southeastern states into a northern region of generally higher elevation, where agriculture is relatively poorer, a region that did not develop a slave-owning economy before the Civil War; and a flatter, lowland Deep South, which is identified with slavery, a historical African American presence, and large-scale agriculture. This primary division of the South into upland and lowland segments also parallels to some extent cultural differences within the Euro-American population, a contrast that has long played a role in regional political economy, between an up-country yeomanry and a lowland planter class. Of the three divisions of the Upland South, the Southern Appalachians best typify the region as a whole.

The Appalachian region is defined primarily in terms of landforms, and it is conventionally subdivided into the Blue Ridge, Great Valley, Ridge and Valley, and Appalachian Plateau sections (map 7.1). The regional names for these divisions as well as their relative size and shape vary over the length of the mountain system. The Blue Ridge, which is a single ridge in Pennsylvania, broadens southward to become the Smoky Mountains along the North Carolina–Tennessee border. The Great Valley is a distinct lowland in northern Virginia where it is drained by the Shenandoah River. The valley briefly disappears near Roanoke but broadens once more southward to become an extensive group of valleys drained by tributaries of the Tennessee River. The Ridge and Valley section is relatively constant in width from Pennsylvania south to the vicinity of Birmingham, Alabama, where the Appalachians end. In New York and Pennsylvania the Appalachian Plateau is drained by the Allegheny River, and the plateau itself is called the Allegheny Plateau. In Kentucky and Tennessee, the same geologic region is known as the Cumberland Plateau.

## The Great Valley (the Valley of Virginia)

The Appalachian region is distinctive culturally because of migration flows in the late 18th and early 19th centuries that followed the grain of the mountains south from Pennsylvania. The Mason-Dixon Line, which cross-cuts the Ap-

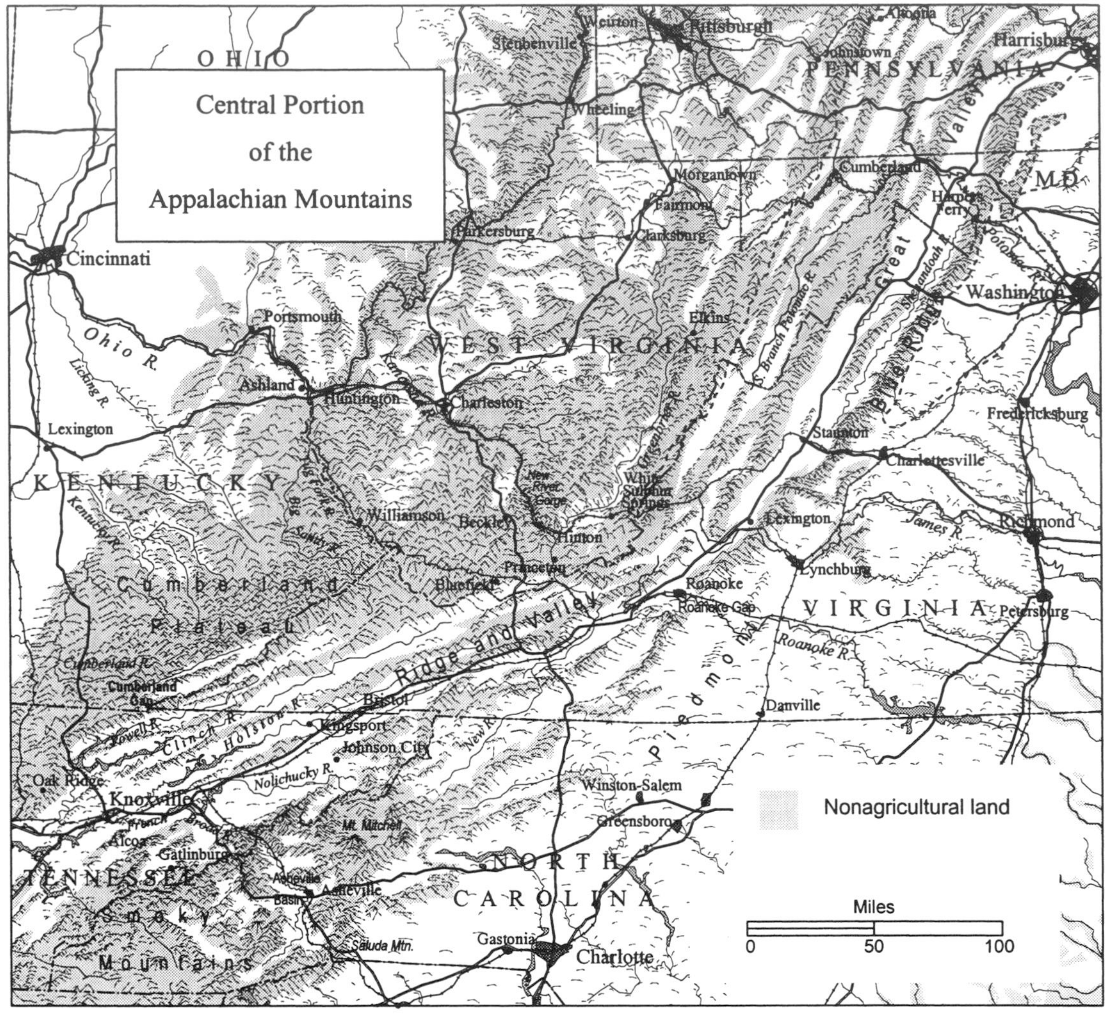


MAP 7.1

palachians, is the northern Maryland boundary and is popularly regarded as the border between North and South in the United States. The Mason-Dixon line had little effect on the migration patterns of early Euro-American settlers, whether northern or southern, but the alignment of the mountains did. Pennsylvanians moved south into Maryland and Virginia; Virginians moved up the Potomac River and settled in southwestern Pennsylvania.

Early-19th-century migrants from Pennsylvania who crossed the mountains to the Ohio Valley provided much of the initial colonizing population for southern Ohio. An earlier generation of Pennsylvanians moved south into Maryland and followed the route of the Great Valley across the Potomac into Virginia. They were the pioneer settlers of the Upland South and much of the Piedmont. Because southeastern Pennsylvania was the common origin of so many migrants, the Upland South, the Piedmont, and the southern Middle West were to share many cultural traits in common in later years. The Great Valley was their common bond.



The Shenandoah River drains most of the Great Valley north of Staunton, Virginia. This portion, called the Shenandoah Valley, is a well-defined lowland, about 25 miles wide, with ridges visible on either side. The Blue Ridge, which is the valley's eastern rim, is not breached by a single through-flowing stream for 175 miles from Harpers Ferry on the Potomac to the James River at Lexington, Virginia. From there it is another 40 miles south to the only other break in the southern Blue Ridge, Roanoke Gap. Folded between the Blue Ridge to the east and the rugged Ridge and Valley section to the west, the Great Valley was an obvious migrant's corridor, and the lack of gaps along it helped channel the flow of settlers southward.

The valley's limestone floor had fertile soils, although the soil cover was thin and the land eventually deteriorated from continuous farming. But in the 1730s, when Pennsylvanians began to move south, the Valley of Virginia offered an attractive environment. Within a decade the flow from Pennsylvania included many Scotch-Irish and German immigrants, some of whom were recent arrivals at the port of Philadelphia. Migrants left the valley at Roanoke Gap and fanned out to the southeast. In the year 1766 more than a thousand wagons moved through the Salisbury district of North Carolina bound for new farms in the Piedmont. By 1780 the tide of southward-moving migrants had settled the Piedmont in South Carolina and northeastern Georgia.

Here was one of the first great thrusts of population redistribution in the United States, a series of southward advances of the frontier that spread a new population into territory that Europeans had not known before. Scotch-Irish, English, and German settlers who moved south from Pennsylvania evolved a folk-cultural system that came to typify the Upland South as a unique region. The Scotch-Irish were Presbyterians, and the Germans were mainly Lutherans; both groups, in the Valley of Virginia and elsewhere, were medium-scale farmers who practiced a diversified agriculture based on livestock and small grains. Few who moved south from Pennsylvania kept slaves, and though they participated in a commercial economy, they were inclined to "do it themselves."

Tidewater tobacco planters also were attracted to the Great Valley, and they brought new customs to the region. These lowland planters were mainly English and Anglican; they raised tobacco with slave labor and were oriented to purchasing rather than making things for themselves. The Great Valley was thus a mixing ground for these two distinct lifeways. South and west of there the two culture complexes were never so distinct.

South of Roanoke, early migrants took advantage of the occasional offset gaps between linear mountains of the Ridge and Valley section. Working westward, valley by valley, they finally reached the edge of the Appalachian Plateau. The sharp front of the plateau against the Ridge and Valley is continuous from Pennsylvania to Tennessee. The only stream that has cut through the middle Appalachian Plateau to reach the Ohio River drainage is the New River, a tributary of the Kanawha, which rises in the Blue Ridge Mountains in northern North

Carolina and flows in a gorge 1,500 feet below the 4,000-foot level of the Plateau in southern West Virginia.

Settlers came to the valleys of the Clinch, Holston, and French Broad Rivers in upper east Tennessee by the 1770s. Early settlements were practically confined to the major river valleys because the grain of the topography was too difficult to cross. The paths of migration later became the routes of trade, first by wagon and then by railroad. Even the modern network of interstate highways in the Appalachians follows the pattern of linear valleys, crossing ridges, east and west, only when necessary.

Tennessee was still a part of North Carolina when Euro-American settlement began, but it was isolated by topography from the main settled territory far to the east. Remote and neglected, upper east Tennessee settlers formed their own government in 1784, proclaiming themselves the state of Franklin. The new state's legality was questionable, and it soon collapsed. Tennessee was admitted to the Union as a state in 1796. Kentucky was first organized as a county of Virginia, but it, too, was difficult to reach from the east. Kentucky was severed from Virginia and admitted to the Union as a state in 1792. The Kentucky-Tennessee border, surveyed in 1779, is simply a westward extension of the colonial boundary between Virginia and North Carolina.

Daniel Boone is the best-known figure involved in Kentucky's early settlement. He was born in southeastern Pennsylvania but moved south to the valleys of western North Carolina as a young man. Boone was hired by a group of land speculators to clear a wagon road across the mountains. The Wilderness Road, as Boone's trail became known, followed an old path through Cumberland Gap, an easy entrance to the Plateau that the native people had long used for hunting and trade (fig. 7.1). At Cumberland Gap (which became the meeting point of the states of Virginia, Kentucky, and Tennessee), tributaries of the Kentucky and Cumberland Rivers diverge westward away from the Plateau. Although the rivers are not navigable here, they can be followed overland, down

FIG. 7.1. *Cumberland Gap, the meeting point of Virginia, Kentucky, and Tennessee, has carried substantial traffic between the Ridge and Valley and Cumberland Plateau regions ever since it became the route of a wagon road in the 1780s.*

to the better lands of the Bluegrass and Nashville Basin. Thousands of settlers bound for the trans-Appalachian frontier took the Wilderness Road through Cumberland Gap in succeeding years and then were guided by either the Kentucky River or the Cumberland River toward the better lands to the west. Later the Cumberland Gap route was used by livestock drovers who walked herds of cattle eastward through the mountains to markets in the eastern states.



## Appalachian Folk Culture

Many of the distinctive cultural traits of Upland South pioneers who moved through the Appalachians have persisted over time. The backwoods frontier was an essentially classless society in which personal freedom and individualism were valued. People distrusted central authority, were skeptical of the value of education, and had little patience with organized religion. Their settlements were small in size and often were organized in terms of groupings based on kinship. Most settlers raised corn and fed part of the crop to cattle and hogs, which they drove to market.

Although the backwoods frontier dweller may have had an aversion to religion, the Great Revival period of religious fervor that began in 1800 transformed the Upland South into one of the most church-conscious sections of the nation. Revivals, camp meetings, and small country churches became a way of life. The old attitudes toward education did not change so rapidly. Public education did not gain a secure footing until after the Civil War, and even the intensive efforts of recent decades have not been enough to close the literacy gap in parts of the Appalachians.

The primary unit of settlement was the family home place, an informal cluster of small buildings that included a house, a stock barn, and a corn crib, all built of logs. The house was a one-room, rectangular log "pen" that could be enlarged by adding a second pen in line with the first. Settlers built corn cribs and barns and enlarged them in similar fashion. Log buildings of this sort appeared wherever Upland Southerners moved: throughout the Appalachians, south to Texas, and north into Indiana and Illinois. The log cabin was much more a part of Upland South settlement than it was of settlement in the North (fig. 7.2).

The only formally platted towns in the Upland South before the arrival of railroads were county seats. Such towns had a substantial courthouse on a central square surrounded by business blocks. Philadelphia and Lancaster were the models for this type of town design. Outside the county seat there were few trade centers with any formal structure. Numerous crossroads general stores served the population. The large number of general stores, each selling the same, limited line of goods, suggests the importance of self-sufficiency in the population. Churches, schools, and other centers of social congregation also were scattered. The dispersal of activities was partly a function of inaccessibility in rugged terrain. The creek, fork, or branch where one lived practically defined the area that was within convenient access.

The list of Appalachian traits is long, and it includes styles in food, music,

FIG. 7.2. *A single-pen log house and lightly used farmland in the Sequatchie Valley of East Tennessee.*

crafts, and folklore as well as patterns of speech and dialect. Notably absent from any checklist of Appalachian traits before the 20th century is poverty. Self-sufficiency and frugality were long-standing habits, but poverty was uncommon before the arrival of coal mining, a development that radically transformed the style of life in the Appalachian Plateau.

## The Appalachian Coal Field

The Appalachian Coal Field produces nearly half of all bituminous coal mined in the United States (map 7.2). The coal field is practically coextensive with the Appalachian Plateau south of the limit of carboniferous rock formations along the Pennsylvania–New York border. Although Wyoming is now the leading coal-producing state, West Virginia ranks a close second. The increased demand for coal to meet industrial, transportation, and heating needs created a westward-moving frontier of coal mining in the Appalachian Plateau during the late 19th century. Periodic spurts of national economic growth during the 1870s and 1880s led to an ever-widening search for new veins of bituminous coal. Western Pennsylvania, West Virginia, and eastern Kentucky were the foci of these new developments well into the 20th century.

Pennsylvania was an established industrial state, but West Virginia was less than two decades into statehood when mining began on a large scale. West Virginia's statehood had not come until 1863, two years after Virginia, of which it had been part, seceded from the Union. Slavery was uncommon in the Appalachian Plateau and the Ohio Valley, but the Ridge and Valley portion of West Virginia had many slave owners. The people living in the counties that were to become West Virginia were divided about 60-40 in favor of the Union at the time of statehood.

West Virginia's population was thinly scattered outside of a few river valleys. Its institutional framework was fragile, land titles were unclear in many cases, and it was possible for entrepreneurs with legal skills to assemble large tracts of coal-rich land. Henry G. Davis and his son-in-law, Stephen B. Elkins, carved out a mining empire, the Elk Garden field, in central West Virginia in the 1880s and built a railroad to haul the coal to market. Henry H. Rogers, a Standard Oil magnate, tapped the Paint Creek and Winding Gulf fields and then financed an expensive engineering marvel, the Virginian Railway, to haul West Virginia coal to tidewater docks at Norfolk. The Chesapeake and Ohio Railroad developed the New River and Kanawha fields; the Norfolk and Western Railway counted the



MAP 7.2

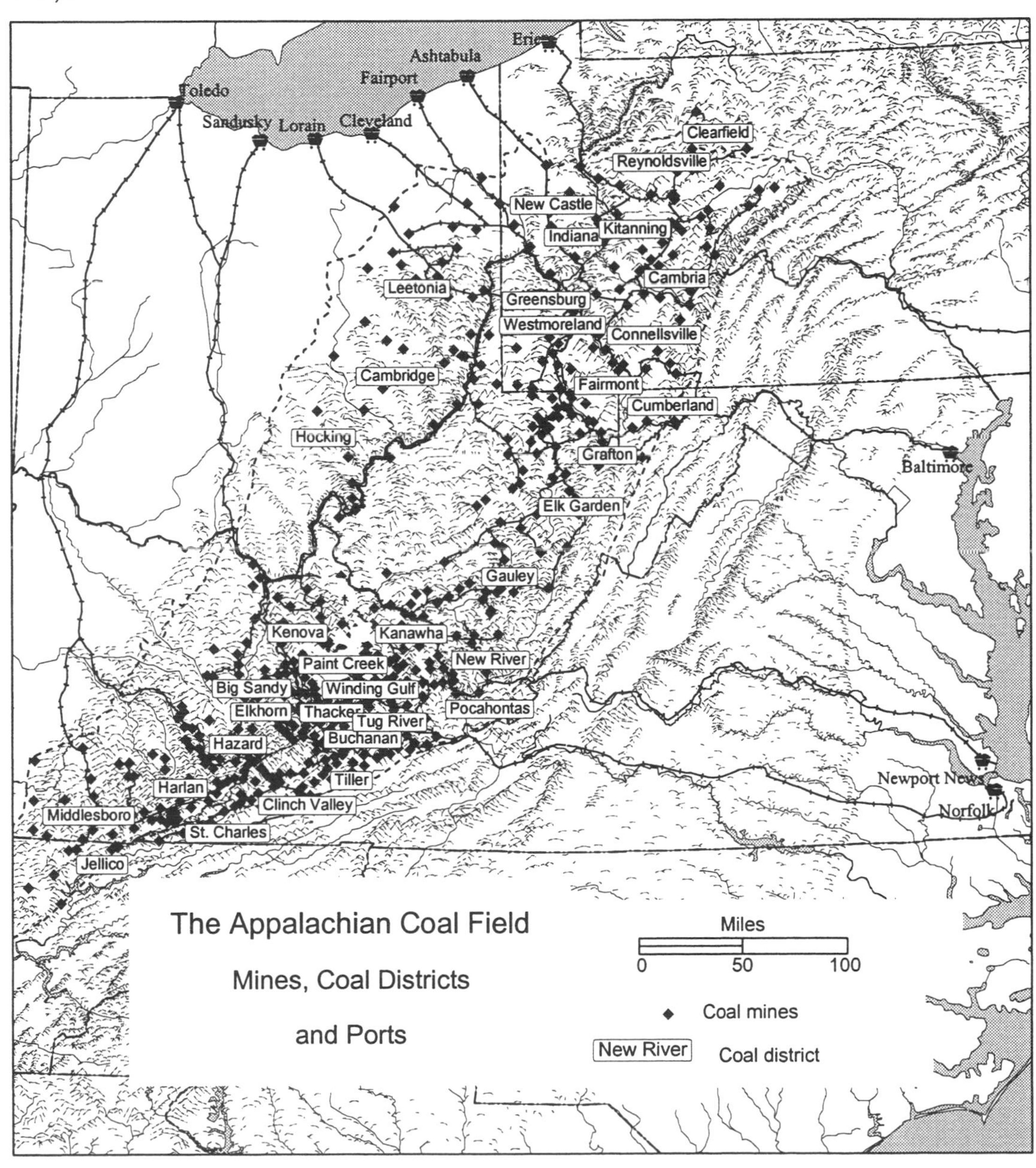

Pocahontas and Tug River fields among its holdings; the Louisville and Nashville Railroad developed the Elkhorn and Harlan fields in Kentucky. Between 1880 and 1910, nearly all the coal in West Virginia and eastern Kentucky became accessible to one railroad company or another.

The result was a glut of coal on the market. Railroads and coal producers tried to solve the problem by restricting output and fixing prices. These attempts failed largely because the West Virginia companies carried coal to Norfolk or Newport News (Port of Hampton Roads) at a cheaper rate per mile than the Pennsylvania-to-New York carriers charged. The purpose was to make their more distant, southern coal competitive in markets of the northeastern states. Because transportation costs commonly amount to more than half the selling price of coal, there always has been a tendency to haul coal by rail the shortest distance possible and to use cheaper coastwise shipping the remainder of the way. Thus, New England–bound coal moves by rail from West Virginia to Norfolk or Newport News and then northeast by ship.

Competition between the coal fields for a share of the northeastern urban market was repeated after 1910 when industrial growth in the Lower Great Lakes region created a new market for bituminous coal. The southern coal fields were again at a distance disadvantage, and railroads once again offered cheap rates to haul coal north across Ohio. Cities along the southern shore of Lake Erie became ports for shipping coal by the Great Lakes to Chicago and other western destinations (see chapter 14, fig. 14.1).

Southern West Virginia and eastern Kentucky coal producers enjoyed two other advantages in reaching markets for which they had a distance disadvantage. One was the high fixed-carbon content of the coal, which is essential in manufacturing coke for use in a blast furnace. The other advantage was the lower wages and generally lower operating costs of the southern fields. Labor union organizers were successful in Pennsylvania, but they were less so toward the south. Though West Virginia's miners are mostly unionized today, only about half of eastern Kentucky's underground coal miners belong to a union. The result is a fairly definite southward-sloping wage gradient and, especially given these other factors, a profitability gradient that slopes upward to the south in proportion.

The farsightedness of investments in coal mines and railroads made by men who developed the Appalachian field was accompanied by their indifference to wages, working conditions, and living standards for the miners and their families. The supply of labor was no problem. Young men who lived in the mountain fastnesses of southern Appalachia were understandably attracted to the mines, given the cash-short, semisubsistence style of life that prevailed there. Sharecroppers from the Piedmont and the Coastal Plain also moved north into the hills where mining jobs were offered.

The traditional pattern of Appalachian settlement—a patchwork of isolated small farms connected by poor roads and served by country general stores—could not support the needs of the coal industry. Making the minimum of investment, coal companies built new towns adjacent to the mines and rented

FIG. 7.3. *Page, a mining community in the coal fields of southern West Virginia.*

houses to the miners (fig. 7.3). They created a single company store at each location that would carry the miner and his family on credit. Wages provided cash, but the miners were forced to exchange cash for company-issued tokens that became the local currency. Coal companies provided little in the way of infrastructure, omitting even the grading of passable roads into the towns, thereby further isolating the people who lived there. Miners and their families were captives, not only economically but geographically as well.

Local and state politics evolved to protect the coal interests. Coal entrepreneurs identified with the new industrial wave, associated themselves with progress and prosperity, and proclaimed their opposition to the old backwoods ways that many were evidently eager to shed. Political cleavages were complex, involving the lingering issues of Union versus Confederacy, the split that created West Virginia and divided Kentuckians as well. Party loyalties even involved local feuding, such as the legendary Hatfields and McCoys who lived in the Tug Fork Valley of southern West Virginia: the Hatfields were Democrats, the McCoys Republicans.

The most effective popular opposition to the status quo was the United Mine Workers Union, which was organized in 1890. Strikes and violence at Harlan, Kentucky, and at Paint Creek and Cabin Creek, West Virginia, came to symbolize life in the coal country. A deep cynicism emerged in local politics, reflected in widespread acquiescence to vote fraud and courthouse gangs. More people were registered to vote than were of an age eligible to vote in more than half of West Virginia's counties as late as 1964. Coal-leasing companies earned millions of dollars from their lands but paid virtually no property taxes in many eastern Kentucky counties, where local officials lacked the ability or inclination to enforce tax laws.

Some of these conditions have changed in the past four decades. The commonly acknowledged date of birth of the modern "Appalachia movement" is

March 1960, when John F. Kennedy, campaigning for the presidency, came to West Virginia with a promise to do something for the people. In 1965 Congress passed an Appalachian Regional Development Act that established the Appalachian Regional Development Commission (ARC), creating a funnel for pouring federal money into the region. The ARC used a growth-center strategy to concentrate investment in the cities and towns large enough to support sustained growth. It has built thousands of miles of development highways to increase accessibility within the region. Critics of the program claim that highway mileage now exceeds need.

Although many problems of the late 19th and early 20th centuries have been eliminated, including, for example, the old company-town system, new issues of dispute have arisen. A wildcat strike led by the Black Lung Association shut down the coal industry of West Virginia in the late 1960s. The Coal Mining Health and Safety Act of 1969 led to a sharp drop in underground mining in some areas and thus had the unintended effect of idling many miners. Public outrage over the damage done by strip mining led to the passage of stringent laws in Pennsylvania and West Virginia. As a result, Kentucky saw an upsurge in surface mining until it, too, finally passed restrictive strip-mining legislation.

The coal industry, which had a clouded future in the 1960s as a result of a general popular shift to cleaner-burning fuels such as oil and gas, also faced potential competition from nuclear power, but the oil embargo crisis of the 1970s and the widespread unpopularity of nuclear installations have restored the coal industry's position in a new business environment. More than two-thirds of the coal produced in the Appalachian field is used for electric-power generation. Current clean-air standards favor the region because it has a substantial reserve of low-sulfur coal. Coking-quality coal remains essential for steel production, and even given recent decreases in U.S. steel output, coals of the Appalachian field are in strong demand by U.S. and Canadian industries. A growing fraction of the Appalachian output is exported, mainly to western Europe.

About two-thirds of the coal is extracted in underground operations. Shafts and tunnels still must be sunk underground, because this is the only way of reaching the remaining coal reserves. The number of men working in underground coal mines has dropped in recent years, however, because of the widespread adoption of continuous mining technology. Machines with rotating teeth shear coal from the seams, and it is removed via conveyor belt. Some mining operations use augers that penetrate the coal seam, making a large underground labor force unnecessary.

The coal mines of the Appalachian field are now medium-size by national standards, given the appearance since 1970 of gigantic strip mines in the western Great Plains. The twenty largest mining companies account for nearly half the output of the several thousand operating coal mines in eastern Kentucky and West Virginia. Petroleum companies purchased coal mines in the 1970s; steel producers own large mines as well. More than four-fifths of the region's coal is sold on long-term contracts to major consumers, especially electric-power utilities.

## Appalachian Industries

In contrast to the landlocked, coal-producing areas of the Appalachian Plateau, the Ohio and Kanawha Valleys developed strong manufacturing bases (map 7.1). The seizure of German patents during World War I and the construction by the U.S. government of nitrates (explosives) plants near Charleston, West Virginia, launched the Kanawha Valley chemical industry. The manufacture of dyes, liquid chlorine, and a variety of caustic chemicals has been the basis for an almost continuous strip of chemical factories in the narrow confines of the Kanawha Valley for thirty miles on either side of Charleston.



On the Ohio River, Wheeling specialized in steel and glass manufacture, using local supplies of coal and natural gas; Parkersburg was based on its oil refineries; and Huntington grew from employment in its large nickel and nickel alloys plant. Ashland, Kentucky, got its start as a steelmaking center in the 1890s. Cheap barge transportation on the Ohio River, coupled with numerous nearby sources of steam coal, have made the Ohio and Kanawha Valleys desirable sites for electric-power generation in recent years, even as some of the older, local industries have declined.

Metamorphosed rocks of the Ridge and Valley section contain some mineral resources, but this section has developed other industries, especially those based on the region's abundant forests. The production of rayon, a synthetic fiber manufactured from wood or cotton cellulose, was the basis for east Tennessee's fiber industry, which, in turn, supplies the textile industry of the Carolina Piedmont as well as the large textile manufacturing industries at Knoxville. Kingsport, Tennessee, a planned industrial city laid out in 1917, was intended as a model for the Appalachian region. Methanol, cellulose acetate fibers (yarns, plastics), and photographic films are produced here. Kingsport also attracted Massachusetts-based cotton spinning mills, paper companies, and printing plants, eventually growing into one of the largest industrial cities of the Appalachian region.

## The Smoky Mountains

Only three rivers—the Potomac, the James, and the Roanoke—cross the Blue Ridge Mountains, and no crossing is south of Roanoke, Virginia. The inaccessibility of the country west of the Blue Ridge and the lack of deepwater natural harbors along Chesapeake Bay and the North Carolina coast (Norfolk–Newport News is the lone important exception) prevented the development of an east-west pattern of ports and tributary areas. The Blue Ridge Mountains diverted settlement, separating the thickly settled Piedmont to the east from the parallel valleys of Tennessee on the west.

Linear topography in the northern Blue Ridge gives way to a broad plateau in southern Virginia that becomes mountainous southward to reach a width of eighty miles in the vicinity of Asheville, North Carolina. Here the Blue Ridge becomes the Smoky Mountains. The Blue Ridge–Smoky Mountains south of

Roanoke have forty-six peaks exceeding 6,000 feet in elevation. The loftiest, 6,684-foot Mount Mitchell, near Asheville, is the highest elevation in eastern North America. The high peaks of the Smokies have a cover of spruce, pine, and hemlock forest. Mount Mitchell's summit recently lost most of its forest cover, probably because of acid rain deposition from eastward-moving air masses containing sulfur dioxide pollutants from industrial smokestacks.

Tree growth is stunted by climatic factors on the high summits of the Smoky Mountains generally, however, because of the cool, moist summer climate. Some Appalachian Mountain summits that are barren of trees are known as *balds* and probably resulted from deliberate burning to remove the forest cover. The balds were recorded as treeless by early Euro-American settlers, who continued the native practice of burning. Wild game are attracted to these grassy openings at high elevation, and some balds are used for grazing. The Appalachian Trail, which follows the highest ridges of the Smoky Mountains, passes through several such bald summits, which today are maintained by mowing or periodic burning.

Lowlands in the Ridge and Valley section have been important agriculturally, but steep slopes are more important than elevation in limiting land-use options in the southern Appalachians. Agriculture benefits from the mild climate, the long growing season, and the abundant moisture, but it is severely limited by slopes so steep that they erode naturally at about the same rate as soils develop. Narrow tobacco patches that wind along creek bottoms, scattered hillside clearings planted in corn, and the vegetable gardens wedged between houses and hillsides typify agriculture in the mountains.

A roughly circular basin nearly fifty miles across surrounds the city of Asheville, North Carolina (fig. 7.4). The Asheville Basin is drained by the French Broad River, a tributary of the Tennessee. Only a short, steep grade over Saluda Mountain separates Asheville from the Piedmont. The French Broad route through the Asheville Basin was an important early migration route and a live-

FIG. 7.4. *The city of Asheville, North Carolina, is in a highland basin of the Smoky Mountains.*

stock drovers' trail. Today it is a major rail and highway route through the mountains. Asheville's mountain climate began attracting wealthy northern tourists in the 1880s. The most prominent was George Washington Vanderbilt, who built a mansion there, purchased a large timber acreage in the mountains, and helped foster modern practices of commercial forestry. Tourism and forest-product manufacturing (including furniture, paper, and cellulose fiber industries) are the basis of the regional economy.

The democratization of tourism that accompanied the growth of automobile ownership and the construction of paved highways has led to a nearly complete penetration of the Blue Ridge–Smoky Mountains of North Carolina, Tennessee, and Georgia by recreation developments. Summer homes for residents of Atlanta and Miami, retirement villages, and outdoor amusements of various sorts line the highways and dot the mountain slopes. The Pigeon Forge–Gatlinburg strip on the Tennessee side of the Smokies ranks as one of the largest concentrations of tourist facilities in North America.

## The Tennessee Valley

The valleys of upper east Tennessee were settled early in the history of westward expansion. The Cherokees, who dominated the region, soon learned that the white advance would not be stopped without force. A militant, breakaway band of Cherokees, the Chickamaugas, divided themselves into smaller groups and occupied well-fortified towns along the Tennessee River. For a time they were an effective force in driving out new settlers.

Following a disastrous defeat in 1794, the Chickamaugas rejoined the main body of Cherokees and, with them, began to adopt Euro-American traits. This was especially true of those with mixed blood, of whom there were many, including the visionary Sequoyah (George Gist), who devised an eighty-six-character syllabary for the Cherokee language. Missionaries provided the Cherokees with a printing press. The Cherokees adopted a written constitution, yet they kept the tribal practice of common ownership of lands within their reservations.

The Cherokees' rapid adaptation to white society ultimately gained them no favor. The state of Georgia refused to recognize their land treaties. During the administration of Andrew Jackson (1829–37), Georgians and others pressed for the total removal of the Cherokees from the Southeast. In 1835 the remaining Cherokees signed a treaty of removal and in 1838 left the Tennessee Valley on a long forced march, the Trail of Tears, which led them to new lands in northeastern Oklahoma.

White settlers began arriving in larger numbers as soon as the removal had taken place. Two major northeast-southwest-trending valleys, the Sequatchie and the Coosa, attracted settlers from the same Virginia and Carolina Piedmont areas that had supplied the initial population stock to the more northern valleys earlier (map 7.3). Cotton was raised in scattered upland clearings by the mid-1830s, and iron furnaces were established where local ores were available.

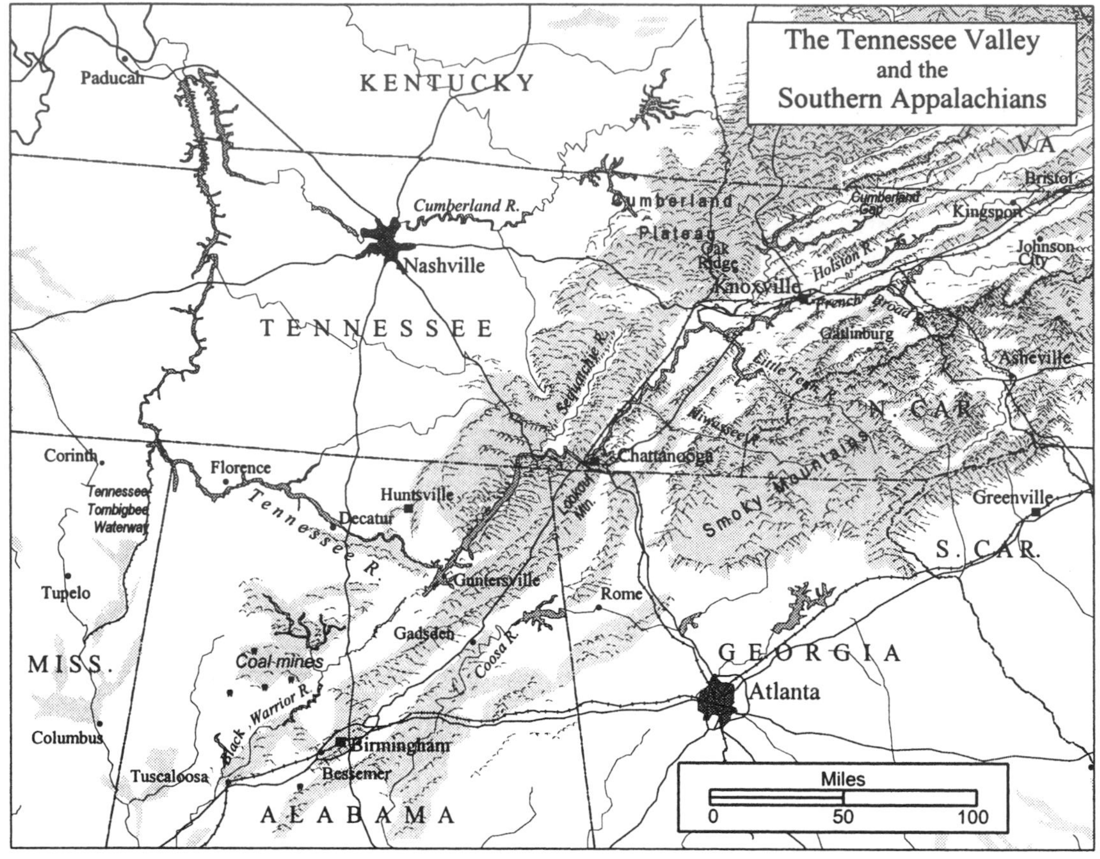


MAP 7.3

Chattanooga is one of the mid-South's oldest centers of heavy industry, with iron furnaces, steel mills, and a variety of metal-fabricating industries. Its site has been strategic for centuries, dating well back into prehistoric times, because it occupies a gap in the Appalachians where the Tennessee River leaves the low ridges of this portion of the Ridge and Valley section and enters a deep gorge in the Appalachian Plateau. Chattanooga's location was strategic during the Civil War because Atlanta and the Georgia Piedmont are within easy access to the south.

Two other industrial complexes of the southeastern mountains suggest the varied character of the industrial base. During World War I the Aluminum Company of America constructed a bauxite processing plant at Alcoa, Tennessee. Three hydroelectric power dams built on tributaries of the Tennessee River provided the enormous amount of electricity needed to process the raw materials used in making aluminum. Oak Ridge, Tennessee, now an important center for research and development, was built in a then-isolated area of the Cumberland Plateau by the federal government in 1943. More than seventy-five thousand people worked there during World War II manufacturing atomic bombs and other war materiel. The very existence of Oak Ridge, then known as the Atomic Bomb City, was not acknowledged by the federal government until 1945. That a manufacturing city of such size could remain an official secret for several years suggests the role that isolation has played in shaping the course of development in the Appalachians.



A remarkable coincidence of natural resources occurs at the southern tip of the Appalachians, where the Plateau and the Ridge and Valley sections meet. Coking-quality coal outcrops along ridges that face, across limestone-floored valleys, other ridges that contain hematite, a fairly high-grade iron ore. All three of steel's major ingredients are in a single locality, making it an obvious location for a steel industry. The city that grew here was Birmingham, Alabama, beginning in 1871. A United States Steel subsidiary purchased the largest works at Birmingham in 1907, and since that time the city has produced iron and steel in various forms, predominantly iron pipe. Birmingham's iron ores were rich but not very plentiful, and its iron and steel industry is now supplied with ores from South America. Barge transportation on the Tombigbee and Black Warrior Rivers through the port of Mobile provides access to ocean shipping for the iron-ore imports.

The Tennessee River and its major tributaries—the Clinch, the Holston, the French Broad, and the Little Tennessee—form a crescent-shaped drainage course. Rising in the Ridge and Valley section of Tennessee and North Carolina, the tributaries flow south, away from the higher ground of the Cumberland Plateau. The Tennessee River itself begins at Knoxville, where the French Broad meets the Holston. At Chattanooga the river cuts the high plateau on which Lookout Mountain is formed, abruptly shifts valleys, and flows southward into Alabama in an extension of the Sequatchie Valley. At Guntersville, Alabama, the river turns back north and eventually reaches the Ohio River at Paducah, Kentucky, making a broad arc around the Highland Rim and the Nashville Basin (fig. 7.5).

Navigation on the Tennessee River was hazardous because of shoals where the river flows from one geologic formation to another. The first hydroelectric power dam was constructed at Hales Bar near Chattanooga in 1913. Wilson Dam, designed to generate electricity and flood the dangerous Muscle Shoals near Florence, Alabama, was begun under U.S. Government auspices in 1916 to produce electricity for nitrates used as ammunition during World War I. The plant was intended for conversion to fertilizer production (also based on nitrates) when the war ended, but Muscle Shoals was still under construction when the war ended, and attempts were made to sell it. Automaker Henry Ford offered the government six cents on the dollar for the entire property.

U.S. senator George Norris of Nebraska saw the possibility for public power development at Muscle Shoals and prevented the Ford purchase, although he was unsuccessful with numerous attempts to establish a government-owned power authority. Norris finally succeeded when Franklin D. Roosevelt was elected president in 1932. The new president was a firm believer in public power, flood control, and regional planning. The Tennessee Valley Authority (TVA) became an important part of Roosevelt's first New Deal.

Cheap power brought new industries to the Tennessee Valley, especially near

FIG. 7.5. *Grain barges and pleasure boaters share the Tennessee River at Guntersville, Alabama. Here, at the river's southernmost bend, corn and soybeans shipped in from the Middle West are manufactured into poultry feed for the southern broiler industry.*

the Muscle Shoals nitrates complex, where a major aluminum-processing facility was established. Flood-control dams and reservoirs incidentally provided new recreational facilities. Barge traffic on the Tennessee River grew steadily. Chattanooga's manufacturing base expanded to include flour milling, using grain brought by barge from Kansas City. In Alabama, Guntersville and Decatur became feed-processing centers for the southern poultry industry, also on the basis of cheap transportation of grain from the Middle West by Tennessee River barges.

The TVA eventually suffered from its success, coming under attack in recent years for pollution from its steam coal plants (power demands have long since required more electricity than the dams can provide), its uncertain attempts with nuclear power, and the Tellico Dam project of the 1970s, which environmentalists opposed because it threatened the existence of the snail darter. Tellico became famous as the first major construction project halted by an endangered species. In some respects the TVA today is just another bureaucracy, but it has had positive effects of lasting consequence on the mid-South.

## *References*

Curry, Richard Orr. *A House Divided: A Study of Statehood Politics and the Copperhead Movement in West Virginia.* Pittsburgh: University of Pittsburgh Press, 1964.

Glassie, Henry. *Pattern in the Material Folk Culture of the Eastern United States.* Philadelphia: University of Pennsylvania Press, 1968.

Harvey, Curtis E. *Coal in Appalachia: An Economic Analysis.* Lexington: University Press of Kentucky, 1986.

Jordan, Terry G. and Matti Kaups. *The American Backwoods Frontier: An Ethnic and Ecological Interpretation.* Baltimore: Johns Hopkins University Press, 1989.

Kniffen, Fred B. "Folk Housing: Key to Diffusion." *Annals of the Association of American Geographers* 55 (1965): 549–77.

Lilienthal, David E. *TVA: Democracy on the March.* New York: Harper, 1944.
Mitchell, Robert D. *Commercialism and Frontier: Perspectives on the Early Shenandoah Valley.* Charlottesville: University Press of Virginia, 1977.
Newton, Milton B., Jr. "Cultural Preadaptation and the Upland South." *Geoscience and Man* 5 (1974): 143–54.
Shifflet, Crandall A. *Coal Towns: Life, Work, and Culture in Company Towns of Southern Appalachia, 1880–1960.* Knoxville: University of Tennessee Press, 1991.
Williams, John Alexander. *West Virginia and the Captains of Industry.* Morgantown: West Virginia University Library, 1976.

## CHAPTER 8

# The Interior Low Plateaus

An irregular, west-facing escarpment marks the western edge of the Appalachian Plateau from southern Ohio to Alabama. West of this escarpment lies a series of plateaus at lower elevation than the Appalachians but higher than the Coastal Plain. This section is called the Interior Low Plateaus (map 8.1). It stretches from the Tennessee Valley of north Alabama to the limit of Pleistocene glaciation in the Ohio Valley. The Low Plateaus extend farther north, but they have little topographic expression because they are buried beneath glacial materials. Topography north of the glacial limit in the Middle West is more a function of glaciation than of control by the underlying bedrock.

Within the Interior Low Plateaus are several broadly uparched limestone plateaus or domes that were grassy or savanna-like when the first Euro-Americans saw them. Two such areas, the Bluegrass region of Kentucky and the Nashville Basin of middle Tennessee, were so attractive to the first trans-Appalachian migrants that they became the favored sites of early habitation at the end of the 18th century. A third such gently rolling limestone surface, the Pennyroyal Plateau along the Kentucky-Tennessee border, was settled soon thereafter. These three areas attracted the first permanent Euro-American settlement on the trans-Appalachian frontier, and all three developed prosperous agricultural economies.

Fringing these areas of good land are several belts of rugged topography, formed mostly on cherts and sandstones, that were avoided by the main body of early settlers. Each upward arch of the underlying bedrock created the opportunity for differential erosion, exposing the edges of the resistant rock layers on which the uplands were formed. The Highland Rim surrounds the Nashville Basin. The Dripping Springs Escarpment (or Chester Escarpment) is a fringe of broken topography that bounds the Pennyroyal Plateau on the north; Muldraugh's Escarpment marks the western edge of the Bluegrass. In southern Indiana the Mitchell Plain is a limestone lowland similar to the Pennyroyal; it is bounded by the Knobstone Escarpment and the Crawford Upland. These alternating sections of smooth lowlands and rugged uplands were correctly perceived as areas to settle versus areas to avoid by early migrants from across the Appalachians, and thus the underlying geology influenced settlement patterns.

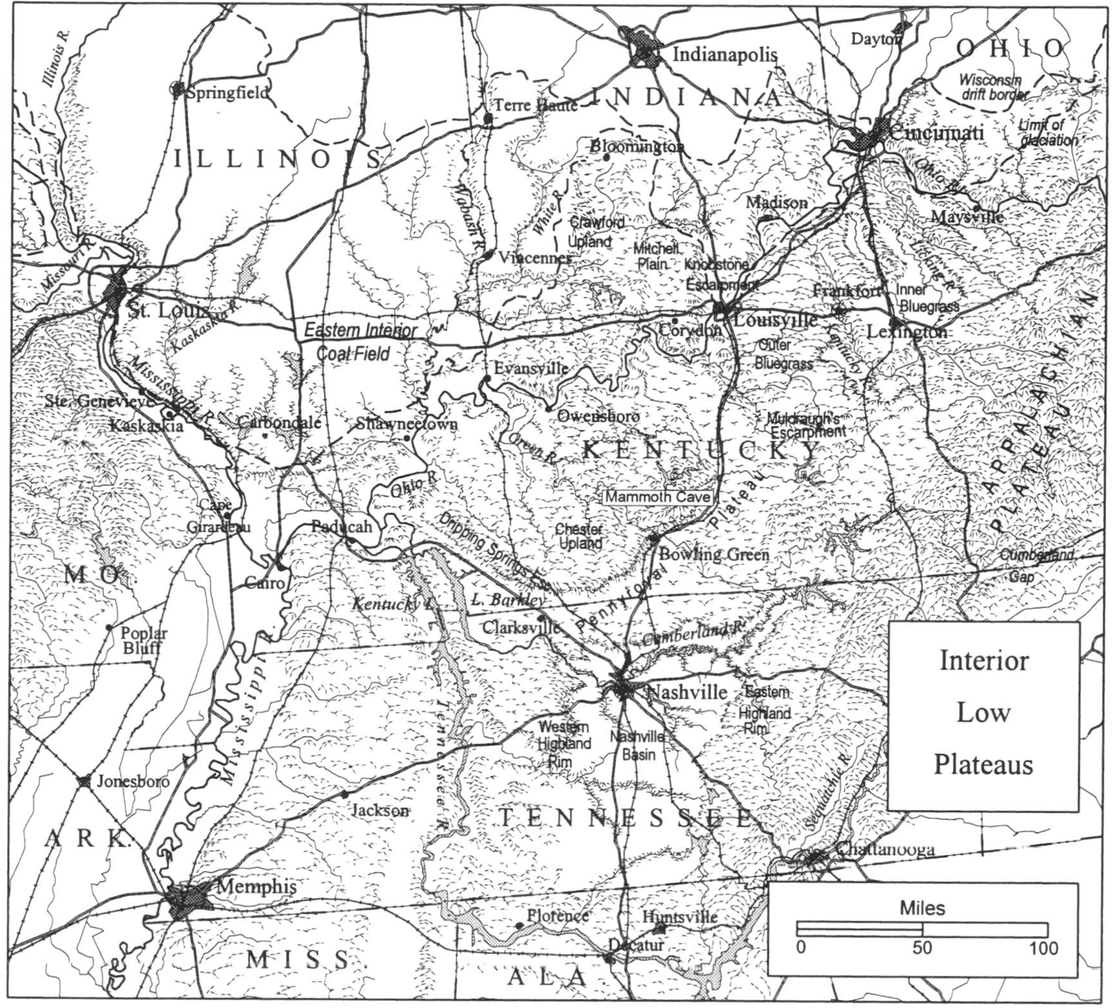


MAP 8.1

## The Nashville Basin

The principal structure determining the arrangement of plateaus and ridges within the Interior Low Plateaus region is the Cincinnati Arch, a broad anticline that stretches south from Cincinnati through the Bluegrass, the Pennyroyal Plateau, and the Nashville Basin (map 8.2). The crest of the arch is near Nashville, where the oldest rocks are exposed at the surface and where the Cumberland River flows across the basin. In 1779 a company of settlers from North Carolina founded the settlement that became Nashville on the banks of the Cumberland River. In later years many more migrants came to the Nashville Basin, and most of them followed the same route that the initial colonists had taken, through Cumberland Gap and then down the Cumberland River.

The early Nashville Basin had rolling expanses of grassy meadows fringed

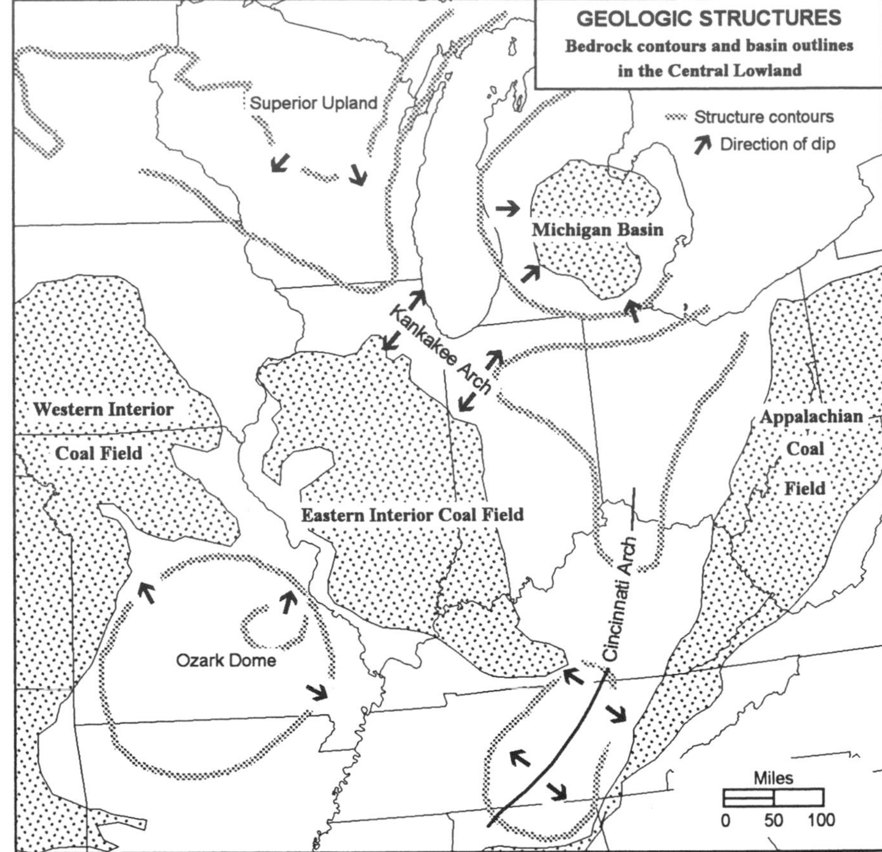


MAP 8.2

by cedar glades on the surrounding uplands. It was an excellent area for grazing when the first herds of cattle were driven there, because it had been grazed by bison for years before that. Its environment had been modified to a state of enhanced productivity through its apparently long history of human use. French hunters and traders from Illinois had been coming to the Nashville Basin for some time before 1770 to slaughter the bison that roamed freely through the Nashville Basin's grassy openings. Native people had used it as a hunting grounds for an unknown length of time before.

North Carolinians named their first settlement Fort Nashborough. The name was changed to Nashville in 1784, a reflection of the anti-British and pro-French sentiments of the Revolutionary War period. Cattle and hogs raised in the Nashville Basin made it one of the most important meat-producing areas of the nation by 1850. Cotton and tobacco grown on Nashville Basin plantations employed more than ninety thousand slaves in 1860. Middle Tennessee's proslavery majority joined the secessionist cotton-slave counties of West Tennessee to produce a statewide margin in favor of secession from the Union in June 1861. East Tennessee remained staunchly pro-Union. These divisions lingered for more than a century after the Civil War, and they were reflected in the large votes given to Republican office-seekers in the Cumberland Plateau and Ridge and Valley counties of East Tennessee.

As the only important city on the Cumberland River, Nashville dominated local trade in the early years. Navigation improvements on the Cumberland made Nashville accessible to barge traffic from the Ohio-Mississippi system. Nashville also became an important insurance and financial headquarters city for the mid-South. Its greatest fame, however, began with a country music radio program, the Grand Old Opry, which has been broadcast live every Saturday night from Nashville since October 1925. Nashville's Music City U.S.A. label of recent years is based only partly on its country music role, however. The city also is the headquarters of the gospel music industry and is one of the most important music-recording and audio disc–manufacturing centers for all types of music transcribed and impressed in the United States.



## The Pennyroyal Plateau

Transportation arteries heading north from Nashville ascend the short but steep grade of the Highland Rim to attain the level of the Pennyroyal, a rolling plateau developed on soluble limestones that extends from the Highland Rim north to the Dripping Springs Escarpment of Kentucky. The same limestone formation extends northward, across the Ohio River, and is known as the Mitchell Plain in southern Indiana. Both the Pennyroyal Plateau and the Mitchell Plain are karst plains, dotted with numerous solution-depression features where the relatively soluble limestone has dissolved into shallow sinkholes. Surface waters flow only a short distance before entering one of the sinkholes (fig. 8.1).

The more insoluble sandstones of the Chester Upland act as a roof over the limestone layer, but the streams continue their work of erosion underneath this roof, flowing in subterranean passages where the dissolving action of streams

FIG. 8.1. *The puddle in the cornfield marks a sinkhole leading to subterranean drainage in the Pennyroyal Plateau near Park City, Kentucky. The higher ground on the horizon is the Dripping Springs Escarpment, the "roof" over Mammoth Cave.*

continues. As the region underwent geologic uplift, streams like the Green River continued to cut downward, creating deep gorges. Surface streams that enter sinkholes in the Pennyroyal Plateau see the light of day once again where they discharge into the Green River. The combination of a karst plain, a protective, resistant rock layer over part of the soluble limestone, and a drainage outlet (the Green River) have combined to create Mammoth Cave.

Like the Nashville Basin, the Pennyroyal was largely a grassland when the first Euro-Americans saw it. Fires set by the native people for hunting had created this great open expanse within the eastern woodlands at some unknown date in the prehistoric past. Some of the rock formations of the Interior Low Plateaus, which originated as sediments in shallow ocean basins, impart a saline quality to the waters that issue from them as springs. Salt licks of the Pennyroyal, the Nashville Basin, and the Bluegrass attracted buffalo, elk, deer, and other large mammals. Overhunting led to the total demise of most of the large game species within a few decades after Euro-American settlement began.

The Pennyroyal was the first extensive tract of grassland settled by the Euro-American population of the United States. The Pennyroyal's early inhabitants, who were of predominantly Scots and English ancestry from Virginia, labeled it as the "barrens" in the late 18th century, a reference to the lack of trees. At that time the French word *prairie* was not commonly used by Americans to describe grassy vegetation. Eventually the Pennyroyal became a tobacco-producing region, specializing in air-cured dark and burley tobacco (map 8.3).

The Tennessee and Cumberland Rivers flow in valleys narrowly separated by a ten-mile-wide upland along the western edge of the Pennyroyal. When dams

MAP 8.3

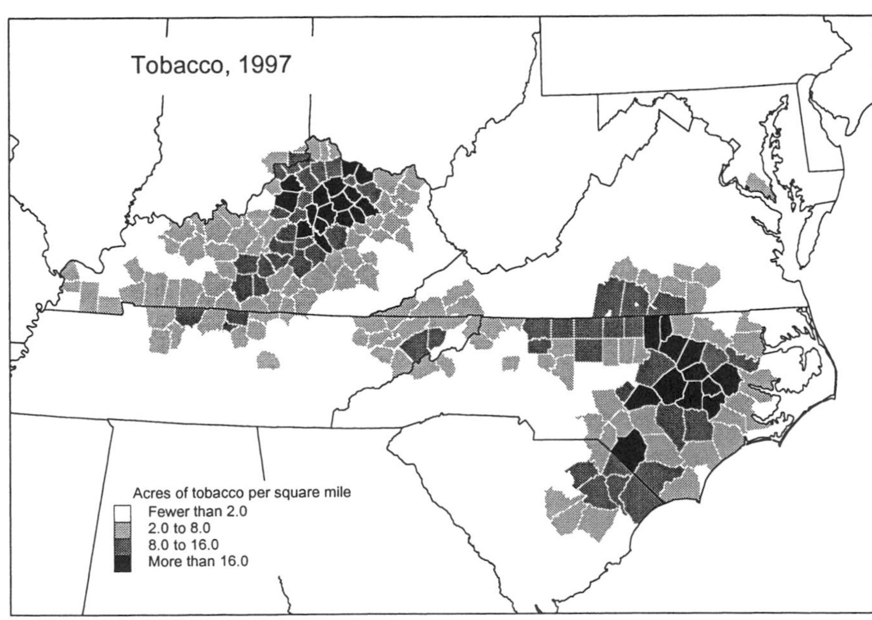

were constructed on the Tennessee and Cumberland Rivers (creating Kentucky Lake and Lake Barkley), a short canal was cut between the two. Together, the lakes have a shoreline more than three hundred miles long and have been responsible for tourism's growth in the local economy. Both the Cumberland and the Kentucky flow into the Ohio a few miles downstream from the reservoirs, and both streams are important avenues of transportation. Coal and grain bound for middle and eastern Tennessee and north Georgia are the major commodities moving on the two rivers.



The most recent navigation improvement is the Tennessee-Tombigbee Waterway, which was completed in the 1980s to link the Tennessee River and tributaries of the Tombigbee River in northeast Mississippi (see chapter 7, map 7.3; chapter 12, map 12.1). The "Tenn-Tom" was designed to emulate the TVA project, but it has not been a success in terms of traffic. An example of self-interest public investment, it was popular with local politicians for many years, but the Tenn-Tom project turned out to be an expensive mistake, largely because there was no great economic need to link the Tennessee and Tombigbee Rivers.

## The Bluegrass of Kentucky

The Bluegrass region of Kentucky was the largest and most populated of the early islands of settlement on the trans-Appalachian frontier at the end of the 18th century. More than a dozen towns were established there during the early 1770s at a time when British colonial authorities prohibited settlement west of the Appalachians. The Bluegrass was accessible by the Kentucky River from Cumberland Gap, but it was reached more easily by the Ohio River. In 1790 the settled area was continuous from Maysville on the Ohio River to the southern Bluegrass margin. By 1800, settlers had spread out sufficiently to create a continuous zone of occupancy from the Ohio River to the Nashville Basin.

Geologically, the Bluegrass is divided into roughly concentric Inner and Outer regions of a dome-like uplift in which the oldest rocks are exposed at the center. The Inner Bluegrass, which surrounds the city of Lexington, has a gently undulating surface; at the time it was settled, it had deep and fertile limestone soils. A belt of poorer, eroded land (the Eden Shale belt) at its margin separates it from the Outer Bluegrass, where the topography and soils compare favorably with the core. Continuous cropping reduced the soils in depth and fertility, but Bluegrass limestone soils still produce excellent grass for pasture.

The grass that gives the region its name was not mentioned by the early travelers. Writing of his reconnaissance of Kentucky in 1782–83, John Filson described a landscape of dense canebrakes and partial clearings with a variety of wild herbage. Other accounts mentioned the open, savanna-like nature of the landscape. Early observations of large herds of bison and the discontinuous forest cover suggest that grazing and Indian burning for hunting probably accounted for the lack of dense forest. The Bluegrass, like the Nashville Basin and the Pennyroyal, was already a well-used country by the time the first whites arrived. So-called Kentucky bluegrass is of European origin and probably was introduced when herds of cattle were brought to Kentucky from the East in the 1790s.

Bluegrass land was acquired in large tracts, especially by Virginia planters who came there to grow crops of tobacco and hemp. Slavery came to the Bluegrass along with the planters, who sought to expand their operations on the better lands west of the Appalachians. Although the Bluegrass region is part of the Upland South, the generalization that the Upland portion of the South did not develop a slave economy applies neither to the Bluegrass nor to the Nashville Basin. When the Civil War came, Kentucky remained in the Union even though its outlook remained distinctly Southern.

The "gentleman farms" of the Inner Bluegrass today are mostly owned by wealthy outsiders who can afford the large investment necessary to breed and train race horses. The local breeding tradition runs back to the 1830s, when the Bluegrass was one of two preeminent centers of cattle breeding in the nation (the Virginia Military District in Ohio was the other). Farmers imported English cattle to improve the quality of herds. The Bluegrass also became a center for sheep and swine breeding. Horse racing, which was part of the sporting, Virginia "Cavalier" tradition brought to Kentucky, fit naturally into the system.

Today's Bluegrass landscape retains its old appearance in places, with mile after mile of board fences, stone walls, and elegant mansions overlooking grassy paddocks (fig. 8.2). The Bluegrass remains the major U.S. producer of burley tobacco, a thin-leafed plant especially desirable for blending with other varieties in manufacturing cigarettes (map 8.3). The elongated barns that are used for air-curing tobacco still dot the landscape, but the Bluegrass also has grown substantially as a result of new manufacturing industries. Urban growth and suburban sprawl have replaced the country estates in many areas.

FIG. 8.2. *Horse farms in the Bluegrass of Kentucky.*

## The Illinois Basin

Bedrock layers of the Interior Low Plateaus dip northwesterly, away from the crest of the Cincinnati Arch, to form a semicircular arrangement of limestone lowlands separated by upland belts. The arc-shaped outline forms the Illinois Basin, a predominantly buried structure that contains deposits of oil, bituminous coal, and natural gas. The Illinois Basin deepened and accumulated sediments when the land surface to the southeast flexed upward along the Cincinnati Arch. Extensive coal deposits formed on a zone of shallow-water marine sediments. The Illinois Basin is coextensive with the Eastern Interior Coal Field, which extends from the Green River of Kentucky northward into Indiana and Illinois, stretching as far west as the vicinity of St. Louis and north to within one hundred miles of Chicago (map 8.2). At one time the Eastern Interior Coal Field was a major factor in the industrial growth of the Middle West.

Coals of this field remain in abundant supply. They lie closer to the surface than those of the Appalachian Plateau, and thus strip mining is the principal means of extraction. Historically, the most important advantage of the Eastern Interior field was its proximity to urban-industrial markets of the Middle West. Electric-power utilities have long been the principal customers, burning more than 85 percent of Eastern Interior production. The proximity of the Ohio River meant low-cost barge transportation to many areas of the eastern United States.

On the negative side is the high sulfur content of the remaining reserves—nearly twice the sulfur content of Appalachian coals and more than five times that of the low-grade bituminous deposits of the Great Plains. Power utilities have canceled their contracts with Eastern Interior coal producers in recent years and have shifted to burning western coal in order to meet clean-air standards. Western Kentucky's portion of the Illinois Basin–Eastern Interior Coal Field once produced two-fifths of the coal mined in that state but now accounts for less than one-fourth. The downward trend of production in the Eastern Interior field probably will continue, because still more stringent requirements for clean-burning fuels have been enacted in recent years.

## The Lower Ohio Valley

Control of the Ohio River Valley was a concern to both the French and the British in their attempts to colonize North America. North of the Ohio, Vincennes and Terre Haute began as outposts founded by French traders on the Wabash River in the 1730s. The British later captured both and held them until the American Revolution. The passage of the Northwest Ordinance in 1787 opened the territory north of the Ohio River to American settlement. The white settlers' frontier advanced westward and northward, crossing the Ohio River, pushed forward by the same population stocks that had settled the Bluegrass, Nashville Basin, and Pennyroyal sections. Settlers occupied the gently rolling limestone plateaus of southern Indiana early in the 19th century, and the better lands of southern Illinois, fringing the Wabash and Mississippi rivers, were taken soon thereafter.

The Ohio River cross-cuts each of the alternating layers of resistant and soluble rocks of the Interior Low Plateaus as it flows toward its confluence with the Mississippi at Cairo, Illinois. The average gradient of the entire river, from Pittsburgh to Cairo, is only about six inches to the mile, but at Louisville there is an abrupt drop of about twenty-five feet—the Falls of the Ohio—where the river passes over the same resistant rock formation that produces the "knobs" of Indiana and Kentucky. Before steamboating on the Ohio, flatboats had to unload their cargoes for portage around the falls, a circumstance that led to the growth of a major settlement.

Louisville, platted in 1779, grew rapidly after steamboating began on the Ohio River. The city's merchants supplied the growing inland settlements of the Low Plateaus region through a trading network that extended south into Alabama. By the 1820s, when a canal was built around the falls, Louisville had grown larger than Lexington. Louisville became a manufacturing center, distilling bourbon whiskey and manufacturing cigarettes from Kentucky-grown products. It was a major meatpacking center as well. Louisville was a southern city, although one with close economic ties to the Middle West and the Northeast, a distinction that the city's merchants tried to turn to their advantage in the years after the Civil War. Railroad lines extending southward from Louisville were an attempt to draw the trade of the mid-South through this port on the Ohio River.

Louisville had two rivals in its attempt to tap the southern trade: the smaller, downriver city of Evansville; and Cincinnati, a larger city upriver. All three became gateway cities for traffic moving between the Middle West and the Southeast. Cincinnati was the more formidable rival, although it was a "northern" city in the opinion of southern merchants who might otherwise have conducted business there. Louisville's manufacturing industries expanded to include automobiles, electric appliances, and a variety of other consumer durable goods.

Cincinnati's and Louisville's trade rivalry continues today in the form of highway development corridors. Cincinnati commands a central location in the I-75 corridor between Atlanta and Detroit, which has especially strong ties to the automobile industry and its component suppliers. Louisville plays a similar role in the I-65 corridor linking Chicago and Indianapolis with Nashville, Birmingham, and the mid-South, another "automotive corridor." Both routes connect suppliers, manufacturers, and wholesalers in a variety of consumer–durable manufacturing industries.

Despite its early date of white settlement and obviously strategic location, the Lower Ohio Valley did not become an urbanized region. With the exception of the Wabash-Ohio confluence zone, the river lacks extensive bottomlands for agriculture. Early settlements such as Corydon, Indiana, and Shawneetown, Illinois, did not develop into urban centers. Frequent floods on the lower Ohio made much of the floodplain unsuited to development. By 1850, when Illinois and Indiana were growing rapidly, railroads were replacing river steamers as the primary means of transportation, and the lower Ohio Valley lost its strategic relevance to regional development.

## The Middle Mississippi Valley



In the steamboat era, the most accessible point in the midcontinent was the southern tip of Illinois, where the Ohio and Mississippi Rivers meet. Cairo, the town that occupies this location, was planned to be a sizable city, but it never developed into the center of commerce that many assumed it would be. Cairo's flood-prone site was eventually enclosed within massive artificial levees to hold back the Mississippi and Ohio Rivers (fig. 8.3). The town's name—as well as the nearly annual floods—gave rise to the regional nickname, Little Egypt, for the southernmost counties of Illinois. Urban investors found better opportunities elsewhere.

Between St. Louis and Cairo the Mississippi is a medium-sized river with extensive floodplains bounded by the rock walls of the Interior Low Plateaus on the east and by those of the Ozark Plateau on the west. The French recognized the strategic location and fertile bottomland soils of this middle portion of the Mississippi Valley. They founded Cahokia, across the Mississippi river from present-day St. Louis, in 1699. Kaskaskia was established by the French in 1703; it became the first capital of Illinois when the Illinois Territory was proclaimed more than a century later.

The west bank of the Mississippi was either French or Spanish territory throughout the 18th century and did not become part of the United States until the Louisiana Purchase of 1803. The east bank of the Mississippi—and all of southern Illinois and Indiana—were liberated from British control by Virginia troops led by George Rogers Clark in 1778. Floodplains on the east bank thereafter became known as the American Bottom, a reference that acknowledged that the territory west of the river still remained under foreign control. Al-

FIG. 8.3. *Cairo, Illinois, at the confluence of the Mississippi and Ohio Rivers, is surrounded by massive artificial levees (that double as railroad embankments) and is protected by a gate that can be closed in times of flood.*

though France transferred the region to Spain in 1763, the Middle Mississippi Valley remained French in language, law, and population until Anglo settlers from Virginia arrived decades later. The French period was not brief, nor was the influence of the French minimal. They surveyed the Middle Mississippi Valley into long-lot settlements, established most of the region's towns (including St. Louis), and opened the early lead mines of the Missouri Ozarks to the west.

St. Louis was established in 1762 as the site of a post that would command the trade in furs from the Missouri Valley to the west and the Mississippi to the north. A city was laid out and farm lots were set aside in a fashion similar to what the French did at Montreal. St. Louis had a role analogous to Montreal's, near the junction of major rivers from which trade would be directed at the city. The name, St. Louis, honored King Louis IX, who was canonized in 1297. However incongruous these plans appear in light of St. Louis's later development, there is little doubt that the French planned that the city would become a major center in the fur trade (which it was, for a time, although not under French auspices). The city languished during the period of Spanish control between 1763 and 1803, although many French settlers who had lived in the American Bottom moved to St. Louis during this period in order to be free of American authority.

St. Louis's ascendancy coincided with the era of steamboating on the Mississippi—the first half of the 19th century, when the city was easily the most accessible trade center west of the Appalachians. The Civil War adversely affected the city in two ways. Its border location perhaps made it inevitable that St. Louis's population would be about evenly divided between Northern and Southern sympathizers, a circumstance that caused bitter division and made it more difficult for the city to pursue a coherent strategy for economic development. The war also produced a temporary blockade of traffic on the Mississippi. Even though St. Louis was a major participant in the railroad-building strategies of the 19th century, it lost influence to Chicago, the preeminent railroad center, which surpassed St. Louis in size by 1880. River access no longer carried the advantage it had had in the past.

Like both Cincinnati and Chicago, St. Louis attracted large numbers of European immigrants, especially Germans, during the second half of the 19th century. It became a major industrial center, although secondary to Chicago. St. Louis became a manufacturer in the transportation field, first of railway equipment, then automobiles, and still later airplanes. Along with its industrial satellite cities on the Illinois bank of the Mississippi, St. Louis also became an important steelmaker, meatpacker, and grain miller. The St. Louis metropolitan area continues to rank among the top ten U.S. cities according to most economic and demographic measures.

## *References*

Belting, Natalia Marie. *Kaskaskia under the French Regime.* Illinois Studies in the Social Sciences, vol. 29, no. 3. Urbana: University of Illinois Press, 1948.

Doyle, Don Harrison. *Nashville in the New South, 1880–1930.* Knoxville: University of Tennessee Press, 1985.

Filson, John. *Discovery and Settlement of Kentucke.* [1784]; Ann Arbor, Mich.: University Microfilms, 1966.
Karan, P. P., and Cotton Mather, eds. *Atlas of Kentucky.* Lexington: University Press of Kentucky, 1977.
Palmer, Arthur N. *A Geological Guide to Mammoth Cave National Park.* Teaneck, N.J.: Zephyrus Press, 1981.
Raitz, Karl B. *The Kentucky Bluegrass: A Regional Profile and Guide.* Chapel Hill: Department of Geography, University of North Carolina, Chapel Hill, 1980.
Sauer, Carl O. "The Barrens of Kentucky." In *Land and Life: A Selection from the Writings of Carl Ortwin Sauer,* ed. John Leighly, 11–22. Berkeley: University of California Press, 1963.
Troen, Selwyn K., and Glen E. Holt. *St. Louis.* New York: New Viewpoints, 1977.
Ulack, Richard, Karl Raitz, and Gyula Pauer. *Atlas of Kentucky.* Lexington: University Press of Kentucky, 1998.

CHAPTER 9

# The Ozarks and the Ouachitas

The Ozarks and the Ouachita Mountains of Missouri, Arkansas, and Oklahoma are the third division of the Upland South. The region is defined mainly on the basis of geological criteria, although culturally it has close ties with both the Interior Low Plateaus and the Appalachians (fig. 9.1). The topography is varied and is controlled in part by underlying geologic structures. All of the highlands between the Missouri and Arkansas Rivers are labeled "the Ozarks," but the region itself contains several distinct subregions of hills, plateaus, and low mountains (map 9.1). The Ouachita Mountains, which resemble the Ridge and Valley portion of the Appalachians, occupy the area between the Arkansas and Red Rivers. Both the Ozarks and the Ouachitas are differentiated from the Great Plains on the west topographically. In some places oak woodlands extend west onto the Great Plains, whereas in others Great Plains grasslands penetrate eastward into the Ozark and Ouachita regions.

As a whole, the Ozark and Ouachita region is an area of sparse settlement and few cities. The mixed hardwood–coniferous forest has scarcely been diminished by human occupancy, despite the importance of logging in some areas. Forests of this type regenerate quickly in the warm, moist climate. The upland areas were largely bypassed by the mainstream, first by aboriginal people and then by early Euro-American settlers. The lowlands of the Arkansas Valley, in contrast, attracted considerable prehistoric settlement and continued in that role in later years when the Arkansas River became one of the important corridors between the Mississippi Valley and the Great Plains.

## The Ozarks

The Ozarks begin abruptly at the western edge of the Mississippi River floodplain south of St. Louis. The region is a dome of upward-arching rocks that is roughly symmetrical around the St. François Mountains, the center of the uplift (see chapter 8, map 8.2). The topography is controlled by the underlying rock formations and varies around the dome according to the layers that form the surface. The oldest rocks are exposed in the St. François Mountains, where Precambrian granites have eroded into a jagged topography of low mountains. Iron-bearing formations are exposed in these mountains and were easy to mine. For a time the St. François Mountains were an important supplier of high-grade iron ores to the St. Louis–area iron and steel industry. Lead and zinc deposits

FIG. 9.1. *A log corncrib with shed wings for machinery storage in the Boston Mountains of Arkansas. Structures of this type once were common in the Upland South—in the Appalachians as well as in the Ozarks.*

around the edges of the St. François Mountains were the basis of another locally important mining industry dating back to early French occupancy of the Middle Mississippi Valley.

Surrounding this mountain core are two limestone plateaus—an inner, more rugged section on Ordovician limestones known as the Salem Upland and an outer, less-dissected plateau of Mississippian rocks, the Springfield Plateau. Most peripheral is a third belt, exposed only on the south, where yet another transition between rock formations produces a change in topography. Rocks of the third belt are the youngest of the Ozark formations. They are associated with the Boston Mountains, an area of rugged topography with elevations above 2,000 feet in a broad section between the White and Arkansas Rivers. Ridges of the Boston Mountains are the outer edge of a belt of north-thrusted strata that also includes the Ouachita Mountains.

Because Ozark topography is largely the product of uplift and erosion, the term *plateau* is more accurate than *mountains* to describe the region generally. The landscape is characterized more by deep valleys cut into the bedrock layers than by prominent ridges or peaks (fig. 9.2). Fairly level to gently rolling uplands separate valleys that are incised several hundred feet below the general level of the plateau. Some of the broader uplands are known locally as prairies, a term that denotes both level land and a grassy, open-woodland vegetation cover at the time of Euro-American settlement. The Ozark prairies probably had a history of human use by prehistoric hunters much like that of the smoother plateaus of Kentucky and Tennessee.

Also like the Interior Low Plateaus, portions of the Ozarks have the necessary geologic ingredients for cavern development. Some caves, such as Meramec Caverns under the Salem Plateau southwest of St. Louis, are popular tourist attractions. Isolated settlements in the heart of the Salem Plateau were clustered around the region's numerous natural springs, which produce an abundant flow of groundwater. Eureka Springs, Arkansas, was among the earliest tourist mec-

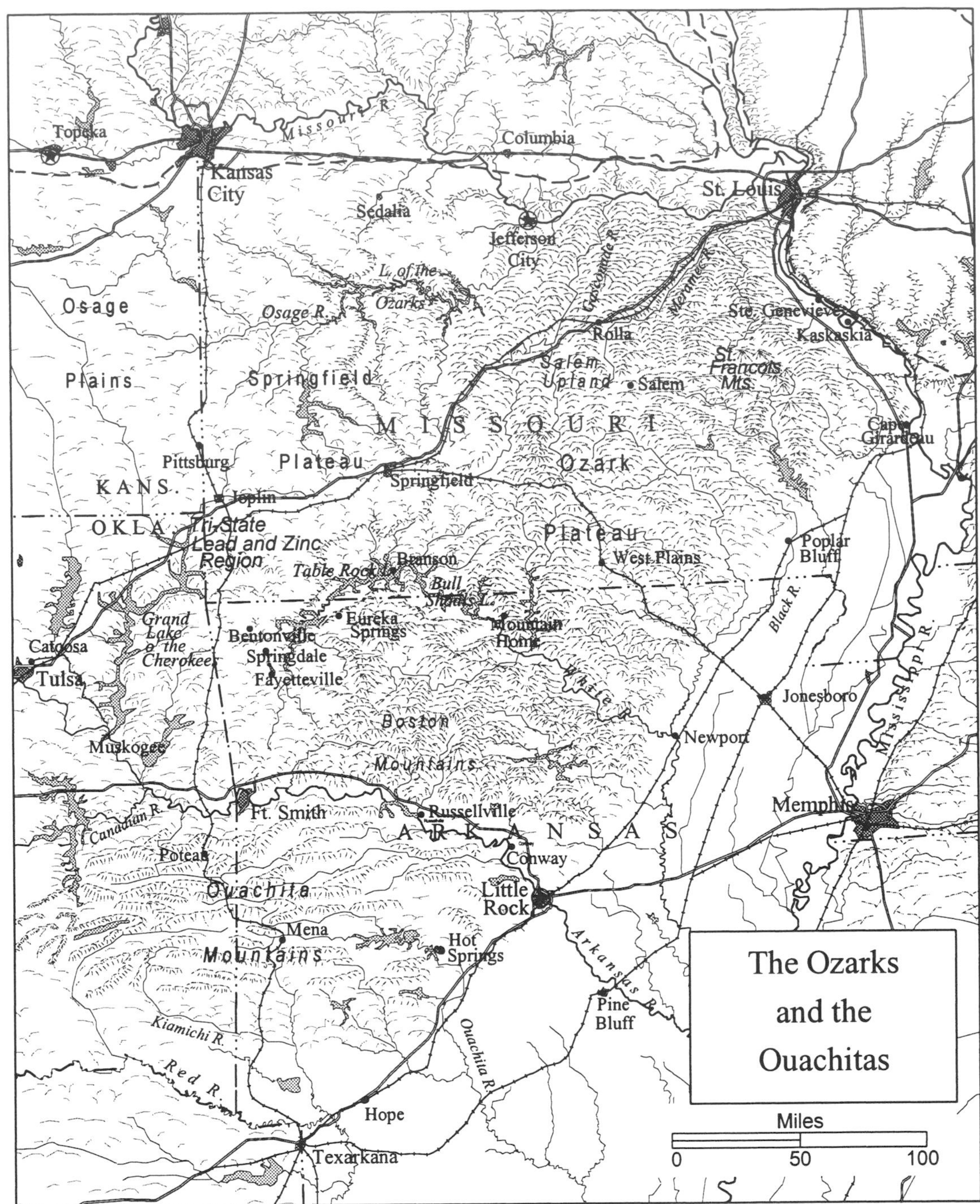
The Ozarks and the Ouachitas
Topeka
Kansas City
Missouri R.
Columbia
St. Louis
Sedalia
Jefferson City
L. of the Ozarks
Osage R.
Gasconade R.
Meramec R.
Rolla
Ste. Genevieve
Kaskaskia
Osage Plains
Springfield Plateau
Salem Upland
Salem
St. Francois Mts.
MISSOURI
Ozark Plateau
Cape Girardeau
KANS.
OKLA.
Pittsburg
Joplin
Springfield
Tri-State Lead and Zinc Region
Branson
Table Rock L.
Bull Shoals L.
West Plains
Poplar Bluff
Black R.
Grand Lake o' the Cherokees
Eureka Springs
Mountain Home
Catoosa
Tulsa
Bentonville
Springdale
Fayetteville
White R.
Jonesboro
Mississippi R.
Boston Mountains
Newport
Muskogee
Ft. Smith
Russellville
Canadian R.
ARKANSAS
Memphis
Poteau
Conway
Ouachita Mountains
Little Rock
Mena
Hot Springs
Arkansas R.
Pine Bluff
Kiamichi R.
Ouachita R.
Red R.
Hope
Texarkana
Miles
0
50
100

MAP 9.1

cas in the Ozarks, dating to the 1880s. Apart from isolated mining ventures and a few resort spas built near the larger springs, however, the Salem Upland attracted little outside investment and long remained sparsely settled.

Elsewhere in the United States, limestone rock layers often are associated with fertile soils, but Ozark limestones are rich in hard, flinty nodules of chert. Upon weathering, cherty limestones produce a loose layer of rubble at the surface. They are generally unsuited to cultivation. The combination of hard rocks and steep slopes plus the lack of extensive bottomlands along the streams combined to discourage crop agriculture. Farming in the Salem Upland focuses on pasturing livestock, especially beef cattle. The low intensity of agricultural land use in the heart of the Ozarks is matched in the Upland South only in the hilliest portions of the Highland Rim and the Appalachian Plateau.



The Ozarks' most important growth sector in recent decades has been the rapid emergence of tourism and outdoor recreation. Natural lakes are few in a topography like this one, where rivers almost perfectly drain the landscape of standing water, but the opportunities for creating lakes behind dams on the Ozarks' rivers were seen to be the means to stimulate a tourist industry. Lake of the Ozarks was created on the Osage River in the 1940s. Later, Bull Shoals and Table Rock Lakes were created behind dams on the White River. The tight pattern of narrow, meandering Ozark valleys produces an equally intricate pattern of winding (although steeply sloping) shorelines when the valleys are flooded, a circumstance that maximizes the amount of recreational front-footage that real estate developers can sell.

Recreational homesite developments appeared throughout the Ozarks as more reservoirs were created behind more dams through the 1970s. Second-home developments, retirement communities, marinas, golf courses, and recreational theme parks predictably followed the creation of water-based recreation sites. The population, which had been stagnant in some Ozark counties for generations, boomed for twenty years beginning in 1960. The Whitewater develop-

FIG. 9.2. *The bluffs along Missouri's Gasconade River typify the rugged topography in the Salem Upland portion of the Ozarks.*

ment, planned for the White River in Arkansas, was to have been a continuation of this sequence, but demand dropped sharply in the mid-1980s economic recession. More recently the entertainment complex built around the once-tiny community of Branson, Missouri, has added another dimension to Ozark tourism. As a result of these developments, the Ozarks are much better known to outsiders than ever before.

The Springfield Plateau is less handicapped by limitations of geology and topography than the Salem Upland. The Springfield Plateau is smoother, it is creased by shallower valleys, and its soils are more productive. Southwest Missouri was settled primarily from Tennessee in the first half of the 19th century, and although many of the original settlers came from the Nashville Basin and other slaveholding areas, they brought relatively few slaves with them to Missouri. The contrast with the comparatively much larger-scale development of slave-based agriculture in the Little Dixie region of Missouri is notable. Southwest Missouri originally developed as a land of mixed corn-livestock farming, although dairying later became the principal form of agricultural activity. Springfield, Missouri, remains as a center of dairying and dairy-product distribution for the south-central portion of the nation. Springfield is Missouri's third largest city and the undisputed "capital" of the Ozarks in the provision of wholesaling, retailing, and services.

Arkansas's portion of the Springfield Plateau became specialized in broiler production beginning in the 1950s. As the nationwide demand for poultry increased, northwest Arkansas developed into one of the major centers of the poultry industry and expanded into the production of preprocessed frozen food entrees. Its central location—near the center of accessibility of the U.S. population as a whole—made it a desirable place for various types of light manufacturing industries and warehouse-based facilities. Some of the nation's largest trucking firms are headquartered here, as are warehouses and shipping facilities for major retail store chains. Located along old Route 66, about midway between New York and California, and on the border between North and South, the Springfield Plateau proved to be a strategic business location that has sustained population growth in recent decades.

Scattered upland prairies within the Ozarks, though resembling similar clearings of the Interior Low Plateaus, also can be interpreted as an eastern extension of the Great Plains grasslands that border the Ozarks on the west. A transition zone known as the Osage Plains occupies the Missouri-Kansas border region and extends southwestward into Oklahoma. In the Osage Plains, precipitation begins to decline noticeably toward the west, and thus small grains such as winter wheat begin to replace more moisture-demanding crops like soybeans.

Two mining areas are in the transition from the Ozark Plateau to the Osage Plains. The Tri-State Lead and Zinc Region, centered on Joplin, Missouri, is the most extensive area of zinc deposits in the world. Nearly two thousand square miles of southwestern Missouri, southeastern Kansas, and northeastern Oklahoma are dotted haphazardly with the waste dumps produced by years of mineral extraction. The industry has declined for some time because of health-related restrictions on the use of lead-based products and a decline in demand for zinc. Many mines have closed, but more than $1 billion of mineral wealth was created during the decades they were in active operation.

The second mineral is a relatively low-grade bituminous coal mined in the Western Interior field (see chapter 8, map 8.2). Like the coal-bearing formations of the Eastern Interior (Illinois Basin) field of which it is actually an extension, most Western Interior coal has a relatively high sulfur content and cannot meet clean-air standards. Western Interior coal was once used locally for steaming in electric-power generation, but it faced strong competition from natural gas produced in Kansas and Oklahoma. Extensive, although shallow, beds of low-grade bituminous coal are widespread in the Western Interior field, which stretches from Iowa to Texas. Coal mines are scattered across this broad region and include several on the Ozark fringe in southeastern Kansas, but as a natural resource Western Interior coal is becoming uneconomical because of air-quality standards.



## The Ouachita Mountains

The Arkansas River Valley in central Arkansas is not an alluvial valley with an extensive floodplain like the lower Mississippi. It is a structural valley that is literally folded between the Ouachita and Boston Mountains. The city of Fort Smith marks the Arkansas River's entrance into this valley. Little Rock is where the river leaves it and enters the much flatter lowlands of the Coastal Plain. The Arkansas Valley is analogous to the Great Valley of the Appalachians just as the Ouachitas are similar in form to the Ridge and Valley section.

The Ouachitas were formed during the same episode of continental collision that produced the Ridge and Valley. Heat and pressure deformed the existing layers of sedimentary rock and pushed them into a series of parallel folds. The Ouachita Mountains occupy much less area than the Ridge and Valley, but their summits rise to equally impressive heights. Geothermal activity accompanied the process of mountain-building and gave rise to numerous thermal springs, the best known of which is at Hot Springs, south of Little Rock. The warm, mineral-rich waters at Hot Springs had long been used by native peoples when Hernando de Soto's expedition (the first exploration by Europeans) happened upon the site in 1539.

The Arkansas Valley was a natural corridor connecting the lower Mississippi Valley with the West. Fort Smith was created as a military outpost on the western frontier in 1817 and became an important supply point and garrison. Fort Smith's site on the Arkansas can be compared with Kansas City's on the Missouri—both were river ports where cargoes were transferred to wagons heading west. West of Fort Smith the land route followed the Canadian River into what is now Oklahoma and continued west toward El Paso or Santa Fe. Fort Smith thus had its own Santa Fe Trail, a southern counterpart of the better-known route that linked Kansas City with the Southwest. In the era of wagon and steamboat transportation, the fledgling city on the Arkansas was a convenient gateway to the Southwest, one reached easily from either the Ohio River or New Orleans.

Fort Smith's development was curtailed when the federal government designated Oklahoma as Indian Territory and established a policy of restricting white settlement west of the boundary line. The Oklahoma border, which lies only a few miles west of Fort Smith, became a wall holding back the tide of westward settlement. Settlers streamed across the Osage Plains to the north, but eastern Oklahoma was a barrier thereafter. Kansas City and Omaha became the principal western gateways, and Fort Smith developed into little more than a regional trade center.

Barge navigation on the Arkansas River was enhanced by the completion of the McClellan-Kerr Waterway in the 1960s. Barges navigate a series of locks and dams that extend the navigation channel as far inland as the port of Catoosa near Tulsa. Grain is the principal southbound commodity, whereas coal, chemicals, and petroleum dominate the northbound traffic. The waterway expanded central Arkansas's and eastern Oklahoma's industrial possibilities, but the volume of traffic has not met the projections made at the time of its construction.

The low ridges of the Ouachita Mountains mark the southern edge of the Arkansas River Valley. Fifty miles south of the river, the topography becomes a maze of plunging anticlines and synclines. Steep-sided ridges here are more than fifteen hundred feet above the adjacent valley floors. Long, straight valleys follow the mountain ridges for miles, then are bent abruptly in a zigzag fold. Rivers abruptly change course to flow into a parallel valley. The Ouachita region is covered by a dense forest of broadleaf trees at lower elevations and conifers (mainly pines) at higher elevations. The forest resembles that of the adjacent Coastal Plain and is cut both for lumber and for wood-pulp production. Extensive tracts of immature forest growing on cutover lands with numerous

FIG. 9.3. *Lightly used woodland and pasture in Pushmataha County, Oklahoma, near the western edge of the Ouachitas.*

clearings used for pasturing cattle typify the Ouachita landscape (fig. 9.3). Little has changed in a century.

Southeastern Oklahoma's portion of the Ouachitas is even more isolated and less developed than the Arkansas side. The Oklahoma portion coincides with the Choctaw Nation, the lands set aside for the Choctaw people when they were relocated from Mississippi beginning in the 1820s. The Choctaws endured a long struggle with the federal government as well as with other native people in their efforts to obtain title to the lands they had been allotted. General farming mixed with forestry is the common occupation of those living in the isolated clearings and villages within this densely forested landscape. The Choctaws' best agricultural land was lost when recreational lakes were constructed and their lowland fields were thereby flooded permanently.



The Ouachita region remains bypassed and thinly populated. The first paved federal highway through the Oklahoma portion of the Ouachitas was not completed until the early 1960s. Unlike the Appalachians, which channeled the flow of migrants southward with the grain of topography, the Ozarks and the Ouachitas stimulated something more like a small detour. St. Louis's main connections with the Southwest were diverted eastward around the Ozark-Ouachita margin and are represented in the series of transportation- and trade-based cities in the arc connecting Cape Girardeau, Jonesboro, Pine Bluff, and Texarkana, all of which developed as railroad centers at the beginning of the 20th century.

Little Rock, which lies where these Coastal Plain routes meet the Arkansas Valley, became the region's largest urban center. In recent years Little Rock has grown in its role as a banking and financial center for the south-central states and has attracted light-manufacturing industries after the fashion of other Sunbelt cities. As the largest city between St. Louis and Texas, Little Rock has attracted an even larger share of the investment capital of the Ozark-Ouachita region in recent decades and has absorbed at least part of the growth that once took place in the regional centers.

## *References*

Bearss, Ed, and Arrell M. Gibson. *Ft. Smith: Little Gibraltar on the Arkansas.* Norman: University of Oklahoma Press, 1969.

Gerlach, Russell L. *Immigrants in the Ozarks: A Study in Ethnic Geography.* Columbia: University of Missouri Press, 1976.

Miller, E. Joan Wilson. "The Ozark Culture Region as Revealed by Traditional Materials." *Annals of the Association of American Geographers* 58 (1968): 51–77.

Rafferty, Milton D. *The Ozarks, Land and Life.* Norman: University of Oklahoma Press, 1980.

Sauer, Carl O. *The Geography of the Ozark Highland of Missouri.* Bulletin of the Geographic Society of Chicago, no. 7. Chicago, 1920.

Schroeder, Walter A. *Presettlement Prairie of Missouri.* Natural History Series, no. 2. Jefferson City: Missouri Department of Conservation, 1981.

*PART IV*

# The Lowland South

CHAPTER 10

# The Southeastern Piedmont and the Coastal Plain

The contrasts between the two principal physical regions of the Southeast, the Coastal Plain and the Piedmont, include their different land surfaces. Swamps, marshes, flatwoods, and other wetlands cover much of the Coastal Plain. Estuaries (former river valleys now inundated by ocean waters) are typical of its outer margins (map 10.1). The term *tidewater* occurs frequently in regional descriptions as a reference to the zone of shallow bays and inlets between the limits of high and low tides.

The Piedmont is a much older land, a maturely dissected complex of metamorphic rocks that are the remnants of a once-prominent range of mountains now worn down to a smooth but rolling surface. The only standing water on the Piedmont is in human-built reservoirs.

The Fall Line, which is marked by waterfalls and rapids on all streams flowing from the Piedmont to the Coastal Plain, was a significant feature limiting settlement in colonial times. Watercraft plying the rivers in Virginia could penetrate no farther inland than Richmond or Petersburg. The rapids limited urban growth to the west even though they did not constitute a barrier to settlement expansion. Fall Line cities became the principal trading and market centers for the backcountry.

In present-day terminology, streams flowing from the Piedmont to the Coastal Plain are called redwater rivers because of the sediment load of reddish-colored Piedmont soil they carry. Blackwater rivers, in contrast, originate in the low-lying swamps of the Coastal Plain and are stained dark because of the high organic-matter content of the swamp waters. Major rivers, such as the James, the Roanoke, the Savannah, and the Altamaha, are the redwater type and have large floodplains in the Coastal Plain, where stream velocities are low. Some redwater rivers pick up blackwater streams as tributaries as they flow to the ocean.

Although the Piedmont is hillier, more eroded, and has soils with a higher clay content than the generally flatter and sandier Coastal Plain, there are no strong landscape contrasts across the fall zone in either vegetation or land use. One reason for the similarity is that the dominant soil-forming process is the same in both regions. The soils are controlled more by climate than by geology. Typical soils of both the Piedmont and the Coastal Plain belong to the order of Ultisols, meaning old soils that have experienced little alteration either chemically or physically for hundreds of millennia. A moist, warm climate affects both regions, and this means that the soils contain a low amount of organic matter,

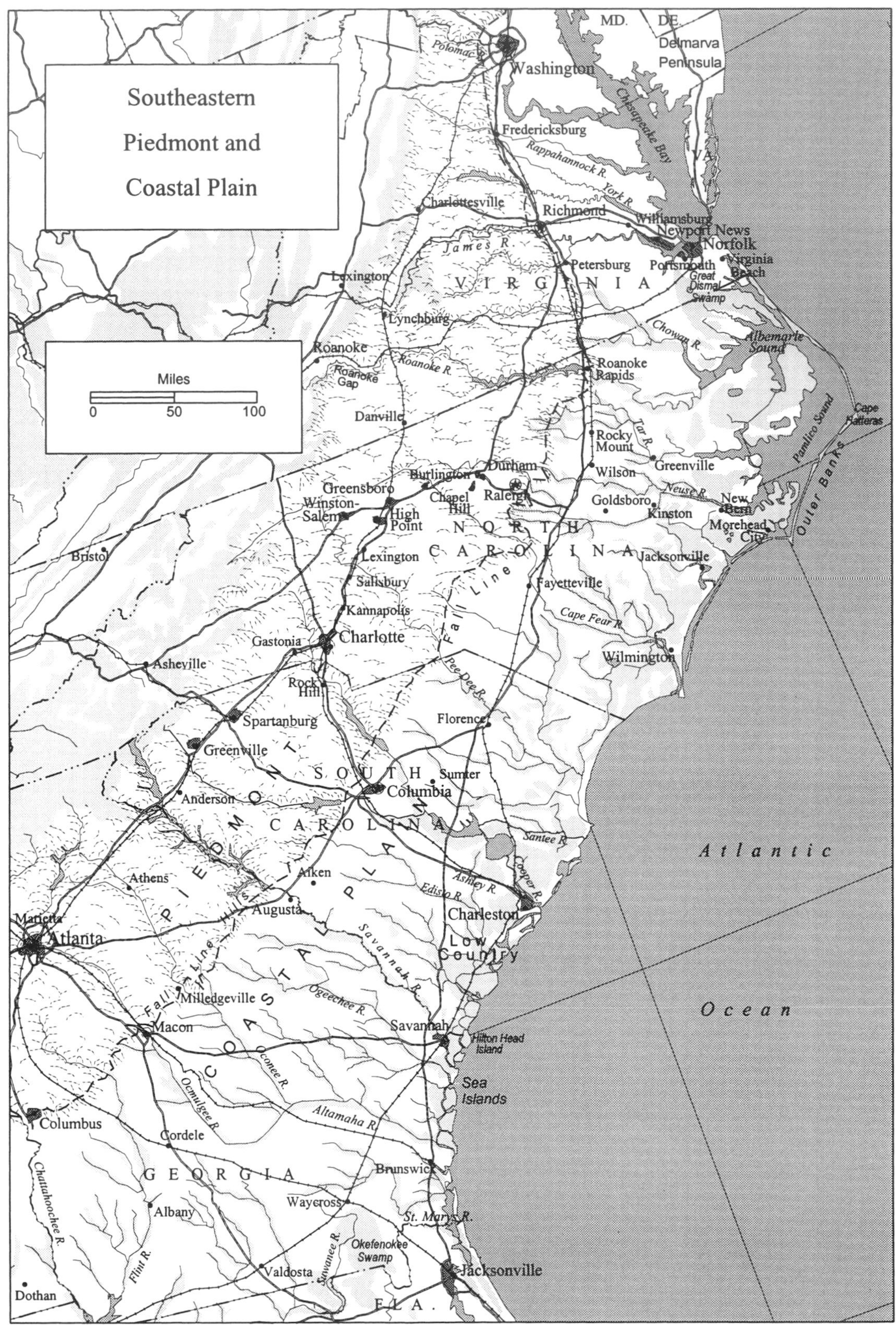
Southeastern
Piedmont and
Coastal Plain
Miles
0
50
100
MD.
DE.
Delmarva
Peninsula
Washington
Potomac R.
Fredericksburg
Rappahannock R.
Chesapeake Bay
VA.
York R.
Charlottesville
Richmond
Williamsburg
Newport News
Norfolk
Portsmouth
Virginia
Beach
Great
Dismal
Swamp
James R.
Petersburg
Lexington
V I R G I N I A
Lynchburg
Roanoke
Roanoke
Gap
Roanoke R.
Chowan R.
Albemarle
Sound
Roanoke
Rapids
Danville
Rocky
Mount
Tar R.
Pamlico Sound
Cape
Hatteras
Outer Banks
Greenville
Durham
Burlington
Greensboro
Winston-
Salem
High
Point
Chapel
Hill
Raleigh
Wilson
Neuse R.
Goldsboro
Kinston
New
Bern
Morehead
City
N O R T H
C A R O L I N A
Jacksonville
Bristol
Lexington
Salisbury
Kannapolis
Fall Line
Fayetteville
Cape Fear R.
Gastonia
Charlotte
Wilmington
Asheville
Rock
Hill
Pee Dee R.
Spartanburg
Florence
Greenville
S O U T H
Sumter
Columbia
Anderson
C A R O L I N A
Santee R.
Atlantic
P I E D M O N T
Athens
Aiken
Ashley R.
Edisto R.
Cooper R.
Charleston
Marietta
Atlanta
Augusta
Savannah R.
Low
Country
Fall Line
C O A S T A L   P L A I N
Milledgeville
Ogeechee R.
Ocean
Macon
Savannah
Hilton Head
Island
Oconee R.
Sea
Islands
Ocmulgee R.
Altamaha R.
Columbus
Cordele
G E O R G I A
Brunswick
Chattahoochee R.
Waycross
St. Marys R.
Albany
Okefenokee
Swamp
Savannee R.
Flint R.
Valdosta
Jacksonville
Dothan
FLA.

MAP 10.1

have a low ability to hold nutrients, are acid, and are low in natural fertility—all common traits of soils in subtropical climates.

Early Euro-Americans followed some agricultural practices that had long been discontinued in Europe but which were taken up as expedients in the new land, perhaps because of the necessity for turning a profit as quickly as possible. These included the continuous production of a single crop, usually tobacco. Continuous cropping soon exhausted the soil. The English, especially, sought to remedy the problem by adding animal manure, which was the common European solution to soil infertility. But England's climate did not match that of the southeastern seaboard. Adding manure only made the sandy soils more acid and had no positive impact on their fertility. It took nearly two centuries for planters to learn that adding lime improved the fertility of Ultisols and simultaneously decreased their acidity.

In both the Coastal Plain and the Piedmont, tobacco was planted in a scheme of land rotation in which fertility was allowed to recover through idling the land, alternating fields in a patchwork of fresh clearings, croplands, weedy openings, and regrowth forest. Some of the early plantations, including George Washington's Mount Vernon, were weed-overgrown, which was not pleasing to the eye but was the only practical means of allowing fertility to recover before the discovery of the beneficial effects of adding lime.

Frontier settlement in the Coastal Plain expanded directly inland from the better harbors. Most of the Euro-American colonization of the Piedmont, in contrast, continued the pattern established by migration through the Great Valley. Migrants from the north entered the Piedmont at Roanoke Gap to establish settlements in Southside Virginia (the counties of southern Virginia). The Carolina Piedmont also experienced a heavy influx from the north, especially Scotch-Irish and Germans who established farms and towns in what eventually became a heavily urbanized zone including Winston-Salem, Charlotte, Greenville, Columbia, and Augusta.

Urban growth was slow in the Piedmont and most of the Coastal Plain. Richmond was founded in colonial times, yet it could count no more than a few thousand inhabitants into the first decade of the 19th century despite having been designated as Virginia's capital (moved there from Williamsburg) in 1779. Cities of the Virginia, Carolina, and Georgia Piedmont were handicapped by inland location. They occupied sites along relatively minor streams, all of which flowed away from the major thrust of settlement expansion and were encumbered by rapids at the Fall Line. Land routes, such as the Philadelphia Wagon Road to the north, offered better access but were a slow and expensive means of transportation. Substantial urban growth in the landlocked Piedmont would wait until investment began to flow into the region during the late 19th century.

## Tobacco

The first permanent English colony in North America was established by the Virginia Company at Jamestown in 1607. Until 1700 a trickle of new immigrants arrived and established themselves, as the Jamestown colony had, on the tidal estuaries bordering Chesapeake Bay. Maryland's settlement began near the mouth of the Potomac River in 1634. Other clusters of villages appeared near the mouths of the Rappahannock and York rivers. By 1700 nearly all the land surrounding the rivers tributary to Chesapeake Bay had been claimed.

Tobacco was the means to wealth in Virginia. It became the common preoccupation of hundreds of Englishmen attracted to the Virginia colony, who acquired a tract of land, cleared part of it, and planted tobacco. A planter aristocracy—including such families as the Washingtons, the Jeffersons, and the Lees—emerged as a dominant influence in the region. In the early years planters procured plantation labor in the form of white indentured servants. African slaves were brought to Virginia as early as 1619, although the total subjugation of Africans did not come about until early in the 18th century, when slave codes were enacted. The so-called three-cornered trade during this period began with the shipment of the tobacco crop to Great Britain, where it was sold. Cargoes of manufactured goods were shipped from Britain and western Europe to Africa, and there the ships received consignments of slaves for transportation to the New World.

The most frequent transatlantic sailings were directly between Virginia and the mother country. The planter consigned his tobacco to a factor (merchant) in England or Scotland who sold the crop and then used the proceeds to procure goods the planter had ordered. Each of the large plantations had its own wharf where sailing vessels docked annually to exchange goods for tobacco. The need for towns and local businesses was reduced because the typical transactions were direct between individual planters and overseas factors.

Eighteenth-century Virginia had few towns. The plantation was the dominant unit of society and the most common form of settlement, and its role expanded to include nearly every aspect of life. Both farm and town, both residence and workplace, encompassing both landlord and tenant, the plantation evolved as a distinct regional type. As tobacco production moved inland, partly in response to the growth of the industry and partly because of the need to find new land, plantations no longer had the direct overseas access they had enjoyed in the tidewater zone; but because the planter society remained essentially intact, there still was little demand for trade-center towns.

Crossroads stores, scattered courthouses, churches, and meeting houses dotted the Piedmont by the early 18th century, but they did not attract one another into agglomerations large enough to be recognized as towns. The markets for tobacco remained at the limit of water navigation, and hence trade of all types was directed there. The lack of trade-center towns in early Virginia later characterized other plantation districts of the South, a circumstance that helps explain part of the long lag in urban growth within the region generally.

By the 1740s new European markets and new trading patterns allowed a greater role for the small-scale tobacco planter, but the most important expansion of production took place up the coastal valleys and into the Piedmont (fig. 10.1). Some tobacco planters went to the Shenandoah Valley. The institution of

FIG. 10.1. *A tobacco patch, Meadow Fork, North Carolina. Tobacco is grown in all sections of North Carolina, from the Coastal Plain to the valleys of the Smoky Mountains.*

slavery spread wherever tobacco production expanded, although small-scale planters typically owned no slaves.

The growth of the American population eventually produced a substantial domestic market for tobacco. Experimentation with new tobacco varieties led to distinct types of tobacco culture, including various methods of drying (curing) the crop after harvest. Maryland specialized in a slow-burning type of leaf that was air cured. Fire curing was the dominant method used in the Piedmont. Around 1840 tobacco planters in the fire-cured belt of Virginia and North Carolina discovered that a superior product resulted if heat was conducted through the curing barn from an outside stove, through a series of pipes (flues).

The flue-curing process was adopted at about the same time that tobacco planters began to move back eastward from the Piedmont to the Coastal Plain. Soil infertility was not a problem so much as an advantage, because the sandy Coastal Plain soils produced a thinner leaf that turned a bright yellow-orange when cured. Tobacco was fertilized early in the season to produce growth but seldom thereafter so as to deprive it of nitrogen and also to control the nicotine content. "Starving" the tobacco plant was easy in Coastal Plain soils of low inherent fertility. This new bright tobacco proved to be especially popular in cigarettes because of its milder flavor. Flue-cured bright-leaf tobacco became the standard of the cigarette industry, and the Coastal Plain emerged as its major area of production.

Machine manufacturing of cigarettes from bright tobacco began on a small scale in the 1880s. Two North Carolina cities, Winston (a new manufacturing town that grew alongside the old Moravian settlement of Salem) and Durham, became centers for cigarette tobacco manufacture. A Virginia-born tobacco peddler, R. J. Reynolds, developed the industry at Winston-Salem; at Durham it was created by a North Carolina–born farmer, Washington Duke. His son, James B. Duke, formed the giant American Tobacco Company trust that controlled nearly 90 percent of cigarette sales in the early years of the 20th century.

Vigorous brand competition emerged following the U.S. Supreme Court's dissolution of the American Tobacco trust in 1911. Sales of Lucky Strike (American Tobacco), Camels (R. J. Reynolds), and Chesterfield (Liggett and Meyers) nearly doubled every five years into the early 1930s.

Unmanufactured tobacco is prone to damage and cannot bear transportation over a long distance, whereas packaged tobacco products can be shipped easily anywhere in the world. Tobacco manufacturing thus became a classic example of a raw-material-oriented industry, with both production of the raw material and its manufacture concentrated in a single region. North Carolina, Virginia, and Kentucky have been the leading tobacco raising and manufacturing states since 1890. These states have had a strong base of political support for both growers and manufacturers ever since. In fact, most of tobacco's "friends" in the U.S. Congress still come from these three states.

Tobacco has long been one of the most labor-intensive crops grown on American farms. Although its introduction into the Tidewater colonies of Virginia and Maryland antedated the use of slaves, most tobacco was grown plantation-style with slave labor until the 1860s. Large landholdings remained after the Civil War, but the practice thereafter was to rent or lease land to tenant farmers and sharecroppers. Each tenant had a piece of ground, usually forty acres, with a crude dwelling (owned by the landlord). The tenant family had a garden, a cow, a few hogs, and a mule. Nearly all of their rented land had to be used for crops, including perhaps eight acres of tobacco, four of cotton, and twenty or more acres of corn and hay for feed. Between one-fourth and one-half of the crop went to the landlord.

Tobacco occupied only a minor fraction of the farm's area, but making the tobacco crop took by far the largest share of the farm family's time. Plant beds were prepared in January, seeded in February, and weeded two or three times before the small plants were transferred to the open fields in May. Then there was watering, suckering, topping, and hoeing in the summer months. The crop was harvested in late summer; hung in the barn for curing; then packed, tied, and carried to market. Even as late as the 1950s, nearly five hundred hours of human labor were required for every acre of tobacco. An eight- or ten-acre plot was about all that one tenant farm family could manage.

During the 1930s the price of tobacco fluctuated, and the market was depressed from oversupply in many years. The federal government began price supports and acreage restrictions on tobacco in 1933. Further refinements led to a stable program that has operated in much the same fashion ever since. The government annually fixes a support price on each grade of tobacco. The supply is restricted by granting production rights only to those lands that were used to raise tobacco in 1933. These "tobacco rights" have been broadened to allow the lease and transfer of allotment away from the original farms, but transfers between counties have been prohibited. The geography of tobacco production has changed only slightly since the 1930s (see chapter 8, map 8.3).

Labor-saving mechanization only recently has begun to change tobacco farming. Even through the 1960s, the tobacco country of eastern North Carolina had the highest rural-farm population density in the United States because a large labor force was needed for this valuable but time-consuming crop. Mechanization was not economical on small tobacco farms; mules were less likely than tractors to bruise the delicate tobacco leaves; and the leaf-by-leaf method of harvest (known as priming) did not seem amenable to machine methods.



Today, however, only the spring transplanting activity remains largely a manual operation. The introduction of chemical fertilizers, pesticides, and automatic primers (machines that harvest leaf-to-leaf) has reduced field time. Even the old flue-cured barn has been replaced by a bulk barn (a metal shed resembling a semitrailer), which reduces labor in the storing and curing phases and eliminates the hand-tying operation. Approximately one hundred hours of labor per acre are required on the larger tobacco farms today—still a high figure in comparison with many other crops, but an 80 percent reduction compared with forty years ago.

## Urban-Industrial Growth in the Piedmont

Some of the fortunes made in tobacco were invested in a wider variety of local manufacturing enterprises. Durham, Winston-Salem, Greensboro, and Charlotte capitalists invested in textiles, an industry that had existed locally on a small scale before the Civil War but which later became the basis for a growth boom that made the Piedmont the most urbanized section of the South by the mid–20th century.

Nearly every city on the Carolina Piedmont began a campaign to finance textile mills during the 1880s. The employment demand drew in young white men from the countryside, who worked 60–75-hour weeks for wages ranging from one dollar to two dollars per day. Women would work for less than half of what men demanded, however, and by 1900 women began to outnumber men in the textile factories. Some of the first "day care" offered by American companies was in the textile mills, where kindergartens were provided to allow women to return to work. Company paternalism was the norm: mill villages to house the workers surrounded the factories, and company-owned stores were commonplace. Education was beyond the grasp of many mill workers, but schools were built in the mill villages nonetheless. Growing demands for textiles fed more industrial growth, which led to even more immigration to the Piedmont's textile towns. North Carolina's population nearly doubled between 1880 and 1920, and by 1930 it had the largest population of any southern state except Texas.

The Piedmont's textile-manufacturing base made it unique as an urban-industrial zone in the largely rural and agricultural South. Some towns focused on yarn and thread mills; knitting mills producing woven fabrics dominated others; but all of the Piedmont's cities had textile factories. Initially, urban growth centered around individual factories or factory complexes located thirty to fifty miles apart, from central North Carolina to the South Carolina Piedmont cities of Greenville, Spartanburg, and Greenwood. New mills and mill towns filled in the gaps between larger cities by the early 20th century. A locally owned railroad that hauled goods and carried people to and from work served this strip of manufacturing cities. Its slogan, "A Mill to the Mile," suggests the dispersed nature of factories and towns in the region. The textile industry eventually spread to the Coastal Plain and the Appalachians as well.

Textiles and apparel were among the first industries to become globalized. Beginning in the 1960s, textile factories moved to cheap-labor locations around the world, away from the Piedmont cities that once had offered a similar advantage in labor costs compared with New England. Textile manufacturing has declined in the Piedmont in the past three decades. The industry would have cut back even more had it not been for the shift to tufted-goods production (especially carpets) in many mills. Numerous mill villages stand abandoned today, many companies have downsized, and others have left the region entirely.

High-tech industries have appeared in the Piedmont in recent years. Metrolina, or Spersopolis, as the Piedmont's dispersed urban form is sometimes called, is now an urban-industrial zone of nearly 4 million people. Research and development industries cluster in the Raleigh–Durham–Chapel Hill Research Triangle. High Point, North Carolina, is the nation's leading furniture market. On the southern end of the region, Charlotte and Greenville-Spartanburg have diversified manufacturing economies. Computer firms, telecommunications industries, and pharmaceutical manufacturers are attracted to the region as well.

Despite the size of this nearly two-hundred-mile-long urban agglomeration that arcs across the North and South Carolina Piedmont, with a population that would place it among the top ten urbanized areas in the United States, there is no single city that serves as its focus. Charlotte is the largest, but it accounts for less than one-fourth of the population (fig. 10.2). Instead, eight separate metropolitan centers, the historical foci of the tobacco and textile-manufacturing industries a century ago, form a series of modest population peaks that give the

FIG. 10.2. *The convention and civic center, Charlotte, North Carolina.*

area its dispersed-metropolis character. Beyond the cities is a densely settled urban-rural fringe, populated by workers who are accustomed to commuting to the next county (or beyond) for work. Today the intermetropolitan sprawl continues south, interrupted by only a few rural counties, to the outer limits of the Atlanta metropolitan area.



The city of Atlanta has only about four hundred thousand inhabitants, but it is surrounded by a densely built-up urban fringe of more than two dozen counties with a total population of nearly 3 million. Atlanta is not an old city by U.S. standards. It was not founded until the 1840s and then was a speculative creation of railroad promoters, who chose it for its relative location. Atlanta straddles the drainage divide between rivers flowing to the Gulf or to the Atlantic. Though a Piedmont city, it is close to the Appalachians. It was the northernmost point where railroads from the lower Mississippi Valley could detour around the Appalachians and bend north, up the eastern seaboard. Adding to its strategic location was a fairly level route north to Chattanooga. The significance of strategic location was brought home forcefully to Atlanta's few thousand inhabitants during the Civil War when Union troops arrived down that same railroad line from the north and burned the city.

Rising from the ashes, Atlanta grew rapidly in the post–Civil War period as it assumed the role of principal financial and wholesaling center for the Southeast. A Georgia-born pharmacist, John Styth Pemberton, invented a locally popular tonic he called French Wine Cola. In 1886 he renamed the syrup Coca-Cola and in 1891 sold the rights to the product for a few hundred dollars. Atlanta capitalists invested in the product, its sales boomed, and the city has remained the headquarters of the company, which is Atlanta's largest corporation.

Atlanta's hub location in the era of rail transportation was duplicated when it became the major air-traffic center of the Southeast. By the 1950s Atlanta also was a major manufacturing city. Its growth in the Sunbelt era has not been continuous, but it has long since surpassed its rivals, Birmingham and New Orleans, as the major economic control center of the South. In recent years both Charlotte and Miami have challenged Atlanta's supremacy as leading metropolis of the region.

## Coastal Plain Land and Land Use

The Coastal Plain formed during Cretaceous times when the land surface was submerged under shallow ocean waters up to and somewhat beyond the present line of contact with the Piedmont. Silt-laden rivers from the land deposited their sediments in shallow, river-mouth bays at the land's margin. A sequence of submergences and emergences of the land surface followed, and the sediments gradually solidified into the rocks of the Coastal Plain. Beaches, reefs, and other ocean-edge environments contributed land-forming materials. Coastal Plain rocks have a feather-edge zone of contact with the older rocks of the Piedmont. They increase in thickness with distance from the Piedmont, to a maximum depth of approximately one mile near the present Atlantic margin. Sea-level changes, upward arching of the earth's crust, and downward tilting of Coastal Plain rock layers have produced numerous episodes of westward advance or eastward retreat of the coastline.

Today's sea-land boundary emerged when the present sea levels were attained after the Pleistocene glaciation ended. About 18,000 years ago, during the Late Wisconsin glacial maximum, sea levels were nearly four hundred feet lower than at present because of the amount of water locked up in the continental ice sheets. Chesapeake Bay was no more than a river, and its drainage course followed that of the preglacial Susquehanna River. The Potomac, James, Rappahannock, and York Rivers were its tributaries. The Susquehanna discharged into the Atlantic seventy-five miles out to sea from the present coastline of the Delmarva Peninsula. Mastodon and mammoth teeth accidentally collected by fishermen over the years plus scattered evidence of human activity show that a wide corridor of emerged dry land bordered the present shoreline of Chesapeake Bay during the glacial maximum. Sea levels began to increase due to glacial melting about 10,000 years ago, and the present levels were reached only within the past 4,000 years.

Notable topographic features of the mid-Atlantic Coastal Plain shore, such as the deep Chesapeake embayment and the barrier islands known as the Outer Banks, were produced by sea-level change. The many peninsulas extending into Chesapeake Bay and into Albemarle and Pamlico Sounds once were swells of higher ground bordering stream valleys but became peninsulas when the sea level rose. The Outer Banks is a former sandy beach ridge at the margin of dry land that became a chain of isolated barrier islands. The islands were reshaped into their present cuspate form through the sculpting of ocean currents. The warm, shallow backwaters of Albemarle and Pamlico Sounds are slowly shrinking in area today as rivers discharge sediment into them and vegetation takes root on their margins. Barrier islands such as the Outer Banks are constantly changing. They are slowly migrating landward from the steady force of onshore winds. They are also among the newest and least permanent of landform features and are subject to drastic alteration by hurricanes; nonetheless, the demand for oceanfront recreation has led to building "permanent" structures there (fig. 10.3).

Standing water covers millions of acres of the outer Coastal Plain all or part of the year. Brackish-water swamps near the coast grade into freshwater swamps farther inland. The site of the Jamestown colony was unfortunately chosen near the contact between brackish (salt) water and freshwater, where the slowing velocity of freshwater streams merges with the inland limit of tidal movements, creating a zone of slack water. Jamestown's own pollution, discharged into the James River, may have led to the outbreaks of typhoid fever and dysentery that contributed to the high death rates in the colony's early years.

From Norfolk to the Georgia-Florida border, the Coastal Plain is a mosaic of wetlands and shallow stream valleys punctuated by tracts of cropland and commercial woodland. The best agricultural lands lie in a belt that begins only a few miles from ocean waters in the north but veers inland to a swath about 75–150 miles inland in Georgia. This is the Inner Coastal Plain, a 600-mile-long

FIG. 10.3. *Nags Head, part of a barrier island in North Carolina's Outer Banks that has long been a popular recreation area.*

strip of reasonably good agricultural land that has been evident on land-use maps for many years (map 10.2). It is not only the best farmland in the southeastern states, but it is also, together with the alluvial lower Mississippi Valley, the best farmland in all of the South today.

The long swath of Inner Coastal Plain cropland does not correspond to differences either in underlying geology or vegetation patterns. The reason these lands are used for crop farming, whereas sandier margins both farther from and nearer to the coast are not, traces back to the era of sea-level rise following the Pleistocene. Streams that flowed more rapidly when the sea level was lower began to deposit sediments over their downstream banks as the ocean's margin penetrated inland. A broad zone of wetlands was created, where flooding was most persistent, in a strip roughly 50–100 miles back from the coast. The wetlands developed a thick vegetation cover under conditions of supersaturation of the soil layer. These soils were more fertile than those of the sandier Coastal Plain that surrounded them, and they accumulated to a substantial thickness.

Two landform features unique to the Coastal Plain—the Carolina bays and the pine pocosins—were formed by this process. Carolina bays (so named because of the typical presence of the bay tree) are elliptical depressions, often underlain by peat, covered with a mat of decomposing vegetation. Pocosins are similar, although their peaty surfaces are elevated as a result of the thickness of peat accumulation. Both have a vegetation cover of shrubs and scrub pine. The lack of heavy forest is apparently due to the very wet conditions and the strong acidity of the soils. Were they species of plants or animals, both pocosins and Carolina bays would appear on the "endangered" list today because, although they are very poor environments for either forestry or agriculture in their natural state, they become highly productive when drained and after their soils are treated to reduce acidity. Commercial timber production has prompted some land conversion, although agriculture has been the most common stimulus (fig. 10.4).

Near the northern and southern limits, respectively, of the Inner Coastal

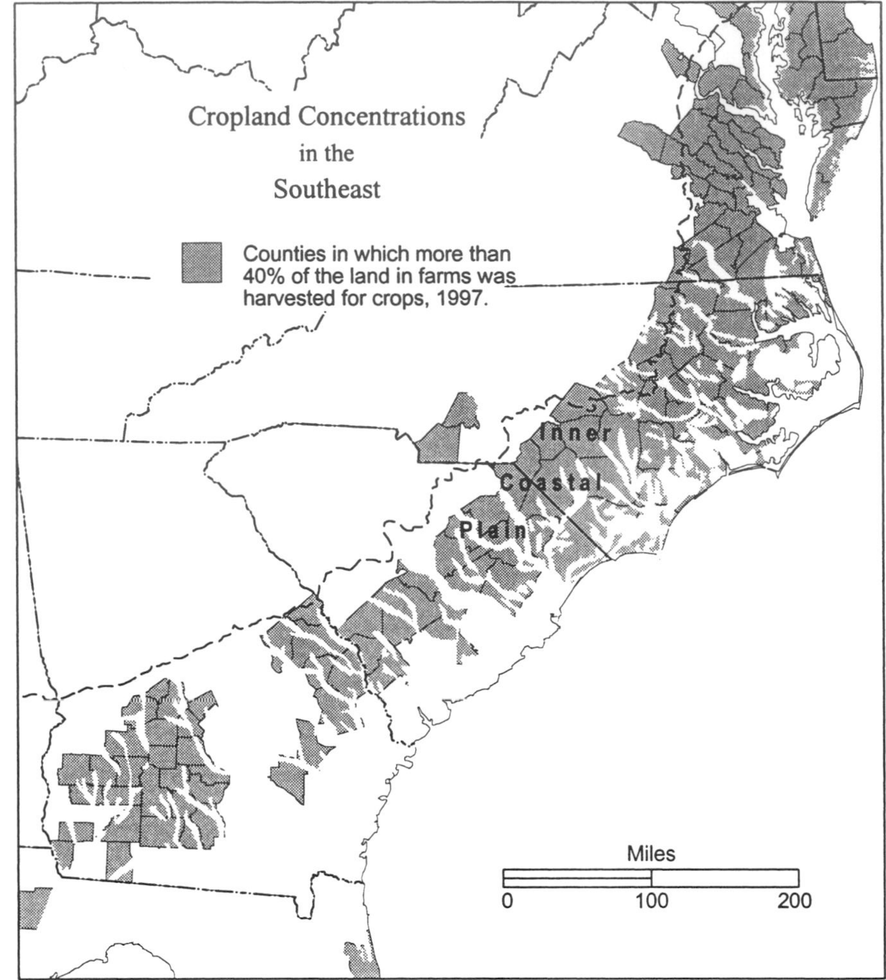


MAP 10.2

Plain are two of the largest freshwater swamps in the eastern United States, the Great Dismal Swamp south of Norfolk, Virginia, and Okefenokee Swamp north of Jacksonville, Florida. Both swamps have extensive areas of peat that have been accumulating for as long as eight thousand years. The drainage of peatlands and other wetlands south of Norfolk began with construction of the Great Dismal Swamp Canal in the 1780s and has proceeded slowly since that time to reclaim wetlands at the swamp's margin for agriculture. Okefenokee is largely a cypress swamp and has been disturbed less for agriculture, although its stands of pond cypress have been lightly harvested for lumber over the years.

Beginning with the federal farm bills of 1985 and 1990, drainage of wetlands has been discouraged and legislation has been enacted to restore wetland functions to existing drained lands. Although conversion to commercial woodland or cropland has been greatly reduced, new types of agricultural intensification have emerged. Fifty years ago soybeans were grown experimentally on Coastal

Plain tobacco lands in an effort to restore soil fertility. The demand for soybeans as poultry feed began to increase in the late 1960s when the broiler industry moved into the Piedmont and the Coastal Plain, and soybeans have become a significant Coastal Plain crop (see chapter 13, map 13.3).

Diversification from broiler production has included the raising of turkeys and, more recently, hogs. North Carolina now ranks second nationally (behind Iowa but ahead of Illinois) in hog production. Coastal Plain farmers began growing larger acreages of both corn and soybeans for livestock feed because of the local increase in demand. Local production has reduced imports of feed grain from the Middle West. Most recently, cotton production has enjoyed a resurgence in the Coastal Plain and is once again being grown as far north as Virginia (map 10.3).



All land uses—commercial forestry, crop agriculture, livestock production, and even wetland reserves—have thus become more valuable in the Coastal Plain because of changes in agricultural patterns within the context of federal farm legislation. With the long-run future of the tobacco industry in some doubt, the trend toward diversification to hogs and poultry will no doubt continue. The Inner Coastal Plain probably will remain as one of the South's most important agricultural regions.

The port of Hampton Roads, Virginia (Norfolk–Newport News–Portsmouth) is the busiest on the south Atlantic coastline. The first transatlantic ships docked there, and wharves were built on both the north and south banks of the James River near its mouth. On the north bank, Newport News grew to become the largest settlement from its roles as a major coal-shipping port and shipbuilding center. Norfolk, on the south bank, also became an important port for shipping Appalachian coal. Norfolk and Newport News, plus Portsmouth, Hampton, and several smaller cities, constitute the largest naval installation in the United States. Norfolk is also a busy container port. Transatlantic container ships carry time-sensitive cargo, and they often make Norfolk their last port of

FIG. 10.4. *Drainage and land clearing are continuous activities on the Inner Coastal Plain of the Carolinas and Georgia.*

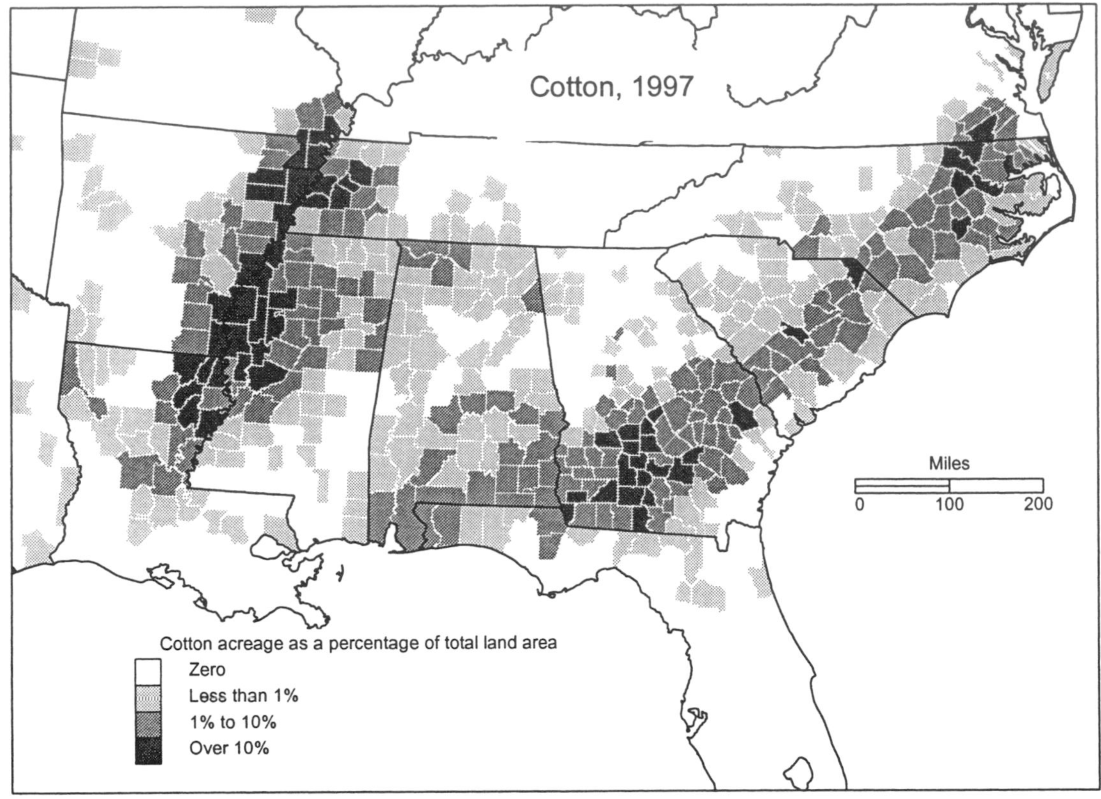


MAP 10.3

call before returning to Europe. This gives Norfolk an advantage in shipping times to Europe compared with the New Jersey ports. Although the more northern ports are closer to London or Hamburg, they have shipping deadlines that are days earlier because of the order of call at ports.

## The Low Country and the Sea Islands

North Carolina's pattern of coastal barrier islands gives way south of Charleston to a myriad of small, tightly packed islands adjacent to the main body of land. This is the Sea Island section of the Coastal Plain, and it extends south to the mouth of the St. Mary's River near Jacksonville. The Sea Islands formed relatively recently as a result of sea-level increases following deglaciation. Large expanses of grassy tidal estuaries backed by coastal forest are separated by former river channels, now greatly widened.

In South Carolina the Sea Island section is roughly coextensive with the Low Country, a reference both to low elevation above sea level and to marshy or swampy ground. The Sea Islands also could be termed the heartland of Afro-America, because it was to the Low Country, especially the port of Charleston, that the largest numbers of African slaves were brought in the 18th century. Until recent times, with the development of resorts such as Hilton Head Island, the

Sea Islands were a largely bypassed area where African Americans had been the dominant cultural group ever since colonial times and where dialects such as the local Gullah were still in common usage.

This small but intensely African portion of the United States was surrounded by a much larger zone where plantation agriculture was established in the early 18th century. It was a development that followed the introduction of tobacco plantations to the Chesapeake Bay region but preceded the establishment of inland cotton plantations across the Lower South. Three crops were introduced into the Low Country, and all three had strong associations with the African slaves brought to labor here. First to be introduced was rice, which was followed by a brief period of specialization in indigo, which in turn was followed by the introduction of cotton. Each new crop marked an advance of agriculture that produced new labor demands and resulted in more slave imports.



Among the varieties of rice brought to the New World, those of West African origin were most often selected for domestication. Natural floodplains of the Niger River in the West African Sudan were used for rice culture at the time of the slave trade. West Africans were the most numerous among the slaves brought to North America, and they likely knew more about the cultivation of rice, indigo, and cotton than did their English masters. Broad river floodplains subject to seasonal inundation were absent in the Low Country, but extensive marshy areas near the streams could be diked and flooded to accomplish the same effect in the production of wet rice (fig. 10.5).

The rice crop brought a good return in the British market, and its production expanded along the coastal rivers within the Low Country during the 18th century. The swamps were unhealthful places, and the mortality rate among slaves was high, but the size of the African American population grew from the importation of more slaves. Throughout most of the 18th century, blacks outnumbered whites two to one in South Carolina's population.

The growth of the plantation economy stimulated a diversification into in-

FIG. 10.5. *Terraces along the Ashley River in South Carolina were used for growing rice during the early years of plantation agriculture in the Low Country.*

digo production at the same time. Indigo, an upland crop, was the principal source of blue coloring used in dyestuffs. Its market was limited, the crop of the Low Country was often of indifferent quality, and its production boom was short-lived. The inland expansion of plantation culture was temporarily halted when indigo declined in the latter decades of the 18th century.

The crop that became most associated with the Sea Islands was neither rice nor indigo but cotton. One of the most prized varieties of cotton is of Egyptian origin. It produces a long-staple fiber that is excellent for spinning into fine cloth. Egyptian cotton requires a long, hot growing season and cannot tolerate cold temperatures. The Sea Islands matched these climatic requirements, and the local soils were adequate for cotton production. By the early 1800s long-staple cotton was leading a new plantation frontier in the Low Country—not back from the coast but rather toward the coast, to the lands where oceanic influence largely eliminates a cold season. Long-staple cotton became so associated with this area that it was subsequently known as Sea Island cotton in American markets. By the early decades of the 19th century, cotton was the principal industry of the Sea Islands.

Savannah and Charleston were the crop's major marketing centers. Both were urban anomalies in the predominantly rural Lower South. Charleston was once counted among the ten largest U.S. cities, although its growth booms of the late 18th and early 19th centuries were followed by a much longer period of relative decline in rank among the nation's cities. Charleston had the dubious distinction of being the capital of the slave trade within the United States. Slaves arrived there from Africa or the West Indies and were sold to traders who supplied slave markets in more inland locations. Part of the city's growth was directly related to the dispersed plantation economy. Many planters constructed their principal residences in Charleston, away from the swampy riverine settings where the actual plantations were located (fig. 10.6). Charleston became the cen-

FIG. 10.6. *The traditional townhouse in Charleston, South Carolina, had large porches to allow air circulation. Porches were built on the long dimension of the house in order to maximize the use of space.*

ter of planter society and the economic focal point of the rice, cotton, and indigo trades as well as the slave trade.

Sea Island cotton culture peaked in the decades before the Civil War, and Charleston attained its maximum influence at that time. The wartime destruction of large segments of the plantation economy could not be overcome by attempts to restimulate the city's growth, however, and Charleston entered a long period of economic stagnation. Its fortunes revived as a naval base and shipbuilding center and, especially during both world wars of the 20th century, Charleston boomed. Like New Orleans, it eventually turned to tourism for economic support. The city's rich architectural heritage remains its most valuable asset.



Savannah began in 1733 as a social experiment of London philanthropists who intended to establish a frontier settlement where England's urban poor might be given a chance to make new lives for themselves. The city was platted in an impressive gridiron, with numerous squares and parks, along the banks of the Savannah River. It became a leading cotton market because of the valuable crop produced in the Sea Islands, and like both Norfolk–Newport News and Charleston, it also became a major naval installation. With the development of the southern naval stores industries (turpentine and resins) Savannah also became a leading center for processing and shipping in that industry. Although it remained smaller in size than Charleston, Savannah was able to capture more of the industrial growth of the late 19th and 20th centuries.

## Piedmont and Coastal Plain Cotton

Nearly all planters who moved inland, "up" the Coastal Plain from the Low Country, chose to plant short-staple, upland cotton. It had a shorter fiber than the prized Sea Island variety, but it grew under a much wider variety of conditions and was a suitable frontier crop as settlement expanded westward. By the mid–18th century, cotton planting had moved into a zone where it blended with the more northern-derived farming population that was expanding southward through the South Carolina Piedmont. The various "cotton belts" of the American South derived from a mingling of early colonial cultures, from a combination of Tidewater plantation traditions and upcountry "yeoman farmer" traditions.

In Georgia the Lower Piedmont is separated from the Inner Coastal Plain by only a narrow sandstone belt, the Fall Line Hills. Apart from this narrow zone, cotton was as productive in the Coastal Plain as it was in the Piedmont. Mid-Georgia's major cities—Augusta, Macon, and Columbus—were founded as water-power sites at the Fall Line, as was Columbia, South Carolina. Until it was moved to Atlanta in 1867, Georgia's state capital was at Milledgeville, also on the Fall Line and midway between the state's most productive Coastal Plain and Piedmont cotton lands. The Georgia Piedmont was the first area to employ cotton gins on a large scale in the 19th century, and its large-scale land holdings, scattered country gins, and plantation stores served as the model for much of the Cotton Belt plantation system that developed across Alabama and Mississippi.

Cotton gins were focal points in the rural economy. The process of ginning separates cotton seed (later pressed for its oil) from the fibers of the boll. Unprocessed cotton is hauled to local gins, compressed into bales, and then shipped to the major cotton markets for local sale or export. Because the plantation remained as the basic unit of the settlement system—both before the emancipation of slaves in 1863 and in the era of sharecropping and tenancy that followed—towns were not important in the local economy. The larger plantations had commissary stores where supplies were issued on credit against the next season's crop. They served the plantations of which they were a part and also the many small-scale planters who lived nearby. The building of railroads gradually impressed a more truly urban pattern of settlement on the cotton regions during the latter half of the 19th century, but as long as cotton remained the major crop, prior to mechanization in the 1930s the overall system did not change.

In the early 20th century, three separate trends converged that gradually led to the demise of King Cotton. One was the appearance of the boll weevil, an insect that first appeared in the Rio Grande Valley of Texas in 1900. It spread rapidly across the South by 1920. Some areas, such as southwestern Georgia, abandoned cotton culture and shifted to the production of peanuts, which are still an important local crop. At Enterprise, in southeast Alabama, citizens erected a monument to the boll weevil because the destruction it caused in the cotton crop prompted local planters to shift into peanut production.

The boll weevil's impact was felt at the same time that a second trend emerged—the first large outpouring of black migrants from the rural South to the urban North. Black out-migration produced a drastic reduction in the number of tenant farmers available to grow cotton. The third trend producing a decline in cotton in the Southeast was the advent of mechanical cotton picking, which was introduced in the 1930s and became the common practice by the 1960s.

To some extent mechanical cotton planting and picking were a response to the general exodus of labor from the region, although the question of whether black out-migration was primarily a cause or a result of mechanization still is debated. Planters in declining cotton districts could not afford to invest in new equipment, and their typical response was to shift out of the business entirely. By the 1970s cotton was no longer a major crop in Georgia or South Carolina, whether in the Coastal Plain or the Piedmont. Cotton returned in the 1990s, following the eradication of the boll weevil and encouraged by high crop prices. Cotton's return thus far has been to the better Coastal Plain croplands; it has not reappeared in the Piedmont.

## *References*

Breen, T. H. *Tobacco Culture: The Mentality of the Great Tidewater Planters on the Eve of Revolution.* Princeton, N.J.: Princeton University Press, 1985.

Earle, Carville. "Environment, Disease, and Mortality in Early Virginia." *Journal of Historical Geography* 5 (1979): 365–90.

Farmer, Charles J. *In the Absence of Towns: Settlement and Country Trade in Southside Virginia, 1730–1800.* Lanham, Md.: Rowman & Littlefield, 1993.

Hart, John Fraser. "The Demise of King Cotton." *Annals of the Association of American Geographers* 67 (1977): 307–22.

Hart, John Fraser, and Ennis Chestang. "Rural Revolution in East Carolina." *Geographical Review* 68 (1978): 435–58.

Hart, John Fraser, and John T. Morgan. "Spersopolis." *Southeastern Geographer* 35 (1995): 103–17.

Kovacik, Charles F. "South Carolina Coast Landscape Changes." *Proceedings, Tall Timbers Ecology and Management Conference,* no. 16, 47–65. Tallahassee, Fla.: Tall Timbers Research Station, 1982.

Littlefield, Daniel C. *Rice and Slaves: Ethnicity and the Slave Trade in Colonial South Carolina.* Urbana: University of Illinois Press, 1991.

Messina, Michael G., and William H. Conner, eds. *Southern Forested Wetlands: Ecology and Management.* Boca Raton, Fla.: Lewis, 1998.

Prunty, Merle C., Jr. "The Renaissance of the Southern Plantation." *Geographical Review* 45 (1955): 459–91.

Prunty, Merle C., Jr., and Charles S. Aiken. "The Demise of the Piedmont Cotton Region." *Annals of the Association of American Geographers* 62 (1972): 283–306.

Wheeler, James O. "Locational Factors in the New Textile Industry: Focus on the U. S. South." *Journal of Geography* 97 (1998): 193–203.

CHAPTER 11

# The Florida Peninsula

Like the rest of the Coastal Plain, the Florida Peninsula receives an average of fifty to sixty inches of rainfall per year. The abundant precipitation falls on a surface that is even flatter than most of the Atlantic or Gulf margins to the north. A slight southward slope of the land surface begins in midpeninsula; and from there on south, the runoff is channeled to numerous small streams entering the Gulf or the Atlantic (map 11.1).

The Florida Peninsula consists of coarse-grained limestones deposited during Tertiary times, 25–50 million years ago, in a warm and shallow marine environment resembling that of the Bahama Banks today. Florida's peninsular outline represents a balance between the total area covered by these limestone layers and the present level of the ocean. A large section of the continental platform on which Florida rests is submerged below sea level. The continental shelf ends abruptly along the east coast near Palm Beach, but it extends nearly two hundred miles westward under the waters of the Gulf of Mexico at the same latitude. Nearly the full peninsula was dry land eighteen thousand years ago, when ocean levels were lower during maximum glacial conditions. Changes in global climate leading to sea-level fluctuations can change Florida's size and shape. A 400-foot drop in the sea level would nearly double its size, but only a 300-foot rise would inundate nearly the entire state. Even a 50-foot increase in sea level would drown southern Florida as well as all of its coastal cities.

Although much of Florida lies barely above sea level, the limestone layers extend down to a depth of 15,000 feet in some places. In the top one-third of the sequence is a series of permeable rock layers saturated with water. These southward-dipping aquifers are exposed at the surface in northern Florida, but they are confined by an overlying layer of less permeable rock in the southern part of the state. The result is an underground reservoir of freshwater that surges up through the surface in the form of artesian springs where the upper, confining rock layer is absent. Artesian springs are most common in the lake region of central Florida. The famous Silver Springs is only one of more than two dozen artesian springs that discharge hundreds of millions of gallons of water every day, making central Florida the most intense concentration of free-flowing springs in the world.

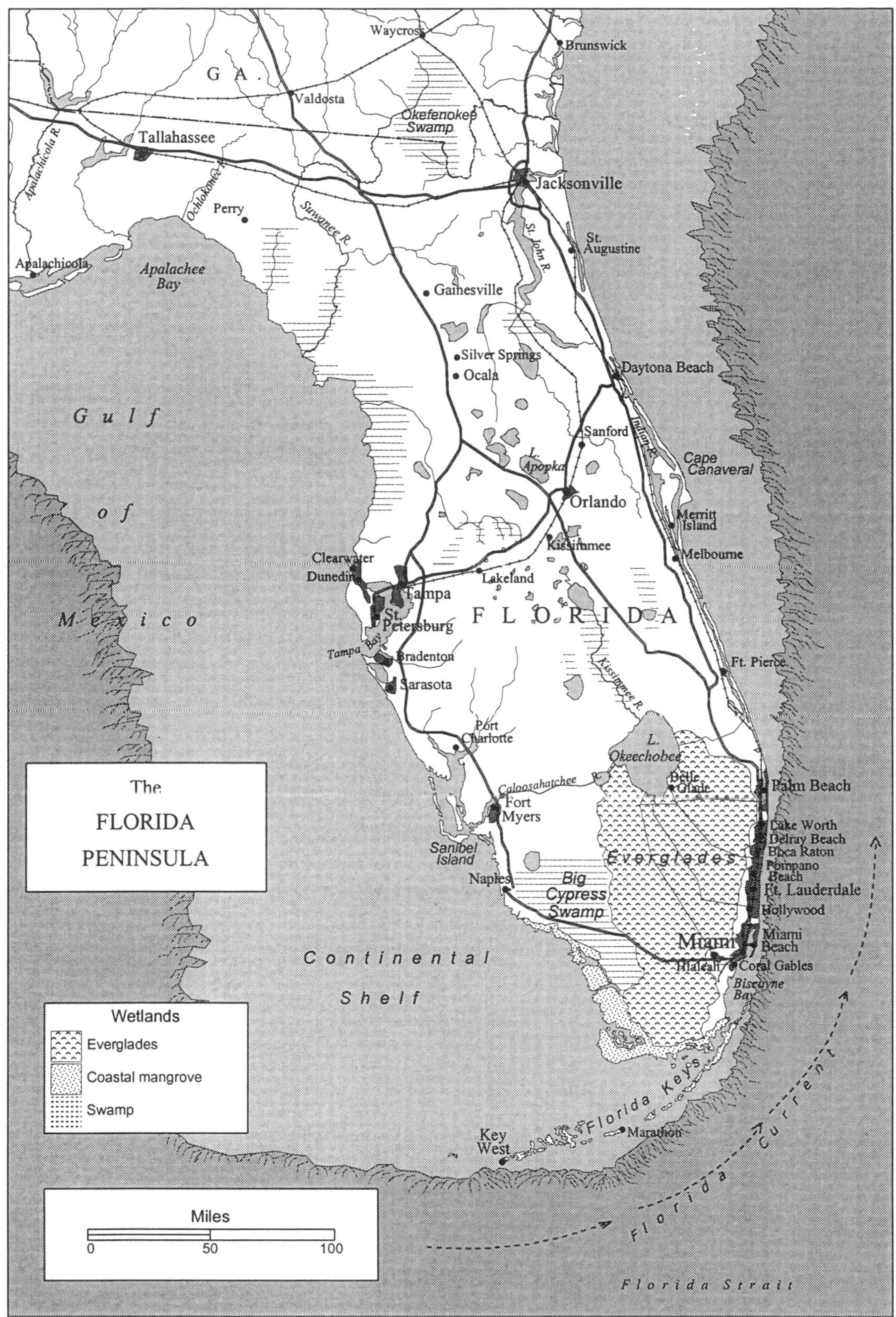
The
FLORIDA
PENINSULA
Wetlands
Everglades
Coastal mangrove
Swamp
Miles
0
50
100
Waycross
Brunswick
G A.
Valdosta
Okefenokee
Swamp
Tallahassee
Apalachicola R.
Ochlokonee R.
Jacksonville
Perry
Suwanee R.
St. John R.
St. Augustine
Apalachicola
Apalachee
Bay
Gainesville
Silver Springs
Ocala
Daytona Beach
Gulf
of
Mexico
Sanford
Indian R.
L.
Apopka
Cape
Canaveral
Orlando
Merritt
Island
Kissimmee
Melbourne
Clearwater
Dunedin
Lakeland
Tampa
St.
Petersburg
F L O R I D A
Tampa Bay
Bradenton
Sarasota
Ft. Pierce
Kissimmee R.
Port
Charlotte
L.
Okeechobee
Belle
Glade
Palm Beach
Caloosahatchee R.
Fort
Myers
Lake Worth
Delray Beach
Boca Raton
Pompano
Beach
Sanibel
Island
Everglades
Ft. Lauderdale
Naples
Big
Cypress
Swamp
Hollywood
Miami
Miami
Beach
Hialeah
Coral Gables
Continental
Shelf
Biscayne
Bay
Florida Keys
Key
West
Marathon
Florida Current
Florida Strait

MAP 11.1

## Florida Settlement

The history of European settlement in Florida begins with the Spanish at St. Augustine in the 16th century. The Spanish (and later, the British) colonial presence in north Florida had few lasting results. The United States purchased Florida from Spain in 1819, the last major area of the eastern United States to be acquired. In 1845, when Florida became a state, it had a population of only seventy thousand. The only large area of continuous settlement lay between the Apalachicola and Suwanee Rivers; it developed as an extension of the South's cotton and tobacco plantation economy.

North Florida would remain distinctly southern: its population came from the southern states, and its political economy was defined by the tensions between plantation-style agriculture and small-scale, backwoods farmers. When they moved on to expand their operations, in the manner characteristic of most development frontiers, north Florida's planters went west to the Mississippi Valley and to Texas rather than continuing the southward drift that had prevailed up to that time.

Peninsular Florida was sparsely settled and remained that way for years. Its flat, swampy landscape offered few attractions to mid-19th-century southerners. Instead, it became a mecca for sun-seeking northerners who had the money to develop it. The foundations for modern-day Florida were laid in the late 19th century when the capital necessary to build the railroads, hotels, and winter-resort communities began to pour into the state. In the meantime there began a series of efforts to change the course of nature in the Florida Everglades by draining and reclaiming the land for agriculture.

## The Everglades

The Seminole Indians, who are sometimes regarded as the aboriginal occupants of Florida, actually derived from several groups of Creek Indians who lived in Alabama until the 1750s. Conflicts between the colonial powers led to schisms within the Creek Nation and eventually to the migration of several Creek bands south into Florida. The British called them Seminoles. British and Spanish authorities were content to leave the Seminoles where they were, but trouble developed quickly once Florida was under the control of the United States. Pressures for Indian removal were intense. A guerilla-style war raged during the late 1830s as the Seminoles resisted government pressure to relocate them to Oklahoma. Some of the Seminoles went west, but others retreated southward and took refuge in the Everglades.

It was the military officers on expeditions against the Seminoles who first recommended that the Everglades' millions of acres of swamp and overflow lands be drained for agriculture. The Kissimmee and Caloosahatchee Rivers had ditching projects in the 1880s. Large dredges deepened waterways, dug drainage ditches, and constructed the levees necessary to channel the flow of water. New channels connected Lake Okeechobee, so that riverboats could reach central

FIG. 11.1. *The conversion of the Everglades from seasonally wet marshland to cropland required the construction of drainage ditches. These ditches border sugar cane fields south of Lake Okeechobee.*

Florida from the Gulf of Mexico. Smaller drainage canals removed excess water from the surrounding lands. By the early 1890s, agriculture had been established in the Everglades. Part of the Everglades drainage scheme involved selling small agricultural plots to new settlers, setting an example for what would become a familiar pattern of land promotion in Florida. Large-scale sugar cane plantations were created around Lake Okeechobee beginning in 1910 (fig. 11.1).

When first drained, the Everglades soils are a deep, black muck—rich in nitrogen with a large component of organic matter in various stages of decomposition. In the United States such soils are known as Histosols. After drainage, and as their organic matter mineralizes, Histosols are well suited for agriculture, although they are prone to shrinkage both from the wind (air drying lowers the water table) and from drainage. They are exceptionally fertile soils for raising vegetables, and the Everglades is one of the largest areas in the United States of agriculturally productive Histosols.

Since the 1920s the principal area of drained Everglades land, south of Lake Okeechobee, has been used for sugar cane and winter-vegetable production. Originally sugar cane was grown here for its value as an industrial fiber, but production was reorganized on a larger scale to supply the crop in its more common use as a sweetener. The "sugar cane parishes" of Louisiana and the Everglades are the two major sugar cane producing areas on the U.S. mainland. The domestic sugar industry has long depended on government price supports and a protective tariff that shields its higher production costs from lower-priced sugar imports.

Winter vegetables such as radishes, eggplant, squash, parsley, celery, and tomatoes also are grown in sizable quantities on Everglades drained land. Production is concentrated near the southeastern margins of Lake Okeechobee and directly west of Miami. Like sugar cane, the winter-vegetable crop suffers from foreign competition (from Mexico, in the case of vegetables). Both sugar cane and winter vegetables demand substantial labor inputs. Despite the long-standing problems of low wages and labor abuse in these Florida industries, profitability in both agricultural sectors is marginal.

The land has been abused here as well. When Histosols are drained of excess water, their bulk begins to shrink and the land surface settles. Subsidence in the vegetable-raising area at Belle Glade on Lake Okeechobee now is more than five feet since the initial drainage—in an area that was only fifteen feet above sea level at the start. Land subsidence has required new efforts to reduce seasonal floodwaters. Sugar cane is burned before harvest to stimulate the flow of sugary juices in the stalks and to remove the leaves and other excess vegetation. Peat in the soils is ignited by the periodic fires, which causes their bulk to shrink. Organic soils that took more than a thousand years to form have been compressed (or burned) back down nearly to the underlying limestone bedrock surface in only seventy-five years.

The flow of water in the Everglades is produced by the slight southward slope of the land surface. Artesian springs to the north feed the system. Current efforts to manage the flow of water began with the construction of new drainage channels and flood-protection levees after World War II, but each new solution has created new problems. The present-day Everglades is much less a natural ecosystem than a massive water-management project directed by the U.S. Army Corps of Engineers.

Lake Okeechobee, which probably originated from the coalescence of several sinkholes in the underlying limestone, was little more than a shallow-water depression before drainage and reclamation. It was a dangerous lake that could whip up sixteen-foot waves when hurricane winds struck. After settlement began, levees were constructed to hold back the surge such storms brought, but they were inadequate. More than two thousand people were drowned in 1926 and 1928 when hurricanes blew Lake Okeechobee's waters out over the surrounding land. The entire lake was then encircled by higher levees to control the hurricane threat, but this prevented the natural flow of water southward from the lake's margin and deprived the Everglades of moisture during the spring dry season.

The Everglades is distinctive partly because it is an open, marshy environment rather than a forested swamp. A high water table discourages tree growth in all but a scatter of isolated islands of slightly higher elevation (known as hammocks) within this "river of grass." Fires were common in the sawgrass marsh during the spring dry season before drainage, and they helped maintain the open landscape, but fires became more frequent and more damaging after drainage ditches were constructed and the spillover from Lake Okeechobee was held back by levees.

About half of the former Everglades, principally the area south and east of Lake Okeechobee, is now devoted to agriculture. An area of about equal size (fourteen hundred square miles) south and west of the lake has been set aside for floodwater retention. The Everglades National Park, created in 1947, represents about 7 percent of the former Everglades, although even this remaining fragment has been much altered by drainage, which turns the grassy wetland into subtropical scrub forest.

A large levee stretches from Lake Okeechobee southward to the edge of the Miami metropolitan area as a protection against Everglades floods that could inundate the east coast. The flood-control problem is matched by an equally serious concern over droughts, which can lower water tables sufficiently to permit saltwater intrusion into freshwater wells of the urban coastal area. The purposes of human intervention in the Everglades have changed markedly over the last century, shifting from drainage for agriculture to flood protection to freshwater supply control for cities. Environmental concerns halted an attempt to create an international air terminal in the Everglades in the early 1970s. But the large population of the adjacent east coast, as well as the steady growth of developments within the state generally, will continue to exert pressure on the Everglades for years to come.



## Urban Growth on the East Coast

Florida's warm winter climate is produced by a fortunate combination of circumstances. January temperatures increase, on the average, about 1°F every thirty miles down the peninsula. The east coast receives additional warmth from the northward-flowing Florida Current, which originates in the still-warmer Caribbean between Cuba and Mexico's Yucatan Peninsula. The warm current flows past the Florida Keys, then turns north along the beaches of Miami, Fort Lauderdale, and Palm Beach before veering out to sea. An abundance of quartz-rich sand—which produces a beach that glistens in the sunlight—was deposited in this zone. A southward drift of the beach sands (opposite to the direction of the Florida Current) originates with river-mouth deposits of sand derived from the hard-rock zones of the Appalachians far to the north.

Henry M. Flagler, who was once a business associate of John D. Rockefeller, was instrumental in attracting wealthy northerners to invest in Florida more than a century ago. He laid his Florida East Coast Railway south from Jacksonville to Palm Beach in 1894 and then extended the line south to the small community of Miami in 1896. Flagler built the elegant Royal Poinciana and Breakers Hotels at Palm Beach and the Royal Palm Hotel at Miami. By 1908 Flagler's various hotels could accommodate forty thousand guests. The heart of Florida's east-coast urban zone was thus created as a product of the winter tourist industry. Because many of the east coast's urban centers—including Miami and Palm Beach as well as Jupiter, Boca Raton, Fort Lauderdale, Delray Beach, and Pompano Beach—began essentially as railroad stations serving resorts, the urban-based tourist industry was spread out over a zone nearly one hundred miles long.

The Great Florida Boom of the early 1920s saw land and housing prices soar as Miami grew and expanded into the new suburbs of Hialeah, Coral Gables, and Miami Beach. When the real estate market collapsed in 1926, Miami and its suburbs entered a period of slow growth. The population then doubled during

FIG. 11.2. *The luxury hotels of Miami Beach rank among Florida's major tourist attractions.*

World War II and doubled twice more by the early 1960s. Miami Beach—a barrier island east of downtown Miami—was transformed into a strip of luxury beachfront hotels (fig. 11.2). Although the early, railroad-centered resort cities concentrated the first waves of growth along the coast, for the past three decades expansion has taken place more along inland corridors served by expressways. The Miami area's growth has substantially leveled off in recent years, however, and its tourist industry has suffered from competition brought by the Orlando-area Disney World development. The former vacation-paradise image has been eclipsed by an image of urban crime and violence. By the 1990s the coastal counties of southeastern Florida contained nearly 5 million inhabitants.

Investors and promoters who began Florida's winter-climate tourist industry in the late 19th century pushed lines of railroad track ever southward down the peninsula, knowing that the winter optimum lay as far south as the tracks could reach. Flagler, who possibly held the greatest ambitions of any of Florida's promoters, built his railroad along the east coast even beyond the limits of land. He constructed a track on pilings across the Florida Keys to Key West in 1912. Hurricanes later destroyed it, but the railroad route was converted into the Overseas Highway during the 1940s.

The reach toward still warmer climates caused Flagler and some of his contemporaries to look even beyond the United States. Cuba, which was taken from Spain by the United States in 1898, also figured in their plans. Although a railway or other land connection between Florida and Cuba was impossible because of the deep waters of the Florida Strait, numerous steamship connections extended the reach of American tourism to Havana. By the 1940s, Cuba was strongly under the influence of U.S. investors and was developing as an extension of the Florida tourist trade.

These connections were severed when Fidel Castro came to power in Cuba in 1958. By then many Cubans had grown accustomed to Americans because of tourism, which helps explain the strong attraction that migration to the United States—especially to the Miami area—held for many Cubans beginning in the late 1950s. The Hispanic population of Dade County (which contains the city of Miami) accounted for only 4 percent of the county's total in 1950, but the proportion has grown steadily until it comprises nearly half of the total at present. Cubans make up roughly two-thirds of the local Hispanic population. Immigration from other Latin American countries has reduced the relative concentration of Cubans during the past several decades, but there has also been a marked second migration to Miami of earlier Cuban émigrés who moved initially to New York City.



South of the Miami metropolitan area, the landscape returns to one based primarily on tourism, with scattered agricultural enclaves. The 150-mile-long arc of elongated islands constituting the Florida Keys is an extension of land within the shallow-water zone of the continental shelf. The Keys were produced by two separate geological circumstances. The more northern islands are the higher ground of a submerged coral reef that grew when sea levels were lower. South of Marathon, in contrast, the Keys are part of a shallow-water limestone formation. Key West, at the southern extremity, is one of the oldest settlements in southern Florida. It was once a center of cigar manufacturing and has been a popular tourist destination for well over a century.

## Florida's West Coast

The west coast of Florida is slightly cooler than the east, lacking the warmth provided by the Florida Current to the east coast. Its beach sands are not as abundant as those along the east coast, because there is no comparable source of sand, nor is there a chain of barrier islands offering continuous ocean frontage and beaches. Nevertheless, southwest Florida developed a winter tourist industry at roughly the same time as the east coast. Henry B. Plant began constructing a railroad from Jacksonville to the small community of Tampa in the early 1880s. The arrival of the railroad brought new hotel construction, and tourists soon began arriving in Tampa. Extensive phosphate beds were discovered in the limestone rock around Tampa Bay in 1883. Phosphatic limestone, which is crushed for producing agricultural fertilizer, was established as a leading local industry. Although the phosphate industry is little known, it is one of the largest businesses in Florida, and it is an integral part of American agricultural production.

Soon after Plant opened the doors of his Tampa Bay Hotel, it became the headquarters for military operations during the 1898 American invasion of Cuba. The growth of Tampa's port facilities was stimulated by the Spanish-American War, and the port was enlarged thereafter as Tampa's role in ocean commerce grew. Tampa also received Cuban émigrés, as well as an immigration of southern Italians, many of whom settled in the Ybor City section of Tampa, where they were employed rolling cigars.

Railroads pushed southward along Florida's west coast after Tampa was founded, reaching Punta Gorda in 1887 and Fort Myers in 1904. No counterpart of the Palm Beach–Miami corridor developed on the west coast, however, because the Gulf side of the peninsula is a series of coastal mangrove swamps separated by river estuaries. Beaches are fewer, and winter resort developments were thereby more isolated on the Gulf Coast side.

Although southwestern Florida is a natural wetland like the Everglades, its vegetation cover consists more of trees than of marsh grass. The largest ecosystem in southwestern Florida is the Big Cypress Swamp, which occupies roughly twenty-five hundred square miles between the Everglades and the coastal fringe south of Naples. It is slightly higher than the Everglades, and better drainage allows a canopy of forest to grow. Isolated patches of higher ground within the Big Cypress region support a growth of tropical and subtropical hardwoods. Palm hammocks are common in this section (*hammock* derives from a Carib word signifying forest growth).

Barron G. Collier, a New Yorker who had made a fortune in the advertising business, purchased nearly 1 million acres of the Big Cypress from timber companies in the 1920s. In the tradition of other northern investors who reshaped Florida, Collier made possible the completion of the Tamiami Trail (U.S. Highway 41) between Miami and Naples in the 1940s, thus creating the first all-weather land route across the southern portion of the state. Collier's heirs still own the majority of the land in this part of Florida; they lease some of it for vegetable or citrus farming.

Coastal mangrove swamps once were common in southwestern Florida (fig. 11.3). The mangrove is not a species but instead denotes several types of semiaquatic tree that produce seeds that germinate while still attached to the tree. The seedlings eventually drop to the mud below and form a dense tangle of roots. The mangrove is an obvious deterrent to coastline erosion, and the sheltered environment offered by its roots is habitat for numerous kinds of aquatic wildlife. But a dense tangle of shrubby growth is incompatible with the type of

FIG. 11.3. *A coastal mangrove swamp, protected as part of a national wildlife refuge, near Fort Myers, Florida.*

beach frontage most land developers want to sell. Southwest Florida's coastal mangrove fringe has been greatly reduced in area because of the demand for ocean-front homesites in the past several decades.

## The Citrus Ridges



Florida's citrus industry, including the production of oranges, grapefruit, and tangerines, has long been associated with the higher beach ridges or inland sand hills of the state. Citrus trees do not perform well in wetlands. Florida's sandy ridges have soils that are practically devoid of natural fertility, but they are well drained and can be irrigated and fertilized to provide ideal conditions for citrus fruit, given the warm winter climate.

The expansion of citrus production in Florida coincided with the growth in national demand for orange juice, especially after the introduction of frozen orange-juice concentrates in the 1940s. Whether the fruit or juice is marketed fresh or as frozen concentrate, rapid access to the national market is critical for citrus-growing areas, and thus Florida's railroad promoters also helped stimulate the state's citrus industry. The orange groves lie more than a thousand miles south of urban markets in the Northeast—much closer than those of California but far enough distant that coordinated, rapid transportation of perishable fruit was required for the industry to be successful. Two areas of intensive citrus production developed in the corridors served by Florida's railroads. One was along the Indian River (which is a lagoon behind a barrier island, not a river) between Cape Canaveral and Fort Pierce on the east coast; the second, a more extensive area, developed in the belts of sandy hills between Tampa, Orlando, and Gainesville.

The long rows of grading and packing sheds that once marked the shipment points for fresh Florida citrus have been replaced by frozen-juice-concentrate plants. In the northern portion of the traditional citrus-growing area, even the crop itself has vanished after a series of freezes in the mid-1980s destroyed more than a quarter of a million acres of orange trees. Urbanization along the east coast has converted many former orange groves to condominium parks. Within the past fifteen years, the Florida citrus industry has migrated southward, away from the danger of winter freezes and toward land less intensively sought by developers. The newest production area is west and south of Lake Okeechobee, where a scatter of sandy uplands provides a suitable environment for the trees to grow.

## Florida Demographics

Florida's average rate of growth has been remarkably constant since statehood in 1845: the state's population has doubled roughly every twenty years since then. Periods of slow growth have separated the booms, yet the long-term trend has remained. The most recent doubling required more than 6 million new Floridians. During the 1980s Florida passed Illinois to become the nation's fifth most populous state, and in the 1990s it passed Pennsylvania to rank fourth. Only California, New York, and Texas are more populous than Florida today.

The state's population is unique in several ways. As a long-term retirement haven for middle-class Americans, Florida has the highest proportion of persons over age 65 (nearly one-fifth) of any state. Its population growth has been based almost entirely on net migration: in a typical year, less than 10 percent of Florida's growth comes from natural increase, the smallest of any state. To a substantial extent, it is Florida's current (rather than past) attractions that cause its population to grow, and its high growth rate has been maintained even as its population has taken order-of-magnitude leaps in size.

Florida receives large numbers of migrants from Caribbean countries and an even larger number come from the northeastern United States, which has long been the most important source of new Floridians. Within Florida, migrants from the Northeast typically move to cities along the east coast, whereas Middle Westerners are proportionally most numerous among migrants to Florida's west-coast cities. Retirees have been most attracted to the west coast and to central Florida. Persons over age 65 account for nearly one-third of the total population in eight Gulf Coast counties centering on Tampa–St. Petersburg. In contrast, none of Florida's thirteen east-coast counties has as much as one-fourth of its population in the over-65 age bracket, reflecting the more diversified economic base and the generally more expensive living conditions of the east coast.

Urbanization itself has a unique expression in Florida. Although the state has twenty-five cities with more than fifty thousand inhabitants, it has no city with a million residents. The municipal limits of Florida's cities never expanded as fast as their populations were growing. Urban growth was dispersed over a scatter of communities fringing the central cities. Until the 1960s, most of Florida's growth was of the low-density, suburban variety that spread the population over a large area. Central and western Florida's heavy concentrations of retirees are indicated by the region's housing stock: more manufactured homes (including mobile homes) are found here than in any other densely settled portion of the United States.

By the 1970s, however, Florida's "house-trailer" image was outdated. A much larger-scale type of planned community was launched by the major development companies in reaction to nationwide concerns over the negative effects of urban sprawl. Developers began to focus on concentrated residential communities in which condominium-type apartments were typical. By the mid-1970s, nearly two dozen planned communities were under construction in Florida, each anticipating an eventual population in excess of fifty thousand. The planned-unit developments were scattered around the state, especially inland from the ocean front along newly dredged "finger canals" that offered water access.

Regional impacts of growth booms have alternated back and forth across Florida in recent decades. The expansion of the aerospace industry created a boom around Cape Canaveral during the 1960s. The construction of the Disney World theme park near Orlando started a boom that led to a doubling of population in the 1970s and again during the 1980s. Since the late 1980s, the Orlando region has become the state's most attractive tourist destination. At the same time, and on opposite coasts, both Fort Myers and Fort Pierce grew at twice the state's average in the 1980s as well. Although growth has slowed along the east coast, resort/retirement areas north of Tampa grew rapidly in the 1990s.



## North Florida

Spain's colonial presence in Florida was limited primarily to a corridor between the Gulf coastal ports and the Atlantic outpost of St. Augustine. Founded by Spanish authorities in 1565, St. Augustine is the oldest city in the United States. Great age does not always mean later prominence for a city, however, as St. Augustine illustrates. The city had an inferior harbor, behind a coastal barrier island, which made it an unlikely place for further developments. It became a "resort of invalids" during the earliest phases of Florida tourism, in an era when a benign winter climate was regarded more as a medical prescription than as a condition for the average person to enjoy.

Jacksonville, which became the major metropolis of northern Florida, also dates from the Spanish period, but it was not platted as a city until 1822. With an excellent harbor near the mouth of the St. Johns River, Jacksonville became an important ocean port and and industrial-manufacturing center as well. Its industries, which have included naval stores, paper, shipbuilding, and chemical products, typify the lower Coastal Plain economy—to which it is most closely tied—rather than the Florida Peninsula. In global-trade terms, Jacksonville lies at the eastern terminus of a strategic land-bridge route between southern California or the Middle West and the east coast of South America.

Between Jacksonville and Tallahassee, the subtropical Florida Peninsula landscape merges with that of the Coastal Plain. Florida's "winter garden" disappears in this zone, where the average number of freeze-free days drops to fewer than 270—a growing season no longer than other parts of the Southeast have. Much of northern Florida's landscape of pine and palmetto is maintained in an open condition because of the herds of cattle that graze there, a practice that dates to Spanish times. Large acreages of pine are a forest crop for the region's paper industry.

## *References*

Anderson, James R. *A Geography of Agriculture in the United States' Southeast.* Budapest: Akadémiai Kiadó, 1973.

Boswell, Thomas D., ed. *South Florida: The Winds of Change.* Washington, D.C.: Association of American Geographers, 1991.

Boswell, Thomas D., and James R. Curtis. *The Cuban-American Experience: Culture, Images, and Perspectives.* Totowa, N.J.: Rowman & Allanheld, 1984.

Carter, Luther J. *The Florida Experience: Land and Water Policy in a Growth State.* Baltimore: Johns Hopkins University Press, 1974.

Craig, Alan K., and Christopher S. Peebles. "Ethnoecologic Change among the Seminoles, 1740–1840." *Geoscience and Man* 5 (1974): 83–96.

Smith, Roger C., et al. *An Atlas of Maritime Florida.* Tallahassee: Florida Department of State, 1997.

Wood, Roland, and Edward A. Fernald. *The New Florida Atlas.* Tallahassee: Trend, 1974.

CHAPTER 12

# The Gulf Coastal Plain and the Alluvial Mississippi Valley

The Gulf Coastal Plain is an almost perfectly concentric series of alternating cuestas (ridges) and lowlands bordering the Interior Low Plateaus, the Appalachians, and the Piedmont (map 12.1). The southward-dipping rock layers comprising each ridge and lowland are exposed in sequence, with the youngest formations at the surface nearest the Gulf. In mid-Alabama the Coastal Plain wraps around and over both the Appalachians and the Piedmont as it bends back toward the heart of the continent, extending as far north as the confluence of the Ohio and Mississippi rivers at Cairo, Illinois. South of Cairo the entire land surface stretching to the Gulf of Mexico is generated by the meandering Mississippi River and its tributaries.

## The Eastern Gulf Coast

The Florida Peninsula disappears into the east-west trend of the Gulf shoreline at Apalachee Bay to become the Florida Panhandle. The coastal zone consists of marshy river deltas and offshore barrier islands extending west to the Mississippi River deltaic deposits of Louisiana. Although the Gulf waters are warm, winter temperatures at this latitude are fifteen degrees cooler than coastal south Florida, which is a major factor limiting winter-season tourism in the Florida Panhandle.

The Panhandle, as well as the narrow, coastal projections of both the states of Alabama and Mississippi, were colonized early by Europeans. Fortified settlements were constructed by the Spanish, the French, and the English during the nearly three centuries that the European powers struggled for control of the Gulf of Mexico. Except for the War of 1812, the Gulf's shores never were considered particularly strategic by the United States. It was not until commercial seaports were developed during the 19th century that urban growth became important to the region.

Biloxi and Mobile were begun by the Montreal-based French between 1700 and 1710; Pensacola was platted by Spanish soldiers a few years earlier. All three passed from Spanish to French and back to Spanish authority prior to the 1763 Treaty of Paris, which awarded control to the British. When the United States acquired the region in 1821, it ended nearly four decades during which no country's laws had demonstrably been in force, so disputed were the claims and ti-

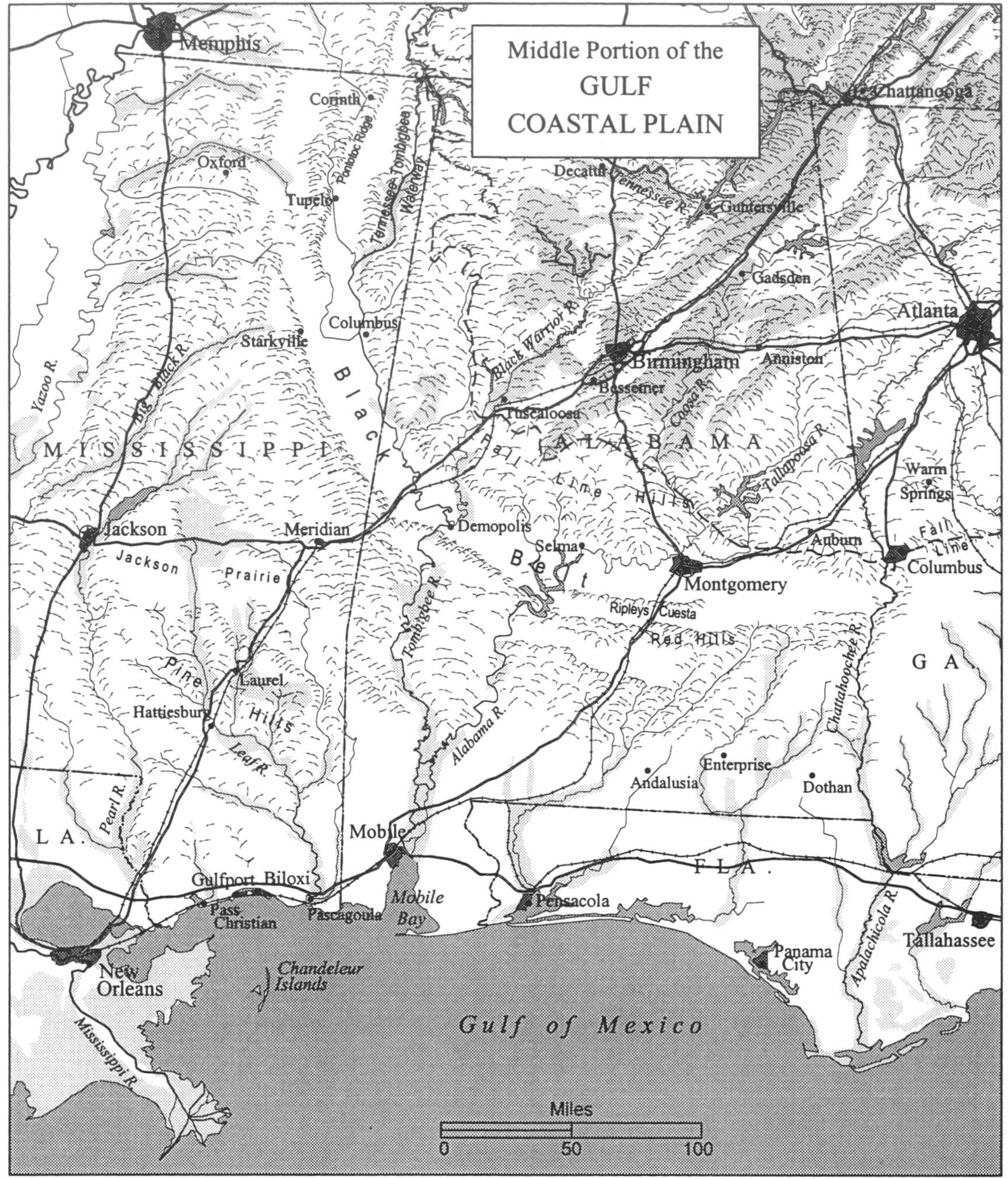


MAP 12.1

tles of various European powers. By then American colonists had moved into the region and established settlements.

Even the larger rivers entering this portion of the Gulf Coast extend no great distance inland. The mouths of the Mobile River, which flows into Mobile Bay, and the Apalachicola (Chattahoochee) River, which enters at the city of Apalachicola, had limited potential as seaports in the era of steamboat commerce.

Only the port of Mobile, which is accessible to most of Alabama by connections to the Black Warrior, Tombigbee, Coosa, and Cahaba Rivers, developed in a manner resembling the great river port of New Orleans. Mobile was a cotton market, a slave market, and a general cargo port during the river era, and it has dominated the external trade of the state of Alabama ever since that time.



The smaller Gulf seaports—Panama City, Pensacola, Pascagoula, Biloxi, and Gulfport—began to grow after railroads linked them to the inland trade. These five, as well as Mobile, became off-loading ports for bananas and other tropical fruit shipped from Central America beginning in the 1900s. The practice continues today, even after the business has shifted to refrigerated trucks. The fruit arrives by ship, it is loaded into trucks, and the trucks begin their northward journeys—all prior to the product's sale. Only after the truckers have traveled some distance north, toward Chicago or Detroit or perhaps Philadelphia and New York, do they learn from a fruit broker who has purchased the cargo and where it is to be delivered.

These same ports have long been the headquarters of the U.S. nearshore shellfishing fleets. Shrimp, scallops, and ocean fish taken in the Gulf of Mexico are brought back to the home ports of the fishing fleet, among which Biloxi and Gulfport are the most prominent. As with tropical fruit, refrigerated truck transportation is used to reach the national market. The general cargo function of Gulf Coast ports is of lesser importance and focuses primarily on paper and wood products or chemicals produced in the region.

Timber exports from Mobile have been stimulated by the opening of the Tennessee-Tombigbee Waterway, which connects Mobile to the Tennessee River, but traffic volumes on the Tenn-Tom have yet to develop to any significance. One of Mobile's major imports is South American iron ore, which is transloaded to river barges at Mobile on its way to steel mills in the Birmingham district. West of the port of Carrabelle, Florida, all of the northern Gulf Coast ports are linked to the Mississippi River system and to the rest of the Gulf by the Intracoastal Waterway.

## The Coastal Plain Piney Woods

Forestry, papermaking, and wood-product manufacturing are the most important industries of the Gulf Coastal Plain, just as forests are by far the most common land cover in the region. These are the famed "piney woods" of the Lower South (fig. 12.1). They are regarded by some as the natural vegetative growth in this region of sandy, acidic soils. Although such limiting conditions generally favor growth of pines (and other softwoods) rather than many hardwoods, the piney woods appear to have evolved as a regional forest type with a good deal of human influence as well.

Travelers who crossed the sandy stretches of pine in the 18th and early 19th centuries reported that the forest was in an open condition. There was little woody undergrowth but rather a luxurious carpet of tall grasses over which pine trees formed a partially open canopy. Wild game abounded and was hunted by

FIG. 12.1. *Two versions of the southern piney woods.* Top: *"quail plantation" pine forest maintained by regular burning, Thomasville, Georgia.* Bottom: *planted pines on former cotton land, Decatur County, Georgia*

the native inhabitants as well as by the recently arrived white settlers. As time passed, the area became more settled, fires were less common, and their effects were more restricted. First the grassy openings disappeared, and eventually so did the pine forest. Without fire, the pines were overtopped by a variety of hardwoods, especially beech and magnolia, which remained as the natural vegetation thereafter.

"Setting the woods on fire" has a long history in the Coastal Plain. Early Spanish settlers grazed cattle in the openings, and so did the Anglos who replaced them. Both groups saw the beneficial effects of fire in maintaining a grassy-pine woodland. Deliberate burning to create browse and enhance habitats was practiced by Native Americans, who realized that it was the best way to increase their supply of wild game. These lessons were relearned in the 20th century when some of the Coastal Plain's abandoned plantations were purchased by northern investors who converted them into hunting preserves ("quail plantations"). Game managers discovered that regular applications of fire enhanced the piney woods, kept out the hardwoods, and created the desired conditions for hunting upland game. The Coastal Plain piney woods is not a natural ecosystem in the sense that it can be maintained indefinitely without intervention.

Southern pine species, such as the longleaf, loblolly, and slash pine varieties, are more resinous than their northern counterparts and were long exploited mainly for their oleoresins in the manufacture of turpentine, paint, tar, and gum. They were considered unsuitable for plywood manufacture until a southern pine plywood mill was constructed at Fordyce, Arkansas, in the 1940s. A wide variety of both hardwoods and softwoods are now used to produce wallboard, paneling, plywood, flooring, fiberboard, and other construction materials. These industries are scattered widely across the Coastal Plain, as are numerous paper mills, which manufacture stock for products ranging from cardboard cartons and paper bags to copier paper.



The decline in agricultural land use in many parts of the former Cotton Belt was followed by the growth of a forest crop on the lands thereby idled. Wood yards, which are local assembly points for trimming pulpwood and saw logs before they are shipped to the mills, now are the focal points of small communities where cotton gins once stood. Because the southern forest industries create a diversified range of products, including chemicals, wood products, and paper, collectively they can exploit almost any type of tree, whether hardwood or softwood, and thus some type of mill can be found nearly everywhere in the Gulf Coastal Plain.

There is little outcry over "endangering" these forests. Perhaps one reason for the lack of concern over forest exploitation in the Coastal Plain is that much of the commercial growth is on land that, less than a century ago, was used for agriculture, especially cotton. The land remained unforested for many decades and sometimes was subjected to severe soil erosion. The South is far more forested and has less soil erosion today than at the height of cotton production.

## The Black Belt

The alternating layers of hard and soft rocks of the Gulf Coastal Plain have a direct impact on land use. Ridges, such as the Fall Line Hills, the Pontotoc Ridge, and the Red Hills, are developed on erosion-resistant sandstones that form a broken, steep-slope topography. Such formations weather to produce inferior, sandy soils that are less suitable for crop agriculture. Most of east-central Mississippi and southern Alabama are marginal for crop production because of such limitations.

These more rugged portions of the Coastal Plain evolved cotton-based economies during the mid–19th century, but because the land was inferior, they were occupied by small-scale farmers more often than by large-scale planters. The lack of large plantations also meant that the Coastal Plain uplands had smaller African American populations. Both socially and economically the small-scale, up-country cotton farmers stood in sharp contrast to the large-scale planters who dominated the more fertile belts where blacks outnumbered whites. The contrast became a major factor in defining local political cultures within Alabama, Mississippi, and Louisiana.

Coastal Plain uplands formed on rock layers that alternate with marine formations that accumulated roughly 100 million years ago and again approximately 35 million years ago when the land surface was lower and the ocean encroached. Among the marine formations are several chalky limestones that erode to produce a smooth, undulating land surface. The chalk layers are almost pure white in their unaltered state; but as they decompose, a heavy, black clay soil forms at the surface.

The most extensive of these layers is the Selma chalk, on which the Black Belt of Alabama and Mississippi is formed. "Black" refers to the color of the soil, not to the color of the people who live there, although the confusion is perhaps not surprising given the development of large plantations in the Black Belt during the mid–19th century. The Black Belt, as well as a similar limestone lowland on black soils known as the Jackson Prairie, were stretches of fertile soil in a region where soils were generally poor (map 12.1).

Cotton planters from the Georgia and South Carolina Piedmont moved west to the Black Belt and Jackson Prairie, where they entered large acreages of public land beginning in the 1830s. The Black Belt had a cover of native grasses that gave rise to the myth that it was a natural prairie, but here, as elsewhere in the South and the East, the prairie lasted only as long as the land was subjected to deliberate burning. It was a landscape like no other the Piedmont planters had known—broad, flat, and for the most part without timber. Planters brought their slaves with them, purchased more in the slave markets of Mobile, and by 1840 had created a new cotton-specialty region. A zone of large-scale cotton plantations, many employing hundreds of slaves each, developed within the crescent of Black Belt land that begins near Montgomery, stretches past Selma, and then arcs northward across northeastern Mississippi before it disappears against the western edge of the Highland Rim in Tennessee.

Black Belt plantations were transformed from the highly centralized operations worked with slave labor that were typical before the Civil War to a more dispersed system based on tenancy thereafter. Beginning in the era of "radical" (northern Republican) rule following the war and continuing through Reconstruction, the old slave cabins were dragged out among the fields, where they served as the one- or two-room tenant or sharecroppers' dwellings. Each tenant house had a patch of garden and a small amount of pasture to keep a cow, both vitally important to a family's food supply. From 1870 until the mid–20th century, the Black Belt was a maze of cotton fields dotted with tenant cabins and subsistence garden patches. Dirt roads connected those who lived on the land to the plantation headquarters as well as to the plantation commissary store, the local cotton gin, and the many religious chapels that were scattered about the countryside. Although not universal, this was the typical landscape inhabited by African Americans who lived in the first half of the 20th century.

The planters and their families often lived in a nearby town; the day-to-day operations on the land were supervised by a white overseer. Although many

FIG. 12.2. *Shotgun houses on Farish Street, Jackson, Mississippi. The shotgun house—one room wide and several rooms deep—was a typical dwelling for tenants on the plantation and also was a common sight in many urban areas of the South.*

plantation headquarters were destroyed during the Civil War, and millions of acres of land changed ownership through tax delinquency proceedings, the plantation itself survived the war. Its layout and method of operation changed to accommodate the shift from slavery to tenancy and sharecropping, but many blacks continued to live on the same plantations where they had been slaves.

Black neighborhoods developed in all of the towns and cities of the Plantation South (fig. 12.2). Although African Americans accounted for more than half the population in some of the towns, it was on the plantation itself where blacks, largely isolated from the world beyond, made up more than four-fifths of the local population. Black people were uncommon and unwelcome in the poorer upland cotton districts, which were identified with the more virulent strains of racism and became strongholds of the Ku Klux Klan. Lynchings of blacks for real or imagined crimes became more common, especially after the 1890s when the oppression of racism grew so intolerable that blacks began to move to northern cities in large numbers.

Life in plantation districts such as the Black Belt was different, both for blacks and for whites. White people constituted a distinct minority in such areas. Black-white relations were constructed around a less violent type of social control that evolved its own mythology based on social separation but which necessarily required physical proximity of the two races. Total disfranchisement of blacks was conceived to be a necessity in the Black Belt and other plantation districts, because allowing the Negro to vote would have meant that whites could not dominate local political and economic affairs.

In the plantation districts, blacks and whites purchased goods in the same stores, but by convention they did so at different times of the day and different days of the week. Schools were totally segregated, with black schools often occupying buildings abandoned by the white school system. Both blacks and whites deposited money in the same banks, but even the small numbers of relatively affluent black people were obliged to stand in the "colored line" when doing so, just as the two races drank from separate drinking fountains, entered restaurants through separate doorways, and rode in different seats on trains and buses.

Racism and segregation thus were not one and the same, and though equally pernicious, each served the interests of different groups of whites. Lynchings and other forms of white-against-black violence were not uncommon in the plantation areas, but planters dominated more through paternalism than by intimidation and violence. Planters looked after what they convinced themselves as being the needs of "their Negroes," who, they rationalized, were thereby freed of the concerns and cares of the white race. Many thousands of African Americans who eked out an existence on the Black Belt's plantations evolved their own ways to cope with these conditions and grudgingly accepted a role in which no challenges to the status quo were deemed advisable.

It was against this set of background conditions that the civil rights struggles of the 1950s and 1960s emerged. Civil rights challenges became more common following the U.S. Supreme Court's decision outlawing segregated schools in 1954. Especially in the plantation districts, whites responded by organizing white citizens' councils to apply economic and social pressure on both blacks and whites to continue rigid segregation. The councils were regarded as "respectable" by local white leaders, who typically avoided identification with the Ku Klux Klan.

When federal courts intervened to enforce school desegregation, a pattern was established in which federal agents and even federal troops were used to carry out antisegregation orders. Dramatic events, such as the Montgomery bus boycott and a series of violent aggressions by local police against mostly younger black people in Birmingham, were portrayed in the national media and started to sway national popular opinion to the side of the African American.

Voting rights had been extended to blacks in many parts of the South by the mid-1960s, and school desegregation was well under way by that time as well, but in the heart of the cotton plantation South, especially in the Black Belt, disfranchisement of Negro voters remained nearly total. Without the vote, black people could not influence the politics of the places they inhabited; and until they gained the vote, they were unable to improve their economic conditions. Dr. Martin Luther King, Jr., was persuaded by local black leaders to focus the attention of his Atlanta-based Southern Christian Leadership Conference on the Black Belt, especially on the community of Selma. Individual challenges to voter registration procedures had produced only a few black voters in Selma and Dallas County before 1965, yet the black population of the area numbered in the tens of thousands.

King brought his strategy of nonviolent confrontation to Selma, anticipating that peaceful rallies of black people would provoke a spectacle of police violence, just as they had in Birmingham. On March 7, 1965, several hundred marchers began walking the fifty miles from Selma to the state capitol at Montgomery, where they planned to petition Governor George C. Wallace for the right to vote. With newsreel cameras rolling and a crowd of reporters looking on, white law enforcement officials suddenly attacked the unarmed marchers with clubs and tear gas on the Edmund Pettus Bridge over the Alabama River in downtown Selma. The march ended there, on what became known as Bloody Sunday, but it was rescheduled and carried out without serious incident on March 21, when King, defying numerous threats of death, led four hundred black and white marchers to Montgomery.



The television coverage of Bloody Sunday in Selma galvanized popular opinion nationwide and led to widespread demand for an end to black disfranchisement. The incident is credited with leading directly to congressional passage of the Voting Rights Act of 1965, which gave blacks the right as citizens to cast a ballot and help determine and operate the government under which they lived. Black voter registration began immediately. The number of black elected officials in the southern states increased from 72 to 2,968 in the two decades following passage of the Voting Rights Act.

Cotton had been on the decline in the Black Belt for decades before these events took place. Fields once used for cotton were used to plant soybeans, converted to pastures, or became tree plantations. The decline, similar to that in other old plantation districts farther east, accompanied the gradual westward shift of cotton production to Texas and California. The legacy of the plantation—and of slavery and tenancy as well—remains in the map of black population in the South, however (map 12.2). Even in the 1990s, the counties with the largest concentrations of black people are those where plantation agriculture was most important: the tobacco-raising areas of Virginia and eastern North Carolina and the cotton districts, including the Inner Coastal Plain and the Lower Piedmont of South Carolina and Georgia and the Black Belt of Alabama and Mississippi.

Nearly all of the other counties with a large percentage of African Americans in the total population are in the alluvial Mississippi Valley between Memphis and New Orleans. These counties fringing the Mississippi River also developed plantation economies, either before or after the Civil War, but unlike the older cotton districts of the Southeast, they are still major cotton producers today. It was primarily in the Mississippi Valley that the cotton plantation underwent yet another modification, beginning in the 1950s, when the production system shifted from tenancy and labor-intensive operation to a system based on mechanization and large capital inputs per acre. It was here that the modern "neoplantation" was developed.

## The Alluvial Lands of the Mississippi Valley

An irregular line of bluffs marks the perimeter of the Mississippi River alluvial lowlands from southern Mississippi north through Tennessee and Kentucky to southern Illinois, then (counterclockwise) back down the western side of the floodplain through Missouri, Arkansas, and Louisiana. The lowland is defined on the basis of sediments that are of Pleistocene or more recent age, which in

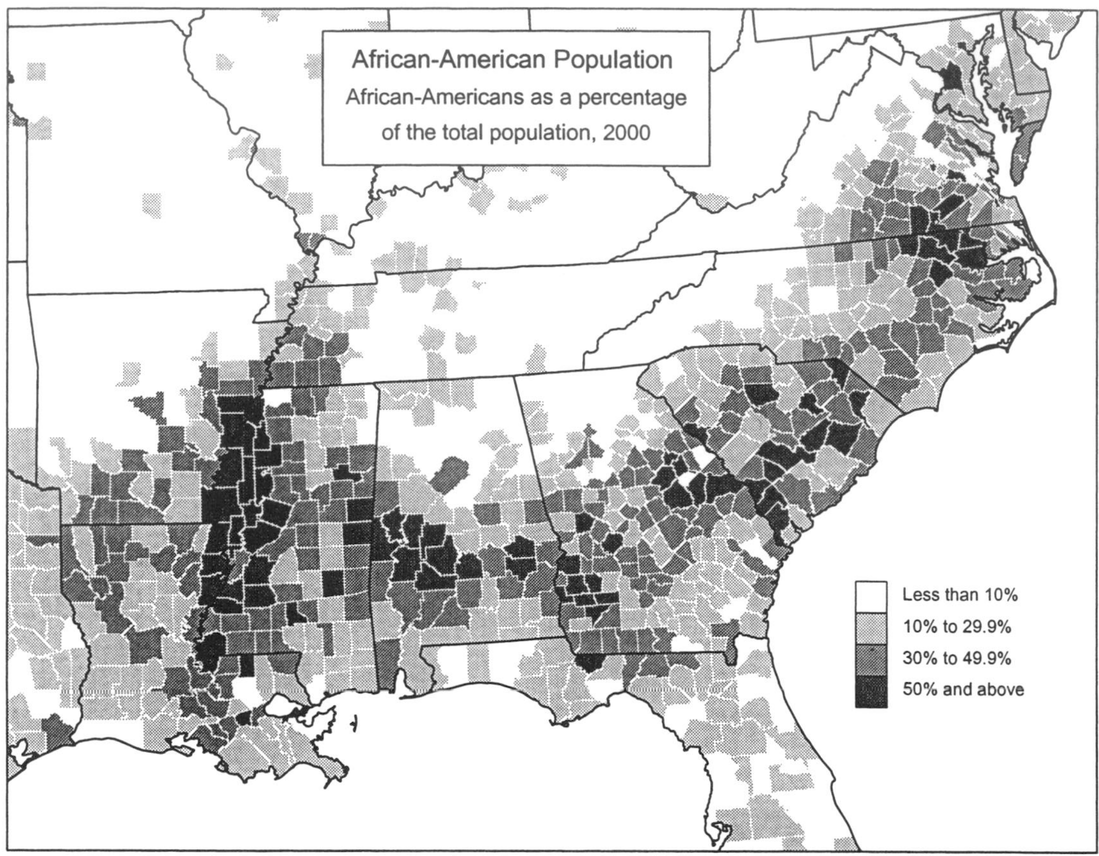


MAP 12.2

this region means nearly the entire land surface. The alluvial valley south of Cairo covers an area of roughly fifty thousand square miles (map 12.3).

The upper basin of the Mississippi River (north of St. Louis) was glaciated at least five times during the Pleistocene. Each time the glaciers retreated, rivers carried meltwaters full of sand and silt away from the glacial margins. The discharge flowed southward through the various drainageways of the Mississippi River and on toward the Gulf of Mexico. The current confluence of the Ohio and Mississippi rivers at Cairo did not exist during the earlier glaciations. The Tennessee River occupied roughly what is now the channel of the Mississippi River as far south as Memphis; the Ohio River ran along the eastern margin of the upland known as Crowley's Ridge in Arkansas; and the Mississippi flowed on the western edge of Crowley's Ridge, beyond which it joined the Ohio near the present city of Helena, Arkansas.

The rearrangement of rivers since that time is only one of numerous changes as the midcontinent's streams adjusted to periodic floods that altered their courses, cut new channels, and built new floodplains. Nearly all of the older Coastal Plain rocks south of Cairo were buried under successive layers of silt and sand deposited by the meandering rivers. Because sea levels decreased dur-

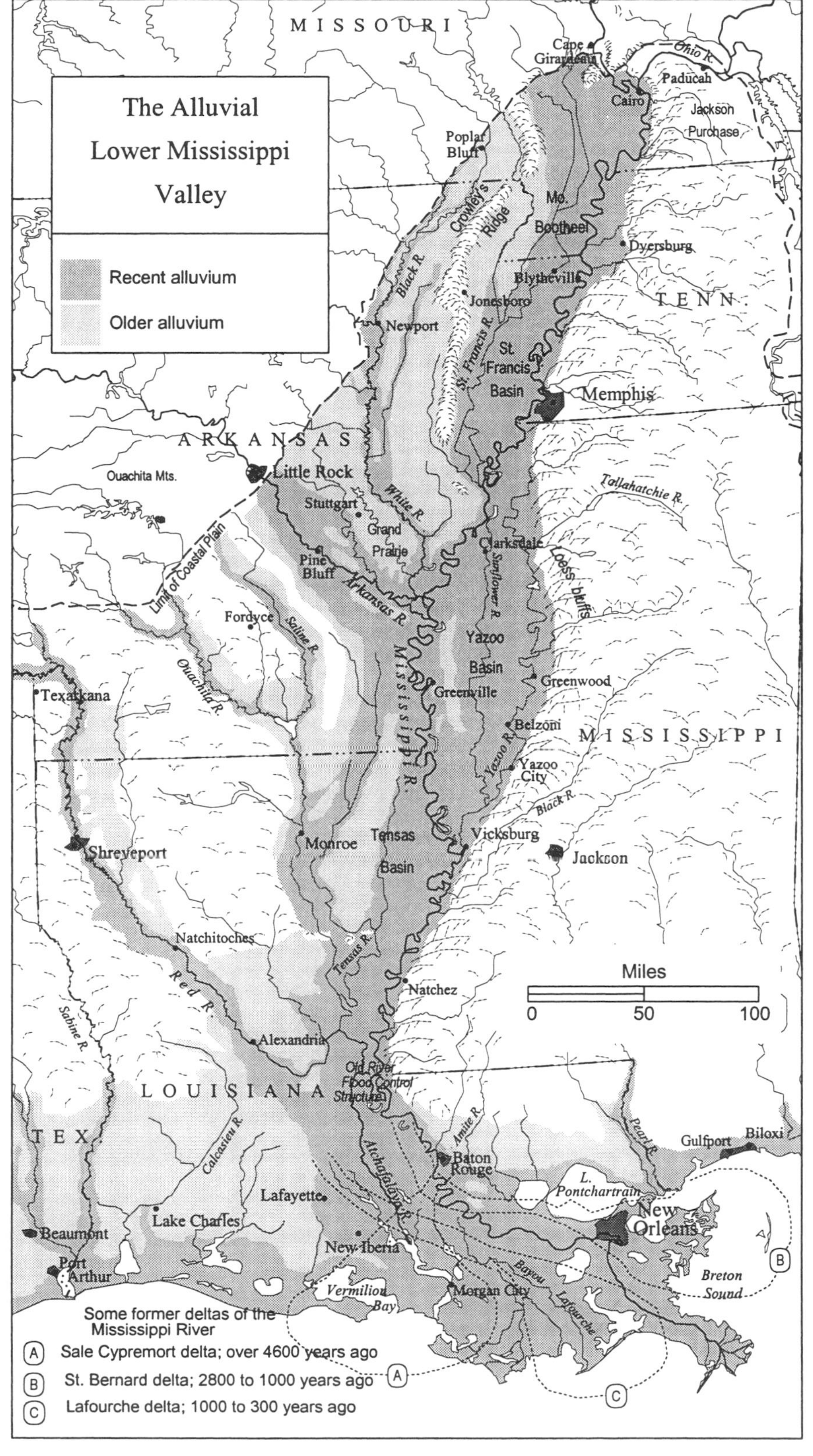
The Alluvial Lower Mississippi Valley
Recent alluvium
Older alluvium
MISSOURI
ARKANSAS
TENN.
MISSISSIPPI
LOUISIANA
TEX.
Cape Girardeau
Paducah
Cairo
Jackson Purchase
Ohio R.
Poplar Bluff
Crowley's Ridge
Mo. Bootheel
Dyersburg
Blytheville
Black R.
Jonesboro
Newport
St. Francis R.
St. Francis Basin
Memphis
Ouachita Mts.
Little Rock
Stuttgart
Grand Prairie
White R.
Tallahatchie R.
Clarksdale
Sunflower R.
Loess bluffs
Pine Bluff
Arkansas R.
Limit of Coastal Plain
Fordyce
Saline R.
Ouachita R.
Mississippi R.
Yazoo Basin
Greenwood
Greenville
Texarkana
Belzoni
Yazoo R.
Yazoo City
Black R.
Vicksburg
Jackson
Monroe
Tensas Basin
Shreveport
Natchitoches
Tensas R.
Natchez
Red R.
Miles
0
50
100
Sabine R.
Alexandria
Old River Flood Control Structure
Calcasieu R.
Amite R.
Pearl R.
Gulfport
Biloxi
Atchafalaya R.
Baton Rouge
L. Pontchartrain
Lafayette
New Orleans
Lake Charles
Beaumont
New Iberia
Bayou Lafourche
Port Arthur
Vermilion Bay
Morgan City
Breton Sound
Some former deltas of the Mississippi River
A Sale Cypremort delta; over 4600 years ago
B St. Bernard delta; 2800 to 1000 years ago
C Lafourche delta; 1000 to 300 years ago

MAP 12.3

ing every glacial episode, then increased again when the glaciers melted, the profiles of the Mississippi River and its tributaries underwent numerous fluctuations, alternating between episodes of erosion when the gradient steepened and deposition when it slackened.



Although the landforms created in these episodes are difficult to interpret on the ground, anyone who flies over the Mississippi Valley south of Cairo can see that the land surface is a maze of former river channels, terraces, backswamps, and natural levees. The main river and all its tributaries continue to meander—and to cut off old meanders just as new ones are created—although today the activity is confined to a zone within a few miles of the rivers. At various times during the Pleistocene, the rivers meandered across the entire alluvial valley.

A secondary result of the creation of vast floodplains during deglaciation was the deposition of silt along the rivers' banks. Winds blew these fine-grained sediments off the floodplains and deposited the material as layers of loess on the fringing bluffs. Loessial soils are generally fertile, and even though loess deposits are prone to gullying, the basis for prosperous crop agriculture is often associated with such areas of wind-blown deposits. A belt of good cropland occupies the high ground along the eastern edge of the Mississippi Valley. Most of it is based on loessial soils deposited by the prevailing westerly winds.

Cotton was once an important crop in more than a dozen counties north of Memphis before the emphasis shifted to soybeans in the 1950s, and in recent years cotton has returned. Loess also covers most of the surface within the eight Kentucky counties west of the Tennessee River. This is the Jackson Purchase, named after Andrew Jackson, who arranged its purchase from the Chickasaw in 1818. The Jackson Purchase contains some of the best agricultural lands in the state of Kentucky and is used now mainly for soybean production (see chapter 13, map 13.3).

The essential fact needed to understand the alluvial Mississippi Valley is its flatness. From Cairo, which has an elevation of approximately 300 feet above sea level, the river meanders more than a thousand miles to the Gulf of Mexico, a gradient of only about four inches per mile. The profile becomes even flatter toward the south, but the volume of flow increases as more tributaries contribute their waters to the Mississippi. The volume of water increases from 67 million acre feet per year just north of St. Louis to 330 million at Memphis and 450 million north of New Orleans. The lower Mississippi River is an enormous sheet of water flowing across a barely sloping surface.

Periodic overbank flooding of the river spreads sediments across the floodplain, usually in greatest abundance along the banks of the main channel. A low ridge, known as a natural levee, is produced, which has a relatively steep slope on the river side but a long, gentle slope across the floodplain. Natural levees form the highest elevations within the alluvial plain. Because the main channel is bordered by natural levees, the flow of tributary streams that drain the floodplain is directed away from the main channel. Local drainage typically follows the low-elevation backswamps at the outer limit of the floodplain. These tributaries, known as "yazoo streams," after the Yazoo River in Mississippi, flow parallel to the main river for miles before entering it. When the Mississippi River floods, the main channel's waters thus back up into the tributary streams, flooding first the backswamps and then, if the volume is sufficient, the entire floodplain.



## The Yazoo Delta

The fourteen counties of northwest Mississippi form an oval-shaped lowland stretching from Vicksburg to Memphis that is popularly known as "the Delta," even though the area is not the delta of a river. The Delta is part of the Mississippi's floodplain, and it is drained by the Sunflower, the Tallahatchie, and numerous smaller rivers. All of those, in turn, are tributary to the Yazoo River, which joins the Mississippi just above Vicksburg. The drainage accomplished by the Delta's rivers is only seasonal at best; and until the 1880s, most of the Delta was seasonally wet everywhere except on the higher ground of the natural levees. Before artificial drainage began, the Yazoo Delta—like most of the alluvial Mississippi Valley—was a cypress and tupelo-gum swamp ruled by snakes and black bears.

Large tracts of cypress land were acquired for a few cents an acre by northern lumber barons under federal swamplands acts during the period of financial chaos that prevailed during Reconstruction. Because the cypress tree has some aerial roots that enable it to survive even when inundated by standing water, it was the typical trcc of swamps and bayous throughout the lower Mississippi Valley, from Cairo to the Gulf (fig. 12.3). Lumber companies dug drainage ditches to make the trees more accessible for cutting, which, in turn, dried the land and removed the cypress's advantage over competitors. Cypress trees still grow in the undrained sections of the Mississippi Valley, but what remains is perhaps less than 5 percent of the former swampy floodplain forest.

FIG. 12.3. *A cypress swamp in the alluvial Mississippi Valley, Lake Providence, Louisiana.*

Once the Delta's lands had been logged, they were sold—again, at low prices—to cotton planters who dug more drainage ditches and expanded their operations away from the high ground where they had previously been confined. The cotton crop of the Yazoo Delta increased decade by decade into the 1920s. Every county in the Delta had a majority black population until at least the 1950s, and many counties were more than three-fourths African American. The Yazoo Delta became the major cotton region of the South, and it contained the heaviest concentration of black sharecroppers and tenants in the United States until the effects of mechanization were felt beginning in the 1950s.

The Delta's trade-center towns—Greenville, Greenwood, Clarksdale, Indianola, Belzoni, and Yazoo City—are busy centers of cotton ginning and cotton oil pressing, but they are only outposts of its "capital city," which has always been Memphis, the major trade center of the region. The rise of blues, rock and roll, and other musical styles in Memphis—including the cross-influences of each on the other—can be attributed to the proximity of the Delta and other nearby cotton districts where both blacks and whites worked and lived. At various times W. C. Handy, B. B. King, Elvis Presley, and Johnny Cash—all of whom grew up in cotton country—brought their distinctive styles of music to the nightclubs and recording studios of Memphis's Beale Street and thereby changed the course of popular culture in the nation.

Although Memphis was built largely on the loess bluffs above the Mississippi's floodplain, it was the river and its commerce that caused the city to grow. It was the major cotton market for the inland South during the era of river commerce, its central location helped make the city a railroad center, and it is now an important air-traffic hub for companies specializing in overnight parcel delivery across the country. Both Vicksburg and Natchez were similarly built on the fringing river bluffs; yet because both of those cities functioned more as satellites of New Orleans, they never attained the size or regional prominence of Memphis. All three amassed wealth from their role in the cotton trade, especially the trade of the Delta.

The continued success of plantation agriculture in the Delta during modern times was made possible by a shift to mechanized cotton planting and harvesting and by a diversification into soybeans, catfish farming, pecans, cattle, and pulpwood on the plantations. The typical Yazoo Delta "neoplantation," like those on the Arkansas side of the river, is an agribusiness enterprise of several thousand acres. Croplands are worked by both white and black employees, who operate a variety of heavy equipment on the flat Delta land. Many plantations contain individual fields of a thousand or more acres. Telecommunications and computers were introduced early into these highly mechanized farms. The shift to catfish production (for fish fillets, fertilizer, fish meal, and by-products) took place during a nationwide decline in the demand for cotton in the 1960s, when polyester fabrics became fashionable. Now that the demand for cotton has returned to levels even beyond those of the prepolyester era, some Delta planters regret flooding so much of their land for catfish ponds. For the time being, at least, cotton is resurging in the Delta.

FIG. 12.4. *Springtime in the Missouri Bootheel. When allowed to flood, the Mississippi River spreads its waters—and its alluvium—over a wide area.*

## The Missouri Bootheel and Eastern Arkansas

A similar sequence of land use—from swamps to logging to cotton fields—took place west of the Mississippi River at roughly the same time that the Yazoo Delta was being developed. The St. Francis Basin, another oval-shaped lowland, covers the bootheel portion of Missouri and the northeastern counties of Arkansas. As a flatwoods swamp, it contained dense stands of hardwoods prior to settlement. The first towns were tiny clearings within an otherwise almost continuous forest. After logging, drainage, and conversion of the land to cotton and soybean fields, the St. Francis Basin is now almost totally without trees (fig. 12.4). Most of the trees are in the towns that once represented the only clearings.

The St. Francis Basin is bordered on the west by Crowley's Ridge, which is a loess-mantled upland belt extending discontinuously from near Cape Girardeau, Missouri, to Helena, Arkansas. Crowley's Ridge is the largest pre-Pleistocene "island" in the alluvial valley, and its surface is hilly. Although the Mississippi once flowed along its western edge and the Ohio on its eastern, Crowley's Ridge remained above the level of floods.

The White River of Arkansas—and its tributaries, the Black and Cache Rivers—drain the area west of Crowley's Ridge, including the backswamp of the larger alluvial floodplain. Much of the White River Basin is slightly more elevated and thus required less drainage to convert it to cropland, although here, too, the land surface is so flat that land-leveling machines must be employed to create a uniform slope to permit drainage. Despite the efforts to drain these lands, and even though the area receives sufficient precipitation for crop production, large tracts of land require supplementary irrigation. The moisture-retention qualities of the alluvial soils are poor in some areas, or the soils are so dense that plant roots cannot extract water from them. The alluvial Mississippi Valley is the most intensively irrigated farming region of the United States east of the Great Plains.

The elevated terrace (or "second bottom") between the White and Arkansas Rivers is known as the Grand Prairie of Arkansas. It was in a grassy condition, almost totally devoid of heavy timber, when white settlers first encountered it. German-Americans who had lived in southern Illinois moved to Arkansas's Grand Prairie and began to grow rice after initial experiments in the late 1890s proved that the crop would grow there. By 1910 the area around Stuttgart, Arkansas, had become specialized in rice production. The heavy clay soils were an advantage for rice because they allowed farmers to hold standing water in the fields longer. The easing of federal acreage restrictions on the crop during the 1970s led to a major expansion of the growing area. Rice is now produced throughout eastern Arkansas.

Fields of cotton, soybeans, and rice continue southward, past the mouth of the Arkansas River, into northern Louisiana. The lowland is part of the Tensas Basin. Neoplantations also dominate the agricultural economy in this section of the river. Blacks outnumber whites in many communities, and both are employed on the plantations. The sharecroppers' shacks that once were ubiquitous on the plantation landscape have all but disappeared and have been replaced by a variety of rural housing subdivisions for plantation workers.

## The Mississippi River Distributaries

Because the gradient on all flowing streams continues to slacken toward the south, bayous and swamps covered in standing water for at least part of the year become more prevalent south of Natchez, where the alluvial Mississippi Valley reaches its narrowest constriction. As more rivers enter the Mississippi and thereby increase its volume of flow, the area inside the river's natural levees continues to increase as well. Artificial levees line the entire lower Mississippi River—atop the natural levees—with only occasional breaks at river junctions or designated floodways. The preoccupation with drainage of wetlands for agriculture in Arkansas and Mississippi is replaced with a necessary concern for flood protection in Louisiana as the great river's flow nears the Gulf of Mexico.

The Mississippi River has built several delta lobe complexes in the Gulf of Mexico during the past 6,000 years as it has meandered back and forth across Louisiana (map 12.3). The present delta, southeast of New Orleans, is only about 500 years old. For at least 500 years before that, the Mississippi discharged through Bayou Lafourche, directly west of New Orleans. Each shift in the delta has responded to a shift in the river's course farther upstream. Seeking the easiest path to the ocean, the river finds a more efficient route, cuts a new channel, and begins a sequence of erosion and deposition. When sufficient sediment builds up in the new channel, and its total length has been increased by meanders, the river once again has to perform an excess amount of work to deliver its volume of flow to the Gulf, the meander is cut off, and the cycle begins again.

Once the river cuts a new path, the old channel rapidly accumulates silt and becomes unsuited for navigation by the types of vessel that now ply the Mississippi. The river-based industries, ports, and flood-protection levees that line the Mississippi River between Baton Rouge and New Orleans today represent an investment of many billions of dollars. More than five dozen oil refineries, petrochemical factories, and grain elevators crowd both banks of the river between Louisiana's state capital and its largest city. Ocean tankers deliver imported petroleum here for refining into gasoline. Apart from the Texas Gulf Coast, it is also the largest concentration of chemical factories in the United States. The New Orleans–Baton Rouge district is now the largest grain-shipping port in the nation, where barges loaded with corn and soybeans grown in the Ohio, Missouri, and Mississippi Valleys are emptied and their cargoes reloaded into ocean ships for export around the world.



Although an eventual shift in the lower Mississippi's course is virtually certain, given the river's past history, it would be an economic and political disaster for New Orleans and the Baton Rouge industrial district. For the past five decades, the U.S. Army Corps of Engineers, which has responsibility for flood control and navigation here as elsewhere on the nation's rivers, has sought to prevent the inevitable from happening.

About five hundred years ago, a westward-meandering loop of the Mississippi broke into the basin of the Red River about fifty miles south of Natchez and captured its flow. Simultaneously it intersected the Atchafalaya River, a distributary of the Red. In 1831 an artificial cutoff channel was dug in the Mississippi just east of the point of capture. The lower limb of the old meander bend, named the Old River, thus became bidirectional depending on whether the Red River or the Mississippi was in flood. The Mississippi began to flow partially down the Atchafalaya River, entering the Gulf at Morgan City, Louisiana, which decreased the river's size at New Orleans. The Atchafalaya route was shorter (142 miles compared with 315), and by the 1950s it became clear that the Atchafalaya would soon become the new course of the Mississippi unless preventive measures were taken.

The Corps of Engineers' tactic was to build a new channel, paralleling the no-longer-functional upper limb of the meander bend, that would divert part of the Mississippi's flow down the Atchafalaya so that the total flow of the Red and Mississippi would be split—30 percent for the Atchafalaya and 70 percent for the Mississippi. The south limb of the old meander was plugged, and a navigation lock was constructed paralleling it to permit barges to enter the Red River as part of the Old River Flood Control Structure. Floodgates and spillways constructed to implement this plan went into operation in 1963.

During the spring flooding of 1973 the structure regulating the intake from the Mississippi was seriously undermined and, although repaired, was further assaulted by floods in 1974, 1975, and 1979. A second, auxiliary channel was constructed in 1981 to permit an additional outlet for the Mississippi into the Atchafalaya. Although the entire complex of control structures surrounding the Old River has been adequate to divert flood waters since that time, the Atchafalaya continues to deepen and flow at a faster rate. This probably hastens the arrival of the date when the Mississippi River, in some still-larger future flood, undermines the formidable earthworks and concrete walls around the Old River and diverts all of its flow down the Atchafalaya.

## Lower Louisiana



The city of New Orleans occupies a strip of natural levee land at the edge of the Mississippi River. No part of the city has an elevation greater than 20 feet above sea level, and massive levees line both the Mississippi and the margins of Lake Pontchartrain as protection for the city. Although its swampy, malaria-infested site inside a broad bend of the river could not have been particularly attractive to the French who founded it in 1722, New Orleans became the principal French settlement of the lower Mississippi. By the time control passed to the Spanish in 1763, the original city—subsequently known as the French Quarter—had become a thriving town. Plantations developed on the natural levees both north and south of the city. New Orleans grew even more rapidly after it became an American city as part of the Louisiana Purchase in 1803. Swamps were drained and new suburbs were added to the old city, flanking the French Quarter and hugging the bends of the riverfront.

The substantial growth of New Orleans during the first half of the 19th century came from its role as a port where cargoes were unloaded from riverboats and then reloaded, either for export or for shipment on coastwise vessels to the eastern seaboard of the United States. The city's trade was stimulated by the exports not only from the plantation districts of the Lower South but also, in the era before railroads, of the agricultural produce of nearly all of the Middle West. When railroads began to reorient the trading patterns of the Corn Belt after the 1850s, the New Orleans traffic from the Middle West plummeted, although it was revived more than a century later, when grain began to move downriver to New Orleans for export once again.

Nearly a century of economic stagnation following the Civil War led New Orleans entrepreneurs to search for a new means of economic support. Beginning in the 1930s, New Orleans became the first large American city to see its future primarily in terms of preserving its past. The rows of balconied wooden buildings lining the streets of the French Quarter had been regarded as a symptom of economic stagnation and thus appropriate for clearance to make room for new businesses and industries. With the growth of tourism, the French Quarter was seen as an invaluable treasure that people from outside the city would come to view and appreciate. Apart from its role as a petrochemical center and a seaport, New Orleans has relied increasingly on tourism to support the local economy for the past five decades.

Historically, New Orleans was also the focus of trade from outlying settled districts within Louisiana. Plantations covered the high ground along the Red River from Alexandria to Natchitoches and Shreveport by 1860. The Red River Valley in Louisiana is flat and swampy like the Mississippi and also became an important cotton-growing region. Like the Yazoo Delta, it is also a region where the crop has made a comeback in recent years. Shreveport, an assembly point

FIG. 12.5. *A version of the Creole cottage, Natchitoches Parish, Louisiana. The house follows the early French* bousillage *style of half-timber and mortar construction. The massive, low-pitch roof and extensive porches can be interpreted as climatic adaptations.*

for traffic in petroleum and chemicals moving from the Gulf Coast to the mid-continent region, is the Red River's major city and the northern limit of barge navigation on the Red.

South of Alexandria the growing season is long enough and the winters sufficiently mild to permit the growing of sugar cane. The crop is produced in more than twenty parishes (counties) in a triangular region extending from Alexandria south to the Gulf Coast and east to New Orleans. Only a fraction of the land within this triangle is dry enough to permit crop production, however; and ever since the first plantations were established in the 18th century, agricultural activity has been confined to the higher ground of the natural levees bordering Bayou Teche, Bayou Lafourche, and the Mississippi River. The eastern and western segments of the sugar cane region are separated by the Atchafalaya Basin, most of which is a designated floodway and is covered with standing water for a portion of the year.

Sugar cane was introduced here by the French, who brought its culture from the West Indies in the mid–18th century. Slaves were imported to work in the cane fields; that led to a direct transfer of African–West Indian people and customs to Louisiana. Louisiana's Creole tradition, meaning a blending of European and African influences, grew out of this experience (fig. 12.5). The original French-stock population came largely from the West Indies, although many who arrived in the 18th or early 19th centuries either had lived in Acadia (Nova Scotia) or were derived from Acadian families. *Cajun,* the corruption of *Acadian,* derives from this ancestry and thus has a different meaning from *Creole.*

Drier lands west of the sugar cane parishes also were settled by Acadians. Rice culture was introduced into southwestern Louisiana slightly before it made its appearance in Arkansas, in the late 19th century. Although many Cajuns became rice farmers, the principal group associated with its introduction to Louisiana were Middle Western farmers, especially German-Americans, the same as were associated with the early rice industry of Arkansas. The coastal prairies of Louisiana proved to be well suited to rice production, as were similar lands in the adjacent coastal fringes of Texas.



## The Coastal Zone

Freshwater streams deposit sediments over a broad area as they enter the Gulf of Mexico. The Mississippi River's succession of deltas accounts for most of the deposition east of Vermilion Bay, but many smaller streams have contributed land-forming sediments to the coastal fringe as well. These processes have left no fixed land-water boundary but rather have created zones of increasingly less firm ground up to the final limits of land. French settlers called the outermost of these environments *flotant* (trembling prairie), a marshy quagmire of matted, decomposing vegetation floating on water. The coastal margins west of Vermilion Bay include sandy ridges that are firm enough to allow trees to take root. This is the chenier plain (*chene,* meaning oak), which is a more terrestrial than aquatic environment.

Left undisturbed, *flotant* accumulated through the natural processes of silt deposition and vegetative growth at the land's outermost margins. The only human inhabitants of such areas were fur trappers, who dug a unique type of canal through the marsh, known as a *trainasse,* to permit navigation with a pirogue, the traditional marshland dweller's canoe that was propelled with a push-pole. As the practice of confining seasonal floods within the limits of artificial levees became more common, less silt was available to spread over the coastal marshes. The accumulation of land-forming materials ceased, and the coastal zone was open to increased erosion. As ocean waters encroached and the old trainasses widened, what had been a mixed land-water environment became open water.

The exploitation of oil and gas deposits in salt-dome structures at the Gulf's margins brought a new threat of destruction to the coastal marshlands. Motorized land-water craft are favored for exploration and well-drilling activities. Although oil was discovered in coastal Louisiana early in the 20th century, production did not reach appreciable levels until the 1940s. The first offshore wells were drilled in 1947, which stimulated the growth of the drilling industry at Morgan City and other coastal towns. By the 1970s tens of thousands of wells had been drilled, and new canals had been dug through the marshes to provide access to them. By the 1990s Louisiana ranked third or fourth in oil and gas production, and although part of the production takes place in the northern portion of the state, the coastal zone is the major focus of activity. The development of hydrocarbon resources also attracted refineries, drilling supply bases, and various oil-field support industries to the area.

The combined effects of the cessation of overbank flooding of rivers, the construction of canals, and the oceanic encroachment into former land areas created a massive land-loss problem in coastal Louisiana. Awareness of the problem led to new restrictions on exploitation of coastal wetlands in the 1990s. The rate of land loss has been slowed, but it has by no means ceased. Predictions that New Orleans will be confined to a narrow neck of natural levee land protruding into the Gulf of Mexico by the year 2020 now seem excessively dire, but Louisiana's coastal zone remains endangered. Combined with the ongoing problem of how to channel the Mississippi River's flow across southern Louisiana, a significant reconfiguration of land and water bodies within the region must remain a possibility.



## *References*

Aiken, Charles S. *The Cotton Plantation South since the Civil War.* Baltimore: Johns Hopkins University Press, 1998.

Brandfon, Robert L. *Cotton Kingdom of the New South: A History of the Yazoo Mississippi Delta from Reconstruction to the Twentieth Century.* Cambridge: Harvard University Press, 1967.

Chestnut, J. L., Jr. *Black in Selma.* New York: Farrar, Straus & Giroux, 1990.

Comeaux, Malcolm L. *Atchafalaya Swamp Life, Settlement, and Folk Occupations.* Baton Rouge: School of Geosciences, Louisiana State University, 1972.

Davis, Donald W. "*Trainasse.*" *Annals of the Association of American Geographers* 66 (1976): 349–59.

Fisk, H. N. *Geological Investigations of the Alluvial Valley of the Lower Mississippi River.* Vicksburg: U.S. Army Corps of Engineers Mississippi River Commission, 1944.

Komarek, E. V. "The Role of the Hunting Plantation in the Development of Game, Fire Ecology, and Management." In *Proceedings, Tall Timbers Ecology and Management Conference,* no. 16, 167–88. Tallahassee: Tall Timbers Research Station, 1982.

Lewis, Peirce F. *New Orleans: The Making of an Urban Landscape.* Cambridge, Mass.: Ballinger, 1976.

Percy, William Alexander. *Lanterns on the Levee: Recollections of a Planter's Son.* New York: Alfred A. Knopf, 1941.

Rehder, John B. *Delta Sugar: Louisiana's Vanishing Plantation Landscape.* Baltimore: Johns Hopkins University Press, 1999.

Whaene, Jeannie, and Willard B. Gatewood, eds. *The Arkansas Delta.* Fayetteville: University of Arkansas Press, 1990.

*PART V*

# The Middle West

## CHAPTER 13

# The Corn Belt

Between southern Ohio and eastern Nebraska lies North America's most productive agricultural region, the Corn Belt (map 13.1). Although agriculture is more intensive in some irrigation districts of the West, no other large area can compare with the Corn Belt's productivity. The Corn Belt is also among the oldest of the continent's agricultural regions. It emerged rapidly in the process of westward settlement expansion during the mid–19th century when, in barely four decades, corn-livestock agriculture advanced from the Scioto Valley of Ohio to the plains of Nebraska (map 13.2).

The term *Corn Belt* refers not only to a region where corn is grown but also to the primary use made of the crop, fattening meat animals for market. Automobile fuel and soft-drink sweeteners are among the many products made from corn today, but historically corn has been consumed by hogs and cattle on the farms where it is grown. Despite the new industrial and food uses and the substantial exports of corn to other countries in recent decades, livestock still consume the bulk of the annual corn crop, which is measured in the billions of bushels.

Corn (maize) is a New World crop. Native people from Canada to southern South America demonstrated maize culture to numerous exploring parties of Europeans, who thereby learned of corn's many varieties and uses and soon began planting it themselves. One of the maize cultivars that spread northward from Mexico during the past several millennia was a productive type that had many rows of dented kernels on a short, thick cob. Known as dent corn, it was first described in Tidewater Virginia in the 17th century. It spread inland as seed carried in sacks, part of the baggage of pioneer settlers who were advancing through the Great Valley and eventually across the Appalachians. Genetically, dent corn crossed freely with the numerous other varieties of "Indian corn," such as the flint types grown in New England and elsewhere.

Dent corn was brought to the Ohio Valley by early settlers who established farms in southern Ohio, Kentucky, and Tennessee. It was raised as a feed grain much more than as a food crop for humans. Its most important quality was the comparative ease with which it could be chewed by animals. Dent corn offered an abundant, ready-made livestock feed that enabled farmers to send larger, heavier animals on the long drives to market and to thus obtain a greater return on their investment. Cattle feeding was a frontier industry in the Ohio Valley, one that many early settlers relied upon for their cash income. By the 1830s cat-

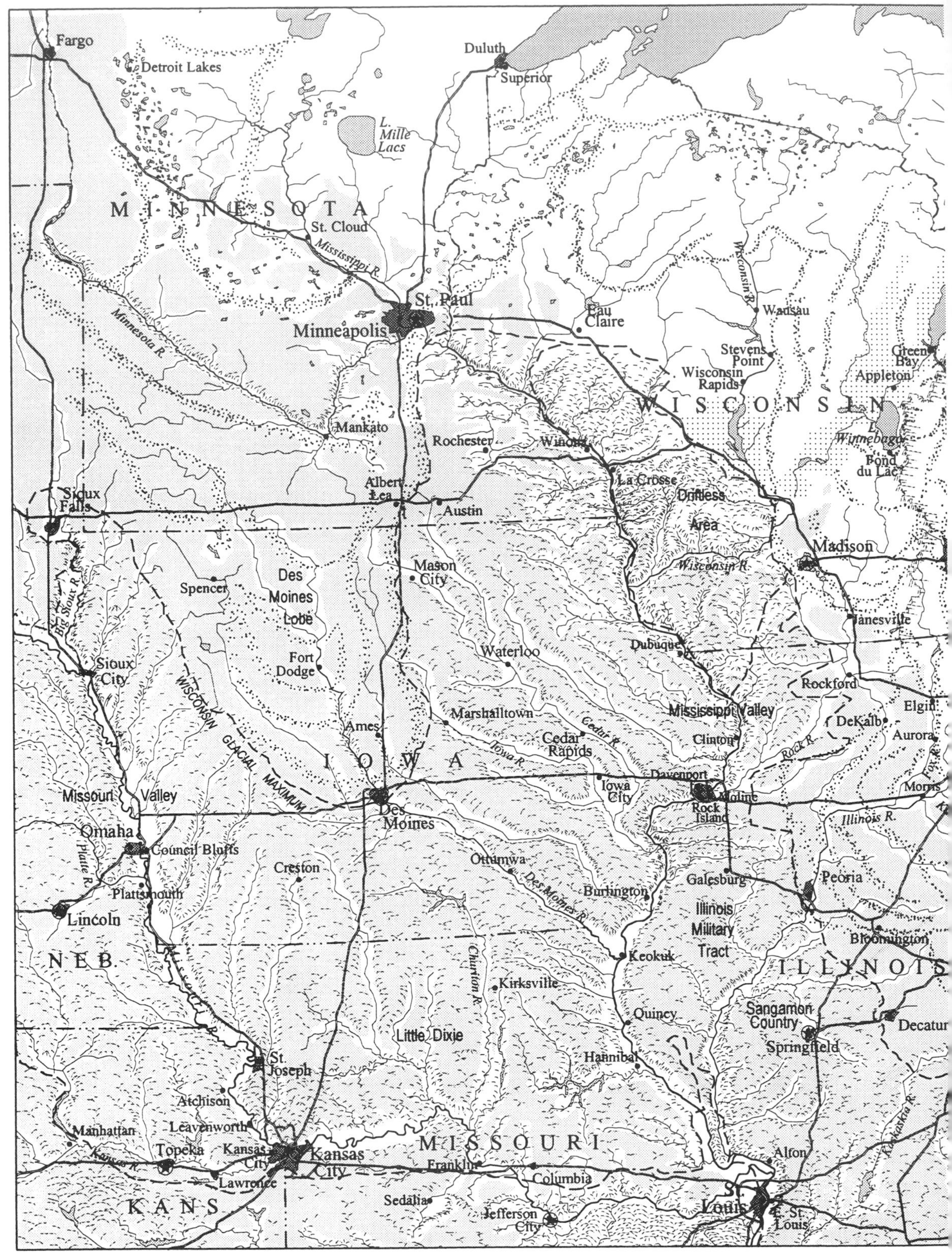

Fargo
Detroit Lakes
Duluth
Superior
L. Mille Lacs
MINNESOTA
St. Cloud
Mississippi R.
St. Paul
Minneapolis
Minnesota R.
Eau Claire
Wisconsin R.
Wausau
Stevens Point
Wisconsin Rapids
Green Bay
Appleton
WISCONSIN
L. Winnebago
Fond du Lac
Mankato
Rochester
Winona
La Crosse
Driftless Area
Albert Lea
Austin
Sioux Falls
Madison
Wisconsin R.
Mason City
Des Moines Lobe
Spencer
Big Sioux R.
Janesville
Fort Dodge
Waterloo
Dubuque
Sioux City
WISCONSIN GLACIAL MAXIMUM
Rockford
Marshalltown
Mississippi Valley
Elgin
DeKalb
Ames
Cedar R.
Cedar Rapids
Clinton
Aurora
Iowa R.
Rock R.
IOWA
Fox R.
Davenport
Iowa City
Moline
Morris
Missouri Valley
Des Moines
Rock Island
Illinois R.
Omaha
Council Bluffs
Platte R.
Creston
Ottumwa
Des Moines R.
Galesburg
Peoria
Plattsmouth
Burlington
Illinois Military Tract
Lincoln
Bloomington
NEB.
Keokuk
Chariton R.
ILLINOIS
Kirksville
Quincy
Sangamon Country
Decatur
Little Dixie
Springfield
Hannibal
St. Joseph
Atchison
Kaskaskia R.
Leavenworth
Manhattan
MISSOURI
Topeka
Kansas City
Kansas R.
Alton
Franklin
Lawrence
Columbia
Sedalia
KANS
Jefferson City
St. Louis
E. St. Louis


MAP 13.1

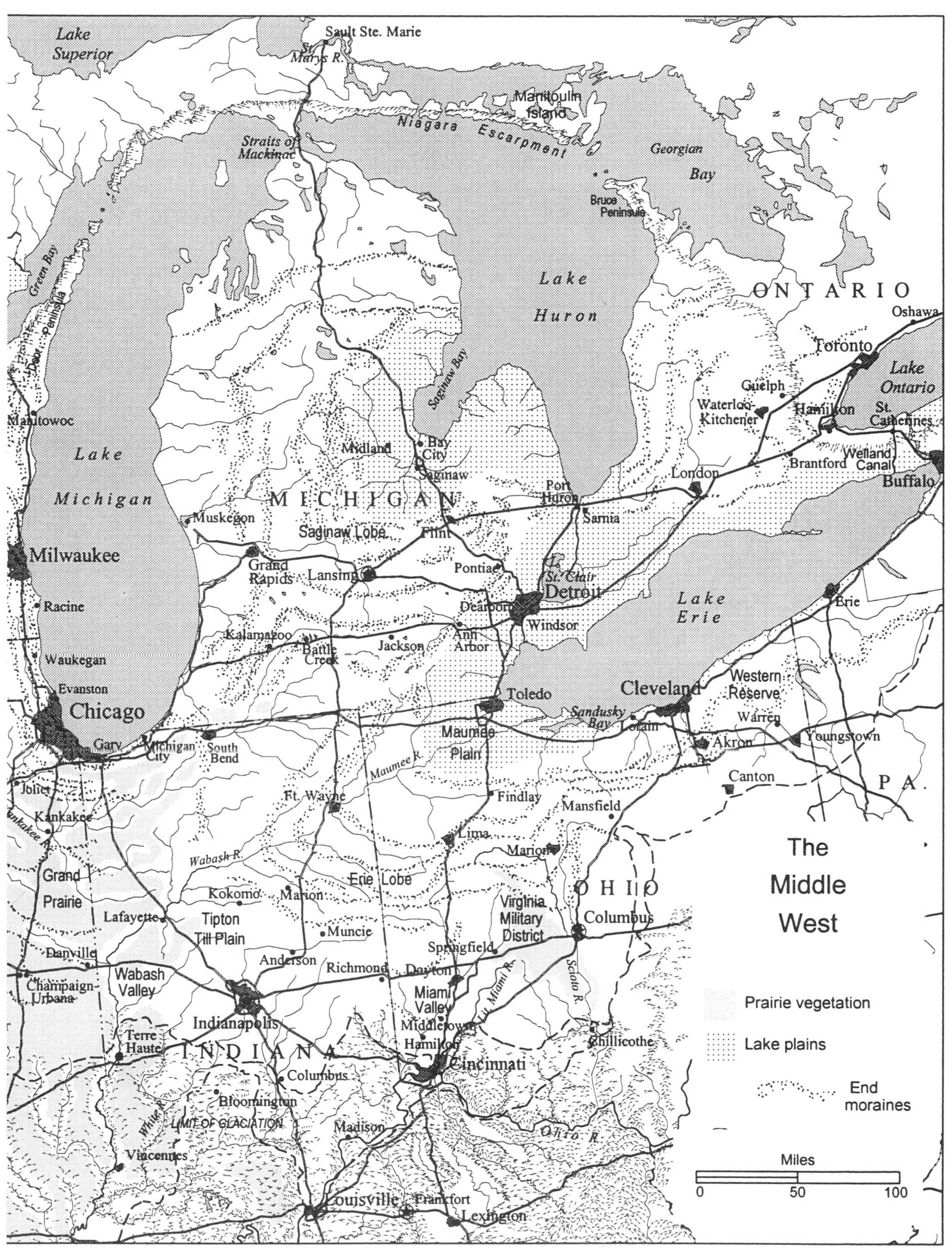
The Middle West
Prairie vegetation
Lake plains
End moraines
Miles
0
50
100
Lake Superior
Sault Ste. Marie
St. Marys R.
Manitoulin Island
Niagara Escarpment
Straits of Mackinac
Georgian Bay
Bruce Peninsula
Green Bay
Door Peninsula
Lake Huron
ONTARIO
Oshawa
Toronto
Lake Ontario
Guelph
Waterloo-Kitchener
Hamilton
St. Catherines
Saginaw Bay
Manitowoc
Lake Michigan
Midland
Bay City
Saginaw
Welland Canal
Brantford
London
Buffalo
MICHIGAN
Port Huron
Sarnia
Muskegon
Saginaw Lobe
Flint
Milwaukee
Grand Rapids
Lansing
Pontiac
Lk. St. Clair
Detroit
Lake Erie
Racine
Dearborn
Windsor
Erie
Kalamazoo
Battle Creek
Jackson
Ann Arbor
Waukegan
Evanston
Chicago
Toledo
Cleveland
Western Reserve
Sandusky Bay
Lorain
Warren
Gary
Michigan City
South Bend
Maumee Plain
Akron
Youngstown
Maumee R.
Joliet
Canton
Ft. Wayne
Findlay
Mansfield
PA.
Kankakee
Kankakee R.
Lima
Marion
Wabash R.
Grand Prairie
Erie Lobe
Kokomo
Marion
OHIO
Virginia Military District
Columbus
Lafayette
Tipton Till Plain
Muncie
Springfield
Anderson
Danville
Richmond
Dayton
Champaign-Urbana
Wabash Valley
Miami Valley
Lit. Miami R.
Scioto R.
Indianapolis
Middletown
Terre Haute
INDIANA
Hamilton
Chillicothe
Cincinnati
Columbus
Bloomington
White R.
LIMIT OF GLACIATION
Madison
Ohio R.
Vincennes
Louisville
Frankfort
Lexington

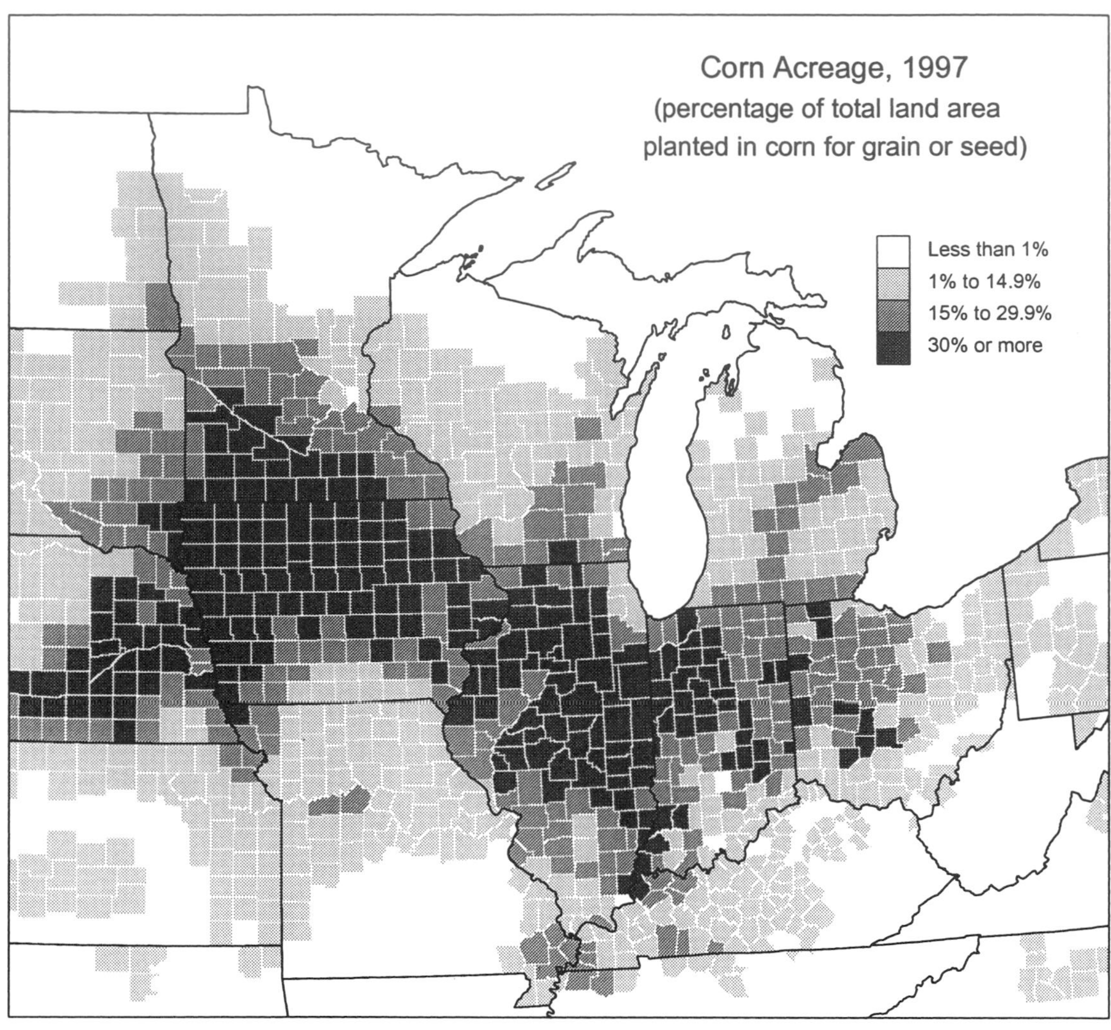


MAP 13.2

tle feeders in the Scioto Valley of Ohio were driving cattle across the Appalachians to the livestock markets of Baltimore and Philadelphia.

## The Virginia Military District and the Miami Valley

Ohio can be subdivided into four land and resource regions, which, in turn, are parts of larger regions that transcend state boundaries. Ohio's northwestern corner is covered by glacial lake plains that had to be drained before they were useful for agriculture. Northeastern Ohio developed as the Western Reserve, a land settled mainly by Yankees from Connecticut and Massachusetts. The state's southeastern corner is part of the Appalachian Plateau and is marginal for agriculture. The best agricultural land in Ohio lies in the central and southwestern portions of the state and includes two smaller regions, the Virginia Military District and the Miami Valley, both of which were important early centers of the Corn Belt.



When they became states of the United States, several eastern-seaboard colonies reserved tracts of land west of the Appalachians to reward their Revolutionary War veterans. Virginia made claims in both Kentucky and Ohio. The Virginia Military District in Ohio included the broad uplands and fertile bottomlands between the Scioto and Little Miami Rivers, a tract that embraced some of the finest prairies encountered by the settlers who moved out of the Appalachian valleys in the early 19th century. Virginia cattlemen grazed their stock on the prairies, raised large corn crops in the bottoms, and fed corn to both cattle and hogs before the stock were driven to market. Here was invented the Middle Western system of range-and-feedlot agriculture that would spread across the entire Corn Belt as succeeding generations moved west.

The regional term *Miami Valley* refers to the area drained by the Great Miami and Little Miami Rivers, which flow into the Ohio River near Cincinnati. The land in the Miami Valley was not so good as that in the Virginia District, because it was hillier and it lacked open prairies when it was settled. With no prairies for grazing cattle, Miami Valley farmers specialized more in hog production, an activity that was especially well suited to an environment in which woodland groves alternated with scattered clearings. Miami Valley farmers achieved great success in hog breeding by 1850, by which time they were driving thousands of hogs to Cincinnati packing plants every year.

Before the invention of the steamboat, the only feasible direction of trade in manufactured goods between southwestern Ohio and the rest of the nation was to ship downriver on flatboats. Downriver was the dominant trade direction even after steamboating began on the Ohio-Mississippi system in the 1820s, because of the difficulty of overland transportation east of Pittsburgh. Cincinnati's packers shipped barreled pork and lard down the Ohio and Mississippi Rivers to New Orleans, where the commodities were reloaded aboard ocean vessels to be shipped to markets in the northeastern United States and the Caribbean.

Despite the circuitous route to market, the Miami Valley and adjacent areas produced enough hogs to make Cincinnati the nation's first "Porkopolis" by 1850. The completion of the Miami and Erie Canal in 1845, which linked Toledo with Cincinnati, gave the Miami Valley packers a market outlet through the Great Lakes as well. The by-products of hog packing, including pork lard and hides, became the basis for Cincinnati's soap, shoe, and leather industries as the city's industrial base diversified. The arrival of thousands of German and Irish immigrants provided labor for more manufacturing. Cincinnati expanded upward and outward from the Ohio River front, from an industrial and warehouse core near the river, through a downtown commercial center built on a higher river terrace, and into residential neighborhoods that occupied the surrounding hills.

By the 20th century, the southern half of the Miami Valley was an industrial region focused on Cincinnati, Dayton, and Middletown, specializing in paper, machine tools, and a variety of consumer goods. Middletown is also home to one of the larger integrated iron and steel plants in the Ohio Valley. Automobile parts manufacturing and automobile assembling are other Miami Valley industries.

In its early years, Cincinnati was a major rival of Chillicothe, the principal city of the Virginia Military District. Whereas Cincinnati was culturally a northern city, albeit one within sight of slave territory on the south bank of the Ohio River, Chillicothe was a southern city in the North. Chillicothe was Ohio's first capital (territorial and state) until Columbus, a site near the geographical center of Ohio, was selected in 1812. Columbus grew from its location where the National Road crossed the Scioto River and was later linked by canals and railroads to the Great Lakes and Ohio Valley trade. It became a diversified manufacturing city during the early decades of the 20th century.

The Virginia District and the Miami Valley were an early proving ground for Middle Western corn-livestock agriculture in a zone where both northern and southern influences penetrated and mingled. As the first state created in the Old Northwest, Ohio's constitutional framers followed the lead of the Northwest Ordinance of 1787 by discouraging slavery in the new state. The same sequence was repeated in Indiana and Illinois, two more states created out of the Northwest Territory, where the first settlers also came west across the Appalachians from both northern and southern backgrounds. The slavery question was projected westward in the migration process.

## The Wabash Valley, Sangamon Country, and Little Dixie

Agricultural settlement north of the Ohio River began to expand rapidly after the War of 1812, when the conditions of frontier life became much safer for Euro-Americans than they had been before. Westward settlement expansion did not take place as a single line of frontier advance. Instead, people left their homes in the Bluegrass of Kentucky and the Nashville Basin of Tennessee to move up the valleys tributary to the Ohio and Mississippi Rivers. Both Illinois and Indiana were thus settled "from the bottom up," and the two states' earliest settlers were overwhelmingly of Upland Southern background. Alluvial soils along major rivers, such as the Wabash and the Illinois, and even minor streams, such as the White in Indiana and the Kaskaskia and the Sangamon in Illinois, attracted Upland Southern farmers, who planted large corn crops, fed the crops to cattle and hogs, and thus extended Corn Belt agriculture farther north (fig. 13.1).

The Wabash River was once a corridor of French trade and commerce between Louisiana and Quebec. Vincennes, which is Indiana's oldest city, was founded by the French in 1727. The sites of both Terre Haute and Lafayette, farther up the Wabash, also were associated with early French habitations. These sites had been the home of Native Americans for some hundreds of years prior to that time. Native people here practiced a maize-based agriculture, and thus prehistoric planting habits were continued in modified form into the era of Eu-

FIG. 13.1. *Wabash Valley, Fountain County, Indiana. Nineteenth-century Corn Belt farmers raised their first large corn crops in the fertile bottomlands along the rivers, fed the corn to cattle and hogs, and drove the fat stock to market.*

ropean settlements. Upland Southerners who replaced the French introduced the common Corn Belt practice of raising corn primarily for animal feed rather than for human food. Corn has been raised in the Wabash Valley for more than a millennium, and it remains as the region's most important product. When soybeans were introduced to Middle Western farms in the 1930s, the Wabash Valley was an early center of production.

*Sangamon Country* is a regional term for the Springfield, Illinois, vicinity and the valley of the Sangamon River, including the city of Decatur, Illinois. *Little Dixie* is a larger region, one that encompasses more than two dozen counties from the Mississippi to the Missouri River in the northern half of Missouri. Both regions were settled by Kentuckians following the War of 1812. Both Little Dixie and the Sangamon Country offered fertile river valleys and expanses of open prairie dotted with groves of hardwoods. It took comparatively little effort to convert lands such as these into productive farms, because only a small amount of land clearing was needed. Clearing the land of trees and brush had already been accomplished by the prairie fires and bottomland agricultural practices of native people in the preceding centuries.

The Sangamon Country and Little Dixie offer a useful contrast: Illinois's constitution severely limited the practice of slavery, whereas Missouri's, although it did not legally establish it, contained no provisions that would discourage the owning of slaves. The backgrounds of the early settlers of the two regions were nearly identical, the Bluegrass of Kentucky being their common origin. The natural conditions for settlement in the two areas were similar as well. Those who wished to move west to establish an agricultural system based on slave labor had Missouri as their option; Illinois offered a setting in which slavery would likely not be established.

Corn-livestock agriculture was the dominant mode of livelihood in both Tennessee and Kentucky, and so it became in both the Sangamon Country and Little Dixie. Crops of tobacco, hemp, and cotton were produced with slave labor, plantation style, in both the Bluegrass and the Nashville Basin during the first half of the 19th century. Little Dixie was too far north to permit cotton production, but it proved to be excellent for both tobacco and hemp. A somewhat scaled-down version of the Southern slave economy evolved in Little Dixie before the Civil War, but tobacco, hemp, and slavery never were established in the Sangamon Country.

The Northwest Ordinance of 1787 had outlawed slavery in the Old Northwest (states north of the Ohio River and east of the Mississippi); but Missouri was added to the United States in 1803 as part of the Louisiana Purchase. Both the French and the Spanish had practiced slavery in the territory that became the Louisiana Purchase, and the institution survived the transfer of the lands to the United States of America. The practice of alternately admitting a slave state and a free state to the Union led to the admission of Illinois (free) and Alabama (slave) in 1819. The admission of Maine (free) in 1820 increased popular pressure to admit Missouri as a slave state, hence the Missouri Compromise of 1820, which extended statehood to Missouri the next year.

Little Dixie acquired its name because it was the center of slavery in Missouri, a regional outlier of the Plantation South that lay hundreds of miles from the main slaveholding areas. The Sangamon Country became associated with the Union cause because of its most famous Kentucky-born pioneer settler, Abraham Lincoln. Glimpses of slave life in Little Dixie, in contrast, were preserved for succeeding generations in the stories written by Mark Twain (Samuel Langhorne Clemens), whose family came to Hannibal, Missouri, as part of the migration from Kentucky.

Unlike other sections of the early Corn Belt, Little Dixie did not evolve into an urban-industrial region. Apart from cities such as Hannibal, St. Joseph, and Jefferson City at its margins, it has remained a sparsely settled region of mixed farming and small towns. As is true in the southern counties of Illinois, northern Missouri's glacial cover is thin, and its soils are not as well suited for corn crops. As a result agriculture is less intensive here than it is farther north. Population has declined in parts of Little Dixie, and northern Missouri generally, ever since the Civil War.

## The Tipton Till Plain and the Grand Prairie

Central Indiana and Illinois proved to be even more fertile grounds for Corn Belt agriculture than older districts such as the Wabash Valley and Little Dixie. By 1850 corn-livestock production was established in a belt from southwestern Ohio across central Indiana into northern Illinois. These lands were more recently glaciated than those of the older Corn Belt to the south, had deeper and more fertile soils, and contained larger expanses of prairie, especially in eastern Illinois and extreme western Indiana.

Central Indiana is part of the Tipton Till Plain, a nearly level glacial plain creased only by shallow stream valleys. The landscape consisted of wooded groves and scattered openings at the time of settlement in the 1820s. Corn and hog farming became the dominant mode of agricultural activity in part because central Indiana was settled by people who came from the Miami Valley. Hog, corn, and soybean producers still dominate the Tipton Till Plain, although today much of the corn is shipped to southern poultry and hog farmers.

Indianapolis was founded in 1820 when its site, in the geographical center of the state, was selected as Indiana's capital. As in Ohio and Illinois, the migration of population northward in Indiana was reflected in a northward shift of the state's capital city. Indianapolis was not on a significant waterway, but in the early years it lay at the intersection of the north-south Michigan Road and the east-west National Road. In later years it became the hub of railway lines and highways, making it the most accessible location in the state of Indiana. Indianapolis, as well as Kokomo, Fort Wayne, Muncie, and Anderson, became important manufacturing centers in middle Indiana, with an emphasis on producing assemblies and components for the automobile industry.

Poor natural drainage is found in most of western Ohio, the Tipton Till Plain, and the Grand Prairie, where ice sheets advanced between fifteen thousand and twenty thousand years ago. The topography is gently rolling, although occasionally it is almost perfectly flat, especially in areas that were covered by glacial lakes at the ice sheet's margins. The Grand Prairie also marks the eastern edge of nearly continuous prairies within the Central Lowland of North America (fig. 13.2). Numerous "oak islands" that formed on higher ridges of better-drained land within the Grand Prairie became favored sites for early towns and farmsteads within the dominant pattern of low-lying, seasonally wet grasslands.

The Grand Prairie is the eastern apex of a region that plant ecologists labeled the Prairie Peninsula, a reference to the prairie's eastward extension beyond the moraines that border Lake Michigan. Burning by humans maintained the Prairie Peninsula as a grassland in the pre-European settlement era, but natural conditions of drought also favored the spread of annual fires. Droughts are more likely here than elsewhere at this longitude, and they enhance the effectiveness

FIG. 13.2. *Corn and soybean fields cover most of the Grand Prairie in east-central Illinois.*

of fire as an agent of landscape change. The past ten thousand years have been witness to several episodes of eastward prairie advance that have coincided with periods of warmer and drier climate.

The Grand Prairie was the last sizable area of Illinois and Indiana to be occupied by agricultural settlers in the 19th century. Its present intensive use for crop farming was postponed until the lands could be drained of seasonal standing water, a development that required years of hard labor, digging ditches to accomplish the drainage that nature has not yet provided. Current federal legislation discourages wetland drainage and promotes the restoration of formerly wet prairies such as these, although the reversal in federal policy has not yet produced a decrease in cropland acreage.

During the mid–19th century, the Grand Prairie was a cattle range that supplied young stock for Corn Belt feedlots to the east and south. It was a center of cattle production even before Chicago became an important meatpacking city. As drainage progressed, and as railroad lines radiated from Chicago across the Grand Prairie, access to the city on Lake Michigan was greatly improved, and the area shifted to the production of corn for cash sale. Today the major cities of the Grand Prairie and its fringes—Decatur, Peoria, Kankakee, Champaign, Lafayette, and Danville—are important agricultural processing centers, where corn and soybeans are converted into livestock feed, edible and nonedible oils, corn sweeteners, ethanol, and a variety of biochemicals.

## Chicago

Lake Michigan's surface lies roughly eighty feet higher than the Illinois River's origin at Morris, where the Des Plaines and Kankakee Rivers meet, sixty miles southwest of Chicago. Southern Lake Michigan's shape mimics a lobe of the Laurentide ice sheet, which pushed southward across the site of Chicago between twenty thousand and fourteen thousand years ago. Moraines constructed by the glacier, as it alternately moved ahead and retreated, ponded the melting ice. When the meltwater at the ice margin was deep enough, it spilled over the low glacial moraines, cascaded through a narrow rock channel, and discharged down the Illinois River. The sequence was repeated each time the continental ice sheets expanded southward and then wasted northward.

Chicago's flat, swampy site was the glacial lake's bottom. The present-day Lake Michigan's outline appeared after the lake's level was no longer high enough to spill down into the Illinois River Valley. The innermost ring of glacial moraines marks the drainage divide between the Great Lakes–St. Lawrence River system and the Illinois River–Mississippi River system. Although the drainage divide continues for hundreds of miles around Lake Michigan, at no point was the portage between waterways easier or more direct than at Chicago, where in wet seasons the canoes of the French traders had to be portaged only a short distance from one waterway to the other in the journey between New Orleans and Quebec.

Until the Illinois and Michigan Canal was completed in 1848, Chicago had

FIG. 13.3. *The Chicago Board of Trade, birthplace of futures trading in agricultural commodities, occupies a prominent position on LaSalle Street in the heart of the city's financial district.*

no waterway access to the tributary area to its south and west. In that same year, Chicago's first railroad to the west was completed and the Chicago Board of Trade was organized (fig. 13.3). With those developments the city became a regional trade and marketing center for the Corn Belt. In the following years Chicago became a meatpacking center, a manufacturer of farm machinery, and a distribution point for lumber cut in the Great Lakes region.

This first Chicago, which was partially destroyed in the disastrous fire of 1871, grew from its role as a broker of agricultural commodities and as a manufacturer of farm and forest products from the Middle West. Chicago rebuilt and grew larger as it received wave after wave of European immigrants who were employed in the meatpacking industries and, after 1900, in the complex of factories that depended on the region's steel mills. Chicago's massive gridiron of city blocks expanded to the north, the west, and the south across the glacial lake plain as residential growth kept pace with manufacturing expansion.

Chicago truly was "hog butcher for the world," from the mid–19th century until well into the 20th. Chicago's Union Stock Yards, on the south side of the city, were the largest ever constructed. The city's one-time concentration of meatpackers, amounting to nearly half the entire U.S. meat industry in 1900, similarly has never been surpassed. As the national population expanded westward, however, so did industries like meatpacking, and by the late 1940s, they were concentrated more in the Missouri Valley than in Chicago.

European immigration began to slow after 1910, yet the need for more industrial labor continued. African Americans, especially from the cotton districts of the mid-South, began migrating to Chicago as conditions in the South grew more repressive for blacks. Chicago's black neighborhoods expanded in number and size, especially to the south of the city's center, until most of the southeastern quarter of Chicago was predominantly African American. Chicago currently has the largest population of African descent of any U.S. city. Eastern European ethnic groups also expanded outward, sectorally, as the city grew. Polish-Americans dominated the northwest axis of growth, Italians the southwest, and African Americans populated another wedge to the west. New influxes of immigrants from Mexico and Asian countries have added new ethnic neighborhoods to the city since the 1960s.

Meatpacking has all but disappeared from the city, but the automotive, food-products, and refining industries have continued to grow. Chicago has retained its position as a national and international center of trade, banking, and finance. The Chicago Board of Trade, which began as a clearinghouse for local corn and wheat crops, became the world's largest commodities exchange. With a total population of just over 8 million, Chicago has declined in rank and now is well behind Los Angeles and New York.

Among its firsts are transportation: Chicago still is North America's most important freight railroad center and claims to have the world's busiest airport. As a center of urban architectural innovation, Chicago's skyscrapers continue to attract worldwide attention. Downtown revitalization has kept pace with the demands of the city's convention business, an especially important part of Chicago's leading role in consumer-goods marketing. Chicago is the command and control center of the Middle Western regional economy, with St. Louis, Kansas City, Omaha, and Minneapolis–St.Paul continuing to act as its satellites. Chicago's influence extends farther to the west than it does to the east or south. It is in the smaller cities of the western Corn Belt that economic ties with Chicago are most evident.

## The Corn Belt in the Mississippi Valley

Corn, soybean, and livestock production dominate the farm economy for five hundred miles west of Chicago, across the valleys of both the Mississippi and Missouri Rivers and across the states of Illinois and Iowa, until the transition to the Great Plains is reached in the eastern half of Nebraska. In the Mississippi Valley the landscape is more rolling than it is flat, the result of interglacial episodes of loess (wind-blown silt) deposition that buried the glacial debris. Loessial soils have steeper slopes and are somewhat more prone to erosion than those that formed on mixed glacial materials. From Illinois westward, the soils are classified as Mollisols, which means that they are dark in color, have deep profiles, and typically are high in natural fertility. Mollisols are associated with historic prairie grassland vegetation from the Prairie Peninsula westward into the Great Plains.

The area between the Illinois and Mississippi Rivers in western Illinois still is known as the Military Tract, a reference to its former status as a land reserve for veterans of the War of 1812. The Illinois Military Tract developed Corn Belt agriculture from the beginnings of white settlement in the 1820s. The cities of the region, including Quincy, Galesburg, Rock Island, and Moline, grew primarily as agricultural processing and manufacturing centers. Eastern Iowa also has the loess-mantled landscape of the Mississippi Valley, with long, smooth slopes separated by low drainage divides. Flat land is rare. The pattern of dispersed farm settlement and the network of small, trade-center towns established more than a century ago still characterize the region.



Eastern Iowa and the Illinois Military Tract remain concentrated on hog production, a local specialty for more than a century. Dubuque, Davenport, Waterloo, and half a dozen smaller cities became major meatpacking centers early in the 20th century by capturing smaller markets within Chicago's zone of general dominance. Although pork production remains important, cattle feeding has declined substantially in the mid–Mississippi Valley, because of the emergence of cattle feedlots in the Great Plains since the 1960s.

Once cattle feeding moved west, Mississippi Valley farmers turned to cash-grain production, an alternative that was especially attractive given the proximity of the Mississippi and Illinois Rivers and the barge transportation facilities they offered. A substantial share of the corn and soybeans exported from New Orleans to markets around the world is produced in the Corn Belt within one hundred miles of the Mississippi or Illinois River barge terminals. The mid–Mississippi Valley thus figures prominently in the foreign trade of the United States even though it lies far inland from the point of export.

North of its confluence with the Illinois and Missouri Rivers near St. Louis, the Mississippi River's flow is inadequate in some seasons of the year to maintain a channel of sufficient depth for barge navigation. The upper Mississippi was transformed into a series of deeper pools impounded by low dams stretching across the floodplain. Twenty-seven barge locks and their associated dams control the Mississippi's flow between St. Paul–Minneapolis (the head of navigation on the river) and Alton, Illinois. Grain is the most important commodity shipped downriver, whereas coal and petrochemicals dominate northbound traffic.

Southeastern Iowa was settled in the first half of the 19th century, roughly at the same time as the Military Tract. Valleys of the Cedar, Iowa, Skunk, and Des Moines Rivers were occupied by farmers from Indiana, Kentucky, Ohio, and southern Illinois who established Corn Belt agriculture in a new, more northern environment. The largest and most easily navigable of eastern Iowa's rivers was the Des Moines, which flows into the Mississippi River near Keokuk at the southeasternmost tip of the state. Steamboats once plied the Des Moines River, integrating Iowa into the national transportation system of the pre–Civil War era.

Despite river commerce, however, Iowa had little time to develop a river-based trading economy before railroads were extended east-west across the state, linking Chicago with the Iowa frontier. Most of Iowa's cities thus developed along railroad lines, even though many are on significant rivers as well. Des Moines, Iowa's largest metropolitan center, began as a frontier stockade in the 1840s, although it remained a small settlement even after the state capital was moved there in 1857. Des Moines's growth in the past half century has been based less on the agriculture for which Iowa is famous and more on the finance and insurance businesses that chose to make Des Moines their headquarters. The city became home for more than fifty insurance companies, earning it the nickname "the Hartford of the Middle West."

FIG. 13.4. *Thousands of carloads of corn move through Clinton, Iowa, every year. Some is shipped south on Mississippi River barges for export, and some is processed here as corn oil and corn sweeteners.*

The economic bases of cities like Cedar Rapids, Clinton, Dubuque, Davenport, Ottumwa, Burlington, and Waterloo are more typical of the Corn Belt (fig. 13.4). Iowa's agricultural-industrial complex is both rural- and urban-based. It produces feed, seed, and livestock on the farm, and its cities fabricate machinery to plant and harvest these products. Food products are Iowa's largest industry. When farmland values collapsed as a result of agricultural overexpansion in the 1980s, Iowa absorbed a disproportionate share of the problem: not only its farms but also its principal manufacturing industries were affected because of their close relationship with the agricultural sector.

## The Missouri Valley

Although corn is now grown under irrigation in many portions of the Great Plains, the traditional western edge of the Corn Belt is the Missouri Valley of Nebraska, Kansas, and western Iowa. The Missouri Valley differs from the rest of the Corn Belt in several respects: it has an even thicker cover of loess deposits than the Mississippi Valley, making it for the most part a hilly region; its agricultural specialty is beef cattle rather than hogs; and because it is the border between wheat production in the Great Plains and corn-livestock production in the Middle West, Missouri Valley cities have specialized both in meatpacking and in flour milling.

The circumstances of national settlement during the mid–19th century also made the Missouri Valley a zone of urban speculation that was prompted by the geopolitical contests of that era. Westward settlement expansion stalled in the Missouri Valley for more than a decade before 1854. In that year the Kansas-Nebraska Act was passed, which opened the area west of the Missouri River to settlement with the provision that the slavery question would be decided by popular votes in the two territories. The Kansas-Nebraska Act was a compromise that grew out of the continuing debate between Northern and Southern factions in the Congress, each of which desired to create a West that reflected the institutions of their respective sections of the country.

The Missouri River flows almost directly southward from Sioux City to Kansas City, where, at the mouth of the Kansas (Kaw) River, it bends sharply eastward toward St. Louis. The Missouri Valley thus represents a noticeable north-south corridor separating east from west near the center of the nation. To this geographical circumstance was added the tension of North-South political division. It seemed that a great city was bound to grow somewhere along this portion of the Missouri River and that its role would be pivotal in the settlement of the West. Numerous towns were founded here, each group of founders hoping that theirs would become the great metropolis.



The "great bend" of the Missouri at the mouth of the Kansas River was the most obvious site for a city. The first wagons to travel the Santa Fe Trail left the river town of Franklin, Missouri, in 1822. The Trail's starting point was moved from Franklin west to Independence, Missouri, and later to Westport Landing at the mouth of the Kansas, the site that became Kansas City (fig. 13.5). Each step edged closer to the great bend, which was the Trail's logical starting point. Steamboat traffic from the east could reach Kansas City easily but could get no closer to Santa Fe because of the Missouri River's course. Thus, Kansas City became the starting point of the Santa Fe Trail, a route that functioned for more than forty years as the main overland link to New Mexico. Kansas City consolidated its position as a transportation center in the 1870s when railroad lines were extended roughly along the same alignment as the Santa Fe Trail.

Missouri Valley cities were well situated to be transportation centers. Kansas City also was the origin of the Oregon Trail, whereas St. Joseph was the eastern terminus of the short-lived Pony Express. The Mormon Trail crossed the Missouri at Omaha–Council Bluffs, a route that was also followed by gold seekers trekking to California in 1849. Other would-be urban centers included Atchison

FIG. 13.5. *Kansas City's skyline, viewed from the south.*

and Leavenworth, Kansas, both in tributary valleys that could be followed by wagons heading west. The smaller cities of Plattsmouth, Bellevue, Brownville, and Peru, Nebraska, likewise began with hopes of becoming great cities of the region. Several of the smaller centers had brief careers as steamboat landings because of the necessity of transloading all goods to freight wagons for overland shipment, but once railroads bridged the Missouri, there was little need for river landings.

With this many cities vying for prominence, it was by no means certain that Kansas City and Omaha would emerge as the largest. Kansas City's role in the Santa Fe trade was matched by the selection of Council Bluffs as the eastern terminus of the nation's first transcontinental railroad. Practically speaking, however, this meant the honor was given to Omaha because it occupied the west bank of the Missouri. Kansas City's and Omaha's access to the east and west favored them as points at which to mass the quantities of wheat necessary to operate a large flour-milling industry. Kansas City's concentration of flour millers has declined from its once preeminent position, but the city is still a milling center.

Both Kansas City and Omaha also developed large stockyards, and both became important meatpacking centers, as did Sioux City and St. Joseph. Beef cattle feeding became concentrated in the Missouri Valley to supply the local packers; the concentration has declined in recent years as the cattle-feeding and meatpacking industries have continued to move west. Grain milling, feed manufacturing, chemical fertilizer production, and agricultural machinery fabrication have long been important industries of the Missouri Valley's cities. Barges navigate the Missouri River as far north as Sioux City, with fertilizer, grain, and chemicals leading the list of shipments.

Away from the Missouri River itself, Lincoln, Nebraska, and Sioux Falls, South Dakota, also became agricultural processing and manufacturing centers. Lincoln, with its base of state government and university employment, developed a more diversified economy. In recent years Sioux Falls has grown as a result of becoming the processing center for financial transactions of one of the nation's largest banks, which relocated there from New York City in search of a low-tax environment and a less expensive workforce.

## The Des Moines Lobe

Loess deposits along the Iowa edge of the Missouri River's floodplain are more than sixty feet thick, enough to create a distinctly hilly topography for the region (fig. 13.6). The great accumulation originated when the Missouri River was a discharge route for glacial meltwaters from the north. When the ice sheets had finally wasted down to nothing, silt and sand in the Missouri's floodplain were subject to wind erosion. The fine-grained loess was carried for miles toward the east in the prevailing winds.

The upper reaches of Iowa's loessial valleys were buried by a series of glacial moraines that arc southward from Minnesota into Iowa and then back north

FIG. 13.6. *Loess bluffs rise steeply at the edge of the Missouri River floodplain in Harrison County, Iowa.*

into Minnesota again. The entire lobe-shaped moraine represents the limit of the late Wisconsin ice sheet and is known as the Des Moines Lobe, after the Des Moines River, which flows across it. The glacier retreated into Minnesota less than twelve thousand years ago, making this the most recently glaciated portion of the Corn Belt. Glacial deposits of the Des Moines Lobe thus lie atop the older loess that is exposed on the western, southern, and eastern fringes of Iowa.

The Des Moines Lobe was the last portion of Iowa to be settled, for much the same reason that the Grand Prairie lagged behind Illinois's general course of development. The land surface was marshy, with numerous sags and undrained depressions. Large-scale ditching did not begin until the early decades of the 20th century. Because of the lag, north-central Iowa did not develop as large a population nor as many cities as other portions of the state. Mason City and Fort Dodge, Iowa; and Albert Lea, Austin, and Mankato, Minnesota, are the region's principal urban centers. All have economies closely linked with meatpacking or grain milling.

Once drained, the Des Moines Lobe became an important crop-producing region. Soybean acreage exceeds that of corn in some counties (map 13.3). Soybeans are crushed for oil, the by-products are used to produce animal feed, and biochemicals derived from both the oil and the fiber are used to construct biodegradable plastics, ink, caulks, and resins. Soybeans, which are native to Manchuria, yield well in the warm but fairly short summers of the upper Middle West. The flat, drained lands associated with the Des Moines Lobe continue northwestward across southern Minnesota and eastern South Dakota. These areas, too, have become important soybean producers in the past several decades. Corn raised on the Des Moines Lobe is a valuable export crop and is hauled by rail to Pacific Coast ports for shipment to the Far East.

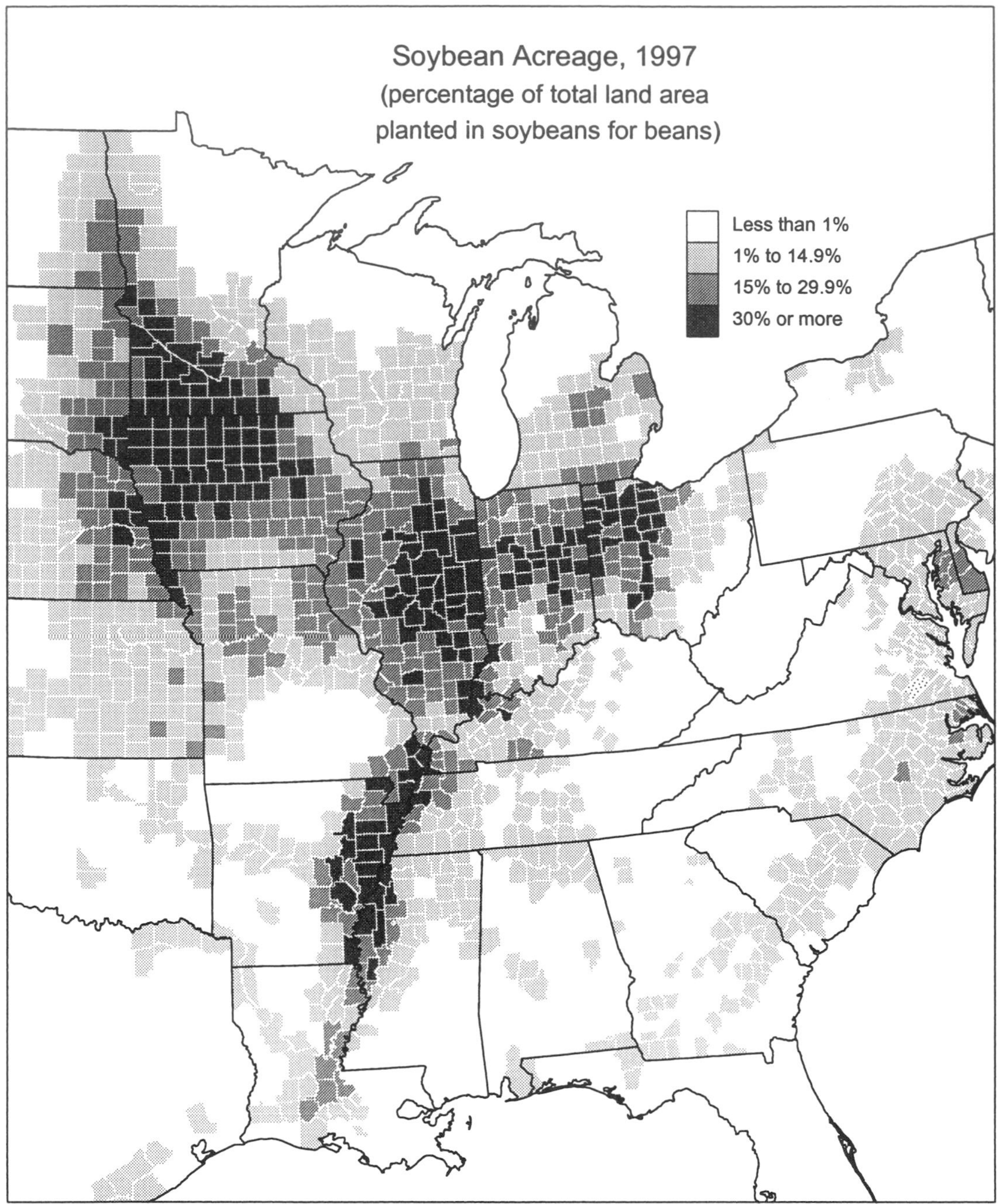


MAP 13.3

## Adjustments in Corn Belt Agriculture

The Corn Belt often is thought of as a region of medium-sized family farms. In fact, the number of Corn Belt farm residents has declined more than 75 percent since the 1920s, and the average size of farm has quadrupled. Accompanying the shift to fewer farms and fewer farmers has been, in percentage terms, an even larger increase in the amount of corn produced on the same amount of land, an intensification made possible by the adoption of high-yielding hybrid seed varieties, the increased use of agricultural chemicals, the employment of larger machinery, and a general increase in capitalization accompanying the decrease in demand for labor.



The increase in scale of Corn Belt farming is by no means a recent development. Throughout the region's history, farmers have borrowed money to expand their operations, invested in the latest technology available, and sought the quickest means to increase their income. At times, such as during the grain-export boom of the 1970s, expansion has been too rapid. Farmers borrowed money against the value of their lands in order to acquire more land, when world demand for corn and soybeans was rising. When the boom collapsed, as it did in the 1980s, the value of farmlands plummeted. Loans were called in by lending institutions, farmers went into default on mortgage payments, and the agricultural economy seemed to be in danger of collapse. Iowa farmers experienced a 55 percent write-down in the value of their assets in just ten years.

Periodic booms and busts have been common in Corn Belt history. New technology to increase agricultural production now seems to lie more in the area of genetic research than it does with larger machines or more powerful agrochemicals. Some argue that Corn Belt agriculture is too productive, perhaps in the belief that its output is larger than the world needs. Though surpluses have been frequent in times past, federal agricultural policy no longer favors massive stockpiling, and the surpluses do not persist for long. Bad weather, acreage restrictions, and policies to curtail surpluses can lead to the near disappearance of corn reserves. A Corn Belt surplus turns out to be far less worrisome than a Corn Belt shortfall when the world's food needs hang in the balance.

## *References*

Baker, Oliver E. "Agricultural Regions of North America, Part IV—The Corn Belt." *Economic Geography* 3 (1927): 447–65.

Bogue, Allan G. *From Prairie to Corn Belt: Farming on the Illinois and Iowa Prairies in the Nineteenth Century.* Chicago: University of Chicago Press, 1963.

Bogue, Margaret B. *Patterns from the Sod.* Springfield: Illinois State Historical Library Society, 1959.

Cronon, William. *Nature's Metropolis: Chicago and the Great West.* New York: W. W. Norton, 1991.

Harl, Neil E. *The Farm Debt Crisis of the 1980s.* Ames: Iowa State University Press, 1990.

Hart, John Fraser. "The Middle West." *Annals of the Association of American Geographers* 62 (1972): 258–82.

———. "Change in the Corn Belt." *Geographical Review* 76 (1986): 51–72.

Hudson, John C. *Making the Corn Belt.* Bloomington: Indiana University Press, 1994.

Prince, Hugh. *Wetlands of the American Midwest: A Historical Geography of Changing Attitudes.* Chicago: University of Chicago Press, 1997.

Transeau, Edgar N. "The Prairie Peninsula." *Ecology* 16 (1935): 423–37.

CHAPTER 14

# The Lower Great Lakes

North America's Central Lowland contains several broad, synclinal basins that are separated by upwarps, or arches, of the bedrock layers (see chapter 8, map 8.2). The basins are filled with geologically more recent rocks of the Carboniferous period, which generally signifies the presence of coal, oil, and natural gas deposits. The Illinois and Michigan Basins, which are separated by the Kankakee Arch near Chicago, contain these minerals as well as others, especially salt; salt is especially abundant in the Michigan Basin. The Illinois Basin (the Eastern Interior Coal Field) is part of the Interior Low Plateaus, whereas the Michigan Basin is the principal structural feature of the Lower Great Lakes region.

## The Niagara Escarpment and the Great Lakes

The most visible feature of the Michigan Basin is its outer perimeter, known as the Niagara Escarpment (or Niagara Cuesta), a single topographic ridge that winds nearly a thousand miles from southern Wisconsin to New York, around Lakes Michigan and Huron, forming the Door Peninsula of Wisconsin, Manitoulin Island and the Bruce Peninsula in Lake Huron, the Niagara Peninsula of Ontario, and the northern edge of the Allegheny Plateau in New York (see chapter 13, map 13.1). The ridge formed on the erosion-resistant Niagara dolomite. The steep, or "scarp," edge of the formation generally faces northward (northwest in Wisconsin, northeast in Ontario). The gentler backslope dips toward the center of the Michigan Basin and is buried under the more recent Carboniferous formations. Midland, Michigan, near the center of the Michigan Basin, is the focus of an important chemical manufacturing complex that exploits underground salt and sulfur deposits.

Oil and gas trapped in buried strata just outside the Michigan Basin fueled industrial growth in northern Indiana and northwestern Ohio during the 1890s, at a time when this was the largest known gas field in the United States. Instead of being used for industrial purposes, however, much of the gas was wasted producing crowd-pleasing pyrotechnic displays of gas fires shooting from the ground. The gas field was depleted by 1910. Northern Indiana's and Ohio's refining industries had to be supplied by pipeline from the more abundant oil and gas reserves of Texas after that time. Canada's major petroleum refineries at Sarnia, Ontario, once depended on Michigan Basin oil as well. Southwestern Ontario still produces oil, although the refineries now rely on oil arriving by pipeline from Alberta or by tanker ship through the St. Lawrence Seaway.

The Niagara Escarpment continues to influence Great Lakes geography, because Lakes Michigan and Huron together form a sizable reservoir of freshwater resting on the Niagara Escarpment's backslope. These two Great Lakes occupy strike valleys behind the escarpment, meaning that their orientation is the same as that of the escarpment itself as it bends around the Michigan Basin. The two lakes actually form a continuous water body that is joined by the Straits of Mackinac at the northern tip of Michigan's lower peninsula.

The present outline of the Great Lakes is a function both of postglacial drainage and of a general rise of the land surface to the northeast since the weight of the glacial ice was removed. The lowest spillway for Lake Huron is its point of discharge into the St. Clair River at Port Huron–Sarnia. For periods lasting thousands of years during the late Pleistocene, the western Great Lakes drained directly to the Gulf of St. Lawrence through the Ottawa River. That outlet is more direct, but, being marginally higher than the St. Clair outlet, it has disappeared. As the land surface rose, the present-day outline of the Great Lakes was formed, about four thousand years ago.

Ridges associated with the Niagara Escarpment also confine the waters of Lake Superior and account for its outlet at Sault Ste. Marie. The rapids of the St. Marys River hold the level of Lake Superior at 600 feet above sea level, or approximately 20 feet higher than lakes Michigan and Huron, which, in turn, are about 10 feet higher than Lake Erie. Lake Erie drains through the Niagara River and over Niagara Falls, cascading some 180 feet down to the level of Lake Ontario, which is the Great Lakes' final crossing of the Niagara Escarpment.

The problem of navigating the Great Lakes is in part the problem of crossing the Niagara Escarpment and thus of gaining access to the successively higher lake levels to the west. French fur traders went from lake to lake by portaging their canoes. In the subsequent era of bulk cargo shipments and large lake vessels, locks had to be constructed to raise and lower heavy ships—up to Lake Erie from Lake Ontario, up to Lake Huron from Lake Erie, and up from the St. Marys River into Lake Superior.

Ontario's Niagara Peninsula is the largest of these steps, and it was where the first canal was built. The initial Welland Canal opened in 1829. A small canal bypassing the rapids at Sault Ste. Marie was constructed in the 1850s when a lake port was established at the western end of Lake Superior. Both of these developments took place before the Great Lakes assumed their present importance in the shipping of industrial raw materials, however, and the Great Lakes–St. Lawrence locks and canals have had to be enlarged several times to accommodate larger vessels.

## The Iron and Steel Industry

Apart from their importance as sources of freshwater, the Great Lakes make a significant contribution as an avenue of cheap, water-borne transportation in

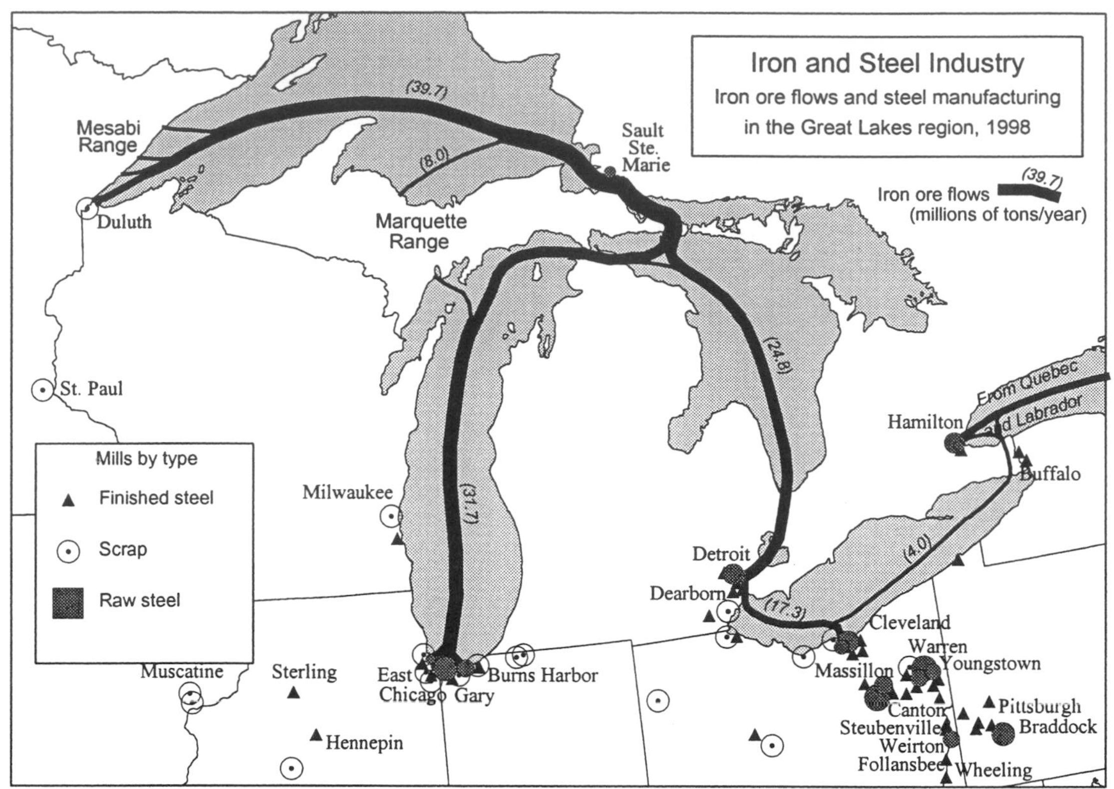


MAP 14.1

the Canadian and American industrial heartlands. The configuration of the U.S. and Canadian steel industries would be difficult to imagine were it not for the Great Lakes. Were it not for steel, industrial growth in the Middle West probably would be limited. The Great Lakes have made it possible to assemble raw materials cheaply and thus to decrease manufacturing costs at sites along their shores, an especially important factor in an industry such as iron and steel where enormous weight is involved (map 14.1).

The largest steel mills of the Lower Great Lakes region are integrated mills, meaning that they produce molten pig iron from iron ore and then, with the addition of alloys, produce steel from the molten iron. Cast, rolled, and forged steel shapes are the mills' end products. The manufacture of pig iron requires quantities of iron ore, coke, and limestone. Coke is an intermediate product, created through the heating of high-volatile bituminous coal in airtight ovens, a procedure that drives out gases and tar and makes a substance that is 90 percent pure carbon.

The blast furnace is charged with coke, iron ore, and limestone. High temperatures are achieved with a blast of air forced into the furnace, which causes the coke to burn. The oxygen in the air combines with the coke to produce carbon monoxide, which acts as a reducing agent to extract the iron from the ore.

Limestone becomes a flux to remove impurities (slag) from the iron as it melts. The dense, molten iron sinks to the bottom of the furnace, while the more buoyant slag floats off through a waste outlet and is hauled away.

Two technological changes in iron manufacture during the second half of the 19th century made the product more successful. The first was a shift from charcoal (made from burning wood) to coke as fuel. The establishment of the iron and steel industry in Pennsylvania, near abundant supplies of coking coal, was part of a westward expansion of the industry that coincided with this shift. A second innovation was the switch from less malleable iron products to steel, a substance that can be stamped, bent, and rolled into a far wider variety of shapes than iron, which can only be cast. The late 19th century saw the shift from iron to steel for use in railway rails, for example, which enabled faster, heavier freight shipments on North American railroads.



The first steel industry was really an iron industry that depended on small deposits of high-grade iron ore of the eastern seaboard states and on local forests for wood to make charcoal. This industry was replaced over the second half of the 19th century by the emergence of coking coal and an emphasis on steel (rather than cast iron) production. During this second period, the efforts of Andrew Carnegie and others made Pittsburgh the center of the steel industry. A third phase, lasting roughly from 1890 to 1910, was marked by the nearly total reorganization of steelmaking as a business, the construction of new steel mills at the southern end of Lake Michigan, and the switch to northern Minnesota as the main supplier of iron ore for Great Lakes mills.

The construction of new steel mills near Chicago was predicated on the use of the Great Lakes for transportation. Iron ore mined in the Lake Superior region moves as short a distance as possible by rail to lake ports, where it is transferred to the much cheaper ore-carrying ships. The "ore boats" can dock at steel mills at Chicago, Gary, Cleveland, and other Great Lakes cities. Limestone is quarried from formations outcropping around the fringes of the Michigan Basin, close to Lakes Huron and Michigan, and it, too, is moved to the steel mills via lake vessels.

Metallurgical-quality coking coal is mined in West Virginia and Kentucky and moves north by rail to the closest lake port, whence it moves westward to the mills, opposite the dominant direction of iron ore flow (fig. 14.1). Lake transportation is so comparatively cheap that eastern Kentucky coal destined for southern Lake Michigan is hauled north to Toledo by rail and there transferred to lake vessels just to move it around the lower peninsula of Michigan. Great Lakes shipping is limited by winter ice conditions, however, and mainly for this reason smaller quantities of iron ore and coal still move directly to the iron and steel mills by rail during winter.

Between 1900 and 1910 a new steel industry was founded on Lake Michigan, the United States Steel Company was formed under the leadership of J. P. Morgan (who purchased the companies that Carnegie owned), and the new industrial city of Gary, Indiana, was created. All of the other major American steel companies of the 20th century were formed at roughly the same time, and most

FIG. 14.1. *Appalachian coal is loaded aboard lake boats and Upper Great Lakes iron ore bound for Ohio Valley steel mills is unloaded at the Lake Erie port of Ashtabula, Ohio.*

of them followed the lead of U.S. Steel by pursuing a policy of vertical integration. Steel companies owned their own iron ore, coal, and limestone mines; their own lake steamship companies; and their own railroads, to connect the lake traffic with mines and mills. All of them constructed mills in the Chicago-Gary region, known as the Calumet district, which became the largest concentration of basic iron and steel production in the United States. Canada's two largest steel companies are at Hamilton, Ontario; they are supplied with iron ore from the Great Lakes region and from the Quebec-Labrador mines. Some iron ore from Quebec also moves through the Welland Canal to reach American steel mills.

World War II and the years immediately following were the high point of the Great Lakes steel industry. Germany's and Japan's steel industries were largely destroyed in the war, whereas those of Canada and the United States were strengthened. U.S. and Canadian producers were slow to adopt the innovations in steel production that appeared in Germany and Japan after the war, however, as those countries rebuilt their steel industries.

One innovation was the Basic Oxygen Process (BOP), which mixes molten pig iron with scrap steel in a furnace into which pure oxygen is blown under high pressure. The oxygen unites with the impurities, and steel is produced as a result. BOP steelmaking has largely replaced the older Bessemer and open-hearth methods of production. Great Lakes steel producers were slow to adopt the basic oxygen process, but the industry has modernized in recent years, even when its level of production was reduced more than one-third during the 1980s.

In other traditional steelmaking areas, production was reduced by more than two-thirds during the same period. Cities of the Pittsburgh–Mahoning–Monongahela Valley district of western Pennsylvania and eastern Ohio once typified the "steel towns" of America. Plant closures there during the 1980s drew national attention to an industry in decline. Western Pennsylvania still makes steel, although now in a smaller, modernized industry that concentrates on specialty products such as rust-resistant, zinc-coated steels demanded by the automobile industry.

Electric furnaces, which are charged only with scrap steel and use electricity as their source of energy, became feasible when technology appeared to supply their massive power needs. Electric steelmaking is not tied to a blast furnace and can operate profitably on a smaller scale. It is the usual means of production in the newer "minimills" that began revolutionizing the steel industry after the 1970s. Numerous minimills have been built in the Middle West during the past two decades. One of the specialty markets they serve is to supply raw steel to automobile factories in the Lower Great Lakes region.



Integrated mills, in which the blast furnace and a series of BOP furnaces form a single production complex, are concentrated in the Cleveland, Detroit, and Chicago-Gary districts, an even stronger localization than in former times because of the loss of productive capacity in western Pennsylvania. After two decades of decline, Lower Great Lakes steel production stabilized in the 1990s. The major customers of the integrated mills are heavy industries that rely on a few types of steel for their manufacturing processes. The heavier the machinery they produce, the more likely it is that they will locate near a steel mill. These linkages have fostered manufacturing centralization in the Lower Great Lakes region over past decades, and they continue to do so at present.

## The Manufacturing Belt

Despite the role of the Great Lakes in cheap transportation of steel's main ingredients, the steel industry is not located primarily with regard to its raw material sources. The market for steel is the most important determinant of steel production. No single factor fully explains this circumstance, although the fact that much of the weight of the raw materials enters the finished product is partly responsible. Another reason for steel's market-oriented location is the great variety of products that a steel mill traditionally produced, including bar, strip, plate, girder, and sheet forms, each of which is demanded by specific customers of the mill. Being as close to as many customers as possible is the optimum location strategy.

The location of the steel industry both determines and is determined by the location of many other types of industrial production. The customers of a steel mill are other factories, those producing machinery, vehicles, and assorted durable goods. The suppliers of each industry desire proximity to their customers, just as each industry desires to be close to its suppliers. It is this effect, sometimes called agglomeration economies, that produces a large, internally connected industrial region. The main such region in North America is the Lower Great Lakes, from Toronto to west of Chicago and generally north of the Ohio River. It was once commonly known as the Manufacturing Belt, although other regions now deserve that label as well. The Lower Great Lakes still is the largest manufacturing region in North America, whether measured in terms of its total industrial output, its number of employees, or the value added by manufacture.

Before 1926 all steel produced and sold in the United States was priced as though it had been manufactured in Pittsburgh. By this infamous "Pittsburgh-plus" pricing system, steelmakers attempted to minimize the inherent geography of their operations and thereby preserve their investment in the Pittsburgh area, simultaneously discriminating against steel consumers farther to the west, who, in fact, obtained their steel from much closer mills. Whatever negative effect the Pittsburgh-plus arrangement had on the dispersion of manufacturing, it could not prevent the continued migration of industries, both to the west and to the south, as markets shifted in response to population growth. The advantages of locating in the traditional Manufacturing Belt were not enough to prevent industries from relocating to pockets of cheaper labor, to low-tax states, or to localities that lacked a tradition of organized labor.

## The Automobile Industry

Motor-vehicle manufacturing was a major industry of the old Manufacturing Belt, and it remains so today (fig. 14.2). Reasons for the automakers' concentration in and around Detroit have long been debated, although early advantage is the strongest explanation. The first automobiles were manufactured in Europe, but by 1903 a cluster of inventors had emerged in southeastern Michigan, including Henry Ford near Detroit, Ransom E. Olds of Lansing, and William C. Durant (Buick motorcar) of Flint. Henry Ford's contributions included an improved design for the internal combustion engine; the concept of a cheap, reliable automobile that anyone could afford; and innovation of the assembly-line method of manufacturing.

Although the industry's early success in southeastern Michigan reinforced its further growth there, other circumstances tended to disperse operations. Another of Henry Ford's innovations was the branch automobile assembly plant, an idea that was copied by General Motors in the 1920s. All of the components for automobiles could be manufactured in and around Detroit and then shipped to distant locations for final assembly. Because of its high import tariffs

FIG. 14.2. *General Motors office buildings, Detroit.*

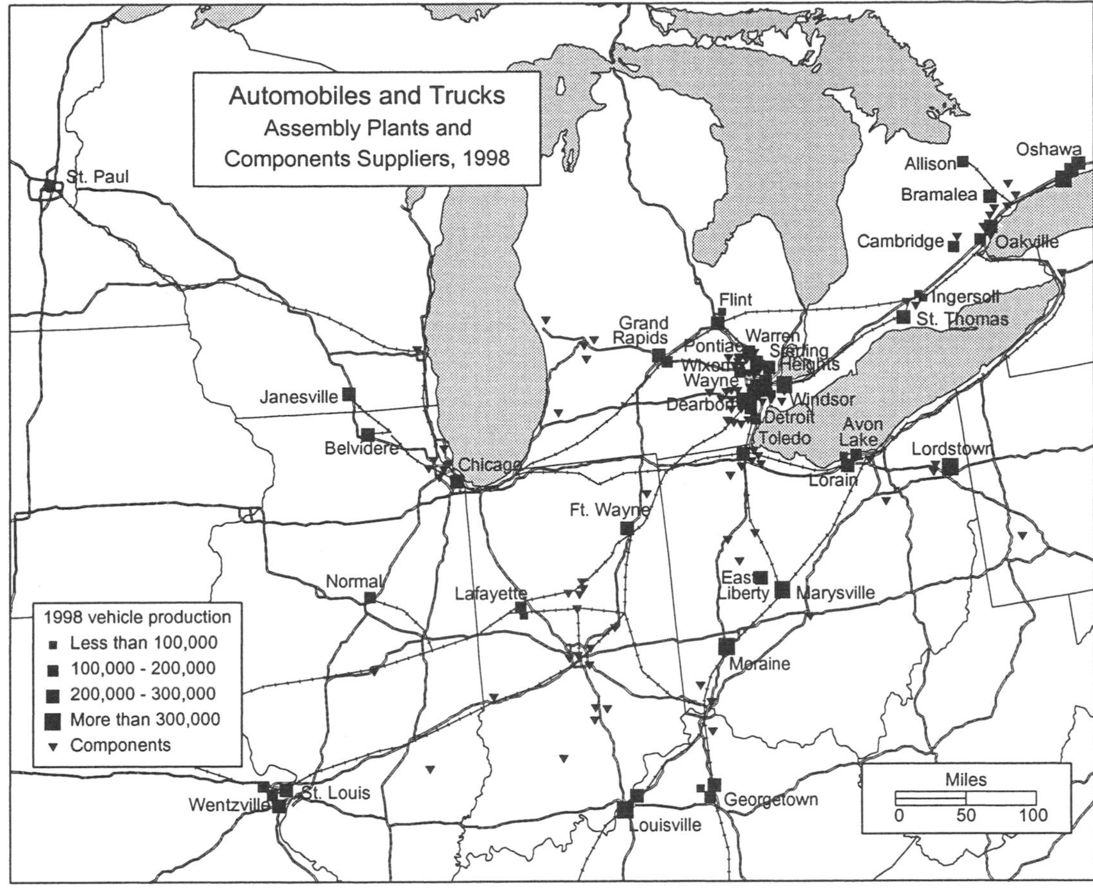


MAP 14.2

on American goods, Canada did not have branch assembly plants but rather had its own counterpart companies of the major American firms, all of which established manufacturing subsidiaries just to serve the Canadian domestic market.

As the major Detroit automakers grew in size, they purchased outright many of the smaller companies that manufactured their specialized parts rather than to engage in patent controversies with them. The component manufacturers were scattered across the Lower Great Lakes region and beyond. Each automaker thus evolved a network of suppliers of parts and components and another of branch assembly plants, both of which were controlled from Detroit. Suppliers of all of those factories, in turn, were other companies in the Manufacturing Belt. Automobile manufacturers depended on efficient transportation to keep their operations functioning smoothly. The industry itself grew to become a sort of megafactory, with divisions, branches, and subsidiaries scattered from coast to coast, all depending on one another for supplies or orders (map 14.2).

Problems of reduced demand brought by foreign competition plagued the U.S. and Canadian automobile industries in the 1970s, at roughly the same time that the steel industry began to experience similar problems. Branch automobile assembly plants on the West Coast fell victim to Japanese imports. The West Coast is nearly as accessible from Japan as it is from Detroit, at least for automobile or component shipments. The major automakers began to reduce their operations in the 1980s, but even as they learned to manufacture vehicles the public wanted to buy and entered into joint ventures with Japanese car builders, they retained much of the product variety for which Detroit was known, with numerous (and competing) lines and models designed for customers of various means to purchase.

Because the number of models was roughly the same, but fewer vehicles of each model were produced each year, there was less need to have several branch plants assembling the same type of vehicle. Former branch plants became specialized factories producing all of the vehicles of one type. Cars and trucks would be shipped to all dealers, nationwide, from every assembly plant, which, in turn, reinforced the desirability of producing cars and trucks somewhere near the center of the market. For most automakers, this means a location within or slightly south of the Lower Great Lakes region.

Canada successfully offered economic incentives to convince U.S. automakers to relocate more of their operations to southern Ontario and Quebec. Many vehicles intended for sale to American customers are now manufactured in Canada. The North American Free Trade Agreement has helped erase the tariff barriers that kept U.S. and Canadian industries artificially separated for many years. Both automobile and truck manufacturing now operate with little regard to the international border.

The value of manufacturing automobiles in the traditional Manufacturing Belt has become greater than ever before. New plants have been constructed from Ohio to Illinois by American and Japanese firms determined to exploit the value inherent in the region's accessibility. Components such as engines, transmissions, body panels, and frames are manufactured in dozens of factories within the Lower Great Lakes region and then shipped as needed to the assembly plants. The Japanese concept of *kanban* (just-in-time delivery) is now totally accepted by U.S. automakers, adding yet another reason for the industry and its suppliers to cluster.

The steel and automobile industries do not represent the totality of manufacturing enterprises associated with the Manufacturing Belt, of course, but many other industries are linked to these two and hence have benefited from the same changes in background conditions. Rhetoric describing the Middle West as a Rust Belt, which was common when the Sun Belt was growing rapidly during the 1970s, has subsided in recent years as both the steel and the automobile industries have regained some of their past strength. Whether the Manufacturing Belt deserves the qualifying label "current" or "former," it illustrates that basic factors of location, distance, and accessibility remain important even through successive episodes of industrial restructuring.

## Yankeeland in the Middle West

Specific regional migration patterns produced the early Euro-American settlement of the Lower Great Lakes region just as other migration patterns characterized settlement of the Corn Belt. The emergent Corn Belt drew people north and west from the Ohio Valley, whereas the settlement of the Lower Great Lakes was based on an outpouring of settlers from New York and New England. The migration lasted from 1810 to the 1840s, when nearly all of the available public land of northern Ohio, southern Michigan, northern Illinois, and southern Wisconsin had been purchased by settlers. Because most of the people in this northern migration stream were Yankees (of New York or New England background), the Lower Great Lakes acquired a different character from the Corn Belt to the south. Yankee backgrounds also predominated in the early settlement of southern Minnesota during the 1850s.

Yankee farmers specialized in wheat and dairy production rather than in raising corn to fatten meat animals. Wheat production moved westward with the frontier. The Genesee Valley of New York was the nation's breadbasket in the 1830s. Then, as settlers from New York took their planting habits westward, northern Ohio had a brief reign as leading wheat producer until the 1850s, when the Rock River Valley of Wisconsin and northern Illinois became the leading area of production.

Unlike corn, which remained the major crop of most regions where it was introduced, wheat raising kept pace with the frontier. Its production declined once the settlement frontier pushed on. Insects, wheat rust, and other plagues typically appeared in areas where wheat was a specialty, and production costs were always lower on the frontier than in longer-established wheat districts. The crop kept moving west with the frontier until the Red River Valley of western Minnesota and eastern North Dakota became the center of wheat raising in the 1880s.

Yankees who moved west sold their farms to newer immigrants, especially to the thousands of arrivals from Germany and the Scandinavian countries who came to Michigan, Wisconsin, and Minnesota during the second half of the 19th century. The population composition changed in these three states as a result. New Yorkers and New Englanders left both Michigan and Wisconsin for Minnesota. Canadians, many of them descendants of Loyalists who had moved to Canada two generations earlier, left Ontario and settled in Michigan as the Yankees moved west. In Wisconsin it was the German-born who replaced the Yankees, and by 1890 Germans dominated eastern Wisconsin. Minnesota's early Yankee settlement was overwhelmed numerically by the arrival of thousands of migrants from Germany, Norway, and Sweden in the 1870s.

Some deliberate matching of environments, between familiar native homelands and new American landscapes, undoubtedly took place as northern European groups sought areas within the Great Lakes region where they could continue to practice forms of livelihood with which they were already familiar. Other factors must be considered as well. Many Europeans disliked the social system of the South and deliberately avoided it in favor of the northern states.

Coincidences of timing also played a role, because much of the Great Lakes region remained unsettled by whites when the great European out-migrations took place between 1850 and 1890. Wisconsin and Minnesota, especially, had large areas of good land awaiting settlers. The upper Middle West thus acquired a strong European ethnic flavor, in its rural farming districts as well as in its cities, more perhaps than any other region of the United States.

The land that these successive groups of domestic and foreign-born settlers chose was the product of repeated glaciations during the past twenty thousand years (see chapter 13, map 13.1). Most of southern Ontario is a former glacial lake, as are the Lake Erie and Lake Huron fringes of Michigan. Northwestern Ohio was known as the Black Swamp in the 19th century, when it was a nearly impenetrable tall-timber wetland on the bottom of former glacial Lake Maumee. The men who built the first railroads across northwestern Ohio found that their construction camps were serenaded at night by the howl of wildcats living in the surrounding woods.

All of these regions became agriculturally productive after they were drained. Southern Ontario's one-time glacial lake bottoms became the basis for the only semblance of Corn Belt agriculture in Canada, the plains between Windsor and London–St. Thomas, where substantial crops of corn and soybeans are produced today. Saginaw Bay and the thumb of Michigan had a similar history. They were wetland forests, then were harvested for timber and drained for agriculture, and now are an important area of cash corn and soybean production. Ohio's Black Swamp (the Maumee Plain) also became a major corn and soybean producer. Both crops are exported to Europe from the port of Toledo by the Great Lakes–St. Lawrence route.

The glacial lake plains were drained early because of their better soils and ease of cultivation. In other parts of the Great Lakes region, standing water still occupies a significant portion of the total land surface. Minnesota's boast of ten thousand lakes is well known, but Michigan also is dotted with small lakes, and so is most of eastern Wisconsin. The region's many moraines (low, discontinuous ridges of glacial debris formed during brief readvances of the glacier) outline the margins of former glacial lakes and also help explain the rather haphazard pattern of streams. Compared with the Corn Belt to the south, the Great Lakes portion of the Middle West has few important rivers, because of the comparative recency of its glaciation.

Alfisols, sometimes labeled as "forest soils," are the dominant soil order of the Lower Great Lakes region. Alfisols are generally fertile, they contain abundant organic matter, and they are typically associated with broadleaf forests (fig. 14.3). Temperatures are warm enough that organic matter decomposes during each year's growing season. Unlike the Mollisols of the prairie, Alfisols occur in humid climates with no marked period of soil-moisture deficiency. Most Alfisols are slightly acid in reaction, although this problem is overcome with applications of lime, as is the practice in the upper Middle West. They are good, all-purpose agricultural soils and are productive of grain crops and pasture grasses.

FIG. 14.3. *Kettle moraine near Whitewater, Wisconsin. The topography was produced by surface collapse following below-ground melting of stagnant glacial ice. Temperate broadleaf forest with a preponderance of oaks typifies vegetation in the Lower Great Lakes region.*

Some poorly drained lands have a different soil order. The Histosol, which contains abundant organic matter, is typically black in color and shrinks in volume when the land is drained. In the Middle West these are often known as muck lands, and they are found in portions of former glacial lake beds, within the morainic belts, and in scattered other sites having poor drainage. The muck lands are highly productive for vegetable crops such as peas, sweet corn, and snap beans. Much of the summer vegetable crop of the United States is produced on Histosols of the Lower Great Lakes region. In still more poorly drained areas of central Wisconsin, where soils developed on sand and are too acid for general agriculture, the major crop in recent years has been cranberries.

## The Driftless Area

Repeated advances of the Pleistocene continental ice sheets covered nearly all of the Middle West. Only one sizable area, the southwestern portion of Wisconsin, escaped glaciation by all of the ice advances. The Driftless Area, as this region is known, shows how the Middle West might have looked had there been no glaciations. Hilltops in the Driftless Area generally are smooth ridge crests of the underlying bedrock. The pattern of drainage is relatively well developed. Isolated hollows in the hills are connected by a twisting network of rural roads somewhat reminiscent of the Interior Low Plateaus, to which the Driftless Area belongs geologically.

Lead deposits attracted early settlers to the Driftless, several decades ahead of the general course of settlement in the upper Mississippi Valley. Despite its early outposts of mining-based settlement, the Driftless Area grew slowly and was bypassed by major urban-industrial developments. Today it is mainly a dairy region with a secondary focus on tourism. The region's major cities lie at its fringes: Dubuque, Iowa; Winona, Minnesota; and La Crosse, Wisconsin are all industrial centers. Wisconsin's capital and second largest city, Madison, lies just to the east of the region.

## The Dairy Industry



The northern limit of Yankeeland in the Middle West is roughly the boundary between temperate, broadleaf deciduous forests and boreal (cold region) forests. Wheat seldom was successful in the poorer soils and shorter growing season of the boreal zone, and immigrants from New York were less attracted to those lands. Dairying is successful in at least the southern fringes of the boreal zone, however, and in central Wisconsin dairying is practiced where white-pine forests once stood. With a few such exceptions, the northern limit of successful agriculture coincides with the southern limit of boreal forest (map 14.3).

The decline of wheat farming in the Great Lakes region over the last half of the 19th century required that farmers find an alternative mode of agriculture. Wheat monoculture had reduced soil fertility, and some believed that the land had become worn out from overuse. Lands no longer suitable for wheat were converted to pastures for grazing. Ohio became a sheep-producing state, and Michigan, Wisconsin, and Minnesota all became major dairy states by the beginning of the 20th century. Vermonters brought sheep husbandry to the Middle West, whereas New Yorkers, especially, were experts in dairying.

Unlike grain crops, which can be stored for long periods of time, milk has to be consumed within a few days. The urban populations in the Lower Great Lakes region grew rapidly between 1890 and 1920, when the shift to dairying was under way, but there were not enough cities, close enough, to permit farmers to market fluid milk (fresh milk) as was the common practice in southern New York and New England. Wisconsin's dairy farmers, especially those at a distance from Milwaukee or Chicago, sold their milk to cheese factories. Minnesota's

MAP 14.3

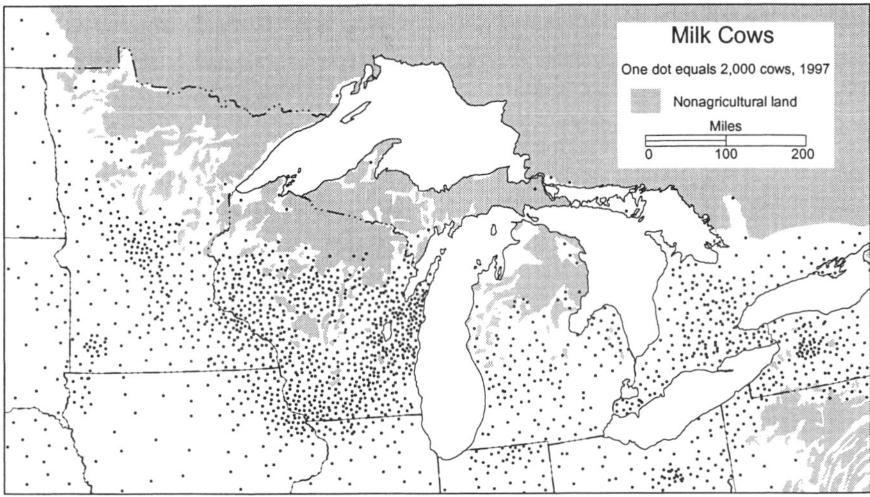

farmers hauled milk to local creameries, where it was manufactured into butter. Cheese and butter can be stored for long periods and are valuable commodities that bear transportation costs to distant markets. The states of Wisconsin and Minnesota, although somewhat remote from the nation's population centers, became the leading producers of cheese and butter.



Dairy farmers grow fodder crops such as alfalfa, and they also raise corn as a feed crop. Short summers with long days once limited the northward extension of corn production, but the limitation was not a serious handicap to dairy farmers, because they use the entire corn plant—which is chopped green and carried to a silo, where it is stored as silage—rather than just the ear of corn, which requires a longer season in which to mature. During the summer season in the Great Lakes states, farmers generally cut two crops of alfalfa, which they bale for winter use. The moisture content of both the hay and the silage are critical for proper storage.

Economies of scale have increased the typical size of Wisconsin dairy herds upward from roughly 20 milking cows per farm in 1960 to 50 cows at present. The increase in average size was accompanied by a decline in the number of dairy farms, although milk production itself is determined by federal milk marketing orders that have kept the supply of milk roughly constant over the past several decades. Even though dairy herds have grown in the Great Lakes region, they have not kept pace with the expansion in the western states—especially California, where the average herd is roughly 400 cows. In 1993 California passed Wisconsin in rank to become the largest milk-producing state. Family-farm dairy operations in Wisconsin and Minnesota now must compete with the larger, industrialized "milk factories" of southern California, a circumstance that is likely to lead to further restructuring of the Great Lakes dairy industry (fig. 14.4).

American consumers are cautious about their consumption of butterfat and have reduced their purchases of butter and whole milk, but the popularity of pizza, yogurt, and ice cream continue to grow and thereby continue to support a strong dairy industry. Cheese factories once dotted the rural Wisconsin landscape. Working usually with only a few employees, they produced a variety of regional cheese specialities: Italian-type (pizza) cheese in eastern Wisconsin; Swiss in southwestern Wisconsin; and cheddar types in all areas. Creameries (butter factories) were another typical "crossroads industry" of Wisconsin and Minnesota, but like the small cheese factories, their production has been consolidated into a few much larger modern dairy production plants. The manufacture of frozen prepared foods, an important industry in Wisconsin, is based on dairy production, the summer vegetable crop, and now even includes imports of seafood from distant sources, all with an emphasis on preparing frozen entrees designed for supermarket sale or restaurant use.

The dairy region of the Middle West extends northwest of Minneapolis–St. Paul in a narrowing wedge between the Corn Belt to the south and the boreal forest to the north until it disappears against the Red River Valley of Minnesota–North Dakota. Turkeys and meat-type chickens are raised in a belt across cen-

FIG. 14.4. *The traditional Wisconsin dairy farm, with barns, silos (for storing chopped green corn), and a milking herd of perhaps two dozen cows, is now economically marginal. Larger dairy operations produce milk at lower cost.*

tral Minnesota, adding to the already large variety of food items produced in this region. As a larger share of the food Americans eat is cooked in factories, frozen, and then shipped for sale—corresponding to a decline in the amount of food cooked at home—the food-processing industries will continue to grow and expand within the areas where the vegetable crops, fluid milk, and meat animals are produced.

## Cleveland, Detroit, Milwaukee, and Minneapolis–St.Paul

Urban growth in the Lower Great Lakes region has been based on a roughly shared set of factors relating to transportation access, the timing of early settlement, and subsequent economic growth followed by restructuring in more recent decades. Although each of the region's major cities has experienced its own unique course of development, these four major metropolitan areas can be considered jointly as a study of urban evolution within the larger region.

The Western Reserve district of Ohio originated as a western land claim made by the government of Connecticut. In 1796 the Connecticut Land Company purchased the entire Western Reserve, a tract stretching some 120 miles from the Pennsylvania line to Sandusky Bay, and began to sell the land. The first settlement was named Cleveland, after Moses Cleaveland, the surveyor who laid off the first lots on the east bank of the Cuyahoga River. The Western Reserve attracted many settlers from Connecticut, Massachusetts, and New York and, as such, became one of the first penetrations of Yankee settlement into the Old Northwest.

Cleveland prospered, especially after the Ohio Canal, linking the Cuyahoga and Ohio Rivers, was completed in 1832. Access to coking coal from Appalachian Ohio fields and iron ore brought by Great Lakes ships enabled Cleveland to become a major steel producer by the late 19th century. The twists and turns of the

Cuyahoga River in Cleveland were improved for navigation to allow large lake vessels to reach the steel mills south of the city's center.

Eastern and southern Europeans were recruited as labor for the mills. African American migrants to the city came later, and they did not enjoy the same access to the better-paying jobs in Cleveland's heavy industries. The disparity became one source of unrest that led to race riots in the 1960s. New leadership, both black and white, eventually brought urban rehabilitation and greater social equality. Cleveland's white population declined by more than half between 1960 and 1990, an example of the "white flight" experienced in other northern industrial cities. During the 1970s Cleveland suffered the humiliation of defaulting on its municipal loans, and it acquired further notoriety when chemicals floating on the polluted Cuyahoga River burst into flame. The sagging fortunes of the steel industry brought new problems in the 1980s. A program of downtown renewal in the 1990s, plus a return to better times in the steel industry, have helped reverse Cleveland's decline (fig. 14.5).



Detroit was founded in 1700 as a fur-trade post and a *seigneury* under the command of Antoine de la Mothe Cadillac, who subdivided the lands in the traditional French long-lot manner. The habitants (farmers) who lived there may have had a life rather similar to that in the smaller seigneuries of the St. Lawrence Lowland during the 18th century. After New France fell to Great Britain, Detroit became part of British North America, but the population remained dominantly French and native until the 1830s, well into the American period of settlement. Detroit did not become an important manufacturing center until the automobile industry began to grow.

Although Detroit was a latecomer to the ranks of American industrial cities, its focus on the automobile industry was largely responsible for making it the fourth largest metropolis in the nation by 1920. Like Cleveland, Detroit's population grew rapidly from the arrival of industrial laborers from eastern and southern European countries between 1900 and 1920. By 1950 African Ameri-

FIG. 14.5. *Cleveland's downtown revitalization of the 1990s included the usual skyscraper construction plus a popular new tourist attraction, the Rock and Roll Hall of Fame.*

cans outnumbered European ethnics in the city. Tensions between blacks and whites over access to jobs in the automobile industry, open housing, and numerous issues involving white authority led to the tragic Detroit riot of 1967. Forty-three people died, hundreds were injured, and nearly $50 million in residential and business property was lost.

Detroit's peak population of 1.8 million in 1950 declined to just over 1 million in 1990. Because the automobile industry never was concentrated entirely in Detroit but rather was dispersed into the suburbs as well as to such nearby cities as Flint, Lansing, and Saginaw, the southeastern Michigan metropolitan region has experienced sustained population growth even as Detroit has declined.

Milwaukee was one among many burgeoning cities of the Old Northwest that attracted migrants beginning in the 1830s. Like Detroit, Milwaukee started as a fur-trade post. In 1833 real estate speculators began to sell lots at the confluence of the Milwaukee and Menomonee Rivers, a short distance from Lake Michigan. The first Wisconsin-bound settlers who disembarked Great Lakes ships at Milwaukee to trek inland purchased goods that had arrived there intended for barter and exchange with the native people, as part of the fur trade, so compressed were the stages of settlement.

During the 1850s Milwaukee became the largest primary wheat market in the United States because of its close access to the Yankee wheat-growing area of southern Wisconsin. Milwaukee's wholesalers and jobbers served a large territory, and its mills and packinghouses received grain and livestock from as far west as Minnesota and Iowa. Like Chicago, Milwaukee also became a railroad center, with lines radiating north to the lumbering and mining districts and west to agricultural areas.

New Yorkers and New Englanders dominated the early settlement of Milwaukee, but by the late 1840s, large numbers of German immigrants were arriving in the city. At first the northern provinces of the German Empire produced the most immigrants, then southern Germany and Austria, but throughout the 19th century Milwaukee was a favored destination. Milwaukee's beer industry was created by men who had learned the arts of brewing in Germany and brought their skills with them. Investment capital flowing to Milwaukee helped establish the city's tool and die industry and later the heavy-equipment manufacturing businesses. It was in these industries especially that later waves of immigrants from Poland, Bohemia, Slovakia, and Slovenia found employment.

By the middle of the 20th century, Milwaukee was a mature industrial city that, like Chicago, Cleveland, and Detroit, attracted African Americans from the Lower South to augment its labor force. Like those cities, Milwaukee also waited too long to recognize the problems of economic and social inequality that divided its population along racial lines. Cleveland's and Detroit's struggles of the 1960s and 1970s became Milwaukee's in the 1980s. All of Milwaukee's industries, including its railroads, foundries, breweries, and machinery manufacturers were under economic duress when the issues of racial equality finally were addressed. Although Milwaukee experienced a less dramatic population loss than Cleveland and Detroit, the city has declined in size for several decades.

Minneapolis and St. Paul did not evolve the same mix of heavy industries as Milwaukee, Detroit, and Cleveland, in part because the Twin Cities lack direct access to Great Lakes shipping. St. Paul originated as a river town, an early outpost of St. Louis, at the northern limit of navigation on the Mississippi River. When Minnesota became a territory in 1849, St. Paul was designated as its capital. In the early years of trade with the Northern Plains and with Canada, St. Paul was a bustling transportation center, the point where cargoes were transferred from steamboat to ox cart and wagon for the long journey to the Red River country. St. Paul had a head start on Minneapolis, its rival seven miles upriver, and was the larger of the two cities until 1880.

Minneapolis grew out of two settlements (St. Anthony's, on the east bank of the Mississippi was the other) built around the only waterfall on the Mississippi River. St. Anthony's Falls had a vertical drop of only sixteen feet; but by running millraces around the main stream channel, a water-power potential of nearly fifty feet of drop was achieved. The first waterwheels to turn at St. Anthony's Falls in the 1850s powered sawmills that converted the logs rafted down the Mississippi River into lumber.

Flour milling was introduced gradually, and by 1870 it had become the dominant use of water power at the falls. Wheat was king in Minnesota in that era, Minneapolis had replaced Milwaukee as the leading wheat market, and millers from New England were devising new methods for grinding flour. Minneapolis entrepreneurs made their first fortunes in flour milling and invested the proceeds in railroads that were built west into the Dakotas and Montana to capture the eastbound grain traffic for their mills. Eventually more than a dozen flour mills were clustered around St. Anthony's Falls. Grain elevators and railroad terminals occupied peripheral districts around the city. Some bore the names Washburn or Pillsbury, both prominent families of northern New England background who were among Minneapolis's early millers.

As the two cities grew in size, they also grew toward each other; and by the streetcar era of the 20th century, the two cities started to become a single urban agglomeration. Flour milling has declined in local economic importance, although Minnesota's smaller cities, including Hastings, Red Wing, and Lake City, still are important milling centers for spring wheat raised in the northern Great Plains. In the Twin Cities new corporations specializing in electronic controls, machinery, and packaging products replaced the older focus on flour milling. The Twin Cities is now a manufacturer of computers, a leading financial center in the Northwest, and a corporate headquarters city chosen by firms seeking a livable urban environment for their employees.

## *References*

Barrett, Donald F., and Robert W. Crandall. *Up from the Ashes: The Rise of the Steel Minimill in the United States.* Washington, D.C.: Brookings Institution, 1986.

Conzen, Kathleen Neils. *Immigrant Milwaukee, 1836–1860: Accommodation and Community in a Frontier City.* Cambridge: Harvard University Press, 1976.

Hart, John Fraser. *The Land That Feeds Us.* New York: W. W. Norton, 1991.

Hudson, John C. "North American Origins of Middlewestern Frontier Populations." *Annals of the Association of American Geographers* 78 (1988): 395–413.

Kaatz, Martin R. "The Black Swamp: A Study in Historical Geography." *Annals of the Association of American Geographers* 45 (1955): 1–35.

Kane, Lucile M. *The Waterfall That Built a City: The Falls of St. Anthony in Minneapolis.* St. Paul: Minnesota Historical Society, 1966.

Lampard, Eric E. *The Rise of the Dairy Industry in Wisconsin.* Madison: State Historical Society of Wisconsin, 1963.

Misa, Thomas J. *A Nation of Steel: The Making of Modern America, 1865–1925.* Baltimore: Johns Hopkins University Press, 1995.

ÓhUallacháin, Breandan. "The Restructuring of the U.S. Steel Industry: Changes in the Location of Production and Employment." *Environment and Planning A* 25 (1993): 1339–59.

Ostergren, Robert C., and Thomas R. Vale, eds. *Wisconsin Land and Life.* Madison: University of Wisconsin Press, 1997.

Rubenstein, James M. *The Changing U.S. Auto Industry: A Geographical Analysis.* London: Routledge, 1992.

Seely, Bruce, ed. *The Iron and Steel Industry in the Twentieth Century.* New York: Clark Lyman, 1994.

Zaniewski, Kazimierz J., and Carol J. Rosen. *The Atlas of Ethnic Diversity in Wisconsin.* Madison: University of Wisconsin Press, 1998.

*PART VI*

# Prairies and Plains

## CHAPTER 15

# Texas and the Southern Plains

Some books about Texas begin with a recitation of superlatives about the state's geography—that it is the largest, broadest, or tallest state (it used to be, but Alaska exceeds it in all of those respects); the most sparsely settled (Texas is the second largest state in area, and it is now also second largest in population); the only state that was once part of a foreign country (all of the Southwest was part of Mexico); or unique because it began as an independent republic (so did Vermont).

Among the lesser known (but true) Texas distinctions is that it is the only state that significantly overlaps both the Great Plains and the Coastal Plain (map 15.1). East Texas—roughly, east of a line separating Dallas and Fort Worth and extending south past Waco, Austin, and San Antonio—is Coastal Plain, from the southern tip of the state north to the Red River and the border with Oklahoma. West of this line lies the Great Plains. Only a fringe of mountains in West Texas belongs to neither of the plains categories.

Because nearly all of the state is flat or gently rolling, distinguishing the various natural regions within Texas might be a difficult exercise were it not for an east-to-west rainfall gradient that roughly parallels the divisions of topography and therefore enhances differences from place to place. Precipitation declines from 55 inches per year at the mouth of the Sabine River on the Louisiana border to 15 inches per year at the mouth of the Pecos, a reduction of .085 inches per mile and one of the steepest rainfall gradients across flatland anywhere in the United States. The Gulf of Mexico is the principal source of moist air masses, not only for Texas but also for much of the interior of the nation. Increased distance from the Gulf means drier air because the influence of a warm ocean is replaced by that of a dry plateau in the interior of Mexico. Ecosystems are compressed into fairly narrow bands across Texas as a result, with a transition from subtropical forests to near-desert conditions spanning no more than a few hundred miles in the southern portion of the state. The transition zones fan out in the northward direction, across the plains, to produce a variety of open-woodland (savanna) environments.

## Hispanic South Texas

Texas also is one of three states (California and New Mexico are the other two) where the first settlement by Europeans was from Spain rather than from

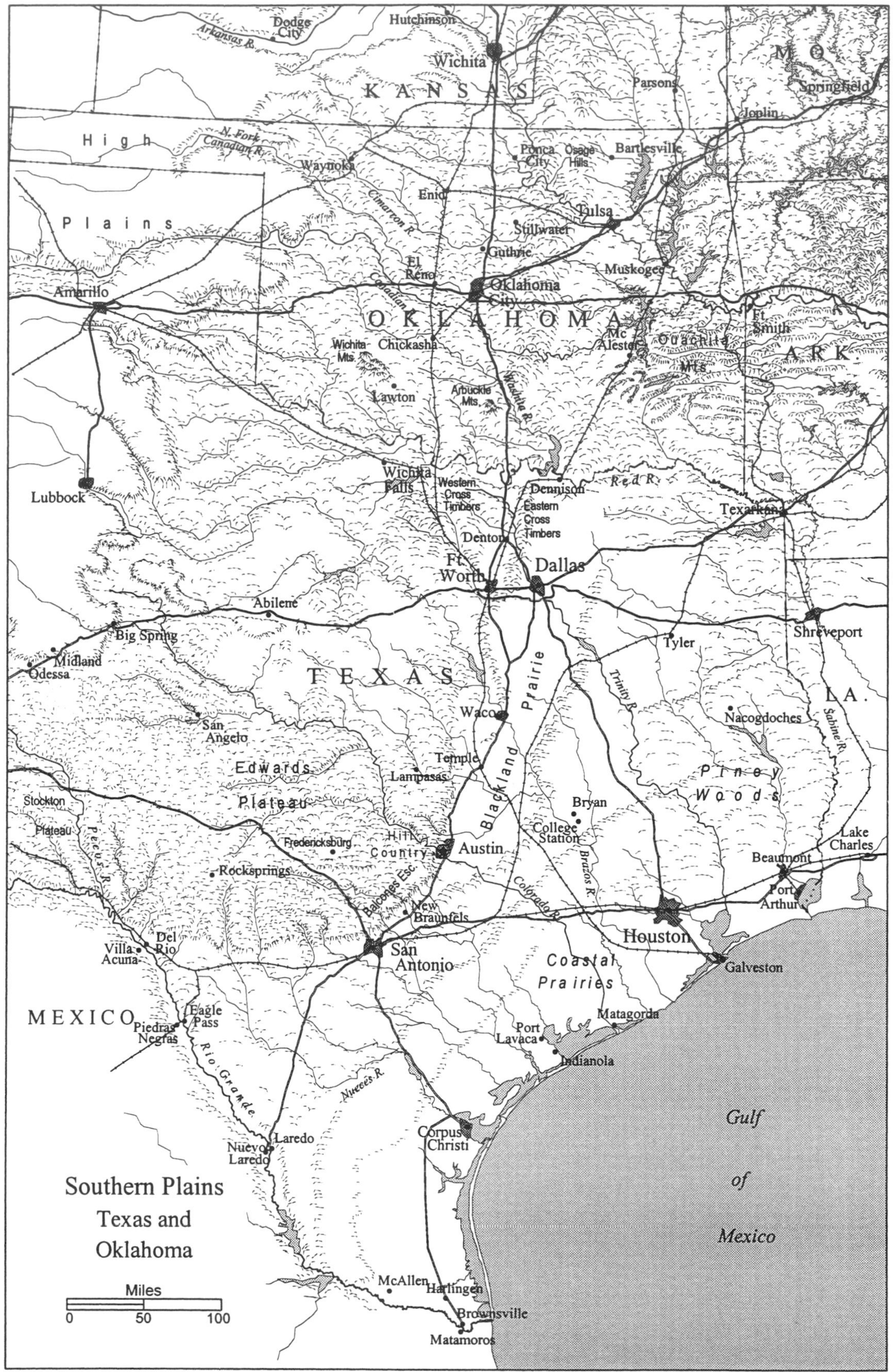
Southern Plains
Texas and Oklahoma
Miles
0
50
100
KANSAS
OKLAHOMA
TEXAS
MO.
ARK.
LA.
MEXICO
Gulf of Mexico
High Plains
Edwards Plateau
Stockton Plateau
Hill Country
Balcones Esc.
Blackland Prairie
Coastal Prairies
Piney Woods
Western Cross Timbers
Eastern Cross Timbers
Wichita Mts.
Arbuckle Mts.
Ouachita Mts.
Osage Hills
Arkansas R.
N. Fork Canadian R.
Cimarron R.
Canadian
Washita R.
Red R.
Trinity R.
Brazos R.
Colorado R.
Pecos R.
Rio Grande
Nueces R.
Sabine R.
Dodge City
Hutchinson
Wichita
Parsons
Springfield
Joplin
Waynoka
Ponca City
Bartlesville
Enid
Tulsa
Stillwater
Guthrie
El Reno
Oklahoma City
Muskogee
Amarillo
Chickasha
Mc Alester
Ft. Smith
Lawton
Lubbock
Wichita Falls
Dennison
Texarkana
Denton
Ft. Worth
Dallas
Abilene
Big Spring
Midland
Odessa
Tyler
Shreveport
Waco
San Angelo
Nacogdoches
Temple
Lampasas
Bryan
College Station
Fredericksburg
Austin
Lake Charles
Beaumont
Rocksprings
Port Arthur
New Braunfels
Houston
Del Rio
Villa Acuna
San Antonio
Galveston
Eagle Pass
Piedras Negras
Port Lavaca
Matagorda
Indianola
Corpus Christi
Laredo
Nuevo Laredo
McAllen
Harlingen
Brownsville
Matamoros

MAP 15.1

England or France. Early Spanish outposts were found from the Carolinas to Florida and around the Gulf of Mexico, but it was in the lands near the Spaniards' settlement base in Mexico that a permanent northward expansion of the frontier took place. Spanish interest in colonizing Texas was kindled in 1682, when the French laid claim to the Mississippi Valley and began exploring the Gulf Coast. Plans were laid for a series of widely spaced outlying settlements that would form a sort of frontier barrier against French encroachment. The most remote outpost was Nacogdoches, in the East Texas piney woods and a long way from Mexico. It proved to be too far distant to be sustained as a functional part of the colonization system.

Several military garrisons (presidios) and missions were built in southern Texas, of which the most successful was San Antonio de Bexar, founded in 1722. San Antonio was a fortified town that was intended also as a center of trade and missionary activity. It was in excellent ranching country and lay close enough to Mexico that it could be kept regularly supplied without difficulty. The territory between San Antonio and the Rio Grande became an extension of Mexico's northern frontier, and so it has remained ever since, culturally if not politically. The counties south and west of San Antonio are some of the most heavily Hispanic in the United States today (map 15.2). South Texas thus became Spanish/Mexican with little effort and has survived as a cultural enclave within the United States despite the fact that the Rio Grande has been Mexico's official northern border for more than 150 years.

As elsewhere in the Southwest, however, the geography of Hispanic populations at present needs more explanation than is provided by Spanish colonial ambitions alone. The Rio Grande Valley is heavily Mexican-American today partly because of irrigation developments after 1900 that transformed the lower Valley near Brownsville, Harlingen, and McAllen into a zone of intensive fruit and vegetable production. It was only then that Mexican agricultural laborers began coming to the United States in large numbers, first on a seasonal basis and later to remain permanently.

Although the Rio Grande Valley's subtropical climate is well suited to winter fruit and vegetable production, it lies more than a thousand miles south of urban markets in the Middle West. Extensions of railways to the Mexican borderlands bridged the gap beginning in the late 1880s. The Rio Grande Valley became one of the first districts of irrigation farming based on migrant labor, and, like both California and Florida, its development required rapid, long-distance transportation to market high-value, perishable crops. Grapefruit, lettuce, and onions are some of the important winter crops that are sent north from the Rio Grande Valley.

San Antonio now has nearly a million inhabitants. Its origins are as thoroughly Spanish as those of any city in the United States, and it remains one of the most dominantly Hispanic of all large American cities. It grew from its role in the Mexican trade and, just as importantly, from the employment provided by the military bases constructed near the city in the 20th century. As trade growth with Mexico continues, San Antonio is likely to become an even more important center of international commerce.

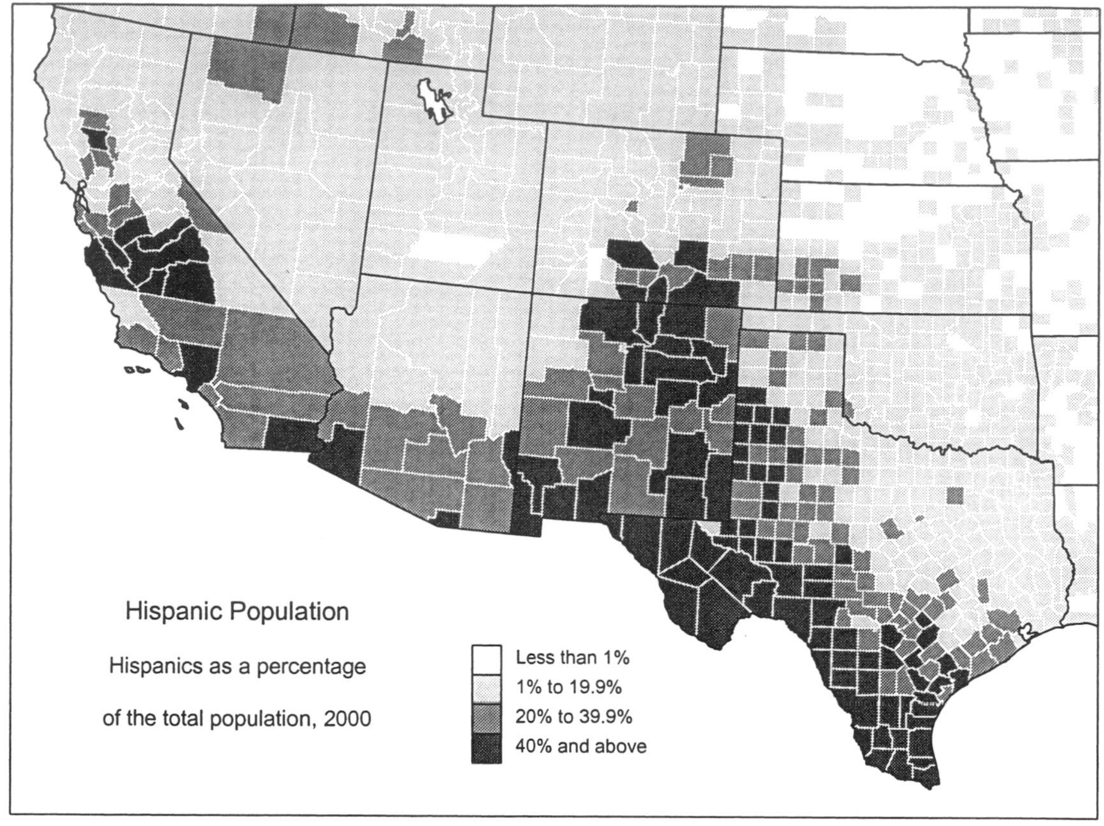


MAP 15.2

South Texas is sparsely populated ranching country, a dry and grassy range that has been invaded from the west by mesquite and other desert shrubs. Apart from the cities of San Antonio and Corpus Christi, nearly all of the population lives in the lowland bordering the Rio Grande River, the irrigated portion of which now has nearly half a million inhabitants. Many who are employed in agriculture reside in dispersed villages within the patchwork of irrigated fields. The Mexican side of the Rio Grande Valley is even more densely settled. Each of the principal U.S. border cities has a larger Mexican counterpart. Nuevo Laredo, Piedras Negras, Villa Acuña, and Matamoros together have more than five hundred thousand inhabitants.

Daily cross-border commuting is common and would be greater were it not for the *maquiladora* system instituted by the Mexican government in 1965. Branch assembly plants of American companies on the Mexican side of the border pay local wages, which are typically far below the U.S. standard. Products manufactured in Mexican *maquiladora* plants enter the United States for sale as though they were domestic goods. Del Rio, Eagle Pass, and Laredo are the principal gateway cities for cross-border traffic in components (southbound) and finished goods (northbound) for products ranging from clothing to automobiles. The North American Free Trade Agreement has increased the volume of cross-border trade, although it has yet to have an impact on the wage disparities that gave rise to the *maquiladoras.*

## The Republic of Texas



At the time of Mexico's revolt against Spanish rule in the second decade of the 19th century, Texas was a frontier consisting of four towns, scattered missions, and a few thousand people. One of the first projects of the newly independent Mexican government in 1821 was to resume the Texas colonization that had begun under Spanish auspices a century earlier. A system of land grants was devised to attract American and European colonists who would develop the territory. Moses Austin, a Connecticut native, and his son Stephen made a colonization agreement with the Mexican government to settle at least two hundred families on a grant along the lower Brazos and Colorado Rivers. The land was subdivided into tracts of several thousand acres each, creating a pattern of large ranches at the outset. Within a decade more than five thousand people had been attracted to the Austin colony.

Although Mexico was more successful in attracting settlers to Texas than Spain had been, the inevitable result of encouraging Anglo colonization was conflict with Mexican rule. The nearest well-populated American frontier at the time the Austin colony was founded was in middle Tennessee, and thus, not surprisingly, Tennesseans were prominent among the early settlers of Texas. By 1836, when the independent republic was proclaimed, Texas had between 35,000 and 50,000 inhabitants, of whom the largest share were Upland Southerners living in the Austin colony and several other land-grant areas inland from the coast and east of San Antonio. Texas was becoming so rapidly Americanized that by 1830 the Mexican government had curtailed the practice of granting land in large tracts and began passing laws to curb Anglo influence. By then confrontation was inevitable.

Battles between Texas and Mexican troops began in 1835. On March 6, 1836, a force of several thousand Mexican soldiers charged a band of 157 Texans barricaded inside the mission of San Antonio de Valero (the Alamo). All of the Texans were killed, including the commander of the forces, Colonel William Barret Travis, and the legendary frontiersmen Davy Crockett and Jim Bowie. A constitutional convention that was in progress even as the Alamo was falling led to the proclamation of Texas as an independent republic. The last battle for Texas independence took place at San Jacinto on April 21, 1836, where General Sam Houston, commander of the Texas forces, decisively defeated the Mexican general (and president) López de Santa Anna.

Between 1836 and 1845 (when the Lone Star republic was admitted as the twenty-eighth state), migrants to Texas came mainly from the Lower South. Though the Mexican government had not favored slavery, the Republic of Texas allowed it and sought to attract planters from the cotton states. Both slavery and cotton culture were established at this time, especially in East Texas, a development that presaged Texas's secession from the Union and joining the Confederacy in 1861. When Texas reentered the Union in 1865, it had been ruled by Spain and by Mexico; had been constituted as an independent republic; and had been a state—first of the United States, then of the Confederacy—all within a span of fifty years.



## The Range Cattle Industry

One consequence of the reduced Mexican influence after 1836 was the abandonment of large herds of native cattle by Mexican ranchers who had lived near the Nueces River in South Texas. The cattle were long-horned Spanish breeds that had evolved, more through neglect than care, over a several-hundred-year period on the Spanish Caribbean islands and in Mexico. These mixed-blood (*criollo*) cattle roamed an unfenced range and were tended only by the infrequent appearance of cowboys when branding was done or when herds were culled to select stock for drives to market. Spanish cattle traditions were implanted in Mexico and also in the Carolinas, Florida, and Louisiana. Anglos were influenced by the Spanish approach to ranching, and when Lowland Southerners moved west through the piney woods of the Gulf Coast, they brought these ideas with them.

It was in Texas that the various pastoral traditions coalesced to form the range cattle system that subsequently was identified with the Great Plains (fig. 15.1). Spanish influence has many manifestations, including words such as ranch, lasso, lariat, corral, and remuda—all Spanish and all common in the subsequent ranching complex of the western United States, although not in countries such as Australia where the Spanish influence is absent.

The semiwild longhorns of south Texas were so numerous that trail drives were organized beginning in the 1860s to direct animals north to market. South Texas was overrun, and its range grasses were severely depleted by the grazing of hundreds of thousands of cattle. The northward drives were organized mainly by Anglo Texans—men with names such as Chisholm, Goodnight, and Loving—with whom the cattle trails were subsequently identified. It was thus a hybrid Anglo-Spanish system, which had already undergone modification in Mexico and was reorganized as a business by Texas Anglo cattlemen.

The early cattle trails north from Texas were aimed at markets in Missouri. In 1867 Joseph G. McCoy, an Illinois cattle feeder, established the first organized market for Texas cattle at Abilene, Kansas, thus creating the first western "cow town." At Abilene cattle that had been driven north from Texas were loaded aboard railroad cars for shipment to feedlots in the Middle West or to meat-packing plants. New cow towns, such as Wichita and Dodge City, Kansas, appeared as more railroads were built across the Great Plains. By the late 1870s the rapidly multiplying Texas longhorns were driven as far north as the Dakotas in search of seasonal grazing. The northward drives spread Texas ranching styles throughout much of the Great Plains.

Although Texas longhorns were abundant, they did not produce the quality of meat that consumers desired, nor were they well suited to fattening on corn,

FIG. 15.1. *Young cattle still roam the wide-open plains of central Texas, as they have for more than 150 years.*

a practice that had already become well established in the Middle West before the era of Texas cattle in the Great Plains. Longhorns had an inbred resistance to tropical diseases that the English shorthorn cattle kept in the Middle West lacked. Quarantines were established to keep Texas cattle out of various northern ranges. Efforts at crossbreeding longhorns with English-Scottish breeds such as the Hereford and Angus, beginning in the late 1870s, resulted in the eventual disappearance of the longhorn as a range animal north of the Texas prairies. The practice of running large numbers of cattle on unfenced public-domain lands of the Great Plains declined after 1886 when droughts and blizzards produced a massive mortality in the cattle herds. This was the end of open-range ranching as such, but the Spanish-Anglo ranching complex survived in tradition and folklore as a permanent part of Great Plains culture.

## The Coastal Plain Piney Woods and Prairies

Environmental regions in East Texas are controlled largely by the pattern of roughly parallel Coastal Plain bedrock outcrops in which the strata dip gently toward (and under) the Gulf of Mexico. Beginning with the oldest rocks, which outcrop farthest inland, each layer produces smooth lowlands or belts of hills depending on its composition and resistance to erosion. Calcareous limestones and marls make good soils on smooth topography, whereas sandstones weather to produce poorer soils and belts of hills. The banded topography of escarpments, hills, and lowlands of Alabama and Mississippi is formed on the same sequence of Cretaceous and Tertiary rocks.

Two land-cover categories, prairies and piney woods, dominate the Coastal Plain of East Texas. Both types reflect human environmental modifications over millennia. The aboriginal practice of firing woodlands to drive game, create browse, and enlarge forest openings for agriculture was as important in Texas as it was in the Coastal Plain east of the Mississippi Valley.

Extensive areas with a heavy growth of long-leaf pines were systematically cut in the early decades of the 20th century after the Great Lakes forests had become severely depleted. Lumber, naval-stores industries (turpentine, pitch, gum), and paper manufacturing have all been important in the Texas piney woods and in adjacent sections of Louisiana and Arkansas. The old pattern of remote sawmills and turpentine camps in the backwoods, connected to the outside world by logging railroads, has been completely replaced by a few—but many times larger—wood- and paper-processing plants. Some of the largest paper mills in the United States are in the Arkansas-Louisiana-Texas pine forests.

If the piney woods were enhanced as the product of human effort, prairies even more surely were. Belts of flat prairie land are separated by stretches of rolling hills that have a savanna or open-woodland vegetation dominated by a variety of evergreen and deciduous oaks, of which Texas has many species, including the jack oak, shinnery, and post oak types. Like the prairies, these savannas also are the product of burning and grazing, a practice that extended well back into the pre-European period when herds of bison roamed the area. The typical Texas landscape of bluebonnets (the state flower) waving in a carpet of grass and studded with massive oak trees owes much to the periodic fires necessary to eliminate shrubby growth.

Nearly all of the prairies are developed on Vertisols, a relatively rare soil order in the United States that is found mainly in East Texas with a small outlier in the Black Belt of Alabama-Mississippi. Vertisols get their name from the tendency of the soil profile to overturn or mix. During prolonged dry spells, deep shrinkage cracks develop that then become filled with loose soil material blowing across the surface. Subsequent wetting of the soils returns the upper layer to a homogeneous state. Vertisols develop in warm climates on soils derived from calcareous limestone bedrock. They weather to produce an almost shiny black, clay-rich soil, the appearance of which is suggested in the term *black waxy prairie*, another name for the Texas Blackland Prairie.

Cotton planters from the southeastern states colonized East Texas in the 1840s, seeking both free land and slave labor. Cotton production was established in scattered localities of the piney woods, savannas, and coastal prairies by the 1850s. The greatest concentration of slaves was between Houston and Corpus Christi, which has remained as the most heavily African American portion of the state.

The Blackland Prairie was not a significant cotton-producing region until after the Civil War. Because African Americans had been freed by this time, their presence no longer automatically followed the appearance of cotton, a change that marks an important break in the nature of cotton production. Blacks did not accompany cotton culture into the Blackland Prairie to any significant degree, nor are cotton and black people associated anywhere in the United States west of there. Among the incidental effects of Emancipation was a severing of the connection between African Americans and cotton production on all frontiers settled after the 1860s. Beginning in the Blackland Prairie, those who raised cotton were known simply as farmers. The term *planter* also disappeared as cotton culture moved west.

The Blackland Prairie, which is bordered on the west by the Eastern Cross Timbers and Balcones Escarpments, stretches nearly four hundred miles from San Antonio to the Red River Valley. Railroads built into the region during the 1880s allowed market access for the cotton crop. By 1910 the Blackland Prairie had the largest cotton acreage of any region in the United States, but its reign was to be brief. Peak cotton acreages were recorded in the 1920s, after which they gradually declined. The impact of the boll weevil, the migration of cotton culture west to the High Plains and California, and a shift to other types of agriculture were responsible for the decline.



## The Texas Gulf Coast

Clay-rich soils such as Vertisols are attractive for certain crops because of their relatively low permeability to surface infiltration. Standing water is absorbed slowly, which is a problem in flood-prone areas but an advantage for crops such as rice, which are grown on flooded land. Prairies in the coastal belt consist of seasonally wet tall grasses growing on almost perfectly flat land, making them suitable for flooding. Beginning in 1900, rice culture expanded from Louisiana westward into the coastal prairies of Texas. The rice crop of Texas, which occupies more than 350,000 acres between Corpus Christi and Beaumont, is the fourth largest in the nation.

The Texas Gulf Coast has a noticeable absence of cities, a condition that is in part a legacy of Spanish colonial policy and in part an adaptation to the environment. The Spanish forced all trade to and from Texas onto a few overland trails that connected it directly with central Mexico. They left the coastal lands vacant as a barrier against outside influence. When Mexican authority succeeded that of Spain in 1821, the policy was maintained in part by making no grants along the coast. The Austin colony's lands, for example, ended twenty-five miles from the Gulf, and its first town, San Felipe de Austin, lay eighty miles up the Brazos River. A few coastal settlements, including Matagorda, were allowed by Mexico, but none of them became major cities. Urban growth in Texas was thus forced away from the Gulf as a matter of colonial policy, and no important cities got a start there during the early, formative years of Texas settlement.

Neither does the Gulf Coast offer a particularly favorable environment for cities. The interrupted series of barrier islands that forms much of the coastline is a product of sea-level change in which former beach lines have become isolated sandbars as ocean levels have risen to flood what were tidal marshes only a few thousand years ago. Wave action and the longshore current sculpted the sand spits even as the force of onshore winds caused them to migrate gradually landward. The mouths of nearly all streams entering the Gulf are blocked by sandbars, and even breaks between the barrier islands require periodic dredging to maintain ocean access. Such is the case at Corpus Christi, for example, which is blocked from the ocean by Padre Island. Corpus Christi was unimportant as a port before the 1880s, when a connecting channel to the Gulf was dredged to permit ocean traffic to reach the city. Corpus Christi's growth came later, after oil and gas discoveries were made nearby and the city developed as a refining and petrochemical manufacturing center.

Texas's largest city, Houston, is near the Gulf Coast but not on it. Houston was founded in the early years of the Texas republic as a river landing on Buffalo Bayou, a sluggish backwater about twenty miles inland from Galveston Bay. It was briefly the capital of Texas and was a cotton-shipping center but was overshadowed by Galveston, on the coast, which eventually became the most important seaport of Texas. Galveston's advantage of actually occupying the seacoast was offset, unfortunately, by the danger of such a location. In 1900 more than six thousand people were drowned in a hurricane that destroyed nearly everything in the city. Galveston was rebuilt behind a seventeen-foot seawall, and its low-lying site was gradually infilled to bring the city up above the level of storm surge (fig. 15.2).

Coastal Texas remains a high-risk zone for hurricane disaster. The pumping of subterranean freshwater and crude oil has produced land subsidence in some areas. The exposure to storm-surge risk along the coast is measured in elevation, and only a few vertical feet separate zones of extreme versus moderate hazard. Limited building is permitted on some of the barrier islands, although probably their best use is for parks or as wildlife sanctuaries where no permanent structures are allowed.

Industrial growth became inevitable in the coastal zone once substantial oil discoveries were made there in the early 20th century. Refineries and petrochemical factories are connected by the Intracoastal Waterway, a system of dredged canals through the coastal marshes and barrier islands that extends the entire distance of the state's coastline from Sabine Pass to Brownsville. The petrochemical industry is the major beneficiary of the waterway, but shipments

FIG. 15.2. *The seawall in downtown Galveston, Texas.*

of industrial raw materials, such as South American iron ore to steel mills near Houston and Carribean bauxite to the aluminum plant at Port Lavaca, also have been made possible by the waterway. Among the major exports are winter wheat and feed grains grown in the southern Great Plains, which move by rail to Houston and Galveston for export.



## The Central Texas Urban Region

The most urbanized portion of Texas is an arc of cities along the Great Plains–Coastal Plain border stretching from San Antonio to Dallas–Fort Worth. Except for San Antonio, whose origins are in the Spanish period, the major cities in this corridor date from the early years of the Texas republic. Their locations were chosen, in part, because of their forward position on the frontier of that time, slightly west of the early Spanish trade routes but east of the semiarid plateaus of the Great Plains.

The line of contact between the Coastal Plain and the Great Plains is a border between more erosion-resistant (lower Cretaceous) rocks on the west and softer (upper Cretaceous) rocks on the east. South of the Colorado River of Texas, the boundary is defined by the Balcones Escarpment, an abrupt topographic break corresponding to geologic faults that separate the Edwards Plateau from the Blackland Prairie. The same line of contact continues northward past Waco and Dallas, where it is marked by the narrow belt of hills known as the Eastern Cross Timbers, although here also the line corresponds to the same division between newer and older rocks.

The Balcones fault zone exposes water-saturated strata underlying the Edwards Plateau and thus gives rise to a series of natural springs at its margin. A large, free-flowing spring near the Colorado River's passage through the Balcones Escarpment made a particularly favored place for settlement. In 1838 it was platted as Austin, the Texas republic's new capital. Although Austin lay many miles inland from the settled portion of Texas at that time, its interior location helped stimulate the westward push of settlement. Austin grew from its twin roles as capital (of the republic and the state) and site of the state's largest university. It did not become a major metropolis until its employment base diversified into high-tech industries, a manifestation of the Sunbelt phenomenon of the 1970s.

Other cities founded at roughly the same time on this Coastal Plain–Great Plains border include Waco, platted on the banks of the Brazos River in 1849, and Fort Worth, founded as a military outpost in the same year. Both cities became trade centers serving the sparsely settled ranch country to the west. Fort Worth grew especially from its role as a cow town, the largest livestock market in Texas, and, beginning in the 1900s, a meatpacking center as well. Fort Worth was the first significant diversification of the beef packing industry away from its established base in the Middle West, and the city became a scaled-down copy of Chicago's stockyards and packing complex. Cattle were shipped from the western ranges to the Fort Worth stockyards and then either sold to local pack-

FIG. 15.3. *Dallas boomed with office-building construction during the oil-rich 1970s.*

ers or shipped to feedlots farther east. Fort Worth ranked among the four or five largest stockyards and packing centers until the industry began to relocate to the High Plains in the 1960s.

In the meantime Fort Worth's economy had expanded and diversified well beyond its cowtown origins, although its growth has always been overshadowed by its larger, nearby rival, Dallas (fig. 15.3). One of Dallas's first industries was a cotton gin. Dallas, which began a few years ahead of Fort Worth, remained larger and grew more rapidly because it attracted more capital investment from outside and in a broader variety of economic sectors. Dallas's prominence as a financial center grew especially after it was designated as a federal reserve bank headquarters in 1913, bringing it up in rank as a regional metropolis to join Atlanta and Minneapolis–St. Paul. Both Dallas and Fort Worth, important transportation and wholesaling centers for the Southwest, are the typical locations chosen for branch assembly plants designed to serve the region.

Growth in the apparel, aviation, and electronics industries (the latter two aided substantially by federal defense expenditures) helped advance the two cities into national prominence by the 1960s. The use of the label *Metroplex* to describe the combined Dallas and Fort Worth metropolitan areas became credible after the completion of the Dallas–Fort Worth Airport in 1974, which stimulated urban growth in the strip between the two cities. Dallas and Fort Worth probably best typified the Sunbelt growth phenomenon of the 1970s when several major U.S. firms relocated their corporate headquarters there at the same time that high-technology industries became a key factor driving urban growth at the national scale and the energy sector was expanding.

The energy boom peaked in 1981. Although Houston depends far more on the oil and natural gas industries, the economic downturn also slowed growth in both Dallas and Fort Worth. Growth and expansion returned in the 1990s, however, with substantial employment increases in the telecommunications, transportation, energy, and financial sectors. The population of the twelve-county metropolitan area containing Dallas and Fort Worth now exceeds 4.5 million.

## The Edwards Plateau and the Plains Border



West of the Balcones Escarpment lies the wide-open landscape of the Southern Great Plains. The contrast is evident immediately west of San Antonio and Austin, where there is an abrupt transition to the higher limestone surface. Geologically speaking, the Edwards Plateau extends eastward to the Balcones Escarpment, although the hilly lands toward its eastern limit are better known in Texas as the Hill Country.

The Hill Country has had a different land-use history from the higher and flatter Edwards Plateau. An organized colonization effort led to the immigration of more than nine thousand Saxon, Hessian, and Alsatian Germans to the Texas Hill Country in the 1840s. Because there was no suitable point of arrival, the German leaders founded their own Gulf Coast port (Indianola—or Carlshafen, as they named it), which later was leveled by hurricanes. New Braunfels, at the edge of the Blackland Prairie, became the largest of the towns they founded, although the focus of German settlement was in the Hill Country itself, where the German settlers created Fredericksburg and numerous smaller communities. Their numbers were comparatively small, but the Hill Country Germans became one of the largest enclaves of non-English-speaking Europeans in the American South, a region that typically held little interest for the thousands who emigrated from Europe to the United States during the 19th century.

The Edwards Plateau has an irregular topography with a thin soil cover over limestone bedrock. Few streams cross its surface because of a system of subterranean drainage associated with cavern development. Collapsed solution features appear on the surface, which further adds to the rugged nature of the landscape. It is not a land well suited for cattle to roam. But these mesquite- and juniper-dotted hills and bluffs are well adapted to the raising of sheep and goats, which are more sure-footed in broken topography.

German graziers in the Hill Country and later the Edwards Plateau hired Mexican shepherds to tend their flocks, a practice that continues today. The other component of the local pastoral economy is the raising of Angora goats, the animal from which mohair for garments such as sweaters and jackets is obtained. Goat herds are concentrated in the vicinity of Rocksprings, Texas, which is also the leading center for goat breeding. The Stockton Plateau, which is the local name for the same plateau west of the Pecos River, is also a major producer of both wool and mohair. The Edwards and Stockton Plateaus are the major area of sheep and goat production in the United States.

At its northwestern limit, the Edwards Plateau is bounded by the High Plains, a higher, flatter, and generally more suitable environment for crop production. Northeast of the Edwards Plateau lies a succession of eroded plains and rugged valleys that stretches across Oklahoma and into Kansas. The term *plains border* is often applied to this section, which lies east of the High Plains but west of the hillier lands of the Ozarks and the Ouachitas. Crops of winter wheat are grown on the better soils, but the broken lands are suited only to grazing. Over extensive areas the soils are reddish in color, having been derived from the Permian "redbeds," which are deep reddish-brown rock layers that formed from continental (rather than oceanic) sediments during the Permian period.

Coal, oil, and natural gas are abundant in the rock layers underlying this bleak landscape, however, and the record of their exploitation forms the essential history of Oklahoma and central Texas over the past century. Coals in this region are part of the Western Interior Coal Field; as elsewhere, they are comparatively high in sulfur content, which restricts their present use, although cities such as St. Louis once relied on Oklahoma coal for their main energy supply.

## The Origins of Oklahoma

Just as the Spanish government viewed Texas as a buffer between Mexico and the United States, so did American policymakers in the early years of the 19th century perceive the lands that would become Oklahoma as a buffer separating American territory from Mexico. The U.S. Army established Fort Smith, Arkansas, in 1817 as an advance outpost to control the international border, which was then just a short distance west of Fort Smith.

Euro-American settlement by then had progressed in the southeastern states to the point where whites and natives found their continued coexistence increasingly difficult. The first steps in making Oklahoma an Indian territory began when the Choctaws of Mississippi asked to be relocated west of the Mississippi River. In 1820 they were granted lands west of the Arkansas border and south of the Canadian River. In 1828 the Cherokees were given 7 million acres between the Canadian and the 37th parallel (the southern border of Kansas). From 1828 until 1837, treaties with the Creeks, the Seminoles, and the Chickasaws awarded them large tracts of Oklahoma land as well. These five native groups—the Five Civilized Tribes—were distinguished by the fact that all had lived in the Lower South, had practiced a diversified form of agriculture, had intermarried to some extent with European and African Americans, and had a system of laws and government patterned after the states from which they had come (map 15.3).

All of the acts that led to removal and resettling reaffirmed the principles that whites were not allowed to live in the Indian territories and that no government—except the Native Americans' own and that of the United States—would exercise authority within their borders. The latter stipulation meant that what had been the common experience up to (and after) that time elsewhere in the United States would not be repeated in the "Indian nations," as they came to be known (fig. 15.4). The practice had been to set off some area as a territory, provide it with a government, and eventually hold a popular election on admission to the Union as a state. The Indian nations were not to go through those

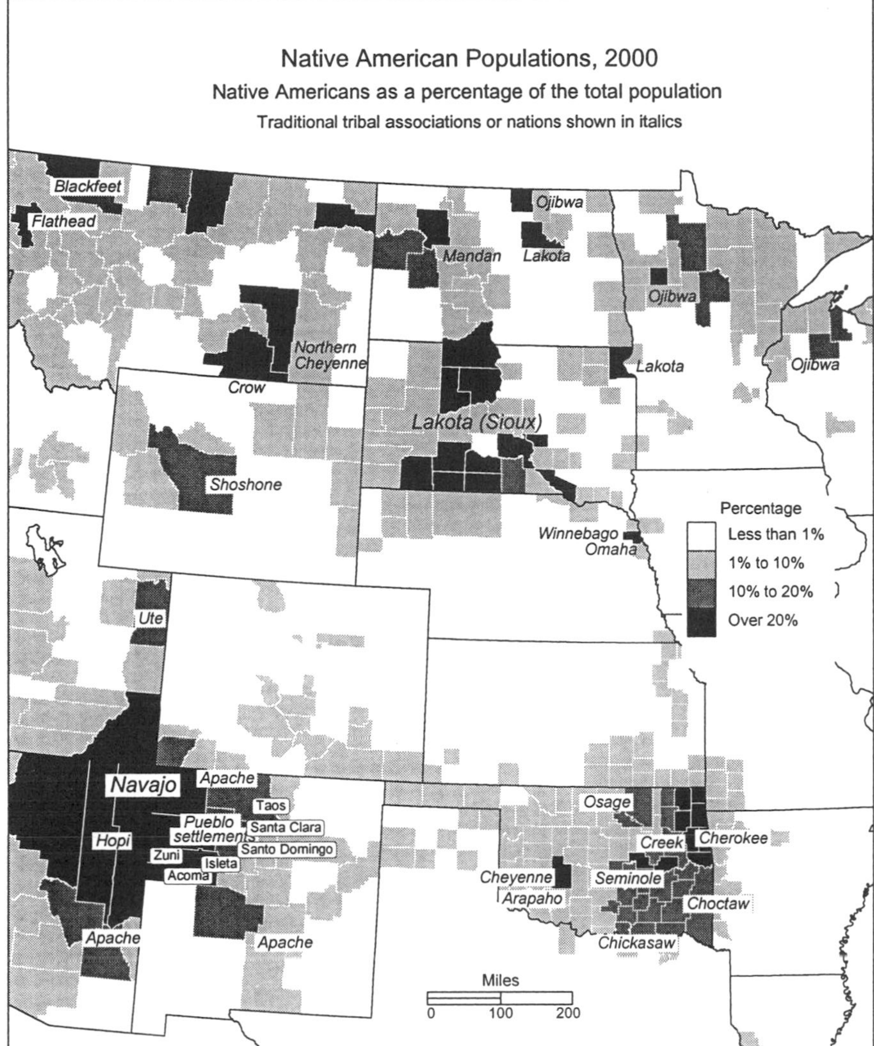


MAP 15.3

steps. What, beyond their own national governments, would be their future was left unspecified.

Between 1830 and 1848 the Indian territory in the West included not only Oklahoma but also nearly all of present-day Kansas and Nebraska south of the Platte River. Whereas most of Oklahoma had been carved up for the Five Tribes, the lands north of the 37th parallel were occupied by the so-called "blanket Indians"—tribes such as the Osage in Kansas—who had a very different lifestyle and no system of laws even remotely resembling that of the Five Tribes.

Among the factors arrayed against the future continuation of Indian territory status for these more northern lands were two important trade and migration routes, the Santa Fe and Oregon Trails, which, by the late 1840s, had be-

FIG. 15.4. *The capitol of the Chickasaw Nation at Tishomingo, Oklahoma.*

come major transcontinental corridors that passed directly through Indian territory. Contention between the slave-state and free-state delegates to Congress postponed any action until 1854, when Kansas and Nebraska were created as separate territories. This left Oklahoma at its current boundaries (minus the panhandle).

The Five Tribes played only a minor role in the Civil War. Popular support was generally for the Confederacy, especially among the most "civilized" persons of mixed blood in the Choctaw and Chickasaw nations, some of whom owned slaves. Much was made of their Southern sympathies after the war, however, and this was used as one justification for taking away the more western portions of their Oklahoma lands to create new reservations for tribes from various parts of the Middle West and the Great Plains. By 1889 Oklahoma's map was covered with the names of Native American groups, some on large reservations, others on small ones, as the process of removal to Oklahoma was broadened.

Whites began to figure more prominently in Oklahoma after the Civil War. The completion of the Missouri-Kansas-Texas Railroad from Parsons, Kansas, to Denison, Texas, in 1871 opened a corridor to outside influence, although whites (other than missionaries and teachers) still were forbidden to settle there. Squatters moved south from Kansas in violation of the law. Efforts to prevent the incursions eventually became overwhelmed by sheer numbers and persistence by those involved. Whites became so much a part of Oklahoma's economy that even the native population was not in favor of total white exclusion. Although bills to create an Indian state (of the United States) were introduced in Congress as early as the 1850s, the motivation behind such proposals nearly always was to open the Indian lands to white settlement by eliminating the Indian nations.

Despite the imitation of other aspects of American law, the Five Tribes (and others) held to the practice of tribal rather than individual land ownership. Beginning in 1887, Congress enacted measures designed to abolish tribal ownership of land in favor of allotment in severalty. Often identified with the work of the Dawes Commission, the policy of granting land to individuals also entailed restrictions on the sale of individual allotments and their associated mineral rights. Step by step, Oklahoma lost the distinctive status that it once had.



The major break came on April 22, 1889, when the so-called Oklahoma District in the central portion of the state was opened to outside settlement. Some twelve thousand hopeful settlers rushed into the district beginning at noon that day. By nightfall the towns of Guthrie and Oklahoma City had been staked out and thousands had claimed homesteads. By 1901 western Oklahoma was "opened" in this same fashion. The lands of the Five Tribes were not subject to these land rushes, but white settlers continued to move into eastern Oklahoma. In 1901 members of the Five Tribes were recognized as U.S. citizens, a precursor to the dissolution of the Indian nations. The final step came in November 1907, when Oklahoma was admitted to the Union as the forty-sixth state.

## The Oil and Gas Industry

Coincident with these developments that produced land rushes and settlement booms was the discovery of oil, not only in Oklahoma but in East Texas as well. No economic activity has had a greater impact on the two states than the discovery, production, processing, and marketing of oil and natural gas (map 15.4). Although the locations of some crude oil springs were known well back into prehistoric times, it was only after 1900, when systematic exploration began in response to a growing market, that a sense of the two states' enormous petroleum reserves began to emerge.

The only area without oil or gas resources is a band that begins in the Ouachita Mountains and extends diagonally southwestward across Texas, where eroded remnants of the folded Ouachita Mountains are buried beneath the surface. During Permian times, materials eroded from these "Ouachita folds" were deposited in a landlocked sea to the north and west, thereby creating the Permian Basin, which is the largest of the inland oil and gas fields (fig. 15.5). During the Jurassic period that followed, nearly the entire state of Texas was submerged under the waters of a shallow ocean that encroached from the southeast. Layers of sediment deposited in this ocean became the Jurassic rocks of the Coastal Plain, which were later buried under the newer (Tertiary) sediments. As the saline layer became pure salt, it expanded and pushed the overlying Tertiary rocks upward, thus creating the salt domes of the Gulf Coast region.

Salt domes are important because they are the structures within which the largest deposits of oil and gas accumulated. There are no salt domes in either Oklahoma or the Permian Basin, where a variety of other structures trap hydrocarbon resources, but salt domes are important in the Gulf Coast field. They occur without regard to the present sea level; hence oil and gas production in the Gulf Coast field take place both on land and at numerous wells in the offshore zone.

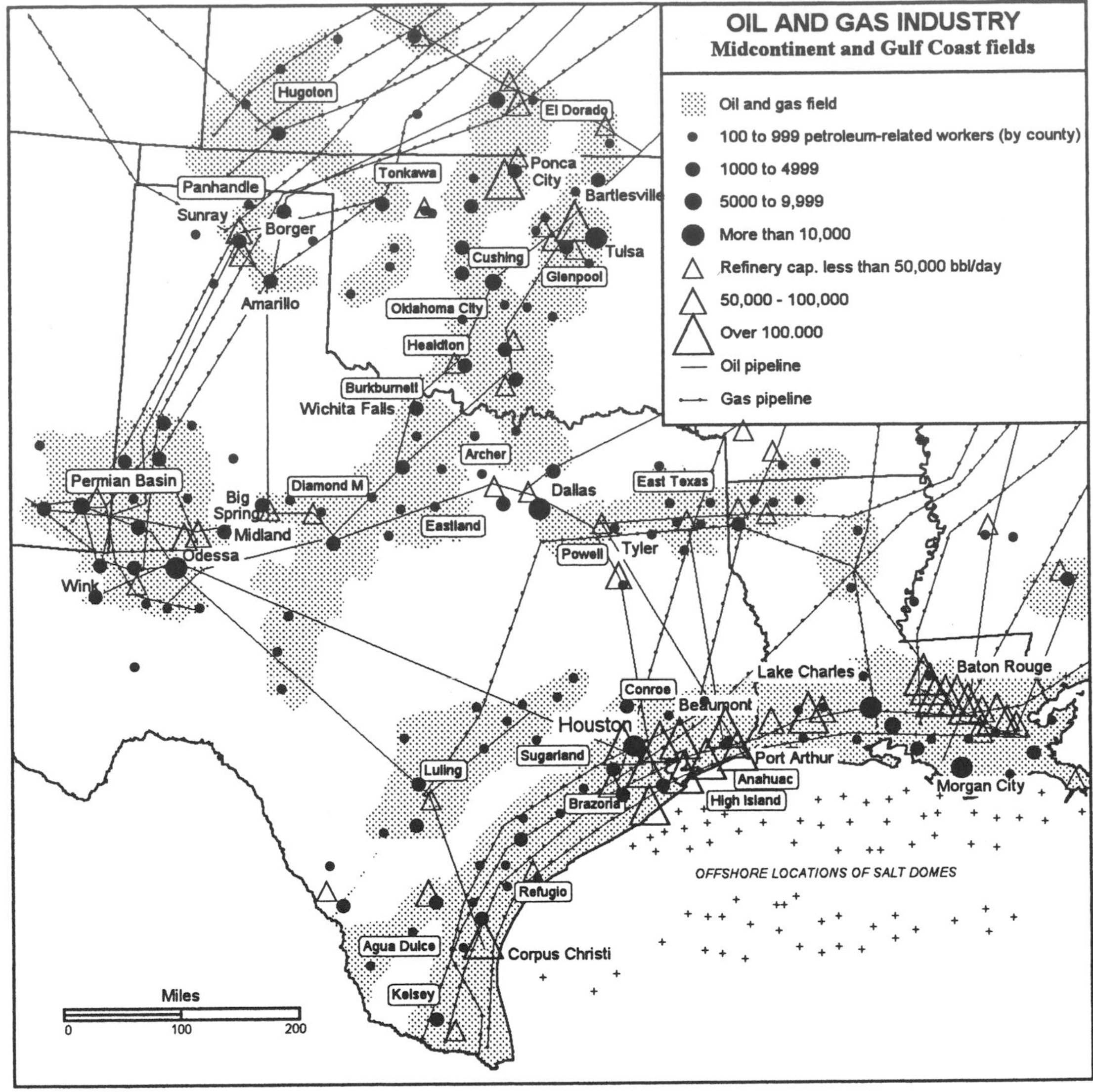


MAP 15.4

The general pattern of exploitation in both the Midcontinent and Gulf fields involved a broadening search for oil that moved from east to west before the 1930s and a movement from dry land to coastal marshes and finally into the Gulf itself since that time. The search for oil dominated the pre–World War II period. The markets for natural gas used, for example, in home heating, awaited the construction of long-distance gas transmission pipelines that took place in the 1940s and 1950s. Employment in connection with oil fields, pipelines, refineries, and oil-land leasing offices grew with each discovery and each new use, usually in boom-bust cycles, until petrochemicals became the leading industry of the Texas-Oklahoma region.

Unlike the Appalachian coal industry, which was largely a backwoods enterprise that involved no local manufacturing and was ruled from distant corporate offices, the oil business totally reshaped the economies of both Oklahoma and Texas, changing them from a rural base of farming and ranching to urban economies focused on the manufacture and sale of petroleum products. Tulsa, Oklahoma City, Bartlesville, Beaumont, Houston, Corpus Christi, Wichita Falls, Midland, and Odessa are cities that achieved prominence because of their home-grown oil and gas industries. Because the oil industry's growth was necessarily linked to the expansion of automobile ownership, Texas and Oklahoma cities are unmistakably products of the automobile era—dispersed, highway-oriented, and generally lacking the central industrial core typical of the railroad era.

The first major discovery was the famous gusher drilled at Spindletop near Beaumont, Texas, in January 1901. Three months later the first tanker ship of crude oil was outbound from Beaumont for Europe, and the city's population boomed. Spindletop was also the first salt-dome oil discovery, setting the pattern for further exploration in the East Texas and Gulf fields. Boom-town growth took place at Wichita Falls in the 1910s, at Dallas and Fort Worth in the 1920s, and at two dozen Gulf Coast locations in the 1920s and 1930s.

Although the first wildcat wells struck oil in the Permian Basin in the 1920s, that basin became even more important as a producer of natural gas after 1950. Midland and Odessa, the office-headquarters and manufacturing centers, respectively, for the Permian Basin, are the most recent boomtowns; both expanded rapidly in the 1950s and again in the 1970s. In other oil-field towns, such as Bartlesville, Oklahoma, one company (Phillips 66) grouped its manufacturing, research, and office operations at a single site, providing a continuing economic base for the city.

Houston was smaller than Topeka, Kansas, before the first oil discoveries in Texas, but by 1970 Houston was the sixth largest city in the United States. Its role

FIG. 15.5. *An oil-drilling rig being readied for operation in the Permian Basin of West Texas.*

as the leading manufacturer of oil-industry equipment brought its steel industry into regional prominence. Houston is the most important headquarters city for U.S. petroleum companies, the center of the largest amount of activity in oil- and gas-land leasing, and the leading manufacturer of both petroleum products and petrochemicals.

The term *refinery complex* used to describe the nature of these industries is not an exaggeration. Raw materials include both natural gases (methane, propane, and butane) and crude oil, including its intermediate processed stages such as cracked gases (ethylene and propylene), aromatics (benzene and toluene), and liquids (paraffins, naphtha, and gasoline). According to Exxon Chemical Corporation, one hundred gallons of crude oil can be used to produce ethylene from which 41 polyester white shirts or 13 garbage cans or 1,155 feet of polyvinyl chloride pipe are manufactured. Or it can be used to produce propylene, which would yield 46 synthetic-fiber sweaters. It can also produce naphtha, from which 910 pairs of nylon pantyhose are fabricated. Automobile tires and carpet backing are some of the other options. Equally, the same raw materials can be used to produce agricultural fertilizers (ammonia and urea), pesticides, and herbicides.

The maze of pipelines connecting catalytic cracking towers and storage tanks necessary to support such an operation requires enormous amounts of land. Chemical industries line the Intracoastal Waterway in the Houston-Baytown district and are scattered as far south as Corpus Christi. Products of the refineries are shipped by rail or barge, depending on their value and bulk, to all parts of North America or are exported directly from the Gulf ports. A Gulf Coast location for the refineries has the advantage that they can receive tanker shipments of crude oil from overseas, which, in Texas, generally means from oil fields in coastal Mexico and Venezuela.

Unique institutional environments in both Oklahoma and Texas have caused the oil and gas industries to contribute more to local economies than might otherwise have been the case. Land ownership rights in Oklahoma were in a state of complexity at the time early oil discoveries were made, because of restrictions limiting sales of individual land allotments. The Indian nations generally benefited from this circumstance because of the large number of oil discoveries made on tribal lands. The poor, rocky lands of the Osage reservation near Ponca City, Oklahoma, for example, turned out to be resting atop one of the largest oil pools in the state. For many years every Osage man, woman, and child received ten thousand dollars per year in oil royalties.

Texas had its own system of land administration. When it joined the United States in 1845, the former Republic of Texas retained ownership of its unsurveyed public lands rather than ceding them to the government of the United States. Substantial amounts of land were still in the state's possession when petroleum discoveries were made. Oil and gas royalties and leasing payments of hundreds of millions of dollars annually accrued to the state's treasury because of the unusual arrangement.

## *References*

Gibson, Arrell Morgan. *Oklahoma: A History of Five Centuries.* 2d ed. Norman: University of Oklahoma Press, 1980.

Gittinger, Roy. *The Formation of the State of Oklahoma, 1803–1906.* Norman: University of Oklahoma Press, 1939.

Jordan, Terry G. *North American Cattle-Ranching Frontiers: Origins, Diffusion, and Differentiation.* Albuquerque: University of New Mexico Press, 1993.

Jordan, Terry G., with John L. Bean, Jr., and William M. Holmes. *Texas: A Geography.* Boulder: Westview Press, 1984.

Meinig, D. W. *Imperial Texas: An Interpretive Essay in Cultural Geography.* Austin: University of Texas Press, 1969.

Webb, Walter Prescott. *The Great Plains.* New York: Ginn, 1931.

## CHAPTER 16

# The Prairie Wheat Lands

The beginnings of Euro-American settlement in Kansas and Nebraska were delayed for a more than a decade after the settlement frontier reached the Missouri Valley while the nation debated the future of slavery in the West. The passage of the Kansas-Nebraska Act in 1854 opened the new territories to settlement with the provision that the slavery question would be decided by a vote of the people. By the time settlers arrived in large numbers during the 1860s, railroads had replaced steamboats as the principal mode of transportation. Congress had passed the Homestead Act (1862), which permitted the acquisition of 160 acres of public land for only a small fee. Both of these developments caused agricultural settlement to expand rapidly once the lands were opened.

Kansas was a meeting place of North and South as well as the border between East and West. The early movement of proslavery forces into the eastern fringes of Kansas in the 1850s was followed by equally determined attempts by antislavery citizens to make Kansas a free state. Despite some early victories, the proslavery minority never had a serious chance of controlling the state, nor was it clear what the purpose of slave labor would be in an unsettled grassland where there was little employment for any class of labor, slave or free. The majority of early Kansans were neither proslavery Southerners nor strongly abolitionist Yankees. Most were Corn Belt farmers from the Middle West. They brought the type of corn-livestock agriculture they knew worked well in their home areas and assumed it could be extended westward into the Plains with but few modifications.

## The Central Great Plains Environment

Definite knowledge of the Great Plains environment—especially its climate—was lacking in the 19th century. Most of what was known had come from military expeditions and travelers' accounts. The trans-Missouri West was known primarily as a "land to get across." Of the several important trade and migration routes, the Santa Fe Trail from Kansas City to northern New Mexico was the most frequently traveled. By the 1860s some five thousand freight and emigrant wagons per year left Kansas City for Santa Fe. Travelers gathered at Council Grove, about one hundred miles west of Kansas City, where the Trail crossed the Neosho River, and there formed into trains for the long journey west (map 16.1). The route followed the Arkansas River into the High Plains, then cut south through low mountain passes to reach Santa Fe.

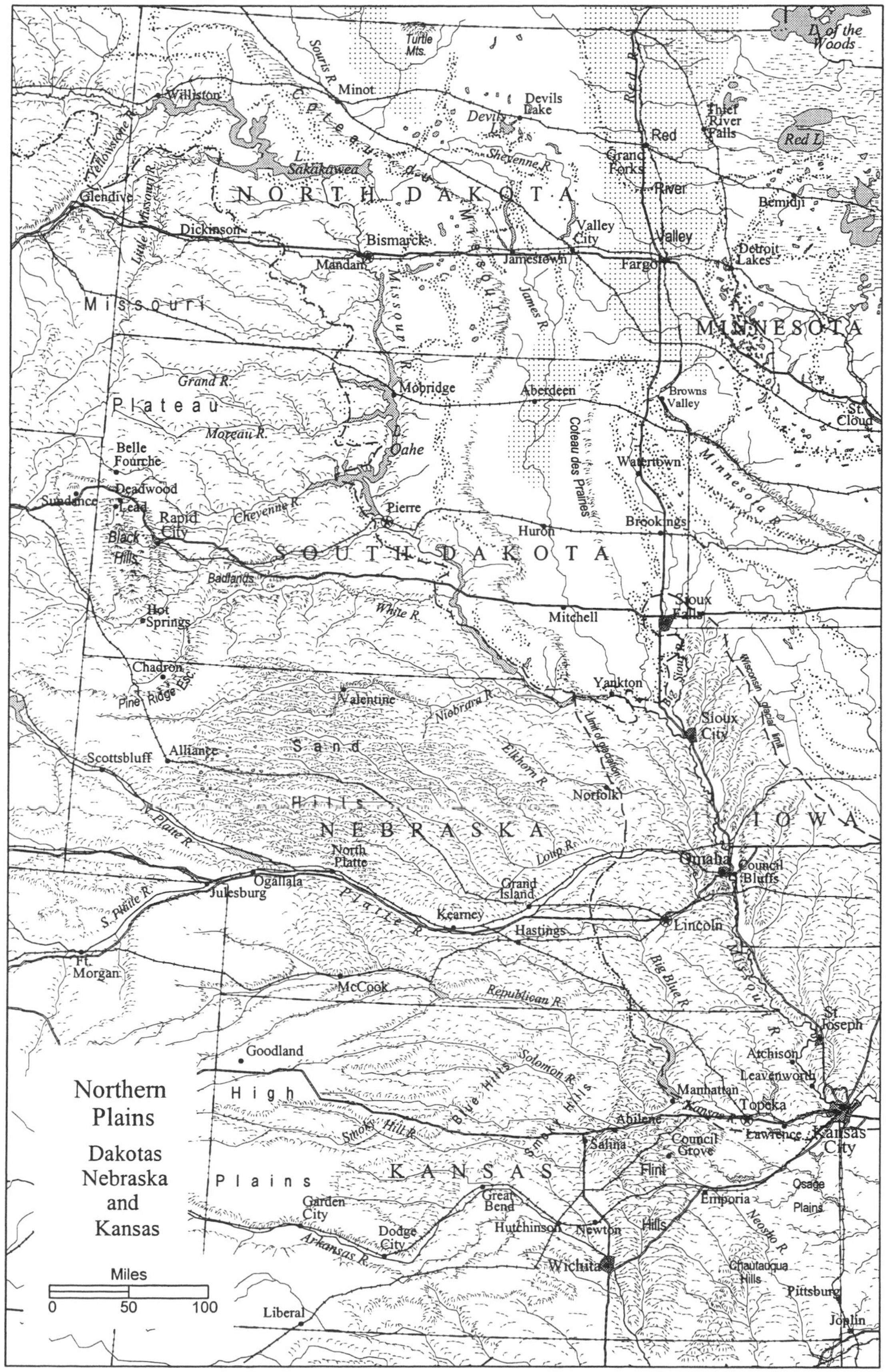
Northern
Plains
Dakotas
Nebraska
and
Kansas
Miles
0
50
100
L. of the Woods
Turtle Mts.
Souris R.
Minot
Williston
Devils Lake
Devils L.
Red River Valley
Red R.
Thief River Falls
Red L.
Grand Forks
Sheyenne R.
L. Sakakawea
Yellowstone R.
Little Missouri R.
Coteau du Missouri
NORTH DAKOTA
Glendive
Dickinson
Bismarck
Mandan
Jamestown
Valley City
Fargo
Bemidji
Detroit Lakes
James R.
Missouri R.
Missouri Plateau
MINNESOTA
Grand R.
Mobridge
Aberdeen
Browns Valley
St. Cloud
Moreau R.
L. Oahe
Coteau des Prairies
Belle Fourche
Watertown
Minnesota R.
Deadwood
Sundance
Lead
Rapid City
Black Hills
Cheyenne R.
Pierre
Huron
Brookings
SOUTH DAKOTA
Badlands
Hot Springs
White R.
Mitchell
Sioux Falls
Big Sioux R.
Wisconsin glacial limit
Chadron
Pine Ridge Esc.
Valentine
Niobrara R.
Yankton
Limit of glaciation
Sioux City
Scottsbluff
Alliance
Sand Hills
Elkhorn R.
Norfolk
N. Platte R.
NEBRASKA
IOWA
North Platte
Loup R.
Omaha
Council Bluffs
Ogallala
Julesburg
S. Platte R.
Platte R.
Grand Island
Kearney
Hastings
Lincoln
Ft. Morgan
McCook
Republican R.
Big Blue R.
Missouri R.
St. Joseph
Goodland
Solomon R.
Atchison
Leavenworth
High Plains
Blue Hills
Smoky Hills
Manhattan
Kansas R.
Topeka
Smoky Hill R.
Abilene
Lawrence
Kansas City
Salina
Council Grove
Flint Hills
KANSAS
Osage Plains
Garden City
Great Bend
Emporia
Hutchinson
Newton
Neosho R.
Dodge City
Arkansas R.
Wichita
Chautauqua Hills
Pittsburg
Liberal
Joplin

MAP 16.1

All travelers, then and now, notice the flatness of the terrain. The surface of the Great Plains appears level to the eye, but it is not quite flat. More accurately, it has an average slope upward to the west of roughly eight feet per mile, from the eastern border of Kansas to the foothills of the Rocky Mountains. The regional slope results from the fact that the rock layers forming much of the Great Plains surface were derived from materials eroded from the Rocky Mountains and then were deposited by wind and water action that carried the sediments eastward.

Drainage patterns are formed by a series of parallel, east-flowing rivers that meander down the gentle gradient, eventually flowing into the Missouri or Mississippi River. The Cimarron and the Canadian are tributaries of the Arkansas River, as is the Neosho. To the north, the Solomon, Smoky Hill, and Republican Rivers join to form the Kansas (Kaw) River, which enters the Missouri at Kansas City. North and south forks of the Platte meet at North Platte, Nebraska, and then are joined by the Loup before the stream flows into the Missouri at Plattsmouth, Nebraska. The Arkansas and both forks of the Platte carry the greatest volume of flow. Unlike the others, these three rise in the Rocky Mountains and carry snowmelt through the summer months when tributary streams originating within the Great Plains often are totally dry.

Millions of years of erosion have stripped the newer sedimentary rock cover from the eastern Great Plains. Transitions between rock formations sometimes form escarpments. Exposed sandstones produce a rough topography in the Chautauqua Hills, a sparsely settled belt in southern Kansas and northern Oklahoma. The land is too broken to be of use to crop farmers, and the soils derived from sandstones are not fertile. The land is hilly, wooded, and used mainly for grazing.

Resistant limestone layers form another north-south-trending belt of low hills between Wichita, Emporia, and Manhattan, across central Kansas. Known as the Flint Hills, uplands in this section are less suited to crop farming than to grazing, and the region has remained as a grassy range for pasturing cattle (fig. 16.1). Crops are confined to valley bottoms, where the soils are deeper. Livestock from the dry ranges of Texas and New Mexico have been shipped to the Flint Hills for well over a century and turned out to graze on bluestem pastures before feeding for market. Because so much of the Flint Hills has never been cultivated, it is one section of the Plains where the grassland of historical times can be seen and appreciated. Several tracts have been set aside as grassland preserves.

Between the Blue Hills and the Smoky Hills another limestone ledge, known as the Post Rock, forms local outcrops and is quarried for building stone. The stone got its name from its use by the area's early settlers, who cut it in post-size pillars and used it for fence posts in a land where few trees were available. Entire communities in this section of Kansas bear the imprint of stone building traditions brought by European settlers who fashioned buildings from the local rock.

Eastern Kansas and Nebraska had numerous wooded valleys when the first

FIG. 16.1. *Flint Hills, Wabaunsee County, Kansas.*

Euro-American settlers arrived. Although people made their early homes in the valleys, eventually they learned that the treeless uplands were fit for settlement. Westward across the central Plains, heavily wooded river valleys give way to lightly timbered ravines and, eventually, to no more than a heavy shrub cover in valley crevasses with totally treeless conditions on dry uplands between valleys.

The precipitation gradient largely explains the dwindling size of trees westward across the Plains, but the presence of the grassland itself cannot be explained entirely by a lack of moisture. Fires, mostly set by native people for hunting purposes, helped maintain the prairies in the treeless state that the first Europeans saw. The grazing patterns of the enormous herds of bison that roamed the Plains also helped maintain grassy conditions. After the bison herds had been decimated nearly to extinction in the late 19th century, and after the land was well enough settled that prairie fires were viewed as a menace to be controlled, trees began to return. The Great Plains is far more wooded today than it was a century or more ago.

Precipitation declines westward across the central Great Plains at a rate of roughly one inch every thirty miles. The gradient is produced by the greater distance that moist air masses must travel from the Gulf of Mexico to reach the interior of the Plains. On the region's western margin, the Rocky Mountains produce a rain-shadow effect for air masses moving from west to east. As air descends the mountain front, it warms by compression, which causes moisture to evaporate. The distance from moist, Gulf air and the rain-shadow effect of the Rockies combine to produce a low level of precipitation in the Plains.

Corn Belt farmers who settled Nebraska and Kansas experimented with a variety of crops and livestock over the last half of the 19th century. Because it was a new type of environment, unfamiliar to them, they had to learn the climate one year at a time. Crops planted in the spring, like corn, took advantage of the region's warm summers and extended growing season, but they were especially vulnerable to summer droughts. Although the average precipitation declines westward across the Plains at a predictable rate, the variation around a given average is larger here than anywhere else in the United States. Weather patterns shift abruptly in the midcontinent region—sometimes several times in a day—but they also are subject to longer-term patterns of persistence, such as sequences of cyclonic storms or long periods of dry, sunny weather. Rainfall and snowfall totals thus fluctuate markedly around the average from one year to the next. The problem of crop farming in the Plains was to be not just marginal precipitation but also unpredictability.

## The Winter Wheat Region

Crops that grow mostly during the cooler fall or spring weather and which depend as much on stored soil moisture as on regular rainfall are well suited to such an environment. Foremost among them is winter wheat, a crop that is planted in the fall, grows to a height of a few inches above the ground, lies dormant over the winter, and then resumes growth the next spring. Winter wheat is harvested in the early summer, following which the ground is idled until fall planting, when the cycle begins once more. Such crops cannot survive the severe freezing conditions of northern latitudes, but they perform well in nearly all of the central and southern Great Plains, from Nebraska to Texas.

Many of the Corn Belt farmers who pioneered the agricultural settlement of Kansas and Nebraska grew winter wheat, although few focused on it as a specialty, because wheat does not yield as much income per acre as does the production of grain-fed livestock. Large corn crops were planted for livestock feed far to the west in Kansas and Nebraska during the 1880s, which was a period of above-average moisture conditions. The wet years were followed by a series of dry years during the 1890s, which reversed the trend and led to a retreat of corn-livestock agriculture from the dry western Plains. Corn production did not return to western Kansas and Nebraska until irrigation came in the 1960s.

Although wheat yields less income per acre, it produces a surer return in an unpredictable environment, mainly because the crop has relatively low moisture demands. Several colonies of German Mennonites who had been living in South Russia were recruited by immigration agents to settle on railroad land grants in central Kansas during the 1870s. Apparently without intending to do so, since they had no foreknowledge of which crops were best suited to the Plains, the Mennonite settlers brought with them the grains of a particularly productive, hard-kerneled winter wheat that grew well in South Russia and the Ukraine, an area with a climate somewhat similar to that of Kansas. Other newly arriving settlers introduced still other strains of winter wheat that were superior to the soft-kerneled varieties, unsuited to mechanized flour production, that were then in common use.

Homesteaders who failed during the droughts of the 1890s packed their wagons and headed back east with signs proclaiming, "In God we trusted, in Kansas we busted." Had they remained, they could have taken part in one of the most rapid transformations of any agricultural region in the United States. Between 1900 and 1920, winter wheat became the overwhelmingly most important specialty crop of the central Great Plains. Kansas City became the leading flour-milling center of the nation. Wichita, Topeka, Omaha, Salina, Abilene, and Newton also became important flour-milling centers and remain so today. Spreading in all directions from central Kansas, winter wheat also became the most important crop of Oklahoma and the northern fringes of the Texas Panhandle as well as eastern Colorado (map 16.2).



Hard winter wheat is used primarily to manufacture bread flour. It is consumed in enormous quantities every day by milling concerns that supply the

MAP 16.2

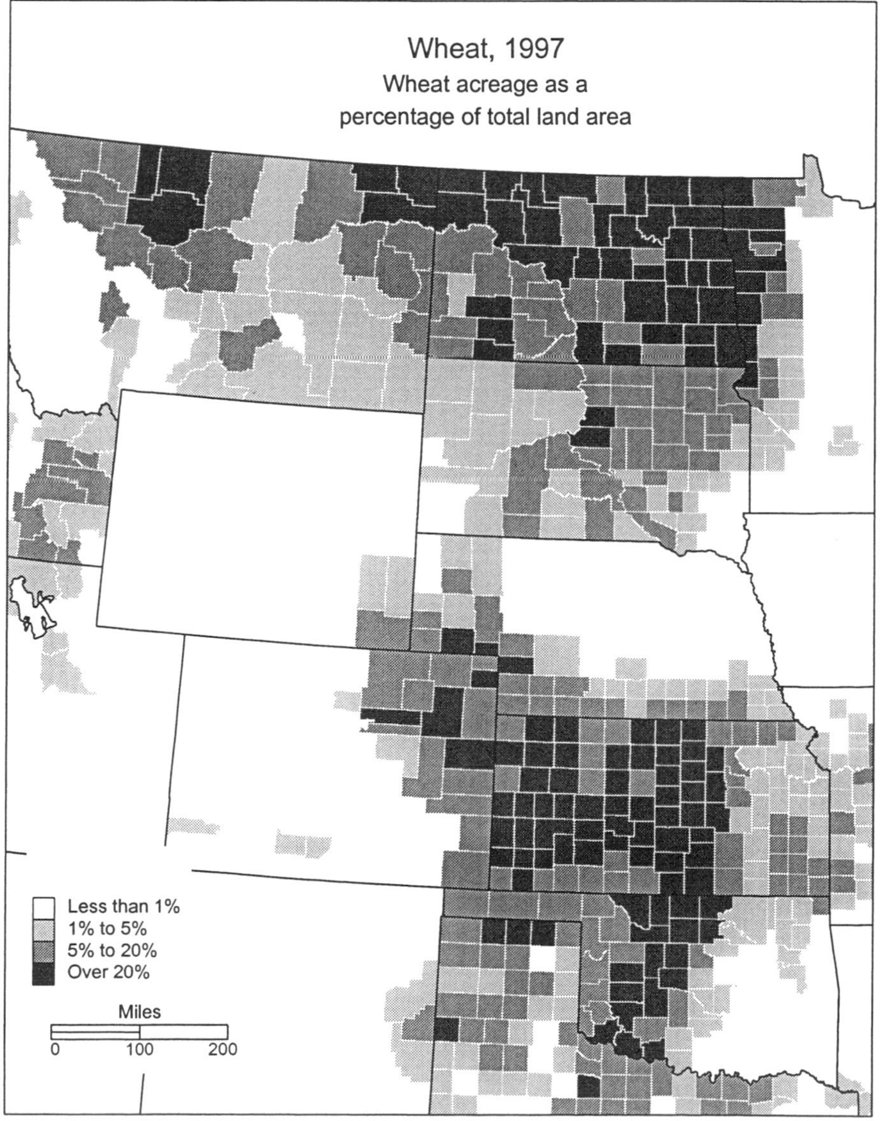

commercial baking industry. Large though the demands of the bakeries are, however, the wheat crop of the United States is so large that domestic demand has required less than half the grain produced in some years. The rest is exported. Winter wheat growers in Nebraska, Kansas, Colorado, Oklahoma, and Texas organized their own marketing cooperative to funnel grain to domestic millers and to overseas markets. Most export wheat from the central Great Plains moves through the Texas Gulf ports, especially Galveston. Asian, African, and Latin American countries are the major customers.

Winter wheat produced in the southern Great Plains is sold in markets where it competes not only with wheat from North Dakota and Saskatchewan but also with wheat grown in France, Australia, Russia, Argentina, and numerous other countries. Because it is a crop of both the northern and the southern hemispheres, large quantities of wheat come on the world market at regular intervals throughout the year. Grain-marketing cooperatives even in small towns of the wheat belt must be aware of up-to-the-minute news of bumper harvests or of crop failures around the world.

The storage of large quantities of wheat is more economical than the storage of flour, and therefore much of the winter wheat crop is stored near the major growing areas until it is sold either to foreign countries or to domestic processors. Gargantuan grain-storage elevators are a hallmark of the winter wheat belt. They dominate the skylines of cities such as Amarillo, Texas; Enid, Oklahoma; and Hutchinson, Kansas. Even smaller towns in the winter wheat belt have grain storage facilities that seem grossly incommensurate with the towns' size. Part of the storage capacity was once used for housing government-purchased wheat resulting from federal price-support programs. Given the withdrawal of the federal government from this activity during the past decade, some of the large grain elevators of the Plains now stand empty.

Wheat farming is highly mechanized today, and it requires only a small, sedentary labor force. Wheat farmers need not live on their farms year round, and many do not live there, except for busy periods during land preparation, planting, and harvest. Rather than owning their own equipment, more than half of Great Plains wheat farmers hire custom threshing crews to harvest their grain (fig. 16.2). A few weeks of the year, the landscape is alive with people and machines, but the rest of the year it appears empty. The semiabandoned look of seasonally occupied farmsteads and small, trade-center towns can be misleading, however. Wheat-farming areas have higher-than-average per capita incomes, and their economies are sound. It is simply a system that has come to rely on fewer and fewer people and requires less in the way of local infrastructure than do most other rural land-use systems.

Winter wheat is the main type planted in Nebraska and southern South Dakota, although the severity of winter temperatures limits the crop northward. Both eastern Nebraska and eastern South Dakota have agricultural economies based more on livestock feeding than on the sale of crops. The northward lapse in winter wheat production is replaced by the appearance of spring wheat culture in central South Dakota, leaving a gap between there and the Platte River Valley where little wheat is grown.

FIG. 16.2. *Wheat harvests in the Great Plains move north with the season, from Texas to Saskatchewan. Custom threshing crews follow the peak of the harvest northward.*

## The Spring Wheat Region

North Dakota and Kansas vie for the title of largest wheat producer, although the two states practice different types of wheat culture. In North Dakota and northern South Dakota, the main varieties are termed spring wheat; they are planted in the spring and harvested in the fall. Spring wheat is more suited to a northern environment, because the crop is in the ground only during the warm season, and it takes advantage of the long summer days of northern latitudes to produce a maximum growth in a short time. A North Dakota spring wheat farmer gets into the fields just as soon as they are ice-free and dry enough to work in the spring, plants the crop, and has it harvested late in the summer before cold temperatures return.

Specialized spring wheat farmers in the Dakotas, like their counterparts in the winter wheat region of Kansas and Oklahoma, often do not live on their isolated farmsteads the year around. Little on-farm work needs to be done except at planting and harvest time. Wheat-threshing crews make their living in the Great Plains by following the harvest northward, beginning in the winter wheat sections of Texas. Then, week by week, they move north through the Dakotas and arrive finally in the Canadian Prairies, where the harvest comes last. The low but seasonally specific labor demands create a slender economic base for most towns in the spring wheat region. Depopulation has resulted from every shift to technologies that have made it easier for people to travel, shop, and commute over longer distances.

Wheat has been a specialty in the Dakotas longer than it has in Kansas. The northern spring wheat region emerged from the successive westward migrations of wheat-raising farmers that began in the Genesee Valley of New York during the early decades of the 19th century. One generation moved from western New York to Michigan, Wisconsin, or Minnesota; subsequent generations pioneered new wheat lands in the Dakotas. By the 1880s the northern Plains had become the nation's principal breadbasket. Short summers with long days and moderate precipitation are ideal for the production of wheat and other small grains. The bitter cold northern Plains winter is not a limitation because no crops are in the ground outside the growing season.



Hard northern spring wheat, like the hard winter wheat of Kansas, is used mainly for manufacturing flour for baking all types of white bread. A closely related variety is durum, which produces an even harder grain that mills into a high-gluten flour that is excellent for manufacturing into pasta. Barley, which looks similar to wheat and durum when it is growing in the field, is the third major crop of the region; it is used for malting in the production of beer. Much of the bread, pasta, and beer consumed in the United States is manufactured from wheat, durum, and barley grown in the northern Great Plains (fig. 16.3).

The production of spring wheat and the related grains have been the principal economic activities of the northern Great Plains ever since the region began receiving Euro-American settlers in the 1870s. During the decade that followed, known as the Great Dakota Boom, thousands of German-, Scandinavian-, and American-born farmers flocked to Dakota Territory, filed homestead claims, and began raising wheat. From the start, Minneapolis was the economic control center for the northern Dakota economy, because that is where Dakota wheat was purchased and milled into flour. Unlike the southern Great Plains, where flour mills were important industries in many small cities, most of the northern Great Plains wheat crop was shipped either to Minneapolis for milling or to Duluth-Superior to be loaded aboard Great Lakes vessels.

Like most Northerners, North Dakotans generally voted Republican for at least three generations following the Civil War. They were scarcely conservative, however. Political life in the northern Plains centered on a continuing struggle with the millers, bankers, and railroaders of Minneapolis and St. Paul. The Republican-controlled U.S. Senate carved the northern Plains into three states in 1889 when North Dakota, South Dakota, and Montana joined the Union as separate states, making a total of six more senators who would be Republicans. Al-

FIG. 16.3. *Spring wheat, barley, and durum are stored before shipment in the grain elevators at Cando, North Dakota. The individual elevators are owned by Minneapolis grain companies or by local farmers' cooperatives.*

though "only one Dakota" might have been a more prudent course, politics dictated division.

Between 1915 and 1921, North Dakota was dominated by the Non-Partisan League, a grassroots political movement that instituted numerous reforms including the operation of a state-owned grain elevator and flour mill, a state-owned bank, and a state-owned insurance program. North Dakota's experiment in state socialism was not widely copied, nor was its existence even much known outside the region, but it has endured nonetheless. The state of North Dakota still purchases some wheat directly from its farmers, mills it into flour, and sells it under the state's own brand label.



## The Red River Valley

The northeastern margin of the spring wheat region is the Red River Valley, an international zone focused on the cities of Fargo, Grand Forks, and Winnipeg that has long produced wheat and other small grains as well as crops of potatoes and sugar beets. In more recent years, substantial quantities of sunflowers, soybeans, and corn have been grown here as well.

The Red River of the North forms the boundary between North Dakota and Minnesota and flows in tight meanders northward across the almost flat plain of former glacial Lake Agassiz. Lake Agassiz covered most of Manitoba but narrowed southward, ending at a ridge of higher ground on the Minnesota–South Dakota boundary near Browns Valley, Minnesota. South of the ridge an extension of the same valley is drained by the Minnesota River. In late glacial times, a torrent poured through the outlet south of Browns Valley that reshaped and broadened the Minnesota River valley, exposing granite formations along its sides (fig. 16.4). Lake Agassiz eventually drained almost completely, although Lakes Winnipeg, Manitoba, and Winnepegosis are its remnants.

The Lake Agassiz Plain/Red River Valley has one of the flattest surfaces anywhere in the United States or Canada. Nearly the entire region requires the artificial drainage provided by a network of ditches and culverts constructed over the past century to remove seasonal standing water. Winter snowmelt comes earlier in the upstream (southern) portions of the valley than it does downstream. In spring, the northern portion of the Red River remains frozen even though its headwaters are receiving a large volume of snowmelt runoff, which creates ice jams and floods during spring breakup. The flat topography adds to the problem because it produces massive flooding over large areas, such as the flood that destroyed portions of the city of Grand Forks in 1997.

Large tracts of Red River Valley land were purchased by eastern investors and operated as "bonanza farms" in the 1870s as an experiment in plantation-scale agriculture applied to wheat. The bonanza farms were not a success, although large wheat crops still are produced in the valley. Red potatoes, which are sold for table stock or used for potato chips, are ideally suited to the black, fertile soils of former Lake Agassiz. In U.S. soils nomenclature, the typical soil of the Red River Valley is a Boroll ("boreal Mollisol"), indicating a deep-profile soil, high

FIG. 16.4. *The rapid drainage of glacial Lake Agassiz produced a torrent that cascaded down the Minnesota River Valley and smoothed these granites near Big Stone City, South Dakota.*

in organic matter content, developed in a cold climate. The Red River Valley is second only to the Snake River Plain of Idaho as a producer of potatoes, and its sugar beet crop is the largest in the United States. Sugar beets are a profitable crop in Minnesota and North Dakota in part because of the U.S. government's tariff on sugar imports, which shelters the domestic crop by keeping the price at a high level.

Some of the most intensive wheat-producing counties in the United States lie at the eastern margin of the Red River Valley, within a few miles of totally nonagricultural, boreal forest lands in northern Minnesota. Although a broad zone of deciduous forest separates prairie from boreal forest over much of the Middle West, the deciduous zone is compressed into a narrow band in northwestern Minnesota. Historical vegetation studies suggest that the prairie once extended farther east into northern Minnesota than it does at present and that the forest has been advancing westward to invade the prairie during the past several thousand years. The result is a compression of ecotones (zones of transition between ecosystems) between boreal forest and prairie that narrows almost to a visible line of trees at the edge of the grassland at latitude 49°.

The 49th parallel was established as the international border west of Lake of the Woods by agreement between the United States and Great Britain in 1818. Although it ranks as one of the world's most peaceful international boundaries, at times issues of contention have arisen between Canada and the United States. Western Minnesota, eastern North Dakota, and the northern fringes of both states are drained by rivers that flow northward into Canada and eventually into Hudson Bay. Three river basins—the Souris (Mouse) in Saskatchewan, North Dakota, and Manitoba; the Rainy in Minnesota and Ontario, and the Red—overlap the international boundary. Schemes to divert irrigation water from the Missouri River into central North Dakota have provoked questions concerning water-quality standards in Canada.

Isolated within this area of Canadian drainage in North Dakota lies a broad, flat depression occupied by Devils Lake, a basin of interior drainage that has no natural outlet to a flowing stream to maintain the lake's level. Devils Lake's level has fluctuated over time, and during episodes of high water its expansion threatens local communities and farming areas. Proposals to connect the lake by an outlet channel to the nearby Sheyenne River (a tributary of the Red) have been opposed by the Canadian government.

## The Parkland Belt

The prairie-forest ecotone broadens northward from the international boundary. Canadians call this region the Parkland, a reference to the open, grassy nature of the landscape, which is dotted with groves of aspen and spruce. The Parkland zone widens to the northwest, covering southern and southeastern Manitoba and central Saskatchewan and extending northward to 59° latitude in Alberta (map 16.3).

Most characteristic of this landscape is the ubiquitous white-trunked aspen poplar, which grows in clumps because it reproduces mainly through its large

MAP 16.3

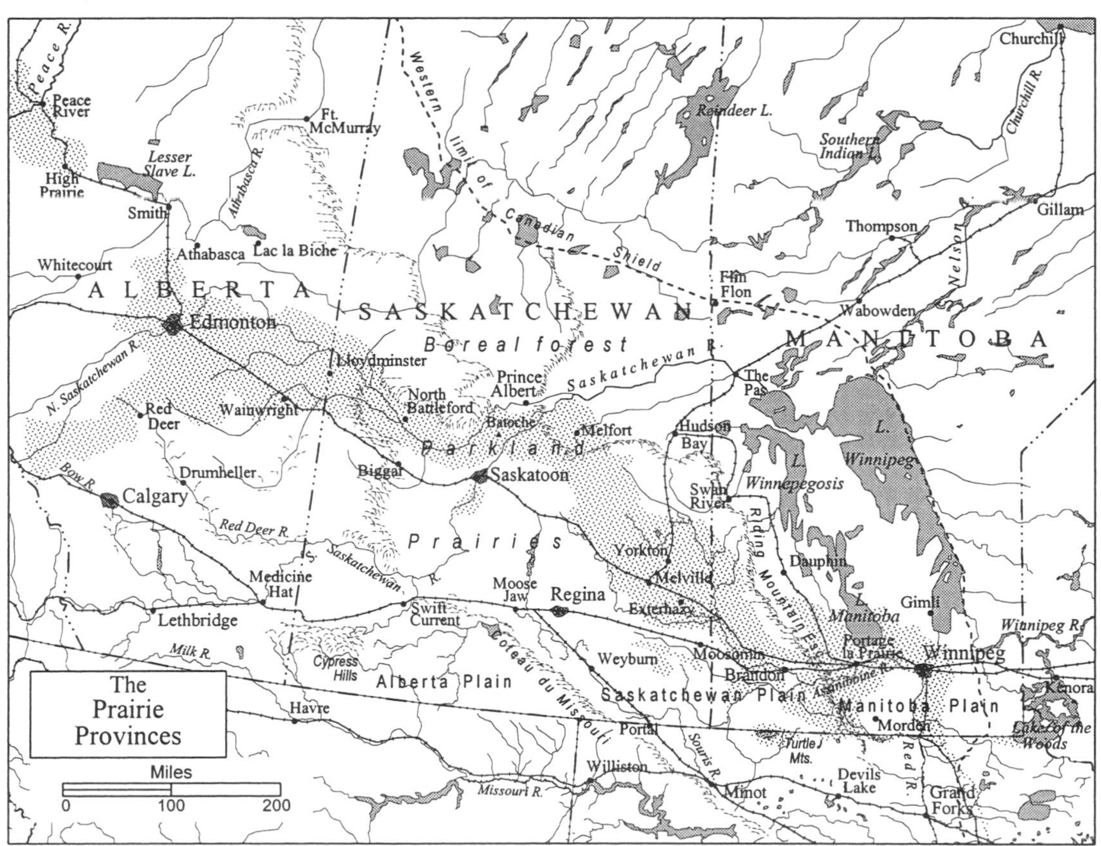

root system. Prairie fires destroy above-ground woody vegetation, but they actually encourage aspen reproduction by causing the damaged plant to extend new shoots and spread. The historic presence of large herds of bison in the Parkland, plus the role of fire in enhancing both grassland and aspen woodland, suggest that this zone may have been occupied by people for most of the period that humans have lived in North America.

Although the bison is often considered a prairie animal, the largest bison reserve in North America is in the Parkland, along the Alberta–Northwest Territories border, just a few degrees of latitude south of the Arctic Circle. The Parkland's open spaces and wooded groves attracted native people because the environment is ideal for game species. White traders, mostly French but also Scots and English, came west to Manitoba with the fur trade. Their contact with the native population produced the métis (mixed-blood) population that figured prominently in the early history of the province of Manitoba. The métis were largely a Parkland people.

Many métis employees from the fur trade remained in the region after the Hudson's Bay Company deeded its former trading territory, Rupert's Land, to Canada in 1812. When the Dominion of Canada was organized in 1867, attention was turned to creating new provinces out of the former Rupert's Land, where there had been little effective government. Lands near present-day Winnipeg were occupied by French métis who, true to the Quebec portion of their ancestry, farmed in long-lot strips stretching back from the Red River. Though long-lot farms had little economic significance, the land system was a powerful symbol of French heritage. When Anglo-Canadians, mainly from Ontario, moved into the Red River Valley in 1869, they resurveyed the land into the more conventional square grid, disregarding métis titles to the long lots.

The first Manitoba was just the southeastern corner of the present, much larger, province. When it was created in 1870, Manitoba's population of roughly twelve thousand was five-sixths métis, the majority of whom had French ancestry. Their leader was Louis Riel, an ambitious, deeply religious, and charismatic French-Chipewyan who had been educated in Quebec. Before the Canadian government could install its first territorial governor in Manitoba, Riel led a popular uprising against the coming tide of Anglo influence. The Red River Rebellion of 1869–70 ended with Riel's métis in control, but they executed one of the Canadian government's soldiers, and Riel fled Canada. Although he was elected to a seat in the Canadian Parliament representing Manitoba in 1873, that body refused to seat him.

Riel was confined to a mental institution for a brief period and then lived in exile in Montana until 1885. By then the arrival of new settlers in Manitoba, mainly from eastern Canada and Great Britain, had overwhelmed the early métis majority. Métis from the Red River and elsewhere drifted west within the Parkland belt and established a new enclave at Batoche, Saskatchewan, north of present-day Saskatoon. Here, again, they laid out their lands in long-lots only to have Anglo-Canadian surveyors ignore their claims and impose what was to the métis an alien system. Riel was enticed to return from his Montana exile to lead a new uprising.

The unfortunate events that followed have cast a shadow over western Canadian history ever since. The métis initially were victorious against Canadian forces, but as news of the second métis revolt spread in eastern Canada, thousands of men enlisted to go west and put down the rebellion—the first military expedition launched by the new government of the Dominion of Canada. Finally, with a force of some eight thousand troops, the Canadians defeated the métis at Batoche on May 12, 1885. As their leader, Riel was tried for treason. He was found guilty and, on November 16 of that year, he was hanged.



It is said that more books have been written about Louis Riel than about any other person in Canadian history. Depending on one's perspective, he could be considered a patriot, a traitor, a prophet, or merely a disturbed individual. Certainly the Canadian government made him a martyr. Just as the uprisings he led were in part symbolic of cultural and ethnic divisions between whites and natives, as well as between Anglos and the French, Riel's memory has been a rallying point ever since for cultural and ethnic groups fighting assimilation to a more dominant culture in both Canada and the United States. Manitoba, like the other Prairie provinces, was to become dominantly Anglo only to have that majority, in turn, overwhelmed by the arrival of thousands of non-English-speaking Europeans, who provided the largest share of new settlers in the prairies after the 1890s.

## The Prairie Provinces

*Prairies,* in Canadian parlance, are the northern equivalent of *Great Plains* in American terms of reference. *Prairie* is a French word indicating grassy vegetation, and *plains* refers, strictly speaking, only to topography, but both the Prairies and the Great Plains are grassy and and have low relief nearly everywhere throughout their extent, from Texas to Alberta. Canada's prairies occupy a roughly triangular region, with Winnipeg at the eastern apex and the Rocky Mountain foothills as the western edge. The Parkland bounds the Prairies on the north.

The Canadian Prairie region often is divided into three sections, or "levels" with respect to elevation. The first and lowest level is the Manitoba Plain, which corresponds roughly to the bed of glacial Lake Agassiz and has an elevation of less than 1,000 feet on Paleozoic formations east of the Manitoba (Riding Mountain) Escarpment. The second prairie level, the Saskatchewan Plain, has an elevation of approximately 2,000 feet and rests on Cretaceous formations that extend westward into Alberta. The Coteau du Missouri marks the transition to the third prairie level, the Alberta Plain, which occupies a more rolling surface 3,000 feet in elevation into which deep stream valleys have been cut. Annual precipitation is largest in the Parkland and Rocky Mountain foothill zones and smallest in southern Alberta, but the Canadian prairies lack the markedly westward-sloping precipitation gradient found in the Great Plains of the United States.

The first western agricultural settlement in Canada took place in the Parkland zone around Winnipeg, but by the late 1880s the frontier was following the course of the Canadian Pacific Railway as it was built westward into Saskatchewan. Winnipeg, which was Manitoba's capital and largest city from the start, became the principal grain-marketing, wholesaling, and transportation center for all of the Prairie Provinces. Winnipeg grew out of the métis settlement and from an even earlier attempt at agricultural colonization led by Lord Selkirk under British auspices in 1812. Regina, the capital of Saskatchewan, was founded by the Canadian Pacific Railway in its efforts to settle the prairies, as were many other towns and cities in the territory west of the Red River.

Realizing that settlement was proceeding so slowly that it would take a century to achieve a sizable farming population, the railways and the Canadian government undertook an active promotional campaign beginning in the 1890s to lure new types of settlers to western Canada. Strong efforts were made to attract Slavic people, especially the farmers of Ukraine, Bukovina, Galicia, the Carpathian Mountains and other prairie-forest zones of east-central Europe. Although such recruitment efforts were important in some parts of the U.S. Great Plains as well, colonization recruitment became the dominant mode of settlement formation in the Canadian Prairies.

Canada was more lenient than the U.S. in allowing colonization parties to settle as groups, in traditional European agricultural villages, rather than just on isolated farmsteads. The population map of the Canadian Prairies is a patchwork of localized ethnic identities as a result. Traditions of endogamy and the relative recency of the settlement have preserved many of these ethnic enclaves intact down to the present day (fig. 16.5).

Wheat was the major crop of the Canadian Prairies from the beginnings of white settlement. The better farming areas were laced with lines of railroad track along which, at regular intervals, trade-center towns were platted as grain-collection points and supply centers for the local economy. Of the three provinces, Saskatchewan became the largest wheat producer; barley, rye, and oats are its important secondary crops. Nearly all wheat is raised on Chernozemic soils (the

FIG. 16.5. *A Ukrainian church near Smuts, Saskatchewan.*

Canadian counterpart of Borolls in U.S. soils terminology), which range from black to brown according the level of precipitation. Black soils dominate in the cooler and moister Parkland; brown soils are most common in the drier portions of the prairie.

In recent years attention has shifted to the production of oilseeds for use as vegetable shortening. The leading crop is canola (rapeseed), a low, bushy plant that produces brilliant yellow flowers. Its seeds are crushed to yield a highly unsaturated cooking oil that enjoys a brisk demand on the world market. Canola can be grown in areas that have a growing season too short for wheat, and thus it is the crop that dominates the northern agricultural frontier into the Parkland belt.

Alberta presents a special case of settlement formation within the Canadian Prairies. It was a late frontier ("the Last Best West," in the promotion literature of the period) where substantial areas remained unsettled as late as 1910. Alberta's early history involved cattle ranching more than farming, and it remained more Anglo both in population and in customs than either Saskatchewan or Manitoba. One of the Canadian government's most successful campaigns to lure new settlers to the Prairie Provinces was aimed at the United States. Nearly six hundred thousand Americans emigrated to Alberta and Saskatchewan between 1898 and 1915. Though not all of them stayed, in Alberta, especially, Americans dominated the early settlement of many rural areas and small towns.

Early Alberta had a livestock-based economy. Grazing operations were concentrated in sheltered valleys of the mountain foothills, especially near Calgary. Cattle and sheep graziers came from the grassy moors and heathlands of England, Scotland, and Ireland. They were experts in the breeding of livestock, but they had little interest in crop farming. Wheat was introduced by the later waves of immigrants to Alberta, who extended the production of this staple crop of the grasslands across the Alberta Plain. Eastern Alberta had a one-crop economy, and its farmers organized cooperatives to obtain higher prices and fairer marketing practices for their crops, just as their counterparts in other wheat-growing regions did.

Like North Dakota, both Saskatchewan and Alberta produced home-grown experiments in political institutions. The Co-operative Commonwealth Federation (CCF) grew out of the farmer-labor coalition in Saskatchewan in the 1930s. In Alberta the United Farmers movement gave rise to a right-of-center popular movement known as Social Credit, which had a philosophy, according to one of its founders, of "Christianity applied to everyday economics." The CCF eventually was absorbed into other political movements in Saskatchewan, but the Social Credit Party has dominated Alberta politics for much of the past sixty years.

The Non-Partisan League, the CCF, and Social Credit were organized to combat the "Eastern" money interests that, in the farmers' opinion, squeezed them unfairly by offering too low prices for their crops, demanding too high interest rates, and setting exorbitant transportation charges. All such prairie populist movements had bona fide grievances. An almost totally agrarian society, where everyone raised the same crop on lands at the outer limits of settled territory, had no alternative but to accept what far-off market centers dictated. That these groups espoused radical economic doctrines but followed conservative social principles that at times amounted to nativism seems to have been in part a response to geographical isolation. Among the prairie social experiments, Saskatchewan's CCF probably was the most successful when, in the early 1960s, it established universal health care in the province, a prelude to the introduction of public health care throughout Canada later in the decade.

Population growth in the Prairie Provinces slowed considerably within a generation after initial settlement. As long as the economy remained agricultural, population growth and urbanization were limited by the relatively low demands of such an economy on the urban-industrial sector. Changes began with the discovery of oil in Alberta and Saskatchewan after World War II. Nearly all of Alberta is blessed with one or more types of hydrocarbon resource. Most of the province lies atop a series of geological basins that, taken together, make Alberta by far the most oil-rich province in Canada. The Rocky Mountain foothills are an important source of bituminous coal. Alberta has more natural gas than any other part of Canada, and it has extensive deposits of tar sands near Fort McMurray that are processed to extract petroleum products. Alberta is to Canada what Texas is to the United States, and in fact it is home to many transplanted Texans who work in its oil and gas industries.

Winnipeg, with its grain exchange, banks, and railway yards, was the unquestioned "primate city" of the Canadian Prairies when agriculture was preeminent, but by the 1980s both Calgary and Edmonton were larger than Winnipeg (which once had been larger than the two Alberta cities combined). The development of Saskatchewan's potash deposits created an important source of agricultural fertilizer, which is sold in North America and is exported to China and other East Asian countries. The complex of developments in Saskatchewan's mineral industries led to substantial population growth in both Saskatoon and Regina. Manitoba had no counterparts of the oil or potash booms, but Winnipeg has grown in recent decades as well, mainly as a result of in-migration from rural areas.

Alberta's oil economy boomed for thirty years until the general downturn in the oil business began during the 1980s. Both Calgary and Edmonton sprouted skyscraper buildings, which, typical of oil boomtowns, stand within a few blocks of wood-framed houses remaining from a not-too-distant past when both cities were little more than regional trade centers (fig. 16.6). Like Atlanta, Calgary has hosted the Olympic games, built an urban mass-transit system, and generally followed the list of what "important" cities do to gain recognition. The rise of Canada's petrochemical industries and its growing trade with the Pacific Rim place Calgary at the center of the national economy, a role that Montreal played a century or more ago. In many respects, Calgary and Edmonton—even more than Dallas and Atlanta—typify the Sun Belt city of the late 20th century.

The Prairie Provinces now depend less on agriculture than on resource extraction and industrial production, although their dominant image remains

FIG. 16.6. *Calgary's evolution from cow town to metropolis happened so fast that wood-frame houses remain standing near the downtown skyscrapers.*

that of the wheat field, the grain elevator, and the isolated farmstead on a rolling glacial plain. Technological and institutional change even have challenged this image. In 1895 the Canadian Pacific Railway accepted a government subsidy in exchange for the stipulation that the railway keep its rates on grain transportation at a low level for the next century. It was called the Crows Nest Pass Agreement, with reference to the government's support of a railroad line built from Lethbridge through Crows Nest Pass in the Rocky Mountains.

So-called Crow rates on grain shipments were seen as a boon to the wheat farmer because they helped stimulate the grain export business. Crow rates helped Canada shift from European to Asian overseas markets for wheat by making it economical to ship to the Pacific ports of Vancouver and Prince Rupert. Canadian railroads eventually lost money because of the rates, and they retaliated by withdrawing service on branchlines operated mainly for hauling grain from country elevators in the Prairie Provinces. When the agreement expired in 1995, wheat shipments to Canada's Pacific Coast ports declined.

By 1995, however, Canadian wheat was penetrating markets in the United States as a result of higher production costs and federal acreage restrictions south of the border. Flour millers in the United States returned to their once common practice of importing Canadian wheat. Trade agreements have enabled both Canada and the United States to sell more wheat to Mexico as well. The trade in wheat thus continues between nations, across oceans, and within trading blocs, causing an almost constant series of adjustments that are felt within the major wheat-producing regions.

## The Peace River District

Where do the wheat fields end? The Parkland–boreal forest border marks the logical northern edge of agricultural territory just as the Rocky Mountains define its western limit. The twin boundaries intersect somewhere northwest of

Edmonton, although the northern apex—theoretically, the northern limit of commercial grain farming in North America—lies farther north than might be expected. At present the northern limit is an outlier of arable land, the Peace River District, along the Alberta–British Columbia border in the Parkland zone some 250 miles north of Edmonton.

Fields of wheat, barley, and canola grow along the rolling uplands bordering the deep valley of the Peace, which is one of the larger tributaries of the Mackenzie River. Although winters at this latitude (56° north) are predictably long and hard, they are softened by the chinook winds off the eastern slopes of the Rockies. Chinooks are warm air currents that sink downslope from the mountains and warm by compression as they settle, producing a rapid although usually brief period of unseasonably warm conditions.

The Peace River District was settled slightly later than southern Alberta, mainly during the late 1920s and the 1930s. Railroads were extended north of Edmonton (and later from interior British Columbia) with the intent of making this yet another wheat-raising region with an orientation to commercial production for the export market. Although wheat failure is too common here to make it an ideal crop, canola performs well in the short, cool summers at high latitudes. Both wheat and canola are hauled by rail to British Columbia ports for shipment to the Far East.

"The Peace" is an established agricultural region, not merely a demonstration of the feasibility of northern agriculture. Despite its remoteness in terms of the settled territory of North America, it is close to the ocean ports where grain crops are loaded for shipment across the Pacific Ocean. The Arctic's two major development highways—the Alaska Highway to Fairbanks, Alaska, and the Mackenzie Highway to Yellowknife, Northwest Territories—both originate in the Peace River District, giving the region something of a strategic transportation role as well. It remains an active frontier of northern settlement.

## *References*

Artibise, Alan F. J., ed. *Town and City: Aspects of Western Canadian Urban Development.* Canadian Plains Studies, 10. Regina, Sask.: Canadian Plains Research Center, 1981.

Drache, Hiram. *The Day of the Bonanza: A History of Bonanza Farming in the Red River Valley of the North.* Fargo: North Dakota Institute for Regional Studies, 1964.

Friesen, Gerald. *The Canadian Prairies: A History.* Toronto: University of Toronto Press, 1984.

Hewes, Leslie. *The Suitcase Farming Frontier: A Study in the Historical Geography of the Central Great Plains.* Lincoln: University of Nebraska Press, 1973.

Hudson, John C. *Plains Country Towns.* Minneapolis: University of Minnesota Press, 1985.

Luebke, Frederick C., ed. *Ethnicity on the Great Plains.* Lincoln: University of Nebraska Press, 1980.

Malin, James C. *Winter Wheat in the Golden Belt of Kansas.* Lawrence: University of Kansas Press, 1944.

McQuillan, D. Aidan. *Prevailing over Time: Ethnic Adjustments on the Kansas Prairies, 1875–1925*. Lincoln: University of Nebraska Press, 1990.
Rees, Ronald. *New and Naked Land: Making the Prairies Home.* Saskatoon, Sask.: Western Producer Prairie Books, 1988.
Shortridge, James R. *Peopling the Plains.* Lawrence: University Press of Kansas, 1995.

*PART VII*

# The Western Great Plains and the Rocky Mountains

CHAPTER 17

# The Southern Rocky Mountains and the High Plains

The Southern Rocky Mountain region includes seven north-south-trending ranges, whose highest peaks are developed on an exposed core of granitic rocks near the center of each range (map 17.1). They begin on the north with the Laramie Range, which broadens southward into Colorado, where it merges with the Front Range. Paralleling the Laramie Range fifty miles to the west is the Medicine Bow Range of Wyoming, which also merges into the Front Range. The Park Range merges with the Medicine Bow and Front Ranges, then splits into the parallel Sawatch and southern Front Ranges. South of these lie the Sangre de Cristo and San Juan Mountains, which extend from Colorado into New Mexico.

Each pair of mountain ranges is separated by a substantial intermontane basin. From north to south, they are the Laramie Basin, which separates the Laramie and Medicine Bow Ranges; the North and Middle Park Basins, which separate the Front Range from the Park Range; the South Park Basin, separating the Front and Sawatch Ranges; and the San Luis Valley, which separates the Sangre de Cristos from the San Juans. The intermontane basins lie at various elevations and have floors that range in topography from flat to broken. Land uses within the basins include grazing and crops of winter wheat, barley, irrigated vegetables, and hay. The San Luis Valley also contains thousands of acres of sand dunes that were produced from a reworking of stream deposits on the valley floor.

## The Formation of the Rocky Mountains and the High Plains

The ranges of today's Rocky Mountains are fairly recent compared with the long series of events that produced the mountain mass itself. More than 100 million years ago, broad basins (geosynclines) formed along what was then the west coast of North America, roughly in the zone occupied by Nevada today. The Gulf of Mexico was connected to the Arctic Ocean by a sea that eventually filled to several miles depth with materials eroded from the land surface. The trough of sediment was subject to pressure from movement on underlying plates, and layer upon layer of rocks were thrust eastward into what are now the Rocky Mountains and the far western Great Plains.

About 65 million years ago, at the close of the Cretaceous period, these sediments were uplifted by the rise of the earth's crust from below. Volcanoes de-

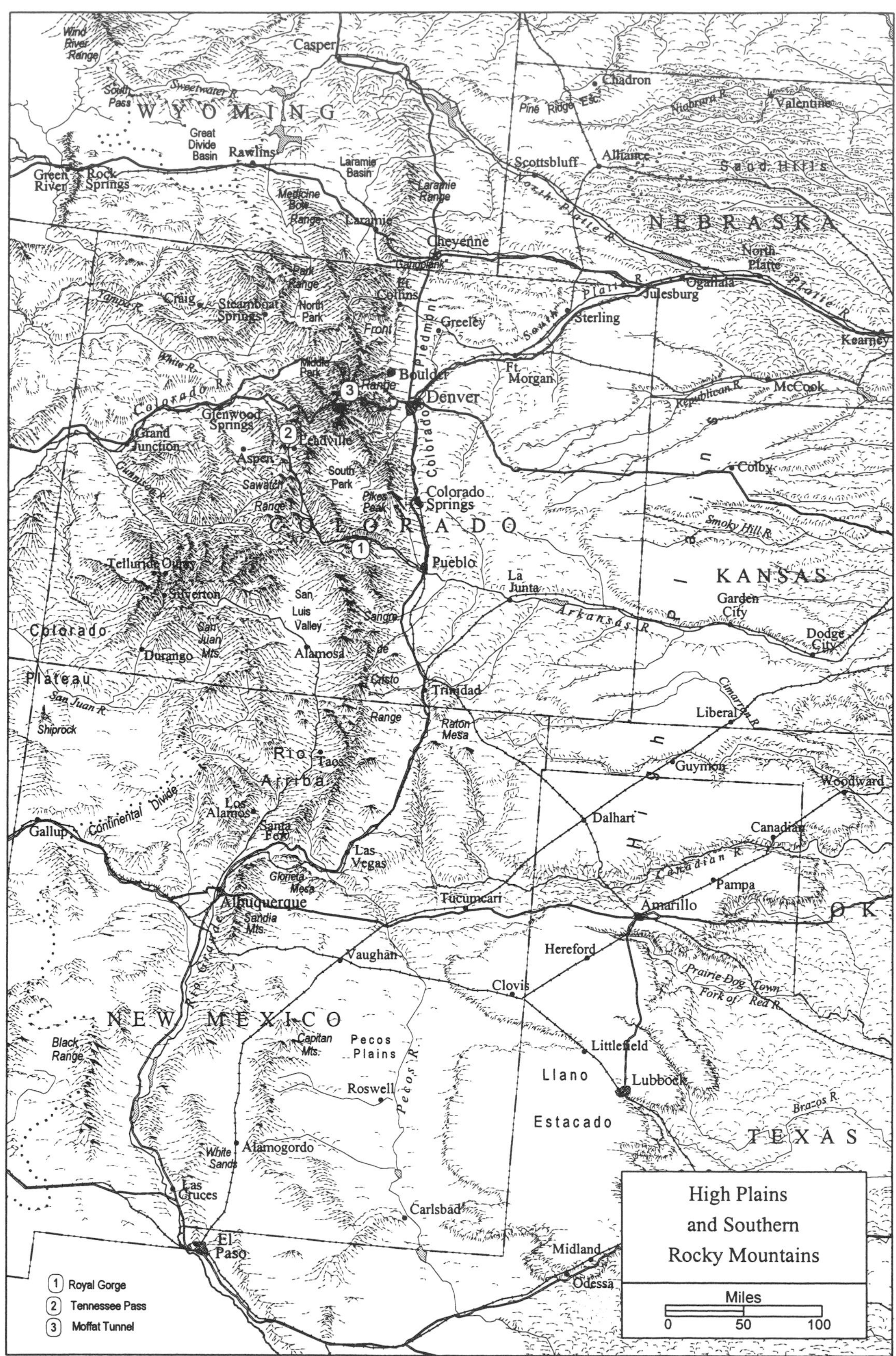
High Plains
and Southern
Rocky Mountains
Miles
0
50
100
1 Royal Gorge
2 Tennessee Pass
3 Moffat Tunnel
WYOMING
NEBRASKA
COLORADO
KANSAS
NEW MEXICO
TEXAS
Casper
Chadron
Valentine
Rawlins
Green River
Rock Springs
Laramie
Cheyenne
Scottsbluff
Alliance
North Platte
Ogallala
Julesburg
Sterling
Kearney
Craig
Steamboat Springs
Ft. Collins
Greeley
Boulder
Denver
Ft. Morgan
McCook
Glenwood Springs
Grand Junction
Aspen
Leadville
Colby
Colorado Springs
Pueblo
La Junta
Garden City
Dodge City
Telluride
Ouray
Silverton
Durango
Alamosa
Trinidad
Liberal
Guymon
Woodward
Taos
Los Alamos
Santa Fe
Gallup
Las Vegas
Dalhart
Canadian
Pampa
Albuquerque
Tucumcari
Amarillo
Vaughan
Hereford
Clovis
Littlefield
Lubbock
Roswell
Alamogordo
Las Cruces
Carlsbad
El Paso
Midland
Odessa
Llano
Estacado
Pecos Plains

MAP 17.1

veloped in the active mountain belt and pierced the surface of the Plains to the east as well. The molten rocks subsequently cooled into granite, but for millions of years the future granitic peaks of the Rockies remained buried. The newer rock formations above the granites were stripped away as uplift took place, and eventually the core of cooled, granitic rock was exposed. Mountains such as Pike's Peak emerged from within the uplifted but eroded mountain mass. The edges of the sedimentary layers that had covered the granites were turned upward and were exposed as a series of hogback ridges bordering the Front Range. In their most vertical position, the hogbacks are known as "flatirons." They are a prominent topographic feature east of the Front Range (fig. 17.1).



Surface rocks of the High Plains are of Tertiary age or younger, whereas the rock formations that underlie them—and whose edges were pushed back to become hogback ridges at the edge of the Rocky Mountains—are of Cretaceous age or older. This suggests that the Rockies are older than Cretaceous but younger than Tertiary, which is roughly true. The period of active mountain-building coincides with the transition from Cretaceous to Tertiary, 65 million years ago, the end of the age of dinosaurs.

The peaks of the Rocky Mountains might be twice as high today, however, had the summits not eroded as the mountains emerged. The erosion of the mountain mass produced sediments that buried the western Great Plains to a depth of more than fifteen hundred feet. The buried section is known as the High Plains, a region stretching from Nebraska to Texas, which is coextensive with the most recent rock layers. Forty million years ago the High Plains was an emerging surface of intertwined stream channels and river deltas as streams originating in the mountains spread their sediments downslope to the east. The High Plains once extended eastward beyond the present Missouri River, but the new surface itself eroded over time until its eastern margin had retreated to the irregular boundary found in mid-Kansas today.

In fact, erosion attacked all of the margins of the High Plains, leaving it as

FIG. 17.1. *Sandstone hogbacks (upturned rock layers) herald the approach of the Rocky Mountains immediately to the west* (left side of photo) *at Golden, Colorado.*

an isolated tableland bounded by abrupt canyons in many places. The uppermost layer of the High Plains, known as the Ogallala formation, has a nearly level surface consisting of consolidated sand and gravel deposits that were laid down by streams. The High Plains surface is so flat in the Texas Panhandle that it acquired the name Llano Estacado, or Staked Plains. It was first recorded by Europeans in the Coronado expedition, which crossed the High Plains in 1541. Coronado's party reported marking their route across the featureless plain ("a land as level as the sea") by using buffalo bones and dry dung, "there being no stones or anything else." Men became lost from the party and disoriented if they were drawn apart by as much as half a league (about 1.5 miles).

In places this Ogallala caprock forms a distinct escarpment, such as along the Prairie Dog Town Fork of the Red River south of Amarillo, Texas. Mazes of eroded canyon slopes covered with piñon (juniper) trees lie just beyond the rim of an almost perfectly flat upland. Other High Plains canyons include those formed along the Cimarron, Canadian, and Pecos Rivers, all of which rise in the mountain foothills of northeastern New Mexico. The expansion of the Pecos River's drainage area in New Mexico probably saved the High Plains from near disappearance as a result of erosion. The Ogallala caprock once extended westward, across the valley of the Pecos, but was eroded as that stream cut headward into the surrounding plains. The Pecos thereby captured what otherwise would be the drainage of the Canadian, Red, and Brazos Rivers and formed a new escarpment line of the Llano Estacado on the west.

## High Plains Settlement

The record of Euro-American occupancy of the High Plains is a story of alternating episodes of misunderstanding, optimism, abandonment, and reappraisal. Expeditions across the High Plains to the Rocky Mountains made by Lieutenant Zebulon Pike in 1806 and by Major Stephen H. Long in 1819–20 reported that the area was a desert. Pike thought the High Plains would be fit only for the "wandering and uncivilized aborigines of the country." Edwin James, the geographer of the Long expedition, extended Pike's disapproving commentary, claiming that the region "was almost wholly unfit for cultivation." Atlases and geography books began to show the label "Great American Desert" written across the lands between the Platte River and Texas.

Two branches of the Santa Fe Trail crossed the High Plains. The main route followed the Arkansas River west to the vicinity of present-day La Junta, Colorado, climbed over Raton Pass at the Colorado–New Mexico border, and skirted the mountains south and west to Santa Fe. The "cutoff" route left the main trail near Dodge City, Kansas, and followed a straight course to the base of the mountains. Despite the substantial traffic along the Santa Fe Trail—which any traveler could see was not a route through the desert—the Great American Desert label did not disappear from maps until the 1880s.

By that time railroads had replaced the Santa Fe Trail. Texas cattlemen had expanded their operations northward and were pasturing enormous herds on the short-grass plains. Outfits such as the Matador Land and Cattle Company, headquartered at Trinidad, Colorado, and the LIT spread near Roswell, New Mexico, were among the large cattle operations in the High Plains. A syndicate of Chicago businessmen received a grant of more than 3 million acres of Llano Estacado land in exchange for building a new capitol for the state of Texas in the 1880s. The land grant formed the basis for the XIT Ranch, the most extensive tract of western range ever enclosed by a barbed-wire fence as a single operation.



By the period 1910–20, the High Plains rangeland had become too valuable to use just for grazing in tracts of such enormous size. One by one, the large ranches were sold in much smaller parcels as farms, and an effort was made to lure crop farmers from other regions who would build a permanent agricultural base in the High Plains. Farmers plowed under the short-grass sod in areas that received too little moisture each year to be cultivated in such a routine manner. Southerners came and planted cotton, and Corn Belt farmers attempted crops of corn and sorghum (milo); nearly everyone who plowed the land tried to raise wheat. Within a decade the former rangelands in western Kansas, eastern Colorado, and the panhandles of both Oklahoma and Texas were plowed under and planted. All of it was a prelude to disaster in the 1930s.

By the middle of that decade the High Plains had acquired a new name, the Dust Bowl, and indirectly had provided a nickname for the decade itself (the Dirty 'Thirties). Improper land-use practices, combined with a sequence of severe droughts, caused the topsoil to blow in billowing clouds away to the east. Scenes of this disaster—both for the land and for those who tried to farm it—were recorded by photographers employed by the federal government and have been passed to subsequent generations as a stirring reminder of that era. The introduction of soil-conservation tillage practices and a withdrawal of the most marginal lands from crop production ended the Dust Bowl by the early 1940s; yet the image remains, and so, in isolated areas, do serious problems of wind erosion.

The solution to the soil-erosion problem advocated by many was to return to the days of ranching. By the 1950s, however, deep-well irrigation systems had been invented that would draw up the stored water of the Ogallala aquifer and sprinkle it or otherwise channel it across those same High Plains fields that had so recently been without water of any kind (fig. 17.2). Sorghum, also known as milo maize, is an African crop once planted in the Plains because it is a drought-resistant cattle feed. After irrigation water was available, sorghum was typically grown under irrigation. Short-staple, upland cotton, the type used for making cheaper cotton goods such as denim, also was reestablished under irrigation. The Texas Panhandle once more became the largest cotton-producing region in the United States.

If irrigated cotton and sorghum could be grown, then so could corn; by the 1970s, cornfields began to appear in various parts of the High Plains. With a base of irrigated feed-grains to draw upon, cattle feeders established large custom feedlots to fatten beef cattle for market. Feedlots attracted the beef packing in-

FIG. 17.2. *Texas County, Oklahoma. Deep-well pumps supply irrigation water for sorghum and other feed-grain crops in the High Plains.*

dustry, which has largely abandoned the Corn Belt and moved to the High Plains since the 1960s. More than half of the beef sold in the United States today is produced in the High Plains. The world's largest meatpacking plants are located here as well. Irrigation farming is now extensively practiced in the High Plains, and the production of grain for Great Plains cattle feedlots is the most important use of irrigated land (map 17.2).

The water on which this irrigated system of agribusiness depends is supplied largely by the Ogallala aquifer. Because surface water seeps into the aquifer at only about one-third the rate at which it is currently being removed, the future of High Plains feed-grains and meat production remains in doubt. The High Plains economy continues to expand, however, not only because of irrigated farming and meatpacking but also because of the abundant local reserves of oil and natural gas. Amarillo and Lubbock, the region's two largest cities, were comparatively late entries into the American urban system, but they have had growth sustained by local petroleum and agricultural developments.

Some irrigation schemes already have been abandoned because of a depletion of groundwater resources. The Oklahoma Panhandle has been the scene of some irrigation withdrawal and offers a microcosm of the larger history of land use within the High Plains. The panhandle itself—three Oklahoma counties—actually is a relict of the slavery question. When Texas was admitted to the Union as a slave state, its northern boundary was limited to latitude 36°30′, the line separating slave and free states stipulated in the Missouri Compromise. Since Kansas's southern boundary was fixed at the 37th parallel, a strip roughly forty miles wide was left as a "no man's land," effectively outside the jurisdiction of any organized government until it was eventually attached to Oklahoma.

The transition from grassy range to dry croplands to Dust Bowl and then to irrigated agriculture required little more than fifty years in the Oklahoma Panhandle. Most recently the region has become the scene of new developments in hog production. The animals are raised indoors, more like poultry than cattle, slaughtered in local packing plants, and the product is shipped to markets in Asia and the United States.



## The Hispano Culture Area

Sharply contrasting with the high-technology farming practices of the High Plains is the Rio Arriba of New Mexico, one of the most tradition-bound cultural enclaves in the United States, in the Rocky Mountain foothills just to the west. The Hispanos of New Mexico and southern Colorado have survived as a distinct culture for four hundred years, although outside influence in recent decades has reduced the area's ethnic isolation and now threatens its cultural distinctiveness as well.

MAP 17.2

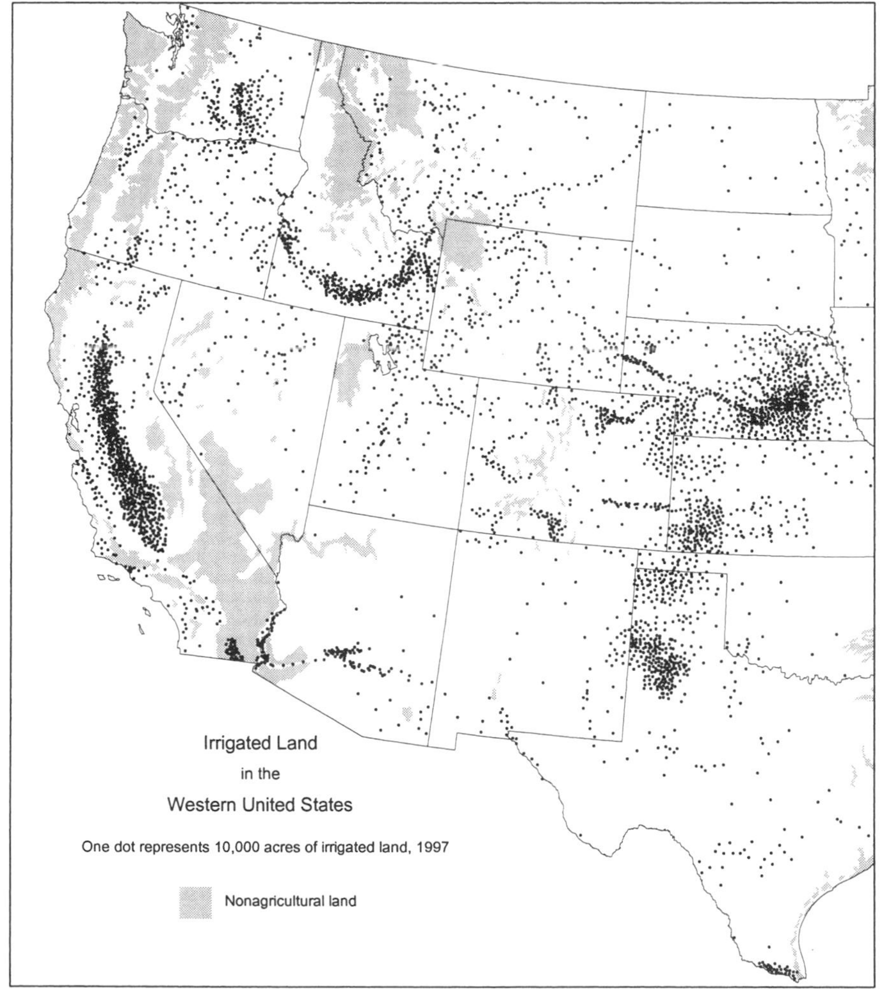

Spanish interest in the region traces to a medieval legend involving seven very large towns of fabulous wealth supposedly some forty days travel north of Michoacán, Mexico. Two expeditions were sent in search of these Cities of Cibola. The second, and most extensive, led by Francisco Vásquez de Coronado in 1541–42, found no cities of gold. Coronado did "discover" the High Plains, however, and what his expedition reported back in Mexico stimulated new plans for expeditions to the north. Coronado's party confirmed the existence of numerous agricultural villages that were settlements of the eastern Pueblo Indians, who occupied the Rio Grande Valley of northern and central New Mexico. In 1598 Don Juan de Oñate and about five hundred followers marched north, conquered the pueblo settlements, and immediately began to transform them into a frontier outpost of Mexico. The conquering Spanish imposed new customs on the pueblos and attempted to Christianize the inhabitants.

Spanish and Indian ways began to mingle almost at once. Adobe construction had been used in the pueblos long before the Spanish arrived, but the Spanish introduced the practice of sun-baking the adobe into bricks. (The Spanish had learned that practice from the Arabs in centuries past). Whereas the pueblo natives were in the habit of constructing multistory dwellings of apartment-like rooms, the Spaniards established the practice of arranging single-story adobe buildings in a rectangular arrangement facing an interior plaza, a design that offered some defensive protection for the village and created a community open space in the center where trade and ceremonial activities could be conducted. Irrigation was a long-established practice in southwestern native agriculture, but the Spanish brought techniques of irrigation farming and water diversion that increased the scale and intensity of the native system. The Hispano culture was thus a hybrid that was unique to New Mexico (fig. 17.3).

Another result of the contact was to create a mixed-blood population. The term *Hispano* is used to describe the descendants of the early Spanish–Pueblo Indian contact, simultaneously setting them apart from the later migrants of

FIG. 17.3. *A traditional sod-and-pole-roofed adobe house in a highland New Mexico village.*

Mexican origin, by whom the label *Chicano* sometimes is used. Anglos, the third ethnic component of New Mexico society, began arriving in larger numbers with the expansion of cattle ranching and railroads into the High Plains during the latter half of the 19th century. New Mexico thus evolved as a triethnic society of native, Anglo, and Hispano components, no one of which was numerically large enough to dominate the interests of the other two completely.



Most of the pueblo settlements were in the Rio Grande Valley at the time of the Oñate party's conquest. The hot, dry valley is less suited for human occupancy than are the piñon woodlands and grassy mesas that lie at a higher elevation. North of El Paso the Rio Grande Valley is not an erosional feature but rather is a 450-mile-long sequence of grabens (fault-block depressions) that were produced by rifting (pulling apart) of the earth's crust. Its northern portion in Colorado is the San Luis Valley, a tributary of the Rio Grande, where the surface is flattest. South of there the Rio Grande River skips from side to side in the long depression through which it flows. Mountainous topography comes close to the river's channel in some places; the river is bordered by steep, sandy alluvial fans in others.

The Spaniards favored sites for settlement that were at higher elevations and had a nearby stream to supply irrigation water. Santa Fe was chosen as the administrative site for the New Mexico settlements in 1609, and it has remained as territorial or state capital ever since. When Mexico became independent of Spain in 1821, the Spanish policy of insulating the colony from outside influence and supplying it with infrequent trading expeditions from Mexico was reversed. Overland trade with Missouri, by the route that became the Santa Fe Trail, commenced almost immediately. Santa Fe's broad, central plaza became the focus of a substantial trade in manufactured goods (westbound) and precious metals (eastbound). For more than twenty-five years, until 1848, when the Mexican war ended, this was an international trade and one that grew to substantial size as the years passed. It ended in the 1870s when railroads arrived from Kansas City. Santa Fe was then largely bypassed in favor of Albuquerque, which was on the Rio Grande and more accessible to the favorable mountain grades that the railroad-builders sought.

Albuquerque was thus opened to outside (Anglo) influence, Santa Fe was spared some of it, and the highland villages of Hispano farmers and sheep herders received almost none. The valleys of the Rio Grande and the upper Pecos, the smaller highland tributary valleys, and parts of the New Mexico High Plains retained a population that was more than 90 percent Hispano as late as 1900. Despite its long history—as long as any European-settled area in the United States—New Mexico's admission to the Union as a state was delayed until 1912, when, with the admission of Arizona, it could be delayed no longer. Hispano-Anglo tensions within New Mexico and fears of how elected representatives of this mixed-ethnic population might vote in Washington had defeated several statehood attempts in earlier years.

The greatest transformation came during World War II, when the federal government created a nuclear weapons laboratory at Los Alamos, near Santa Fe.

Scientists and skilled workers drawn to New Mexico became the basis for a substantial broadening of the economy into defense contracting, electronics, and computer industries in the years that followed. First Santa Fe, then Taos, were "discovered" by tourists who were fascinated by local cultural traditions that had been preserved partly because of prior isolation. Population decline in the Rio Arriba's agricultural villages became common by the 1970s as young people sought employment in the state's larger, growing cities. Highland New Mexico still contains numerous small villages with Hispano majorities, but their aggregate population represents a declining share of the state's total.

## The Front Range and the Colorado Piedmont

The Rocky Mountain Front extends as an almost continuous wall of mountains for 185 miles from the Arkansas River at Pueblo to the Colorado-Wyoming border. Its summits include Pikes Peak (14,109 feet), Mount Evans (14,264 feet), and Longs Peak (14,255 feet). West of the Front Range lies the series of small, intermontane basins, the South Park, Middle Park, and North Park. In this usage the term *park* refers to open, grassy landscapes within a zone of alpine coniferous forest rather than to a designated recreation area.

The Colorado Piedmont lies directly east of the Front Range and includes the hogback ridges at the foot of the Rockies as well as the lowland formed on older rocks of the Denver Basin. After the mountains were formed, erosion cut the connection between the High Plains and the Rockies, and a widening lowland, the Piedmont, appeared between the two. The lowland is drained by the north-flowing South Platte River and its tributaries. Denver, which originated as a small settlement on the banks of the South Platte River, is famous for its "mile-high" status. Because Denver is on the Piedmont, its elevation is actually lower than some land to the east (fig. 17.4).

The Rocky Mountain Front–Colorado Piedmont is the most urbanized re-

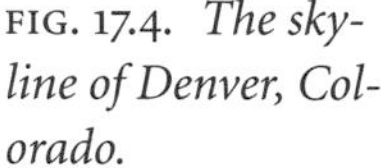

FIG. 17.4. *The skyline of Denver, Colorado.*

gion of the country between Chicago and California. Forecasts have its population of 2.25 million growing half again as large in the next twenty-five years. Its major cities—Denver, Colorado Springs, Pueblo, Boulder, Fort Collins, and Greeley—illustrate the region's multifaceted nature. Denver, the oldest and largest, began as an outfitting camp for the 1850s Colorado gold rush. The site for Colorado Springs was selected by 19th-century railroad promoters for its amenities as a resort and a health spa. Pueblo grew as a heavy-industry center based on its steel mills. Boulder and Fort Collins, both university communities, have grown mainly in the past thirty years as centers of high-tech innovation and amenity-based living. Greeley is in the center of one of the largest cattle-feeding areas of the United States.

The existence of such a large urban concentration at the foot of a major mountain range might suggest the importance of a local mountain pass, but easy routes through the Rocky Mountain Front are scarce. Only the rocky defile known as Royal Gorge, the canyon incised by the Arkansas River as it exits the mountains west of Pueblo, offers reasonably easy access to a crossing of the Continental Divide, although the route (Tennessee Pass, elevation 10,025 feet) is remote and rugged. Pueblo's heavy industries relied on the Tennessee Pass route for the supply of coal from western Colorado; iron ore for Pueblo's blast furnaces was carried in from Wyoming. Pueblo is a small industrial complex by national standards, but it has served the regional economy since 1910, and it helped stimulate industrial growth in the Colorado Piedmont.

Not until the 1930s, when the railroad through Moffat Tunnel was completed to Salt Lake City, was Colorado's largest city part of a direct, transcontinental transportation route. Until then Denver had been a regional metropolis dominating a somewhat isolated economy that was based on the fortunes made in Colorado's gold and silver mines. Denver entrepreneurs had constructed a network of narrow-gauge railroad lines that scaled the canyons into the mountains, linking the great mining bonanzas at Central City, Idaho Springs, Cripple Creek, Leadville, and even the remote San Juan Mountains at Ouray and Durango. Denver was integrated closely with the outlying mineral workings and boomtowns, but the system was confined largely to the state of Colorado and was effectively isolated from the national transportation network.

In addition to the growth stimulated by the state's mineral wealth, Denver became a stockyards and meatpacking center and a wholesale distribution center for the western Great Plains and the central Rockies. The notion that the amenities of a place—climate, scenery, access to recreation—could be a factor in regional growth did not emerge until research and development activities became a significant component of many manufacturing industries in the 1950s. California was much better known for its attractions, but the Denver region began to grow based on the same advantages.

Rapid growth based on amenity factors emerged during the 1960s, and the Rocky Mountain–Colorado Piedmont region entered the growth phase that has characterized its expansion up to the present. Denver is a headquarters city for oil companies, computer firms, and telecommunications industries. Although it has barely half a million inhabitants, the city dominates the dispersed urban zone that extends throughout the Piedmont. As an international air-traffic hub, Denver enjoys close access to cities of the United States, Canada, and Mexico.

Irrigation began on the Colorado Piedmont when settlers diverted water from the Cache la Poudre River near Greeley in the 1870s. The Piedmont's rain-shadow location dictates that water will be in short supply here. The solution was to transfer water from one basin to another, in this case from the much wetter, western slopes of the Rocky Mountains and down to the Piedmont. In the 1950s the Colorado–Big Thompson project was completed to capture water in Rocky Mountain National Park (Grand Lake, an upper tributary of the Colorado River) and send it through the mountains by tunnel to irrigate the South Platte River Valley around Greeley and scattered areas downstream from Fort Morgan. Other interbasin water transfers have been made in Colorado as well, although the diversion from the upper Colorado River to the South Platte basin is the largest.

Colorado River water is thus used to irrigate crops on the Colorado Piedmont. Cattle feeding was established here using wastes from sugar beets, one of the area's traditionally important crops. Irrigated corn and other feed grains now are more important than sugar-beet wastes in supplying the giant cattle feedlots concentrated near Greeley. Other crops, such as pinto beans and onions, are raised in the Piedmont as well, in part by Mexican-American farmers, who are numerous in this region. Here, as elsewhere, agriculture must compete with residential water demands, and the Piedmont's future growth as an agricultural region is in doubt. Pressures to reduce irrigation water consumption will increase in response to the demands of urban and suburban growth.

## The Platte River Valley

The Colorado Piedmont is a north-south corridor flanking the Rocky Mountains on the east, whereas the Platte Valley is an east-west corridor immediately north of the Piedmont. In addition to being a strip of irrigated agriculture across the Great Plains, the Platte River's broad valley is one of the oldest transportation routes in North America. The river itself is not navigable and consists of myriad braided stream channels, none carrying much water, that collectively form a river. Land routes have long followed the nearly level Platte corridor, however, from the Missouri River to the Rocky Mountains. Today the I-80/Union Pacific corridor across Nebraska and Wyoming carries on a role that trails and wagon roads played 150 years ago.

The Platte's alignment made it the most important wagon road to the West. It was followed by all of the long-distance pioneers who traversed the Oregon Trail, the Mormon Trail, and the Forty-Niners (California gold rush) Trail. Beyond the forks of the Platte, the Oregon Trail (as well as the others) followed the North Platte River past the present location of Casper, Wyoming, followed the Sweetwater River west from there, and crossed the Continental Divide at 7,500 feet in the South Pass before dropping down into Wyoming's interior basins.

The Oregon Trail had some severe grades, but elevation mattered little because the trail was not used in winter when snows blocked the South Pass. It was an all-purpose route to the West, from which branches struck in various directions at the Salt Lake oasis.

When the Union Pacific Railroad was built at the end of the Civil War, it also followed a course along the north bank of the Platte River from a point west of Omaha. Because the South Pass was not suitable for the year-around operations of a railroad, a better alternative was found by following the South Platte River to Julesburg, Colorado, and then building directly across the plains.



West of Cheyenne (a city founded by the Union Pacific Railroad in 1867), the Great Plains reaches a maximum elevation of nearly 7,000 feet above sea level. This highest portion of the Plains, the "Cheyenne gangplank," is a natural, bridging grade between the soft sedimentary rocks of the Plains and the hard granites of the Rocky Mountains, which are here eroded down to less than 1,000 feet of additional climb above the Plains surface. The Laramie Range represents the upper end of the gangplank, and it offers the easiest crossing of the Rockies anywhere north of El Paso, standing in especially sharp contrast to the Rocky Mountains of central Colorado. Beyond Laramie the railroad flanks the Medicine Bow Range and crosses the Continental Divide through the Great Divide Basin between Rawlins and Rock Springs.

At present the Platte Valley is one of the most heavily irrigated regions of the United States (map 17.2). Sugar beets are the dominant crop in portions of the North Platte Valley, especially near Scottsbluff, Nebraska. Irrigated corn crops are produced throughout the length of the Platte Valley for local beef feedlots. Both ditch (river diversion) and sprinkler (underground well water) irrigation are extensive. Although some recharge of the Ogallala aquifer takes place through the Platte River's bed, irrigation-water withdrawal produces a net deficit in the water level. Conservation interests, seeking to preserve migratory bird nesting grounds in the valley, have sought legal means to control the withdrawal of water from the Platte, but irrigation continues. Corn crops have spread up both forks of the Platte, into both Wyoming and Colorado, where they constitute the westernmost areas of large-scale corn production in North America.

## The Nebraska Sand Hills

The Nebraska Sand Hills are adjacent to the Platte Valley but contrast with it in nearly every respect. The only crop that can be raised safely in the Sand Hills is wild hay, because it involves no tillage of the land surface. The tall-grass cover of the Sand Hills keeps the soil layer intact, but the removal of the grass sod immediately exposes the sandy surface to deflation by the wind (fig. 17.5).

The one episode of crop farming in the Sand Hills was as brief as it was disastrous. In 1904 the Kincaid Act allowed entry of 640-acre homesteads in the remaining public land of Nebraska, which effectively limited its application to the Sand Hills, the only area that was left unclaimed. Settlers arrived and started cultivating the thinly covered dune sands, only to watch their farms blow away

FIG. 17.5. *Cattle gather around a water tank (kept full by a windmill-driven pump) in the Nebraska Sand Hills.*

in a few years' time. The Sand Hills was primarily a cattle-ranching country before the "Kincaiders" arrived, and it quickly returned to a focus on ranching once they had left. Today the Nebraska Sand Hills is home to some of the nation's largest cattle ranches.

The source of the sand—enough to cover some twenty thousand square miles with dunes hundreds of feet high—remains a matter of debate (see chapter 18, fig. 18.2). The sand may have eroded from local rock formations, or it may have been deposited in glacial lakes that existed within a few hundred miles to the west. The dunes themselves are geologically recent features, resting in some cases on surfaces that are no more than four to five thousand years old, well within the period of human occupancy. Dates such as these are misleading for any given dune, however, because the land surface is constantly being reshaped by wind action even though the deep-rooted, big bluestem grass that grows here anchors the sod from the most severe wind erosion.

The Sand Hills is a great reservoir of fresh water, and its porous surface is a principal intake zone for the Ogallala aquifer. The rolling landscape is dotted with small livestock watering ponds that are replenished with water pumped by windmills. The wind, which is a ubiquitous resource on the Plains, drives hundreds of isolated windmills that accomplish the pumping without electricity. The Sand Hills is cowboy country, where children grow up learning to ride horses and to do the rancher's work of herding and fence-building. This region has one of the lowest ratios of population per square mile in the United States; some county-seat towns have fewer than fifty inhabitants. The Sand Hills looks like people expect the Old West to have looked, with a grassy range extending to the horizon and the occasional cowboy on horseback. It is little known to outsiders, despite lying literally within sight of the busy Platte Valley on its southern margin.

## *References*

Atherton, Lewis. *The Cattle Kings.* Lincoln: University of Nebraska Press, 1961.

Bowden, Martyn. "The Great American Desert and the American Frontier, 1800–1882: Popular Images of the Plains." In *Anonymous Americans: Explorations in Nineteenth Century Social History,* ed. Tamara K. Hareven, 48–79. Englewood Cliffs, N.J.: Prentice-Hall, 1971.

Bunting, Bainbridge. *Early Architecture in New Mexico.* Albuquerque: University of New Mexico Press, 1976.

Carlson, Alvar W. *The Spanish-American Homeland: Four Centuries in New Mexico's Rio Arriba.* Baltimore: Johns Hopkins University Press, 1990.

Gonzalez, Nancy L. *The Spanish Americans of New Mexico.* Albuquerque: University of New Mexico Press, 1970.

Green, Donald E. *Land of the Underground Rain: Irrigation on the Texas High Plains, 1910–1970.* Austin: University of Texas Press, 1973.

Leonard, Stephen J., and Thomas J. Noel. *Denver: Mining Camp to Metropolis.* Boulder: University of Colorado Press, 1990.

McIntosh, Charles Barron. *The Nebraska Sand Hills: The Human Landscape.* Lincoln: University of Nebraska Press, 1996.

Meinig, D. W. *Southwest: Three Peoples in Geographical Change, 1600–1970.* New York: Oxford University Press, 1971.

Nostrand, Richard L. "The Hispano Homeland in 1900." *Annals of the Association of American Geographers* 70 (1980): 382–96.

Vance, James E., Jr. "The Oregon Trail and Union Pacific Railroad: A Contrast in Purpose." *Annals of the Association of American Geographers* 51 (1961): 357–79.

Vigil, Ralph H. "Spanish Exploration and the Great Plains in the Age of Discovery: Myth and Reality." *Great Plains Quarterly* 10 (1990): 3–17.

Worster, Donald. *Dust Bowl: The Southern Plains in the 1930s.* New York: Oxford University Press, 1979.

CHAPTER 18

# The Missouri Plateau

In comparison with the High Plains, the Missouri Plateau, which borders it on the north, is even more sparsely populated. The two regions meet at the Pine Ridge Escarpment, an irregular wall of rocky buttes along the Nebraska–South Dakota border (map 18.1). The Rocky Mountains bound the Missouri Plateau on the west, although various mountain inliers, including the Black Hills, are found in the Plateau region. The Missouri Plateau's eastern border is the Coteau du Missouri, a band of low hills stretching from central South Dakota into Saskatchewan and Alberta. The Missouri Plateau thus extends into Canada, although it is more often identified there with the third and highest prairie level, the Alberta Plain.

The surface rocks of the Missouri Plateau are comprised of Tertiary and Quaternary formations in many places, as in the High Plains, but the Plateau has no counterpart of the Ogallala aquifer. Consequently, irrigation is less common in the Missouri Plateau, and the region contains no major areas of livestock feeding and feed-grains production.

Like the High Plains, however, the Missouri Plateau's newer surface rocks are derived either from sediments originating in the Rocky Mountains or from volcanic activity that accompanied the rise of the Rockies. The Missouri Plateau was a region where dinosaurs roamed before the end of the Cretaceous period (65 million years ago). Volcanic activity associated with mountain-building spread layers of volcanic ash downwind from the Rockies at that time, burying the remains of the dinosaurs. The bones were subsequently rearranged, perhaps during episodes of catastrophic flooding that reshaped the Missouri Plateau's surface during past millennia. Alberta, eastern Montana, Wyoming, and the western Dakotas have become well known for discoveries of megafaunal remains during the past century.

Some of the rock layers comprising the Missouri Plateau weather easily and produce a landscape of intricate gullies ("badlands") on a broad, gently sloping plain. Occasional hard-rock layers in the geologic sequence produce the buttes, bluffs, and scattered tablelands that give the landscape a stair-stepped quality. The low natural fertility of soils produced on rock formations such as these, the vulnerability of exposed surfaces to rainsplash erosion from occasional thunderstorms, and the semiarid climate have combined to limit agricultural development in the region.

This is also a region of mountains-within-the-plains, of broad basins between mountain ranges, and of countless patches of broken ground that can be described neither as plains nor as mountains. The association of grassy vegetation with the Great Plains and of coniferous forests with the Rocky Mountains is complicated as a result. Nearly all sharp topographic breaks in the Missouri Plateau have a cover of ponderosa pine or other drought-tolerant conifers. The coniferous forests increase toward the west in every stepped increase of elevation as the sequence of geologic formations progresses toward the Rocky Mountain foothills.

This chapter on the Missouri Plateau also covers that portion of the Rocky Mountains that lies east of the Continental Divide, including the Yellowstone Plateau and the adjacent mountains of southwestern Montana. Geologically speaking, these mountains are not part of the Missouri Plateau, but they are equally part of the Missouri River's drainage basin, and their human occupancy is related to developments east of the Divide more than to those west of it.

In central Wyoming a gap in the Rocky Mountains connects the Missouri Plateau on the east with the Colorado Plateau on the west. The connection, known as the Great Divide Basin, is a four-thousand-square-mile region of interior drainage between Rawlins and Rock Springs. Streams flow to the Colorado River on the west and to the Missouri on the east, but within the Great Divide Basin they meander slowly downslope and eventually disappear through infiltration and evaporation. This basin, which lies literally within the Continental Divide, permits an almost mountain-free crossing of central Wyoming and has been an important route of travel ever since the Union Pacific Railroad was built across its barren wastes in the 1860s.

## Mountains and Basins

The term *basin* has several geographical meanings. It may refer to a collection zone for interior drainage, such as the Great Divide Basin. It also can refer to surface topography, such as a lowland drained by a river. A third meaning refers to a buried structure, such as a downwarping of rock strata of the sort often associated with coal, oil, or gas deposits. Wyoming has all three types of basins, and they figure prominently in the state's topography, resources, and economy.

The map of Wyoming suggests that its rivers and mountains are somewhat out of adjustment. Rivers like the Big Horn flow straight through the mountains rather than around or between them. The rivers are very old features that are part of a system known as superposed drainage. As the present mountain peaks were exhumed by uplift and erosion, today's streams were already flowing, above the peaks, on a land surface that subsequently eroded away. The rivers cut their way down through the mountains at the same time that the peaks were emerging.

Wyoming's widely separated mountains include the Wind River Range, which has peaks above 13,000 feet in elevation, more than a mile higher than the dry, sagebrush scrublands at their base. The Big Horns are another formidable range, famous for the switchback highway that climbs the hogback ridges on

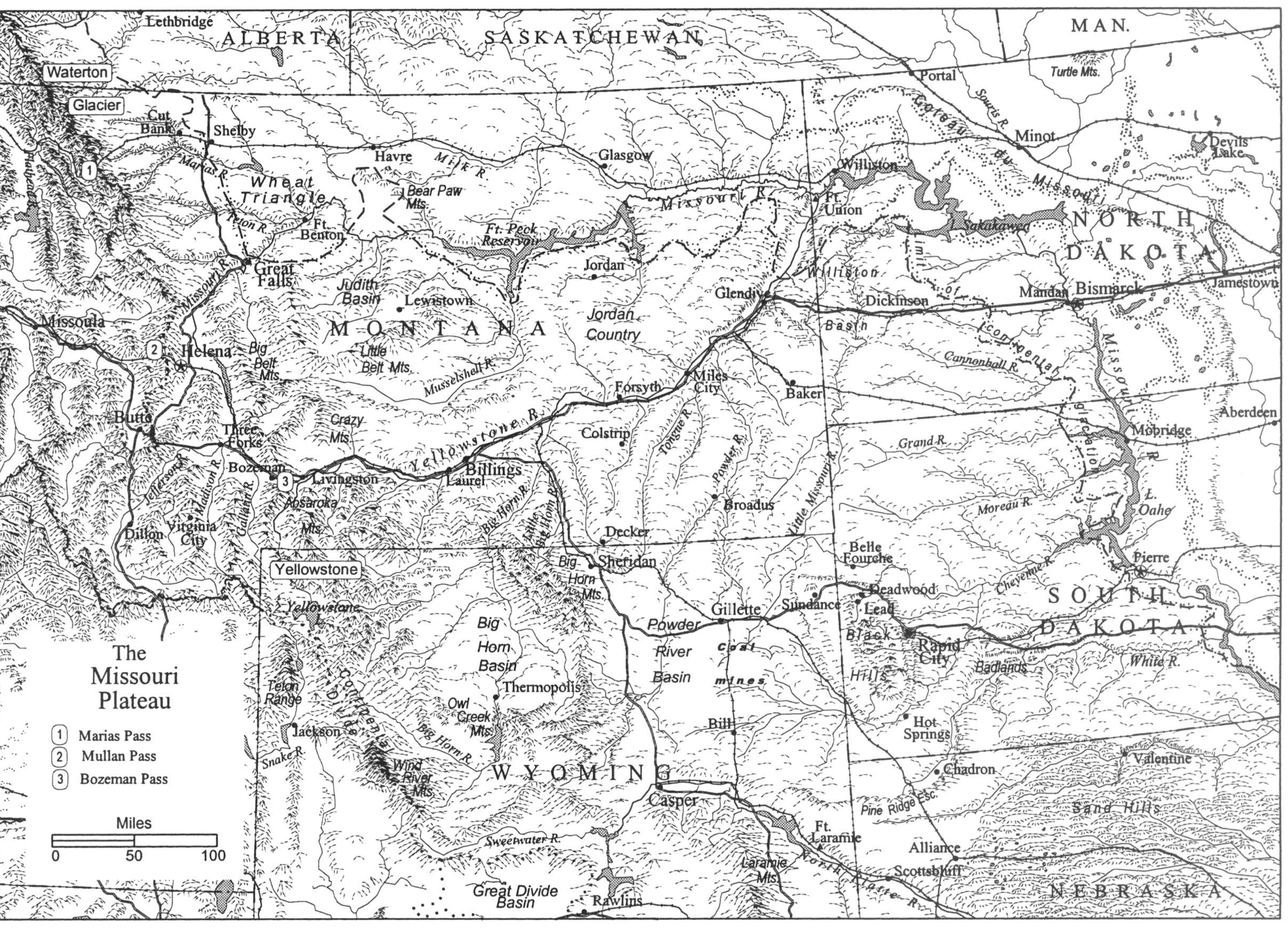

The Missouri Plateau
1 Marias Pass
2 Mullan Pass
3 Bozeman Pass
Miles
0
50
100
ALBERTA
SASKATCHEWAN
MAN.
MONTANA
NORTH DAKOTA
SOUTH DAKOTA
WYOMING
NEBRASKA
Lethbridge
Waterton
Glacier
Yellowstone
Cut Bank
Shelby
Havre
Glasgow
Portal
Minot
Devils Lake
Turtle Mts.
Souris R.
Coteau du Missouri
Williston
Ft. Union
Sakakawea
Missouri R.
Milk R.
Marias R.
Wheat Triangle
Bear Paw Mts.
Teton R.
Ft. Benton
Ft. Peck Reservoir
Jordan
Great Falls
Judith Basin
Lewistown
Missoula
Helena
Big Belt Mts.
Little Belt Mts.
Musselshell R.
Jordan Country
Glendive
Williston Basin
Dickinson
limit of continental glaciation
Mandan
Bismarck
Jamestown
Cannonball R.
Forsyth
Miles City
Baker
Aberdeen
Mobridge
Grand R.
Butte
Three Forks
Crazy Mts.
Yellowstone R.
Colstrip
Tongue R.
Powder R.
Little Missouri R.
Bozeman
Livingston
Billings
Laurel
Jefferson R.
Madison R.
Gallatin R.
Absaroka Mts.
Big Horn R.
Little Big Horn R.
Broadus
Moreau R.
L. Oahe
Dillon
Virginia City
Decker
Belle Fourche
Pierre
Sheridan
Big Horn Mts.
Deadwood
Cheyenne R.
Gillette
Sundance
Lead
Powder River Basin
coal mines
Black Hills
Rapid City
Badlands
White R.
Big Horn Basin
Thermopolis
Teton Range
Continental Divide
Owl Creek Mts.
Jackson
Snake R.
Big Horn R.
Wind River Mts.
Bill
Hot Springs
Chadron
Valentine
Casper
Pine Ridge Esc.
Sand Hills
Ft. Laramie
Alliance
Sweetwater R.
Laramie Mts.
Scottsbluff
North Platte R.
Great Divide Basin
Rawlins


MAP 18.1

their ramparts. The margins of both the Wind River and Big Horn Mountains are partially buried by alluvial fans that rest on ancient erosion surfaces, an arrangement characteristic of mountains in a desert environment. Both of these Wyoming ranges, as well as the Black Hills of South Dakota, also resemble the Rocky Mountains in that their highest peaks are formed on the granitic rocks that comprise the central, uplifted core of the mountain mass (fig. 18.1).

Both the eastern and western flanks of the Big Horns are zones of irrigated agriculture in a region where only sparse hay crops grow without a water supplement. Most of Wyoming is public land rather than privately owned cropland. The public domain is used as a range for cattle and sheep rather than for crops. As is typical of the Mountain West, ranch headquarters are generally at low elevations, along streams, where irrigated hay meadows provide winter range for livestock.

Stock are driven to higher pastures in summer, through a maze of federally controlled lands. Most grazing lands at low elevation are administered by the U.S. Department of the Interior's Bureau of Land Management. At higher elevations, where trees and grass replace sagebrush, national forests take over, and control passes to the U.S. Department of Agriculture's Forest Service. Water for irrigation is controlled by the Interior Department's Bureau of Reclamation. Western ranchers, though sometimes seen as irrationally hostile to the federal government, probably have more experience dealing directly with it than do most people.

Domes or uplifts are the natural counterparts of the underground basins. Buckling and stretching of rock layers rearrange the hydrocarbon resources trapped within the sedimentary layers. Wyoming's most famous such structure is the Teapot Dome, an oil-producing region north of Casper, whose name became synonymous with political scandal after illegal oil-land leases were made there during the administration of President Warren G. Harding in the 1920s.

FIG. 18.1. *Eroded granite peaks in the center of the Black Hills uplift west of Rapid City, South Dakota.*

Casper's petroleum-refining industry is an important part of that city's economic base, and Wyoming is the fifth largest oil-and-gas-producing state.

The Williston Basin, which underlies most of the western Dakotas and eastern Montana, also has abundant oil and gas resources. Production began in the 1950s and is large enough to support refinery complexes at Billings and Laurel, Montana, and at Bismarck and Mandan, North Dakota. Western North Dakota also has substantial reserves of lignite—sometimes called "brown coal"—a substance that is midway in its evolution from peat to coal and which consequently has relatively low heat-giving qualities. It is abundant, and several attempts have been made to exploit lignite as a resource on a larger scale, including its conversion to natural gas.

## Powder River Basin Coal

In a profound reversal of resource-use patterns and transportation logistics, Wyoming's formerly little-known reserves of subbituminous coal have become one of the most important natural resources of the United States during the past thirty years. A series of federal clean-air acts have mandated reductions in sulfur dioxide emissions for industrial smokestacks. The legislation coincided with an era of increasing distrust of the nuclear power installations that once were thought to be the preferred supplier of heat for future electric-power generation. High-sulfur coal, though remaining abundant in the Interior Coal Field of Illinois, Indiana, and Kentucky, for example, could no longer be used for electric-power generation without the installation of expensive pollution-abatement technology. Since the first Clean Air Act in 1969, the national air-quality standards have been raised several times, which further reinforced the desirability of burning low-sulfur coal.

The nation's largest reserve of clean-burning coal is in the Powder River Basin of Wyoming and adjacent portions of Montana. As a surface basin, the region is drained by the Powder River, although the river basin itself is unrelated to the coal deposits. As an underground structure, the Powder River Basin is a broad downwarp that connects the rise of the Big Horn Mountains on the west with that of the Black Hills on the east. The underground basin contains billions of tons of subbituminous, low-sulfur coal. (The designation *sub* refers to its inferior heat-giving qualities, measured in BTUs per ton, as compared with, say, the high-grade bituminous coals of eastern Kentucky.) Once standards were mandated by the Environmental Protection Agency for sulfur dioxide emissions from smokestacks, however, the desirability of low sulfur content more than overcame the disadvantage of lower heat-giving properties.

Powder River Basin coal is cheap, abundant, and ideally suited for steaming, that is, for heating water to generate electricity. The annual production of Powder River coal increased from 70 million tons in the early 1970s to 190 million tons by the early 1990s. Wyoming is now the leading coal-producing state, ahead of Kentucky and West Virginia, the two leading Appalachian producers. The main coal-bearing formation consists of a seam roughly one hundred feet thick,

FIG. 18.2. *A mile-long coal train, one of more than two dozen loaded each day in the Powder River Basin, winds eastward through the Sand Hills. Some trains may travel a thousand miles to reach the single electric-power plant the train serves.*

which is buried under only a shallow layer of overburden that is easily stripped away. The reserves, which are measured in the thousands of quadrillion BTUs, can last a century or more even at the current rate of mining, which is roughly 300,000 tons per day. The Black Thunder Mine, near the remote hamlet of Bill, Wyoming, produces more than 35 million tons of coal per year and is the world's largest coal mine.

Coal is hauled by rail to electric-power plants, to barge terminals on the Mississippi River system, and to Superior, Wisconsin, where it is poured into Great Lakes vessels. Approximately 250 coal trains, consisting of one hundred or more cars each, are in daily use hauling the output of Powder River Basin mines (fig. 18.2). Electric-power plants from the Pacific to the Gulf through the Great Lakes region now rely on western, low-sulfur coal. It is the most important source of energy used to generate electricity in the United States today.

The 1970s were boom years in the Powder River Basin, but thus far, at least, no substantial population growth has followed. Modern coal strip mining is capital intensive. There are no grubby "coal towns" in Wyoming, nor are the ugly scars of strip mining as visible, because of environmental regulations requiring restoration of the land surface following mining. For the next century and more, the production of electricity in the United States will depend on Powder River Basin coal.

## Dry-Farmed Wheat on the Missouri Plateau

Ice sheets advancing toward the west overrode the eastern edges of the Missouri Plateau, then retreated back east to form the range of morainic hills known as the Coteau du Missouri. The Missouri River lies west of the Coteau but slightly east of the limit of glaciation. Two types of landscape evolved on the Missouri Plateau—a smoother, glaciated, somewhat moister zone on the north and east and a more dissected, unglaciated, drier area on the south and west. The first became a wheat- and barley-growing region; the second, which is marginal for grain crops, has remained livestock-oriented ever since the first large herds of cattle were pastured there in the 1870s.

Most of the better wheat lands of the northern Great Plains had been settled by 1900. General economic prosperity in the years that followed saw a new movement toward the remaining uncultivated lands to the west as land prices advanced in older, settled areas. The world's demand for wheat continued to increase. One consequence was the emergence of the winter wheat belt in the southern Great Plains. Between 1900 and 1920 wheat farming also advanced in the north, into the better lands of the Missouri Plateau, especially its glaciated sections. Northwestern North Dakota, northeastern Montana, and the "wheat triangle" of northwestern Montana (bounded by the cities of Shelby, Havre, and Great Falls) became spring wheat–growing areas at that time.

The main limitation to successful wheat culture, even where the soils and topography permitted it, was the lack of precipitation. The solution was a set of land-use practices known as dry-farming, which, employed together, permitted the extension of grain production into the semiarid climate zone. Those who have flown over the western Plains have seen dry-farming's most visible aspect, the parallel strips of alternating cropland and fallow that are actually an adaptation to climate rather than to the land (fig. 18.3). The fallow land absorbs moisture during the year that it is not used for a crop. It is cultivated to remove moisture-robbing weeds but otherwise lies idle. The following year it is planted in wheat, and the previous year's cropland is then idled so that it, in turn, can absorb moisture. Two years of precipitation are used to grow one year of crop. Farmers alternate their crops and fallow in a series of parallel fields, giving a striped pattern to the landscape.

Wheat production is decreased by half in such a system, but it has proved to be a sustainable method for using lands of the subhumid and semiarid Plains—

FIG. 18.3. *Dry-farmed wheat fields, with alternating strips of cropland and fallow, also can be oriented to minimize wind erosion.*

a compromise between grazing (not cropping at all) and conventional farming, which would be unsuccessful in this zone of marginal precipitation. Other dry-farming adaptations include limiting the frequency and depth of soil tillage to reduce erosion. Great Plains wheat farmers rarely plow their fields because it opens the surface to erosion. They disturb the ground surface as little as possible and leave a residue of plant matter on the soil surface. Critics who have argued that the western Great Plains should not be cropped at all have underestimated the value of these fairly simple expedients to cope with the marginal nature of precipitation. Wheat farming is sustainable on the Missouri Plateau with such conservation practices.

Dry-farmed wheat and barley areas of the northern Plains have even lower population densities than other grain-growing districts because of the relatively low intensity of land use. The region has no large urban centers, and those cities that do have more than a trade-center function, such as Minot, Billings, and Great Falls, have economic bases dependent on a narrow range of economic activities. In recent years, conservation reserve programs of the federal government have withdrawn a substantial amount of dry-farmed wheat land from production, especially on rolling to hilly portions of the Missouri Plateau. Acreage reductions have produced a new cycle of infrastructure decline and population loss, especially in the zone just south of the U.S.-Canadian border.

## The Great Sioux Reservation

The juxtaposition of geologically complex mountain systems with the dry, western Great Plains had the incidental effect of intensifying native-white conflicts in the latter half of the 19th century. The cause derives on the one hand from the nature of mountain-forming processes. Rocks subjected to intense heat and pressure rearrange their physical structures and chemical properties; those actions form gold, silver, and other precious-metal deposits. Gold and silver were discovered in a succession of strikes, beginning with California in the late 1840s, Colorado in the 1850s, Montana and Idaho in the 1860s and 1870s, and the Black Hills also in the 1870s.

Each gold discovery drew a stream of Euro-American fortune seekers through the Plains, which prompted the U.S. Army to build forts for their protection. Meanwhile, the western Great Plains, unwanted by the majority of white Americans, had also become one of the last refuges of the native people, a region where vast acreages still were being awarded "in perpetuity" to some native group. A treaty concluded at Fort Laramie in 1851 set aside most of the land between the Missouri and Powder Rivers as a reservation and hunting grounds for the Sioux. Every trail that crossed Indian hunting grounds and every fort constructed by the military to protect white travelers became an issue of contention between natives and the federal government.

The near demise of the bison accompanied the arrival of white hunters, just as protracted military conflict, disease, and starvation eventually amounted to near genocide against the Plains Indians. The Missouri Plateau was the scene of this conflict. Much of the area set aside under the Fort Laramie Treaty of 1851 became known as the Great Sioux Reservation, only to be subsequently whittled down in numerous treaties.

The Sioux (the Dakota/Lakota) are recorded as living in woodland environments on the upper Mississippi River in the late 17th century. The French from Montreal traded with them for furs. Eventually, as the westward push of European settlement from the eastern seaboard increased, the Sioux moved westward, first into the Minnesota River Valley (from which they were later removed after bloody conflict) and eventually into the Missouri Plateau. They gave up their woodland ways, adopted the horse and the rifle, and began to live a life on the open plains that was based on the availability of bison for food and clothing.

The Sioux were by no means the only native people in the region, however, and their history through the 19th century is punctuated by conflicts with other Plains groups. The pressure intensified as white settlement moved westward. First wagon trails, then railroads were projected across the short-grass Plains toward California or Montana. In 1868 Red Cloud, chief of the Oglala Sioux, persuaded the government to close the Bozeman Trail, which connected the Oregon Trail with Montana, because traffic along it had grown substantially during the Montana gold rush. In exchange, the Sioux agreed not to disturb an even more permanent intrusion, the route of the new Union Pacific Railroad then being built across Wyoming.

Despite occasional swaps such as this one, the clear direction was to decrease Indian territory in every treaty. After gold was discovered in the Black Hills in 1874, there was immediate pressure on Washington to extinguish Indian title to southwestern Dakota Territory. When this was done in 1876, the Sioux mounted their strongest effort, one that culminated in the death of General George Armstrong Custer and 250 of his men at the Battle of Little Big Horn, Montana. "Custer's last stand" was the end of significant Indian victories. In treaty after treaty into the 1900s, the Great Sioux Reservation was reduced to roughly eight counties in the driest portion of West River South Dakota, which became the permanent reservations for more than half of the tribes of the Sioux nation (see chapter 15, map 15.3).

After 1900 white settlers began demanding that all of the Missouri Plateau be opened to homesteading. In 1907 two railroad lines were constructed across South Dakota's West River country, linking Rapid City to the Missouri River. Thousands of land-seekers crossed the Missouri and filed claims on land so marginal that more than half of the farms failed within five years. That the fringes of the South Dakota Badlands could once have seemed so attractive to farmers, railroad builders, or national policymakers is hard to imagine today, but in the early 1900s, an era when the last free land in the West was being claimed, the pressures to open all land to farming—regardless of title, quality, or previous treaties—was intense.

Dry-farmed wheat did succeed on the better uplands of the Missouri Plateau, especially on the northern, glaciated portion, where the soils are better. The weathering of the softer surface rocks of the unglaciated Missouri Plateau produces a sticky clay soil that is not only drought-prone but also impassably slippery when wet. "Gumbotil," as it is known locally, proved to be totally unfit for agriculture. Soils, rather than rainfall, turned out to be the best predictor of agricultural success on the Missouri Plateau.



Rapid City developed into an important trading and financial center for the Plateau region. Deadwood and Lead, the gold rush towns of the Black Hills, survived the initial boom and became urban centers as well. The Homestake gold mining operation at Lead consolidated its holdings and grew to become the largest straight-gold mining operation in the Western Hemisphere. Production has declined in recent years, but the Homestake mine remains as an important reserve of gold for the United States. Although the extent of the Black Hills forest is small in comparison with the vast acreages of pine timber farther west in the Rocky Mountains, the Black Hills continues to supply logs for lumber manufacture in this region where timber resources are scant.

Family-based automobile tourism is the leading industry of the Black Hills region. The state of South Dakota was an early promoter of tourism and employed advertising copywriters beginning in the 1930s to boom the state's attractions to tourists who would drive there for a vacation. As an outlier of mountain scenery hundreds of miles closer to the more densely populated Middle West than the Rocky Mountains, the Black Hills and the Badlands became favorite stopping points on the way west. Recent demands that all or part of the Black Hills be returned to the Sioux tribes have stimulated intense discussions locally, given the impact that such a transfer could have on the tourism economy of the region.

## Missouri Basin Water Projects

Two large rivers drain the Missouri Plateau: its namesake, the Missouri, on the north; and the Yellowstone, a major tributary that joins the Missouri near the old fur-trading post of Fort Union on the Montana–North Dakota border. The Missouri River was once navigable by steamboats from St. Louis as far upstream as Fort Benton, Montana, even though that trip took weeks to complete through the meanders and point bars of the Missouri's twisting, sand-choked channel.

Today, however, only pleasure boats can navigate the Missouri River above Sioux City, because the river has been permanently impounded behind five enormous earth-fill dams built between 1940 and the 1960s mainly for flood-control purposes. The U.S. Army Corps of Engineers, which is responsible for flood control and navigation on the inland waters, desired the large dams on the main river. The Bureau of Reclamation, which controls irrigation, sought to build more dams, but on the Missouri's tributaries, in order to expand the amount of water available for irrigating crops. In the late 1940s Congress approved the Pick-Sloan plan, which basically added the two agencies' plans together and appropriated funds to build all of the dams. Only relatively short stretches of the Missouri, immediately below the giant dams, resemble the river as it once existed.

The Missouri River rivals the Tennessee in the extent to which it has been permanently dammed, but the Yellowstone River has no dams anywhere from its mouth to its source in Yellowstone National Park. Irrigated lands along the Yellowstone, which produce crops of sugar beets and hay, receive water from some of the Yellowstone's tributaries rather than from the main river. Also unlike the Missouri River, which never had a city on its banks for four hundred miles from Williston, North Dakota, to Fort Benton, Montana, the Yellowstone River Valley contains some of Montana's larger cities.

The Yellowstone was a natural corridor across the rough topography of the Missouri Plateau, which, though it is rarely mountainous, has many buttes and other topographic breaks that make the uplands difficult to cross. When the Northern Pacific Railroad was built west across Montana in the 1880s, the company chose to follow the Yellowstone River for 350 miles from Glendive to Livingston. In so doing, the railroad founded Billings and most of the other cities of the Yellowstone Valley, establishing it as the "Main Street" of the Northwest.

Eastern Montana between the Yellowstone and Missouri Rivers is sparsely settled even by standards of the Missouri Plateau. Its eastern portion, sometimes called the Jordan Country, after the isolated community of Jordan, its only trade center, is inhabited by widely scattered sheep and cattle ranchers. Ranch families are so few that school-age children live in a dormitory adjacent to the high school in Jordan when classes are in session. The Jordan Country is bordered on the west by the Musselshell River Valley, a zone of dry-farmed wheat culture and cattle ranching. North of the Musselshell Valley's pine-clad slopes lies the Judith Basin, an intermontane lowland set off by the Little Belt Mountains. Wheat and barley grown in the Musselshell Valley and Judith Basin are shipped to the Pacific Coast for export.

## The Rocky Mountains East of the Continental Divide

The Yellowstone Plateau separates the continuous ranges of parallel fault-block mountains typical of the Northern Rocky Mountains from the pattern of widely separated mountain found in Wyoming. The Yellowstone Plateau is primarily a volcanic feature, a massive series of lava flows that surrounds a collapsed, extinct volcanic crater (caldera), Yellowstone Lake. More than 2 million acres of the lava plateau were set aside by Congress in 1872 as Yellowstone National Park.

The Yellowstone lava plateau is bounded by the Absaroka Mountains on the east and north. On the south it connects with the Wind River and Teton Ranges, which are separated by the Jackson Hole Lowland. The geysers that have made Yellowstone Park famous are caused by water seeping into the ground and coming into contact with the semimolten rocks above a "hot spot" of the earth's upper mantle. Hot springs and geysers abound where intense heating takes place not far beneath the earth's surface.

Yellowstone was the first national park in the world and continues to rank as one of the most popular scenic attractions in any country (fig. 18.4). In addition to hosting millions of visitors each year, Yellowstone National Park is one of the largest wildlife preserves in the United States. It is home to substantial

FIG. 18.4. *The falls of the Yellowstone River in Yellowstone National Park, Wyoming.*

herds of elk and bison; it is a natural preserve for the grizzly bear and, more recently, the timber wolf. Yellowstone's role as a wildlife sanctuary may become its most important one in the years ahead.

The mountains within Montana's portion of the Missouri Plateau include the Snowy, Big Belt, Little Belt, Crazy, and Bearpaw Ranges. Several of these resulted from thrust faults that carried a sheet of folded rocks eastward under the impact of continental collision, creating isolated mountain ridges. These isolated, fault-block ranges of west-central Montana were avoided, first by the railroad surveyors and then by the highway engineers, who simply detoured around them. The mountains' low silhouettes, above the otherwise unbroken Great Plains horizon, help give meaning to the "big sky" image of Montana.

Montana's mountain ranges, which strike in various directions of the compass, also give the state's settlement pattern a somewhat haphazard quality. The tighter the fabric of mountain ranges and valleys, the more important is the influence of the location of favorable mountain passes on settlement patterns. In Montana, where the arrangement of mountain ranges returns to a pattern more like that of Colorado, a solid front of mountains marks the edge of the Great Plains. Although the term *front range* is not used in Montana and Alberta, the topography more closely resembles that of Colorado than it does Wyoming.

When the Northern Pacific Railroad was built across Montana in the 1880s, its line exited the Yellowstone Valley west of Livingston, crossed the Absaroka Mountains near Bozeman, and descended to the Missouri River Valley at Three Forks, where the Madison, Gallatin, and Jefferson Rivers join to form the Missouri. Like Lewis and Clark before them, the railroad builders were faced with the problem of finding a mountain pass that would take them across the Continental Divide. Lewis and Clark struck west and south from the three forks of the Missouri but found no easy route to Idaho. The railroad builders chose to head north from this point by following the Missouri downstream to Helena, which lies near the approach to Mullan Pass, the site of an old wagon road over the Continental Divide that is low enough to permit construction of a railroad. Helena began as a gold-mining camp in the 1870s but amounted to little of permanence before the railroad over Mullan Pass was constructed and the city then was chosen to be Montana's state capital.

The other major city in this portion of Montana is Great Falls, a former center of copper-smelting and copper-refining activity and once the state's largest urban center. The Great Northern Railway was extended west to Great Falls in 1886 to tap the Montana copper traffic, but no mountain pass existed west of the city that would permit an easy crossing of the Continental Divide. Great Falls prospered nonetheless, both from its role in the copper industry and from being the principal center of trade and milling activity for the nearby wheat triangle region.

At 5,900 feet, Mullan Pass was a relatively low-level crossing of the Continental Divide, although it is not the lowest pass in Montana. In 1889 the Great Northern Railway secured a route through what has remained ever since as the easiest known crossing of the Divide in the northern United States. Marias Pass, at an elevation of 5,280 feet, had been known to Euro-American traders in earlier times, although it was not used as a transportation route because of the reputed ferocity of the Blackfeet Indians who occupied its eastern approaches. Marias Pass, which is a slight niche in the mountain wall at the upper end of a tributary of the Flathead River, lies almost within sight of the rolling, grassy Great Plains just to the east.

The Great Northern found Marias Pass, built a railroad through it, and also recognized the tourism potential of the area. The railroad promoted the idea of setting aside a national park, which Congress did in 1910. Glacier National Park became a northern counterpart of Yellowstone and grew into nearly as popular a tourist attraction. Canada created an adjacent reserve and designated it as Waterton National Park.

Both of Alberta's major cities, Calgary and Edmonton, also grew as railroad centers along transcontinental lines aimed at favorable passes in the Rocky Mountains. The Canadian Pacific extended its line from Calgary up the Bow River Valley through Banff and crossed the Continental Divide (also the Alberta–British Columbia border) at 5,400 feet through Kicking Horse Pass (see chapter 19, map 19.1). Another transcontinental line, through Edmonton, crossed the Divide at Tete Jaune (Yellowhead) Pass near Jasper at an elevation of only 3,720 feet, although it was a route requiring a substantial northern detour to achieve. Both of these lines stimulated the growth of cities at their eastern approaches, and both Calgary and Edmonton prospered in their roles as cities along the transcontinental passage.

Paralleling the American example of railroad-promoted tourism, both Canadian lines sponsored the creation of national parks at their Rocky Mountain crossings as well (fig. 18.5). Jasper and Banff National Parks in Alberta are, taken together, the most visited mountain parks in North America. Tourists from all parts of the world, especially from Asia, are in the throngs gathering at the two parks during the brief summer season. As elsewhere in the Rockies, year-

FIG. 18.5. *Moraine Lake, Banff National Park, Alberta.*

round developments based on skiing have more than doubled the number of tourists visiting the parks each year.

## *References*


Brown, Dee. *Bury My Heart at Wounded Knee: An Indian History of the American West.* New York: Holt, Rinehart & Winston, 1971.

Clements, Donald W. "Recent Trends in the Geography of Coal." *Annals of the Association of American Geographers* 67 (1977): 109–25.

De Voto, Bernard. *The Journals of Lewis and Clark.* Boston: Houghton Mifflin, 1953.

Hargreaves, Mary Wilma M. *Dry Farming in the Northern Great Plains, 1900–1925.* Cambridge: Harvard University Press, 1957.

Hart, Henry C. *The Dark Missouri.* Madison: University of Wisconsin Press, 1957.

Martin, Albro. *James J. Hill and the Opening of the Northwest.* St. Paul: Minnesota Historical Society Press, 1991.

Nelson, Paula M. *After the West Was Won: Homesteaders and Town-Builders in Western South Dakota, 1900–1917.* Iowa City: University of Iowa Press, 1986.

———. *The Prairie Winnows Out Its Own: The West River Country of South Dakota in the Years of Depression and Dust.* Iowa City: University of Iowa Press, 1996.

Popper, Deborah Epstein, and Frank J. Popper. "The Great Plains: From Dust to Dust." *Planning* (December 1987): 12–18.

Schell, Herbert S. *History of South Dakota.* 3d ed. Vermillion: University of South Dakota Press, 1975.

*PART VIII*

# The Intermountain West

## CHAPTER 19

# The Northern Rocky Mountains and the Columbia-Snake Plateau

The Continental Divide asymmetrically divides the Northern Rocky Mountains into a relatively narrow fringe of foothills and low mountains east of the drainage divide and a much larger region of mountain ranges and basins to the west (map 19.1). The asymmetry is expressed in terms of elevation, because the eastern foothills lie at a comparatively high level in the Missouri Plateau, whereas valley floors west of the Divide are at lower elevations. Because the wagon trails and railroads built through the Rockies were the work of a westward-looking people, and because the objectives of trade were concerned largely with hauling goods in the westbound direction, transcontinental routes were most often built through passes favorable from the eastern slopes. The western destinations for the routes were less important. There were only a few Pacific Coast urban centers to choose among at the time the trans–Rocky Mountain surveys were made, and it was assumed that new cities would be built as needed.

The greater range of elevations on the Rocky Mountains' western slopes also means a greater range of ecosystems. Valley bottoms in the western mountains are warm and dry. The temperature is controlled more by altitude than by latitude. The single transition from grassy, wind-swept plains to forested mountains on the eastern flanks of the Rockies has no counterpart on the west, where a variety of temperature and moisture regimes produces vegetation ranging from desert shrubs to dry pine forest in the lower elevations. Correspondingly, the western slopes have a greater range of land uses and support a greater variety of crops.

The evergreen-clad slopes of the Northern Rockies are mostly a part of the public domain, notably the national forests. No region of the United States has a larger proportion of land in this category. Twenty separate national forest jurisdictions administer cutting, grazing, and recreational land use. The familiar sequence of grazed or cropped valley bottoms, grazed and forested midelevations, and densely forested high country is repeated on every mountainside. Controversies that result from this mixture of uses include the demands by nonresident groups that neither timber-cutting nor grazing be permitted on public land. Forests are dynamic entities, however, and they cannot simply be fenced off in the belief that their composition will remain constant over time. Management strategies must evolve continuously to meet the competing needs of the various groups. During the past decade urban-based nonresidents have won notable victories restricting logging and grazing on public lands.

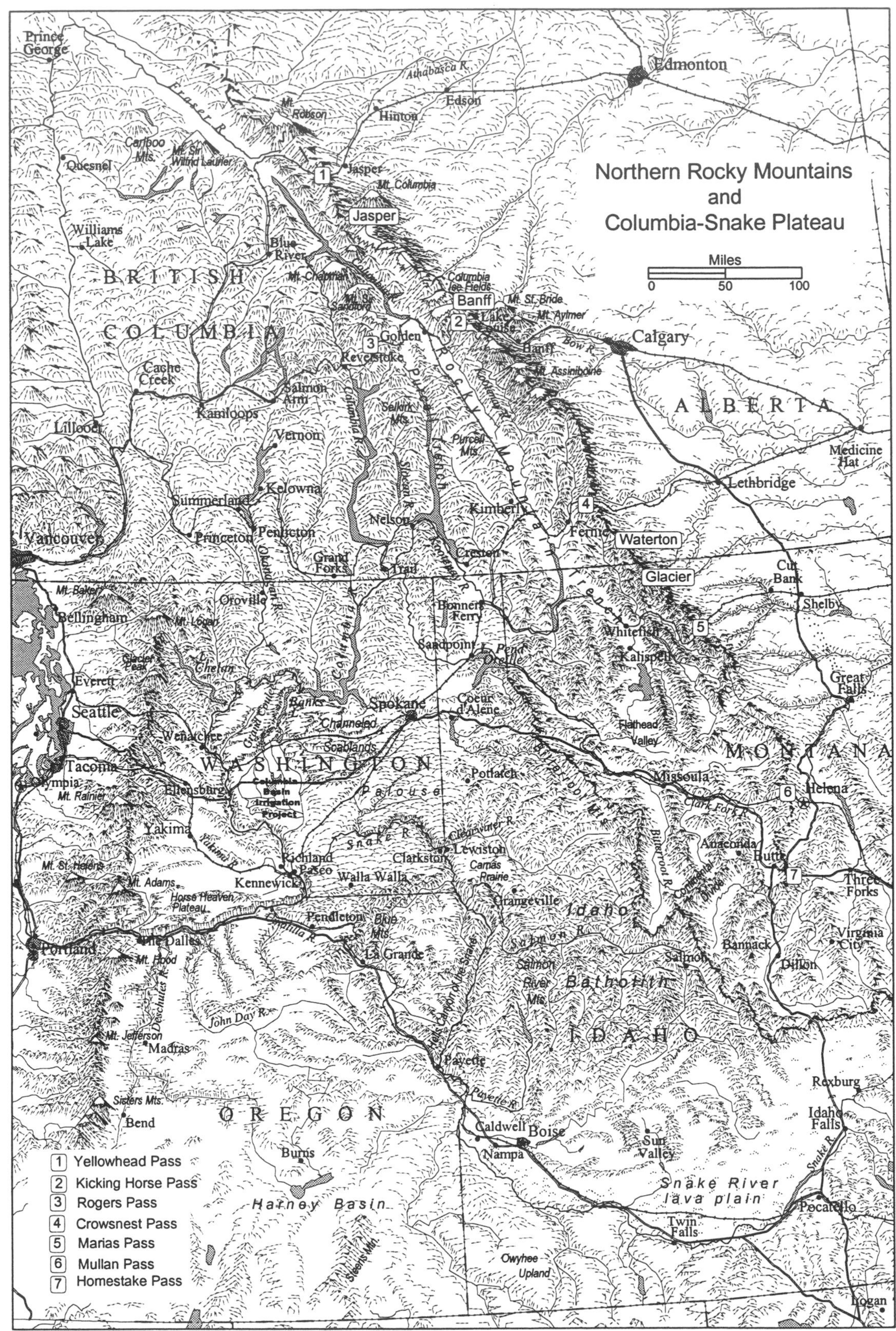
Northern Rocky Mountains
and
Columbia-Snake Plateau
Miles
0
50
100
1 Yellowhead Pass
2 Kicking Horse Pass
3 Rogers Pass
4 Crowsnest Pass
5 Marias Pass
6 Mullan Pass
7 Homestake Pass
BRITISH
COLUMBIA
ALBERTA
WASHINGTON
OREGON
IDAHO
MONTANA
Prince George
Edmonton
Athabasca R.
Edson
Mt. Robson
Hinton
Cariboo Mts.
Mt. Sir Wilfrid Laurier
Quesnel
Jasper
Mt. Columbia
Jasper
Williams Lake
Blue River
Mt. Chapman
Columbia Ice Fields
Mt. St. Bride
Banff
Mt. Aylmer
Lake Louise
Golden
Revelstoke
Banff
Bow R.
Calgary
Mt. Assiniboine
Cache Creek
Salmon Arm
Kamloops
Lillooet
Vernon
Selkirk Mts.
Purcell Mts.
Medicine Hat
Lethbridge
Kelowna
Summerland
Kimberly
Princeton
Penticton
Nelson
Fernie
Waterton
Vancouver
Creston
Grand Forks
Trail
Glacier
Cut Bank
Shelby
Mt. Baker
Oroville
Bonners Ferry
Bellingham
Mt. Logan
Whitefish
Sandpoint
Pend Oreille
Kalispell
Glacier Peak
L. Chelan
Everett
Great Falls
Seattle
Spokane
Coeur d'Alene
Channeled
Scablands
Flathead Valley
Wenatchee
Tacoma
Olympia
Potlatch
Missoula
Helena
Mt. Rainier
Ellensburg
Columbia Basin Irrigation Project
Palouse
Clark Fork R.
Yakima
Snake R.
Clearwater R.
Lewiston
Anaconda
Yakima R.
Richland
Pasco
Clarkston
Camas Prairie
Butte
Mt. St. Helens
Kennewick
Walla Walla
Mt. Adams
Three Forks
Horse Heaven Plateau
Grangeville
Idaho
Pendleton
Blue Mts.
Bitterroot R.
Continental Divide
Virginia City
Portland
The Dalles
Salmon R.
La Grande
Salmon
Bannack
Mt. Hood
Dillon
Salmon River Mts.
Batholith
Deschutes R.
Hells Canyon of the Snake
John Day R.
Mt. Jefferson
Madras
Payette
Rexburg
Payette R.
Sisters Mts.
Bend
Idaho Falls
Caldwell
Boise
Sun Valley
Nampa
Burns
Snake R.
Snake River
lava plain
Harney Basin
Pocatello
Twin Falls
Steens Mt.
Owyhee Upland
Logan

MAP 19.1

In the northern United States and adjoining portions of Canada, the western slopes of the Rocky Mountains are drained by the Columbia River and its tributaries. Although the Columbia Basin of Washington and Oregon has a very different topography and resource base, lands bordering the Columbia in those states were affected by glaciers that formed in the Rocky Mountains. Settlement in the Columbia Basin reflects adjustments subsequently made to these conditions, especially to the episodes of deglaciation toward the close of the Pleistocene epoch. Even within the Rocky Mountains, the lowlands now used for agriculture were shaped by glacial and periglacial activity. Before agricultural settlement, a series of gold, silver, and other precious-metal discoveries drew Euro-American settlers to the region.

## The Idaho and Montana Mining Frontier

In 1862–63 gold was discovered almost simultaneously near Boise and on tributaries of the Salmon and Clearwater Rivers in Idaho and near Bannack and Virginia City, Montana. Even during the Civil War, the mining frontier attracted thousands from the eastern states (as well as from California). Lawlessness plagued all of the gold camps, a condition that was made worse by the added tensions of the war period. Congress's first attempt to control the vigilantes who acted in lieu of an effective government was the creation of Idaho Territory (out of Oregon) in 1863. It was to encompass most of the present states of Idaho, Montana, and Wyoming and had a provisional capital at Lewiston, Idaho. But it was both a political and a geographical impossibility. The Montana settlements were hundreds of miles across the mountains from Lewiston.

Various proposals to create a new territory out of the eastern portion of Idaho included plans to use the 113th meridian, the Continental Divide, or the 116th meridian as the basis for division. Montanans maneuvered this border westward and then proposed that their main boundary with Idaho follow the crest of the Bitterroot Mountains between the Continental Divide and Lake Pend Oreille, north of which the 116th meridian would be the border. This plan, which was adopted and eventually became the Montana-Idaho state line, reduced northern Idaho to its familiar panhandle shape. Not only was Idaho reduced in size, but also the northern panhandle was almost impossible to reach by any route from the southern part of the state without detouring into Montana or Oregon. Idaho was effectively divided into two settled areas separated by a wide swath of uninhabited territory, the Salmon River Mountains.

Boise was chosen as Idaho's permanent capital in 1864. Bannack, the provisional capital of Montana, was soon succeeded by Virginia City, which, in turn, was succeeded by another mining town, Helena, in 1875. People clustered around the mining camps, wherever they were, and only the discovery of mineral ores could change the population map. The greatest mining bonanza in all the West was found at Butte, Montana. The focus on Butte began with silver and gold strikes in the early 1870s, which were later overshadowed by the discovery of a large vein of copper ore. Almost immediately Butte gained a reputation as

FIG. 19.1. *Butte Hill copper-mining district, Butte, Montana.*

a mining mecca where fortunes could be made not only in minerals but also in the processing of ores (fig. 19.1). Among the legendary "copper kings" of Montana were William Andrews Clark, a young Pennsylvanian who came to Bannack in search of gold in 1863; and Marcus Daly, an Irish immigrant who arrived at Butte in 1876. Daly, backed by San Francisco investors, developed the Anaconda mine at Butte.

Because copper ore is found in rocks that contain only a small fraction of pure copper, the ore must be concentrated to 25–30 percent purity at the site in order to reduce the amount of waste material transported away from the mine. The next step is smelting, which converts copper concentrates into blister copper, an intermediate stage that is roughly 97 percent pure. Smelting requires an assured supply of water, but because it involves less waste it can be done away from the mine itself. The final stage in copper processing, known as refining, converts blister copper into copper wire and other products; it produces little waste but requires a substantial amount of electricity.

In 1884 Daly's company erected its smelter and built a company town at a site they named Anaconda, twenty miles west of Butte. Cheap electric power was available at Great Falls, the site of the only major waterfall on the Missouri River, and several refineries and smelters were built there. Thus three of Montana's early cities—Butte, Anaconda, and Great Falls—owed their livelihood to different phases of the copper industry. Daly promoted his company town of Anaconda to be the state's capital after statehood was achieved in 1889. Clark, who invested heavily in copper mines and had extensive investments at Helena as well as elsewhere in the state, was determined to keep the seat of government there. Corruption charges surrounded the election in which Helena narrowly won the decision to remain as the capital. Money from outside investors, especially Boston and New York capitalists, flowed into Montana, where it mingled with the enormous profits that Anaconda and other copper companies were earning. The state acquired a reputation as one where money could buy power and where corporations could dictate their own terms to the government.

With copper as its base, western Montana became one of the most industrialized regions of the West by the beginning of the 20th century. The demand for copper kept the mines producing at capacity for decades as electricity became the preferred source of light and power across the nation. Eventually the Montana copper industry declined as the great open-pit mine at Butte gradually became depleted of its higher-grade ores. The smelters and refineries have operated only occasionally since the early 1980s, and the Butte mine itself is now operated primarily to rework the waste material from previous ore extraction.



Idaho lacked a single spectacular mining development like the Anaconda, but it had a wide variety of commercially exploitable minerals. Gold-bearing quartz mines, which also produced silver and lead, were developed in Idaho soon after the 1860s gold rush began. By the time a second mining boom took place, in the Coeur d'Alene district during the 1880s, base-metal mining had become a more carefully engineered procedure of sinking shafts and tunnels, reducing the metals, and marketing the product. Mining was transformed from an individual activity into a corporate undertaking, and the wide-open life of the frontier mining towns gradually disappeared. Mining continues in Idaho today, although like Montana, the state derives more income from agriculture and forestry than it does from mining.

## The Idaho Batholith

Many of the precious-metal discoveries in Idaho were made around the perimeter of the Idaho Batholith, a sixteen-thousand-square-mile region of rugged mountains that forms the central core of the state. A batholith is a large mass of once-molten rock. Central Idaho's numerous hot springs, similar to those of the Yellowstone region, indicate that the bottom of this mass of rock is still warm. Batholiths cool gradually under the earth's surface and are exposed only when the rock layers above them are stripped away by erosion. The result is a maze of low but sharp-peaked granitic mountains in which the peaks form no pattern of alignment. Valleys wind and twist with little direction and typically lack flat bottomlands. Streams cascading through the zig-zagging hard-rock canyons are full of rapids and sometimes are lethal to those navigating them in wooden boats. A batholith also presents an almost impossible topography to cross overland because every routeway must be blasted out of the extremely hard granites.

The Idaho Batholith posed problems such as these to Meriwether Lewis and William Clark (evidently the first Euro-Americans to see Idaho) in 1805; they could find neither a suitable land route nor a safe water route to the west once they crossed the Bitterroot Mountains. Their guide, Sacagawea, was Shoshoni by birth and a native of the Idaho mountains. After much consultation with her people, the party found a passage by land along tributaries of the Clearwater River, but it exacted a toll of injuries to the travelers and their pack horses. Later

FIG. 19.2. *The main street of Yellow Pine, a gold-mining town in the Idaho Batholith.*

in the 19th century, railroad builders searched for feasible routes across the Idaho Batholith but found none. The region has yet to have a paved road constructed across its breadth.

Isolated gold mines still are worked in the remote mountains of Idaho (fig. 19.2). The only practical way in or out of the mining camps is by primitive roads or helicopter. Much of the area is officially designated as wilderness, a category of national forest administration that protects it from certain types of human use. But its formidable topography probably is the best guarantee that the Idaho Batholith will remain in its present, largely unspoiled state for the indefinite future.

## The Rocky Mountain Trench

The Bitterroot Range forms the eastern boundary of the Idaho Batholith. Between the Bitterroots and the Continental Divide is a hundred-mile-wide swath of north-south-trending mountain ranges separated by broad valleys. In British Columbia this pattern of valleys and ranges expands into a zone of more than three hundred miles' breadth and constitutes the typical terrain of the province's interior. The north-south-trending ranges are upraised edges of rock formations that were thrust eastward in the process of forming the Rocky Mountains. Parallel ridges and valleys developed on rocks of different age and composition. The number of separate ridges is difficult to determine, because they appear and disappear at occasional gaps, but there are at least a dozen ranges of aligned peaks separated by smooth-floored valleys that intersect the U.S.-Canadian border west of the Continental Divide.

Two of the valleys are especially noteworthy because of their length. The larger one is the Rocky Mountain Trench, which parallels the front range of the Rockies about fifty miles to the west and stretches more than nine hundred miles from the Bitterroot Valley of Montana to northern British Columbia. The

Rocky Mountain Trench is a structural (rather than an erosional) feature that resulted from differential movement along parallel fault planes at the base of the mountains. The Rocky Mountain Trench in Montana is drained by the Bitterroot, Clark Fork, and Flathead Rivers and in southern British Columbia by the Kootenay River; all are tributaries of the Columbia.



The Columbia River originates in the middle portion of the Rocky Mountain Trench. It first flows north, past the city of Golden, British Columbia, but then bends back south to pick up the Kootenay and Pend Oreille Rivers as tributaries as it enters the United States. Still farther north, the Fraser River flows in a segment of the Rocky Mountain Trench, and north of that the Peace River occupies a portion of it. Because the trench is a natural corridor for rivers, highways, or railroads to follow, it has been important in short-distance segments. But the nine-hundred-mile length of this great, smooth-floored valley that stretches from southwestern Montana nearly to Yukon Territory has had little significance for people. Its alignment is not one that humans, whether prehistoric or modern, have felt much need to follow.

A second such structural valley, the Purcell Trench, lies about fifty miles to the west. Extending from the vicinity of Coeur d'Alene to near Golden, British Columbia, where it joins the larger Rocky Mountain Trench, the Purcell Trench also carries the flow of several rivers, including portions of the Kootenay and the Clark Fork–Pend Oreille. Microclimates are an important resource of these mountain lowlands. Air settling down the adjacent slopes warms by compression. The greater the descent of the settling air, the greater is the warming, and the milder the climate on the valley floor. The Kootenay Valley in the Purcell Trench at the northern tip of Idaho is only 1,700 feet above sea level (fig. 19.3). Hops are grown here, as they are in similar valley-bottom settings as far north as Kamloops, British Columbia. The Okanogan Valley, which cross-cuts the U.S.-Canadian border, is another warm lowland, despite its northern latitude. Summerland, British Columbia, on the Okanagan (the name is spelled differ-

FIG. 19.3. *Cattle graze the banks of the Kootenay River, in the Purcell Trench of Boundary County, Idaho.*

ently north of the border), claims to have the mildest winters in Canada. Peaches, apples, and cherries are grown with irrigation in the warm lowlands. The broadest of the Rocky Mountain structural valleys is the Flathead Valley of Montana, an extensive plain that specializes in livestock farming and orchards.

The Rocky Mountains often are described as a barrier or as an obstacle to travel, and indeed they are, but only for certain directions of movement. Nearly all transportation routes were built across this region east to west, to reach the Pacific Coast from the interiors of Canada and the United States. Although valleys between the mountain ranges are numerous, they trend north and south, across the transportation routes. Most routes built through the Northern Rockies represent a compromise between directness and ease of crossing. Nearly all of the major valleys contain portions of transcontinental railroads and highways, even though the valleys do not strike in the desired direction of travel.

Westbound travelers who crest the Continental Divide through Homestake Pass at Butte, Montana, having traversed less than one hundred miles of mountainous topography, may be unaware that they have reached the maximum necessary elevation on their journey and that, except for occasional redundant grades, they will travel downslope the entire distance to the broad plain of the Columbia near Spokane. The avenue that accomplishes this downhill, cross-mountain travel is the Clark Fork River, a route used by the fur traders, the first northern transcontinental railroad, and, over most of its length, the interstate highway. The Clark Fork route offers a great detour past Lake Pend Oreille in terms of westward travel, but it is one that avoids both the steep slopes of the Bitterroot Range and the Idaho Batholith.

## Interior British Columbia

At times this topographic grain also has had geopolitical implications. The valleys of British Columbia are more easily accessible from the United States than they are from either Vancouver or eastern Canada. The international boundary (at the 49th parallel) cross-cuts the mountains, leaving southern British Columbia relatively open to American entry. Canada's territorial integrity was compromised by this circumstance, not from any danger of military invasion but rather because the natural resources of southern British Columbia were thereby open to easy access from the United States.

The first railroads crossed the Continental Divide about four hundred miles apart, on Mullan Pass in Montana and Kicking Horse Pass in Alberta–British Columbia. Both routes, constructed in the early 1880s, were a long way from the international border. In 1889 the Great Northern Railway built its transcontinental line through Marias Pass, along an alignment that took advantage of the Kootenay Valley and both the Rocky Mountain and Purcell Trenches. This railroad, which reached as far north as Bonners Ferry, Idaho, became the front line for American economic penetration of southern British Columbia.

Large quantities of high-grade bituminous coal suitable for industrial uses were discovered in several British Columbia districts during the 1880s and 1890s.

Copper and silver were found in the Kootenay and Slocan Valleys in the late 1880s and in the Boundary District near Grand Forks, British Columbia, in 1897. F. A. Heinze, another of the Butte "copper kings," built a copper smelter at Trail, British Columbia, in 1895. With these developments American entrepreneurs began to show interest in extending railroad lines north of the border in order to tap the province's resources and perhaps to encompass the British Columbia mining and smelting complex within the sphere of operations at Butte.

The Canadian government countered by subsidizing construction of the Crowsnest Pass railway in exchange for an agreement by Canadian Pacific to maintain low rates on grain shipments (the so-called Crow Rates, which remained in effect until 1995 and helped the Canadian prairie wheat farmers sell their grain at Pacific Coast ports). Meanwhile, smelters were constructed at Grand Forks, Nelson, and Revelstoke, British Columbia, all of which served local copper and silver mines. American capital competed with British for control of the region's mines, smelters, and railroads. Although Canada's sovereignty was not directly threatened, easy cross-border access remained inviting for American capitalists.

The first mining boom in the mountains of British Columbia was a long way north of the international boundary. In 1860 gold was discovered in the Cariboo district, more than four hundred miles north of Vancouver. The "Cariboo Road," roughly paralleling the Fraser River, saw many would-be miners making the trip north. The gold was abundant, although much of it required sinking deep shafts to gain access to the ore. Most men who went to the Cariboo ended up working for wages in the mines rather than striking it rich on their own. The mixed grass and Ponderosa pine landscape of the Cariboo offered an attractive environment to cattlemen, however, and was well suited to livestock grazing. Beginning in the 1880s, its warm and dry valleys became the scene of extensive ranching activity. The area remains as a zone of livestock ranching in British Columbia.

Coal from mines in the Elk River district near Fernie still is an important Canadian export and in fact has grown in importance during the past two decades. The coal moves by rail to Vancouver for shipment to Asian steel mills, and it also moves by rail to the United States, where it has partially replaced Appalachian coal in the blast furnaces of the Gary steelmaking district of Indiana. Farther north, coal mined in the Tumbler Ridge development west of the Peace River District also is hauled by rail to Pacific seaports for shipment to industrial customers in Asia. Along with Alaska, British Columbia is the principal supplier of coal in the westbound trade of the Pacific Rim.

Nearly all of southeastern British Columbia's population lives in the isolated but linked mountain valleys. Although most of the valleys were deepened and widened by glaciation, some—especially those containing lakes or reservoirs—are too narrow and steep-sided to permit highways or railways at the water's edge. Ferry service still connects highway fragments in some of the interior valleys where there is no room for a road. Travel across southeastern British Columbia generally involves circuitous routes between the widely separated population centers.

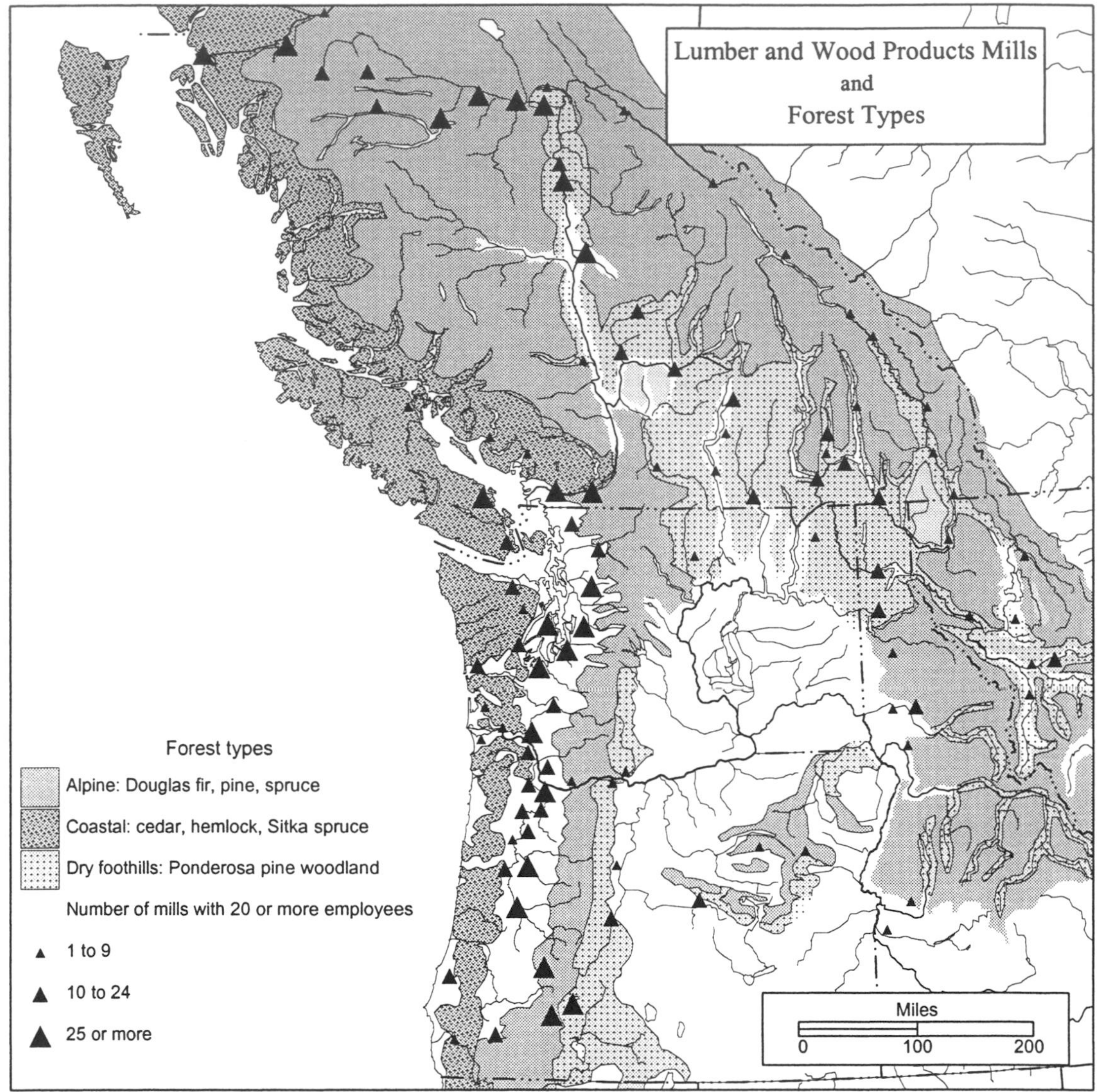


MAP 19.2

## Forest Industries

The pattern of mountains, valleys, and rain shadows largely determines the map of forests in the Northern Rockies and adjacent regions (map 19.2). Rain shadows are especially dry areas that, in the middle latitudes, are on the east-facing slopes of major topographic barriers. Air masses move predominantly from west to east and are cooled as they are forced to rise against the mountain slopes, producing rainfall or snowfall depending on elevation and season. When those same air masses descend the east-facing mountain slopes, they are warmed by compression as they settle, which transforms liquid precipitation back to water vapor stored in the air. As a result of this process (known as orographic precipitation), mountainsides not only have different vegetation formations according to altitude but also different types based on slope aspect (east-facing or west-facing).

Interior British Columbia is covered with coniferous forests between the alpine tree-line (cold) and valley-bottom (dry) limits of forest growth. Both of these plant-physiological limits migrate upslope with distance southward into Idaho and Montana until, in southern Idaho, trees occupy only the highest peaks or the shaded north slopes. Nearly all species of needleleaf evergreens in the Great Lakes forest realm also reproduce spontaneously in the Northern Rocky Mountains.



The repetition of species dominance made it especially easy for Great Lakes lumbermen to relocate their operations to the Northwest after they had cut the forests of Minnesota, Wisconsin, and Michigan beyond the point of diminishing returns in the early 1900s. Men like Frederick Weyerhaeuser of St. Paul, who had made a fortune in Wisconsin white pine, bought large tracts of Idaho white pine land between 1900 and 1910. The Weyerhaeuser family and other Wisconsin lumber barons organized the Potlach Lumber Company in 1903 and built the company town of Potlach, Idaho, where millions of board feet of lumber have been processed in the years since.

The drier foothills of the Cascade and Blue Mountains of Oregon produce ponderosa pine, which, along with Douglas fir, is the typical tree cut for construction lumber. Hemlock, red cedar, Douglas fir, and ponderosa pine are the basis for interior British Columbia's lumber industry, which produces a substantial portion of the construction lumber used in the Canadian (and a portion of the U.S.) homebuilding industry. The largest concentration of lumber mills and plywood factories in North America occupies the valleys of the Fraser River and its tributaries near Prince George, British Columbia.

Plywood and wood-veneer industries are less important in the U.S. Rocky Mountain region than they are farther west, although Idaho is home to several large plywood mills. Despite the scale of lumbering in the western Rocky Mountains, the forest growth is not so heavy as it is on the lower but moister west-facing slopes of the Cascade Mountains and Coast Ranges in Washington and Oregon, a fact reflected in the much smaller lumber industry of the mountains compared with that of the Pacific Coast region.

Unlike the experience of the Upper Great Lakes states, logging in the Northern Rockies has yet to reach extinction of the commercial species. The large acreage of forested land, the inaccessibility of much of it, and the relatively restricted use for wood today compared with a century ago combined to preserve the forest from total demise. Forestry is the most widespread rural land-use in the Northern Rockies, however, and especially outside the major urban centers, it is the largest employer. Lumber prices are extremely sensitive to fluctuations in the home-building industry, whereas timber-cutting allowances are subject to interpretations by federal courts charged with upholding endangered-species laws that indirectly protect old-growth forests. The loss of a lumber mill often means the loss of the entire economic base for the more remote valleys of western Montana and the Idaho panhandle. Attempts to find alternatives to logging employment continue as the controversies over cutting allowances rage on.

## The Columbia-Snake Plateau



*Columbia-Snake Plateau* refers to a region that takes in most of southern Idaho and the western two-thirds of Washington and Oregon. Although this large area contains numerous geological subdivisions, its defining characteristic is its surface rock, which is the product of volcanic activity. The volcanism of the Columbia and Snake region was extrusive, involving massive flows of molten lava that moved upward through the earth's crust during various episodes from Cretaceous times until only a few thousand years ago. The lavas issued from vertical passages in the existing rocks, building layer after layer to a thickness of thousands of feet and covering a total area of two hundred thousand square miles.

Extrusive volcanic rocks such as those of the Columbia-Snake Plateau weather and disintegrate, grain by grain, from the action of ice, wind, and running water. When eroded, they form deep gorges and steep slopes, but because the rock is porous, surface water infiltrates the ground layer before it is removed as runoff. The result is a topography of steep-sided canyons but fairly smooth slopes that show little dissection by streams (fig. 19.4).

The blackish-colored basaltic lavas constitute a surface like no other in the mainland United States; the topography formed here is unique to the region. The Columbia-Snake's two major subdivisions are the Snake River Plain of Idaho and the Columbia Plateau (or Basin) of Washington, which are connected by the mountains and basins of eastern Oregon, another region of primarily volcanic landforms.

FIG. 19.4. *Steep but largely uneroded slopes, formed on volcanic rock, border the Salmon River near Riggins, Idaho.*

## The Snake River Plain

The Snake River rises in the Teton Range of Wyoming and flows in a broad arc south and west of the Idaho Batholith before emptying into the Columbia River at Pasco, Washington. At their southern limit, the mountains of the Idaho Batholith become a series of low ridges with wide valleys (such as Sun Valley) opening out to the Snake River Plain. The defining characteristic of this plain is its covering of fairly recent basaltic lavas that were extruded from fissures and vents in the earth's crust perhaps only thousands of years ago. The lava surface formed recently enough that in some places it still remains without a vegetation cover.



Because the lava plain is porous, it absorbs the flow of the smaller streams entering from the north. They become "lost" rivers that reappear as natural springs along the southern margin of the lava plain where the Snake River flows. About 90 percent of the Snake River Plain is covered with lava; and although not all of the lava plain is irrigated, nearly all of the irrigated land is on the lava formations (see chapter 17, map 17.2). As one of the first large-scale irrigation efforts in the West, the Snake River Plain served as a model for other irrigation districts established in later years.

Agricultural settlement began in 1860 with a few Mormon families who moved north from Salt Lake. Initially the Mormon settlers lived in compact villages that were defensible against possible Indian attacks, but over time the settlement pattern grew more dispersed. Like all Mormon villages, those in Idaho were organized around institutions such as the tithinghouse (a sort of community bank) and various agricultural cooperatives. They had a strong degree of centralized decision-making authority, in which the church's role was preeminent. Irrigation was practiced experimentally almost from the beginning and grew as an organized, community effort within the framework of institutions that were part of Mormon village life. Although Mormons were officially discriminated against in territorial Idaho (they were not allowed to vote), the anti-Mormon laws were relaxed after statehood in 1889, and southeastern Idaho became "Utah extended."

Irrigation is necessary for agriculture in the Snake River Plain because of the arid climate. The Carey Act of 1894 and the Newlands Act of 1902 created federal programs to foster irrigation developments in the West, transferring to state control the authority to organize irrigation districts, divert water from streams, and construct reservoirs. The projects were financed from the sale of irrigable land, which was offered in small parcels to encourage family farms. The Snake River Plain was ideally suited to such developments. Soils on the lava plain are well drained and fertile when irrigation water is applied. Long summer days with cool nights are especially suited to root crops such as potatoes and sugar beets. The Mormons were irrigation experts, and they also mastered the techniques for organizing production as a community enterprise. Eventually, Snake River agriculture evolved on an almost industrial scale.

Reservoirs were built to provide water for three major irrigation districts.

The largest irrigated tract, near Pocatello and Idaho Falls, was supplied with water from the upper Snake and its tributaries, beginning at Jackson Lake in the Teton Range, which is the first reservoir in the system. Sugar beets were introduced to Idaho by Mormon farmers, and by the early 1900s beets were the major crop of the eastern Snake River Valley. The Mormon church was heavily involved in the Idaho sugar industry, from the level of irrigation farming to the processing of beets and the refining of sugar. Potatoes, which were also introduced by the Mormons, were grown on the same lands and also performed well under irrigation on the sandy, volcanic soils. Idaho's reputation as a potato producer was enhanced by aggressive advertising, and its crop eventually became the largest in the nation.

A second irrigation district, near Twin Falls, also specializes in potatoes and sugar beets; it is supplied with water from the Snake River as well as from tributary streams flowing north from Nevada. The third irrigated area is western Idaho, which is an almost continuous maze of irrigated fields between Boise and Caldwell. It developed a more diversified agricultural base including orchards, hops, and onions as well as sugar beets and potatoes. Large cattle feedlots and aquaculture farms are found here as well.

The Boise irrigated area is heavily urbanized today, and urban sprawl has consumed some agricultural land, but the area remains an important irrigated oasis. Boise is an amenity-rich city, where life is enhanced by the proximity of ski slopes, trout streams, whitewater rapids, and uninhabited mountains—all within easy driving distance of the city.

The Snake River Valley changes character markedly beyond Payette. The river flows in a six-thousand-foot-deep canyon (Hell's Canyon of the Snake), which is also the Oregon-Idaho boundary. Hell's Canyon is as deep as the Grand Canyon and just as inaccessible. No cities are found along the river for 250 miles north to Lewiston (the lowest elevation in Idaho), where the Snake River bends west and flows through the rolling Palouse hills. Dams completed on the lower Snake River in the 1970s gave it sufficient depth for all-year barge navigation downstream from Lewiston. Wheat and barley crops grown in western Montana and even in the Canadian Prairies are trucked over Lolo Pass, through the Bitterroot Mountains, to take advantage of the low rates on barge shipment of grain, much of which is bound for export from the mouth of the Columbia River.

## The Palouse and the Channeled Scablands

All of the north-south-trending Rocky Mountain valleys were occupied by ice sheets at various times during the Pleistocene. Ice filled the Rocky Mountain and Purcell Trenches, blocking the northward outlet from Lake Pend Oreille. The Clark Fork drainage was thereby dammed, resulting in a long, deep, many-fingered lake, known as Glacial Lake Missoula, which extended southward into the Bitterroot Valley and north into Montana's Flathead region, completely drowning the present Clark Fork Valley to a depth of hundreds of feet. When the ice dam failed, the waters were suddenly released and cascaded across the lowlands south of Lake Pend Oreille, past Spokane, and spread out over the plain to the west.

Disagreements exist as to the number of times this may have happened, but there is little doubt that catastrophic flooding did take place. The result was the Channeled Scablands, the scarred and grooved lava plain between Spokane and Pasco, where dry waterfalls, isolated lakes, and stream-cut valleys without streams form the landscape. The surface gouged out by the cascading waters soon became one of calm pools that, in turn, evaporated to expose layers of silt. Winds blowing from the west and north picked up these fine-grained sediments (loess) and transported them to the southern and eastern fringes of the Scablands. Reworked by the wind, the loessial surface became a dune-like landscape of low but steep-sided hills. The resulting region of rolling, convex hills, known as the Palouse, begins in the Channeled Scablands and extends across eastern Washington into the bounding mountain valleys of the Idaho panhandle.



Slopes as steep as those of the Palouse could not be cultivated without risk of erosion were it not for the rapid infiltration of precipitation on the loess-mantled volcanic surface (fig. 19.5). But the lack of runoff here limits gullying and makes a surface ideally suited for the production of grain crops. The soils are Xerolls (a summer-dry type, otherwise similar to the Mollisols of the Great Plains wheat areas). Winter wheat culture is especially suited to the precipitation regime, which has a maximum in the winter months.

Agricultural settlement began in the 1870s near Walla Walla and Lewiston, spreading slowly westward across the Palouse and eventually onto the higher ground between canyons of the Channeled Scablands. By the 1880s the Palouse had become an intensive wheat-producing region and was a major factor in making Spokane the economic capital of the region. Spokane's site, at a substantial waterfall on the Spokane River, was chosen with waterpower in mind. The city grew primarily as a trading center for the wheat region, which publi-

FIG. 19.5. *Palouse wheat fields at Central Ferry, Washington, in the gorge of the Snake River. Much of the grain is shipped to Portland for export.*

cists touted as the Inland Empire, a label chosen to contrast it with coastal Oregon and Washington.

Producing and marketing the wheat posed special problems. Because the area was farther from the national market than grain-growing districts in the Great Plains, there was a greater reliance on overseas exports through Pacific Coast cities. The hilly topography required high-horsepower machinery to work the fields. Because the rolling wheat fields lie at elevations in many cases more than a thousand feet above the river valleys, the harvested grain was fed by special chutes down to warehouses along the Snake River, where it could be loaded aboard river barges.

Outliers of the Palouse, less hilly but equally suited to the production of grain crops, extend into the mountain valleys of Idaho. The largest of these rolling grasslands is the Camas Prairie, which occupies a high benchland between the Snake and Clearwater Rivers north of Grangeville, Idaho. The area takes its name from the camas plant, a bulbous root that was processed into a bread flour by native inhabitants of the region, especially the Nez Perce Indians, whose traditional lands included the Camas Prairie. Other native people also visited the Camas Prairie each year to dig the roots of this blue-flowered plant that was an important source of food. The Nez Perce were forcibly resettled in Oklahoma for a time, but they eventually were given back a portion of the former Idaho homeland, including the Camas Prairie. Today the region produces legumes such as dry, edible beans, as well as winter wheat.

Grain farming also extends westward from the Palouse and occupies the smoother benchlands along the Columbia as far west as The Dalles, Oregon. The main crop is winter wheat; also as in the Palouse, wheat culture emerged here with the beginnings of settlement in the 1870s and 1880s. Deep gorges cut by the John Day and Deschutes Rivers create a discontinuous zone of dry-farmed wheat crops resting on thick volcanic layers high above the south bank of the Columbia River. Although volcanic rock is porous, it is also hard and sharp. Farmers in this region do not even attempt to drive fence posts into the ground where the soil cover is thin.

## The Columbia Basin Project

Precipitation declines westward across the Columbia Basin from an average of more than twenty-five inches per year in the eastern Palouse to less than ten inches in the Big Bend region south of Wenatchee. For several decades after wheat farming was established in eastern Washington, the central portion of the state remained very thinly settled. Proposals to irrigate it culminated in plans for the Columbia Basin Project, a combined hydroelectric-power and irrigation scheme that was launched by the U.S. Bureau of Reclamation in the late 1930s.

The centerpiece of the project is Grand Coulee Dam on the Columbia River, a mile-long concrete structure that is now the largest single source of electric power in the United States (fig. 19.6). Some of the power generated by the dam's turbines is used to pump water out of the impoundment (Franklin D. Roosevelt

FIG. 19.6. *Grand Coulee Dam. A fraction of the electricity generated by the dam is used to pump water across the ridge* (right center) *and into reservoirs. From there the water is fed by gravity across the Columbia Basin's irrigated acres.*

Lake) behind the dam, over the drainage divide, and into Banks Lake, a reservoir that was created in the formerly dry bed of the Grand Coulee. The Grand Coulee was cut by the Columbia during a brief period of the river's diversion by ice blockades to the north, then abandoned when the Columbia sought the more western course it now follows.

From Banks Lake, irrigation water is channeled across farm fields in the Big Bend region and eventually back to the Columbia, into which the excess water drains, near Pasco. The Grand Coulee irrigation project thus allows a portion of the Columbia's flow to take a shortcut over higher ground, passing through irrigated fields before it returns to the river. Of the nearly 1 million acres of land that were considered irrigable when planning began, nearly three-fourths are now under irrigation. Crops of potatoes and onions are the most productive of those that have been tried, but hops and fruit orchards also have succeeded.

The Grand Coulee project was both costly and a long time in the making. The dam was constructed in the early 1940s, and the power turbines were added in stages until the present generating capacity was reached in the 1970s. Irrigation was planned in the 1930s but did not actually begin until 1953. Part of the long planning process that guided the Grand Coulee project was the government's determination to attract small-scale irrigation farmers. In the 1940s planners believed that a population of at least 350,000 could be supported on the farms and in the towns and cities that were bound to grow in the region. People were encouraged to settle in groups, to form cooperatives, and to practice a fairly modest scale of farming.

The Columbia Basin Project was one of the federal government's last attempts at social engineering by the process of rural land settlement. Not even one-fourth of the envisioned population has been realized. Urban growth was stimulated more by the federal government's plutonium-processing facility at the Hanford Works, which for several decades produced a population boom in the Tri-Cities (Richland-Kennewick-Pasco) area. Grand Coulee was conceived in the era before small farms became economically marginal and before the great waves of off-farm migration as a result of technological change began in the 1950s. The agrarian and communitarian values espoused by the project's planners were modeled, in part, after the successful system of irrigation farming found in the Snake River Plain, although without the religious aspect. The long-held belief that the West was best occupied by small-scale irrigation farmers had its last test in the Grand Coulee Project.

The project's obvious success, instead, derived from the electricity it provided. Following the completion of Bonneville Dam on the Columbia east of Portland in 1937, Grand Coulee added evidence that adequate power would stimulate industrial growth in the Northwest. One industry that was able to locate in the region because of the electricity supply was aluminum manufacturing. Washington became the leading aluminum-producing state, and that led directly to the growth of aircraft manufacturing in the Pacific Northwest. The flow of the Columbia and Snake Rivers is now interrupted by more than a dozen hydroelectric power dams that have turned both rivers into a stairstep of reservoirs upstream from Bonneville.

Unconnected to the Grand Coulee project itself, but sharing a similar location, are the irrigation districts based on the Columbia's west-side tributaries, principally the Wenatchee and Yakima Rivers. Irrigated apples, pears, peaches, and plums are grown along the narrow Wenatchee River Valley west of that city and in the more extensive lowlands of the Yakima River near the cities of Ellensburg and Yakima. This is the largest apple-producing area of the United States and an area of intensive hops production. All crops are grown under irrigation. Without the added water, livestock grazing probably would be the most intensive land use that could be sustained.

## *References*

Arrington, Leonard J. *History of Idaho.* 2 vols. Boise: Idaho State Historical Society, 1994.

Eagle, John A. *The Canadian Pacific Railway and the Development of Western Canada.* Kingston, Ont.: McGill-Queen's University Press, 1989.

Malone, Michael P., Richard B. Roder, and William L. Lang. *Montana: A History of Two Centuries.* Seattle: University of Washington Press, 1976.

Meinig, D. W. *The Great Columbia Plain: A Historical Geography, 1805–1910.* Seattle: University of Washington Press, 1968.

*Pattern of Rural Settlement.* Columbia Basin Joint Investigations. Washington, D.C.: Bureau of Reclamation, U.S. Department of the Interior, 1947.

Twining, Charles E. *F. K. Weyerhaeuser: A Biography.* St. Paul: Minnesota Historical Society Press, 1997.

Wyckoff, William, and Lary M. Dilsaver, eds. *The Mountainous West: Explorations in Historical Geography.* Lincoln: University of Nebraska Press, 1995.

CHAPTER 20

# The Great Basin

Most of Nevada, the western half of Utah, and portions of California, Oregon, and Idaho make up the Great Basin, a 220,000-square-mile zone of interior drainage bounded by the Wasatch Mountains, the Colorado Plateau, the Sierra Nevada, and the Columbia-Snake Plateau (map 20.1). It is not a single drainage basin in which all rivers drain toward a common low point of topography but rather a collection of dozens of such basins, few of which are tributary to any other.

Although the Great Basin is a drainage-defined region, its most important —and most limiting—characteristic is its arid climate. It occupies the entire rain shadow of the Sierra Nevada and a portion of the rain shadow of the Cascade Mountains (fig. 20.1). Aridity is the major factor limiting land-use options in the region, it is the basic control on vegetation and soil-forming processes, and until recent years it has served as a strong deterrent to population growth. Even some of the region's nonmetallic surface minerals are climate controlled, because they accumulated under conditions of low precipitation and rapid evaporation.

California ranchers established large cattle- and sheep-grazing operations in Nevada during the latter decades of the 19th century. Immigrant Basque sheepherders found a suitable environment there as well. They ran their stock on the public domain and made few improvements to the vast acreages over which their animals ranged. Some of the Great Basin's valleys have grassy meadows that support larger numbers of animals, but most of the landscape has a discontinuous cover of sagebrush and other desert plants of little value to graziers. The passage of the Taylor Grazing Act in 1934 reduced the number of animals to protect the environments against overgrazing.

Most of the land in the Great Basin remains part of the public domain. The patchwork of federal land jurisdictions covering the Great Basin today includes substantial areas administered for grazing, numerous military bombing ranges and test sites, and other public lands that see little private use and over which the region's herds of mustangs (feral horses) and other critters roam. Except for the Canadian Shield and the Far North, this is the least intensively used land in North America. As is typical of other desert and semidesert environments, nearly all of the Great Basin's people live in cities and towns, because there is little need for anyone actually to live "on the land."

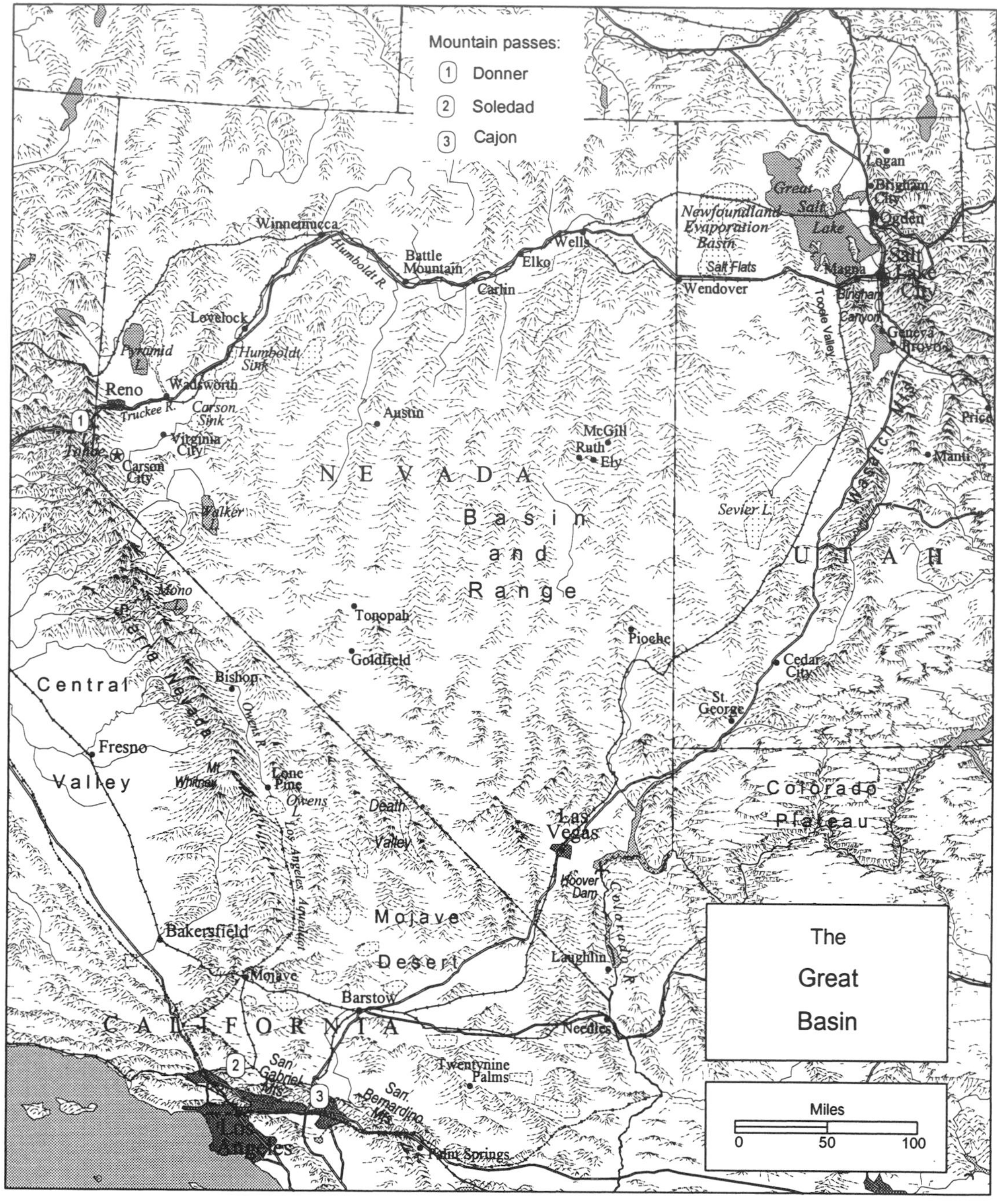


MAP 20.1

## Basin and Range Topography

The Great Basin covers roughly the northern half of the Basin and Range province, a region that is defined by its distinctive geologic structures. Basin and Range topography has evolved over hundreds of millions of years, beginning with a series of geosynclines (large basins) that formed during Paleozoic and

Mesozoic times. They gradually were filled with materials eroded from fringing volcanic mountains or deposited on the sea floor during periods of marine inundation. These sediments were compressed, folded, and, during the past 100 million years, subjected to intrusion by volcanic activity. The widely separated low mountain ranges characteristic of the Basin and Range resulted from subsequent episodes of block faulting and a gradual infilling of the basins with materials that have been eroded from the mountains.

It is said that geologist C. E. Dutton remarked that a landform map of the Basin and Range resembles "an army of caterpillars crawling northward out of Mexico." The pattern begins in the Mexican states of Coahuila and Chihuahua and extends northward across southern Arizona and California, covering all of Nevada and western Utah before it ends against the Snake River Plain of Idaho and the Blue Mountains of Oregon. There are more than 150 separate ranges of low mountains in the Basin and Range region, although they are not strictly countable because they coalesce and diverge in places to form almost a honeycomb of small basins flanked by mountains.

The processes of erosion in an arid region favor the lateral retreat of mountain slopes, with high slope angles being maintained even as the mountains diminish in size. Lower mountain flanks are covered with materials accumulated by downslope movement, whether the result of gravity or of infrequent episodes of rapid surface runoff produced by thunderstorms. Collectively, such long, smooth slopes of boulders, rocks, and sand are known as pediments or, when deposited by the action of stream flow, alluvial fans. In the Southwest they are known as bajadas, a term the Spanish explorers gave them.

Where water is scarce, these processes only rarely produce a system of integrated river valleys that drain a landscape so that stream flow is directed outside the region. Depressions at the foot of the bajadas, known as bolsons (or playas, another term of Spanish origin), are basins of interior drainage where surface waters evaporate because there is no natural discharge outlet. The basin

FIG. 20.1. *The Nevada slope of the Sierra Nevada is the rain-shadow side of the mountains, with a vegetation cover of grass and sagebrush.*

aspect of Basin and Range topography thus has a special meaning: in many cases the basins accumulate water that can only evaporate. Salts and other residues remain after evaporation and sometimes accumulate in substantial deposits.

John C. Frémont bestowed the name Great Basin on this region when, on exploring trips made in 1843 and 1845, he could find no large rivers; even those streams that flowed within it discharged into small lakes and salt flats, where their waters evaporated. In a cool, humid region such as the Canadian Shield—which also has many small streams that flow into isolated lakes—interior drainage produces wet, boggy conditions and acidic soils. In the Great Basin, interior drainage leads to rapid evaporation and excessive alkalinity or salinity of the upper soil layer, a condition that only the hardiest of desert shrubs can endure.

During Pleistocene times the basins of Utah and Nevada received far more water from the bordering mountains than they do today. When the climate was cooler and rainier and when mountain glaciers were melting, the basins became filled with water until, one after another, they coalesced to form temporary lakes of substantial size. Between 25,000 and 10,000 years ago, more than one hundred closed basins in the West contained these so-called pluvial lakes. The largest was Lake Bonneville, which covered nearly twenty thousand square miles and had a maximum depth in excess of 4,300 feet. Its principal remnant is the Great Salt Lake in Utah. Another, Lake Lahontan, covered much of northwestern Nevada. Mountain ranges formed islands and peninsulas within the pluvial lakes, whose shorelines fluctuated in response to climatic change. The former shorelines of glacial Lake Bonneville are visible as a series of "bathtub rings" on the Wasatch Mountain foothills at Salt Lake City.

One river—the Humboldt—flows more than two hundred miles across the Nevada desert before it disappears in the Humboldt Sink near Lovelock. It is the Great Basin's longest river, and it provided a route that was easy to follow, first for wagon trails and then for railroads and highways threading their way between the isolated mountain ranges. Lovelock, Winnemucca, Battle Mountain, Carlin, Elko, and Wells—the "roadhouse" towns along the I-80 corridor across Nevada—were created by the Central Pacific Railroad as it built eastward along the Humboldt River and across the desert in 1867–69.

Those who drove freight and emigrant wagons across Nevada learned that only one stretch of the trail, the Forty Mile Desert west of the Humboldt Sink, was totally without water. Near present-day Wadsworth, Nevada, the wagon trail reached the Truckee River, a small stream flowing out of the Sierra Nevada immediately to the west. But streams flowing down the Sierra's eastern slopes fare no better than their west-flowing counterparts like the Humboldt. The Truckee is swallowed by Pyramid Lake a few miles north of Wadsworth. To the south, the Carson and Walker Rivers, also of Sierra Nevada origin, are swallowed in the Carson Sink and Walker Lake. The eastward flow of all waters from the Sierra disappears within fifty miles after entering the state of Nevada.

The massive Sierra Nevada, which here loom as an immense wall against the desert, have eastern approaches so difficult that the earliest travelers disassembled their wagons, built windlasses, and hoisted both wagons and oxen up into the mountain passes. From Donner Pass, California (where members of the ill-fated Donner party perished in the winter of 1846–47), the elevation declines 2,700 feet down to the level of the Truckee River at Wadsworth. An air mass traveling east over the Sierra and sinking down the Nevada slope to this level would experience a compressive warming of approximately 15°F, which is sufficient to change condensation back into vapor. The wet, west-facing Sierra slopes give way to dry slopes under descending air on the eastern front. Clouds disappear as air temperatures rise.



Nevada's low mountains produce numerous episodes of lifting as the air masses travel toward the east, but few of the mountains are sufficiently high to cool the air to the level of condensation once more. Not until the slopes of Utah's Wasatch Mountains are encountered, more than five hundred miles to the east, does the air again rise to a height where the vapor condenses to fall as rain or—more likely—snow. Skiing in the western mountains is primarily a west-slope phenomenon because of the atmospheric processes involved.

## The Mormon Settlement of Utah

No region of the United States is as closely identified with a single religious group as the Great Basin is with the Mormons (the Church of Jesus Christ of Latter-day Saints). The Mormons were the first group of Euro-Americans to settle in the Great Basin. Ever since their arrival in 1847, they have played a leading role in the region's development (fig. 20.2). In addition to their well-known presence in Salt Lake City, Mormons colonized most of the better mountain valleys of Utah's Wasatch Range. Beyond that, they extended a scatter of outposts to the fringes of the Great Basin and, by 1849, held the ambition of creating a Mormon-dominated state, to be known as Deseret, that would have included within its borders all of the Great Basin as well as the Colorado Plateau.

The main body of Mormons had known four temporary homes between

FIG. 20.2. *A reconstructed Mormon village at Salt Lake City. The buildings are typical of those that would have been found in a Mississippi or Ohio valley town of the 1840s.*

the time of the church's creation in western New York in 1830 and their arrival in Utah in 1847. At Nauvoo, Illinois, Joseph Smith, the Mormons' founding prophet, had become convinced that the group should move west, beyond the margins of settled territory. Even before Smith was murdered by an angry mob of gentiles (the Mormon name for non-Mormons) near Nauvoo in 1844, plans had been laid to seek a home in the Great Basin. Under the guidance of a new leader, Brigham Young, they abandoned their homes and property at Nauvoo, and in July 1847 the first party of Mormon settlers arrived at the Salt Lake Valley.

Brigham Young's famous proclamation upon seeing the Wasatch Mountain foothills overlooking the Salt Lake ("This is the place") had been preceded by careful study of possible settlement sites. Traders, explorers, and Indian scouts had advised the Mormons, who sought a location that would support a growing, agricultural-based population eventually expected to number in the tens of thousands. Within two days of their arrival, the Mormons had dug an irrigation ditch, the first of many that would channel and divert meltwater flow from the high peaks of the Wasatch down through the foothills and into valley-bottom fields.

The site of Salt Lake City was staked out, following Joseph Smith's design for a biblical City of Zion, as a gridiron of spacious blocks with each block divided into four smaller properties that would contain houses, orchards, and gardens. This design was repeated in numerous other Mormon villages, where, for purposes of religious congregation as well as defense, farm families lived in a compact village, near one another, and traveled outward each day to work the surrounding tracts of cropland belonging to the village, in the manner of European or New England agricultural villages.

Mormon strategies for colonizing the Great Basin called for a repetition of these plans in every settlement. At the core was the church and its hierarchy of officers, committees, and responsibilities. The villages were scaled-down copies of Salt Lake City, with large farm lots in town, an ambitious plan for irrigation, and a general emphasis on the creation of institutions to support the welfare of the entire group. By 1851 Mormon outposts had been extended northward to Brigham City and Ogden, Utah; south to Provo, Manti, and Cedar City, Utah; and west to San Bernardino, California, and Carson Valley, Nevada.

The Mormons flew the American flag (even when Utah was not yet part of the United States); they sent a battalion of troops to assist in the 1846–48 war against Mexico; and the civil institutions they dreamed of creating, including the state of Deseret, were understood to be included within the general framework of American laws. Were it not for their practice of polygamy, they likely would have been considered good citizens—a valuable, hardworking group anchoring the far western frontier. But not only was polygamy against the law; it seems also to have been extremely offensive to most gentiles. Travelers who visited Salt Lake City typically found evidence of polygamy's immorality; and although they often admired Mormon thrift and industry, they condemned the group and its leaders for the unconscionable practice of encouraging men to take more than one wife.

It was this peculiar Mormon trait, more than any other, that had produced conflict with their neighbors back east and had led them to seek a domain far beyond the settled frontier where they would be left in peace. As Salt Lake City grew, it became an attractive stopping point for the growing numbers of travelers heading west. Mormons traded with the trains of emigrant wagons moving past the Salt Lake and eventually built a brisk business that brought money into the local economy. Early attempts by church leaders to discourage all interaction with gentiles and to rely on Mormon self-sufficiency gradually shifted to a view that emphasized economic (but not social) integration.

The change in outlook was especially evident when, in 1867, the Union Pacific Railroad was building west across Wyoming to meet the tracks of the Central Pacific, which was building eastward across Nevada at the same time. Brigham Young tried to convince the railroad builders to lay their tracks through Salt Lake City and follow a route just south of the Great Salt Lake. The railroad chose a more northern route, through Ogden. When the two lines of track met at Promontory, Utah, to create America's first transcontinental railroad in May 1869, Salt Lake City had been bypassed.

From that time forward, church leaders and other local authorities planned to ensure that Salt Lake City would grow as the economic focal point of the Great Basin. The hierarchy of religious and bureaucratic functions centered on Salt Lake City provided a framework for the organization of satellite settlements and intermediate centers in the Wasatch Mountain zone. Ridges of the Wasatch are arranged *en echelon,* meaning that they are parallel and separated by intervening valleys with occasional offset gaps that allow passage between the valleys. The arrangement was ideal for the founding of new settlements, one per valley segment, each of which was about the right size to support the population of a small farming community.

Irrigated sugar beets, dry-farmed wheat, sheep grazing, and other typical western agricultural activities augmented the Mormon village economy of irrigated gardens and orchards. At St. George, in the "Dixie" region along the Virgin River in southern Utah, Mormon pioneers successfully introduced the planting of cotton. In all, the comparatively fertile and well-watered valleys the Mormons chose offered a good support for farming communities. Mormon successes reflect their own foresight and careful planning more than a later Mormon rhetoric that emphasized near heroism by the early pioneers who tamed a parched and hostile land. Utah had been a good place for them to settle, and through their own industry they had made it a better place.

## The Evolution of Utah

The federal government did not allow the Mormons to establish their own state of Deseret in 1849. Nor, even, was the much-scaled-down successor territory, known simply as Utah, allowed admission to the Union as a state for almost fifty years thereafter. In the meantime, Mormon leaders established new outposts as future places of refuge in both Canada and Mexico. In 1857 federal troops appeared ready to attack the Mormon settlements on the excuse that Mormon leaders planned secession, but the separation of church and government functions in Utah led to an easing of tensions. The admission of Utah as a state waited until 1896—six years after the Mormons finally renounced their practice of polygamy.

Utah became an industrial state following the discovery of large porphyry copper deposits at Bingham Canyon in 1905. Copper concentrates were produced near the mine, and a large smelter was constructed at Magna on the southern margin of Great Salt Lake. The Bingham Canyon–Magna complex became the largest copper mining and smelting operation in the world. The modernization of facilities during the 1980s permits the continued extraction of copper ores, which now average only 0.6 percent copper.

Deposits of coking-quality coal are mined along the western margins of the Colorado Plateau in central Utah, and a small but high-grade deposit of iron ore was once mined intensively near Cedar City in the southwestern corner of the state. The coal and iron ore became the basis for a small steel industry at Geneva, Utah. The industry developed during World War II, when it supplied steel to shipyards and other industries of the Pacific Coast. It continues in operation, although now the iron ore must be brought by rail from northern Minnesota. The steel industry considerably expanded the range of manufacturing industries in the Salt Lake City–Provo region.

Although Utah has a basis for heavy industries, its population has grown in the past several decades more as a result of employment in aerospace and other defense-contracting industries. Like both California and Colorado, Utah offered an amenity-rich environment that attracted a white-collar workforce. Related developments include numerous ski resorts in the Wasatch Mountains. Gradually, Utah's image has shifted away from the Mormons and pioneer settlement of the Great Basin to outdoor recreation and urban-based, amenity-oriented living.

FIG. 20.3. *Two railroads and an interstate highway are crowded between the copper smelter, the foot of the Wasatch Mountains, and the fluctuating shoreline of Great Salt Lake at Magna, Utah.*

Even the Great Salt Lake has not been a constant. Because it is a basin of interior drainage, the lake's level fluctuates in response to short-term climatic variations that produce a greater or lesser flow of streams entering the lake. Heavy snows during the early 1980s produced rising lake levels that, by the mid-1980s, began to threaten the lower-lying elevations around the lake's margins (fig. 20.3). The specter of a latter-day Lake Bonneville rising to drown the Salt Lake Valley became a concern. To solve the problem, engineers expanded the lake by creating a new drainage outlet to the west, permitting the Great Salt Lake to spill over into the desert. This "second" Great Salt Lake, which is needed only in times of high water, is called the Newfoundland Evaporation Basin (named after the nearby Newfoundland Mountains), a name that indicates its function.



## Nevada

If the physical environment is held to explain a people's customs, outlooks, and habits, then no better demonstration of the weakness of such an explanation could be found than the adjacent Great Basin states of Utah and Nevada. The Mormon theocracy in Utah appears even more anomalous when contrasted with the saloons, casinos, and brothels that have been part of Nevada's image ever since frontier days. In fact, Nevada was once a part of Utah Territory, and the early settlements founded near present-day Reno and Las Vegas were the work of Mormons.

Discoveries of gold and silver in 1859 brought the first rush of would-be miners to the Carson City–Virginia City district of Nevada. Many who went there objected to living in "Mormon territory," and this was used as one justification for establishing a separate Nevada Territory in 1861. The federal government's relative treatment of Nevada and Utah is interesting. Nevada, which consisted of a few mining camps and had a population less than half that of Utah, was rushed into admission as a state in 1864 when its reliably pro-Union votes were needed in the U.S. Senate. Tiny Nevada thus became a state more than thirty years ahead of Utah, and it remained the nation's least populous state for a century thereafter.

Virginia City lay at the heart of the Comstock Lode, which attracted thousands of fortune seekers in 1859–60. Many came east over the Sierra from California's Mother Lode, the scene of a gold rush only ten years before. The Comstock was first mined for gold but was found to be richer in silver, a metal only slightly less precious than gold and one that the U.S. government experimented with as a currency standard during the latter half of the 19th century. Nevada became a political base for the Silverites in the West, but ultimately to no avail. When support for silver currency waned as a national issue, so did the price of silver, and Nevada's mines went into decline. The state was so tied to silver mining that it lost one-third of its people between 1880 and 1900 and entered the 20th century with a population no larger than it had during the Civil War.

Most of the rest of Nevada's towns and cities are associated with mineral deposits as well. Austin, Pioche, Tonopah, and Goldfield were gold and silver mines. Ely and Ruth are based on copper mines, for which the nearby town of McGill serves as the smelter. The two most significant exceptions to this generalization are Nevada's largest cities, Reno and Las Vegas, neither of which began as mining camps. Reno was created by the Central Pacific Railroad in 1868 as the company extended its tracks across northern Nevada. It became the principal trading center for the Sierra Nevada foothills region and soon surpassed the old Comstock Lode cities in size.

Reno also was the first to gain nationwide attention when, in the 1930s, the state of Nevada began experimenting with its laws regulating gambling and divorce, two "industries" that before that time rarely had been thought capable of stimulating economic development. Nevada progressively relaxed its laws governing the residency period for divorces, it made divorces easy to obtain, and it also removed most residency and health restrictions on marriage. Several well-publicized divorces in the 1930s (including one granted to the wife of actor Clark Gable) began to draw attention first to Reno and then to Las Vegas as places to divorce and remarry in the minimum possible time.

More significantly, in 1931 Nevada changed its laws so that the state's long-flourishing gambling casinos would no longer be outlawed but would be regulated and taxed in a manner that brought money into the state's treasury. Reno boomed from the growth of gambling, styling itself as the "Biggest Little City in the World." The growth of its entertainment industries soon spread to nearby Lake Tahoe, which developed its own combination of resort, gaming, and second-home developments.

The major beneficiary of Nevada's relaxed approach to the gaming industry was a city of fewer than twenty-five thousand inhabitants as late as 1950. Las Vegas, the site of a Mormon outpost during the church's Deseret expansion phase, became a railroad service center in the early 20th century, but until Hoover Dam (originally, Boulder Dam) was constructed on the nearby Colorado River in 1936, Las Vegas was anything but a boomtown. The large labor force brought in for the dam's construction and a temporary influx of military personnel during World War II caused Las Vegas to grow and allowed its gambling casinos to expand.

Las Vegas's growth, which is largely responsible for having propelled Nevada from 48th to 35th in population rank among the states in four decades, is remarkable by any standard. Jet air travel made it possible to reach Las Vegas from anywhere in a conveniently short time. A shift in the national economy toward a professional workforce caused a rapid expansion in the number of organizations holding national conventions. What has proved to be an almost insatiable demand for gambling in the American population continues to produce growth in Las Vegas even as the number of casinos nationwide continues to increase as well.

Although Nevada's gaming industry made the transition to a legalized, regulated status in the 1930s, gambling remained illegal nearly everywhere else. The attention of bookmakers and others identified with organized crime was drawn to Nevada. Las Vegas's first casinos were the work of gamblers, oil millionaires,

FIG. 20.4. *Las Vegas casinos.*

and various other entrepreneurs willing to bet that a fortune could be made operating a gambling casino in the desert. The proximity to southern California stimulated an interest among Hollywood investors and led to the presence of nationally known entertainers, who began performing in theaters built in conjunction with the newer Las Vegas hotels. Casinos, show lounges, and hotels were packaged into multimillion-dollar entertainment complexes that began to line Las Vegas Boulevard ("the Strip") by the early 1960s.

An important but unseen change took place in 1969 when the state dropped its requirement that all casino stockholders be licensed, a stipulation that had been designed to regulate organized crime (which by then was an accomplished fact despite the law). The requirement also had prevented casino ownership by large corporations, including the nationwide hotel chains and Hollywood entertainment conglomerates. From that time forward, Las Vegas entertainment complexes grew to twice the size they had been before and the pace of construction even increased (fig. 20.4).

Gambling reached its peak in Las Vegas in the 1980s. Entertainment complexes created since that time have increasingly followed a theme-park orientation with an emphasis on family entertainment. As a metropolitan area of more than three-quarters of a million people, Las Vegas also has diversified its economic base. The construction of still larger, more spectacular facilities designed to dazzle the tourist and attract more millions of visitors to the city each year continues.

## The Mojave Desert

The southern end of the Great Basin is the Mojave Desert, an arid landscape of partially buried mountains separating small drainage basins. Near its center, about one hundred miles west of Las Vegas and 282 feet below sea level, lies

Death Valley, the lowest elevation in the United States. Death Valley's extreme climate is a product of its rain-shadow location as well as the surrounding block-faulted topography. Elevations below sea level are in basins centered on downfaulted blocks of the earth's crust. The Mojave Desert, which is adjacent to the major fault zones of southern California, has many such fault blocks, and they form basins of interior drainage.

The added heat to produce an even greater evaporation of surface waters comes from the extra distance air masses have to sink down to the bottoms of such basins, which produces an additional temperature increase from compressive warming. The Mojave Desert contains numerous alkali flats or "soda lakes," which owe their existence to this condition. Among the mineral residues produced by rapid evaporation in such environments are borax, soda ash, trona, and other salts that are refined for industrial processes. Alkali chemicals are used in the manufacture of cleansing agents and certain types of paper.

Some lake beds at higher elevations in the Sierra Nevada foothills would contain more water were it not for their diversion for urban uses. A celebrated case is that of the Owens River, which once supplied irrigation water for farms in the Owens Valley between Bishop and Lone Pine, California. The city of Los Angeles purchased large tracts of Owens Valley land and, in 1913, constructed the first aqueduct to convey water across the mountains to Los Angeles. The city leased back its Owens Valley land to farmers and ranchers but allowed only a small amount of irrigation water to be used for local crops. The project was expanded to include diversion of water from the Mono Lake basin in 1940, and a larger aqueduct was added in 1970. Bitter controversies over Owens Valley and Mono Lake have raged from time to time, but the city of Los Angeles remains in firm control of the water.

Environmental gradations within the Mojave Desert are based mainly on elevation. The lower elevations near the Mojave's eastern border are all but devoid of vegetation as well as human habitation. The higher elevations lie toward the western fringes of the region, and because the high desert is both cooler and slightly less arid, it has been more favored for settlement. The foothills of the San Bernardino and San Gabriel Mountains within the Mojave Desert are part of the great southern California conurbation. The Cajon and Soledad Passes through the mountains already were well-worn trails in pre-European times and have long carried the routes of the principal highways and railroads linking Los Angeles with the rest of the United States. They now play an additional role of channeling the heavy volume of commuter traffic into the Los Angeles Lowland from the Mojave Desert each day.

## *References*

Arrington, Leonard J. *Great Basin Kingdom: An Economic History of the Latter-day Saints, 1830–1900.* Cambridge: Harvard University Press, 1958.

Francaviglia, Richard V. *The Mormon Landscape: Existence, Creation, and Perception of a Unique Image in the American West.* New York: AMS Press, 1978.

Hulse, James W. *The Nevada Adventure.* 6th ed. Reno: University of Nevada Press, 1990.

Jackson, Richard H. "Mormon Perception and Settlement." *Annals of the Association of American Geographers* 68 (1978): 317–34.

Meinig, D. W. "The Mormon Culture Region: Strategies and Patterns in the Geography of the American West, 1847–1964." *Annals of the Association of American Geographers* 55 (1965): 191–220.

Moehring, Eugene P. *Resort City in the Sunbelt: Las Vegas, 1930–1970.* Reno: University of Nevada Press, 1989.

Nelson, Lowry. *The Mormon Village: A Pattern and Technique of Land Settlement.* Salt Lake City: University of Utah Press, 1952.

Smith, George I., and F. Alayne Street-Perrott. "Pluvial Lakes of the Western United States." In *Late Quaternary Environments of the United States,* ed. H. E. Wright, Jr., and Stephen C. Porter, 190–214. London: Longman, 1983.

Starrs, Paul F. *Let the Cowboy Ride: Cattle Ranching in the American West.* Baltimore: Johns Hopkins University Press, 1998.

Symanski, Richard, "Prostitution in Nevada." *Annals of the Association of American Geographers* 64 (1974): 357–77.

———. *Wild Horses and Sacred Cows.* Flagstaff, Ariz.: Northland Press, 1985.

*CHAPTER 21*

# The Colorado Plateau and the Desert Southwest

High peaks of seven prominent mountain ranges dot the perimeter of the upper Colorado River's drainage basin (map 21.1). The Wasatch, Wyoming, and Wind River Ranges flank the Colorado Basin on the northwest; and the Medicine Bow, Front, Sawatch, and San Juan Ranges border it on the east. Other mountains, including the east-west-trending Uintas of northern Utah, lie largely within the basin.

A gap between the Uintas and the main body of the Rocky Mountains marks the drainage area of the Colorado's major tributary, the Green River, which extends northward into Wyoming. Only the gently sloping surface of the Great Divide Basin separates the Green from the headwaters of the North Platte. The gap has long been important to transportation routes crossing the West, connecting the forks of the Platte with Salt Lake City, and it might have been exploited as an easy route into the Colorado Basin as well.

But no transportation route ever was constructed south along the Green River and down the Colorado. The rivers follow deep, winding canyons, they lack bottomland, and their flow is turbulent, as Major John Wesley Powell and his exploring party discovered on the first recorded trip made down the Colorado River in 1869. Ever since Powell wrote about his experience, the Colorado has attracted mainly those who seek adventure. Except for its lower reaches, the river never has been a route for transportation or trade.

The Colorado is, however, the single most important source of water for urbanization and agriculture in the arid Southwest. Like the Nile in Africa, the Colorado is an exotic stream, gathering its water in one region and then flowing through a desert on its way to the ocean. Its many upper-basin tributaries—including the Green (which receives the flow of the Yampa and White Rivers in Colorado), the Gunnison, the Dolores, and the San Juan—are the beneficiaries of orographic precipitation from air masses that are forced to rise against the mountains one more time as they are carried toward the east. Snowfall totals average more than 150 inches per year on the higher elevations of the Sawatch, Front, and Medicine Bow Ranges.

Also prominent as a source of water are the high, west-facing slopes of the San Juan Mountains, where snowmelt feeds numerous streams flowing toward the Colorado. The San Juans are an uplifted maze of volcanic peaks and glaciated mountain valleys that pose an extremely difficult topography to cross and effectively isolate southwestern Colorado from the rest of the state. Early

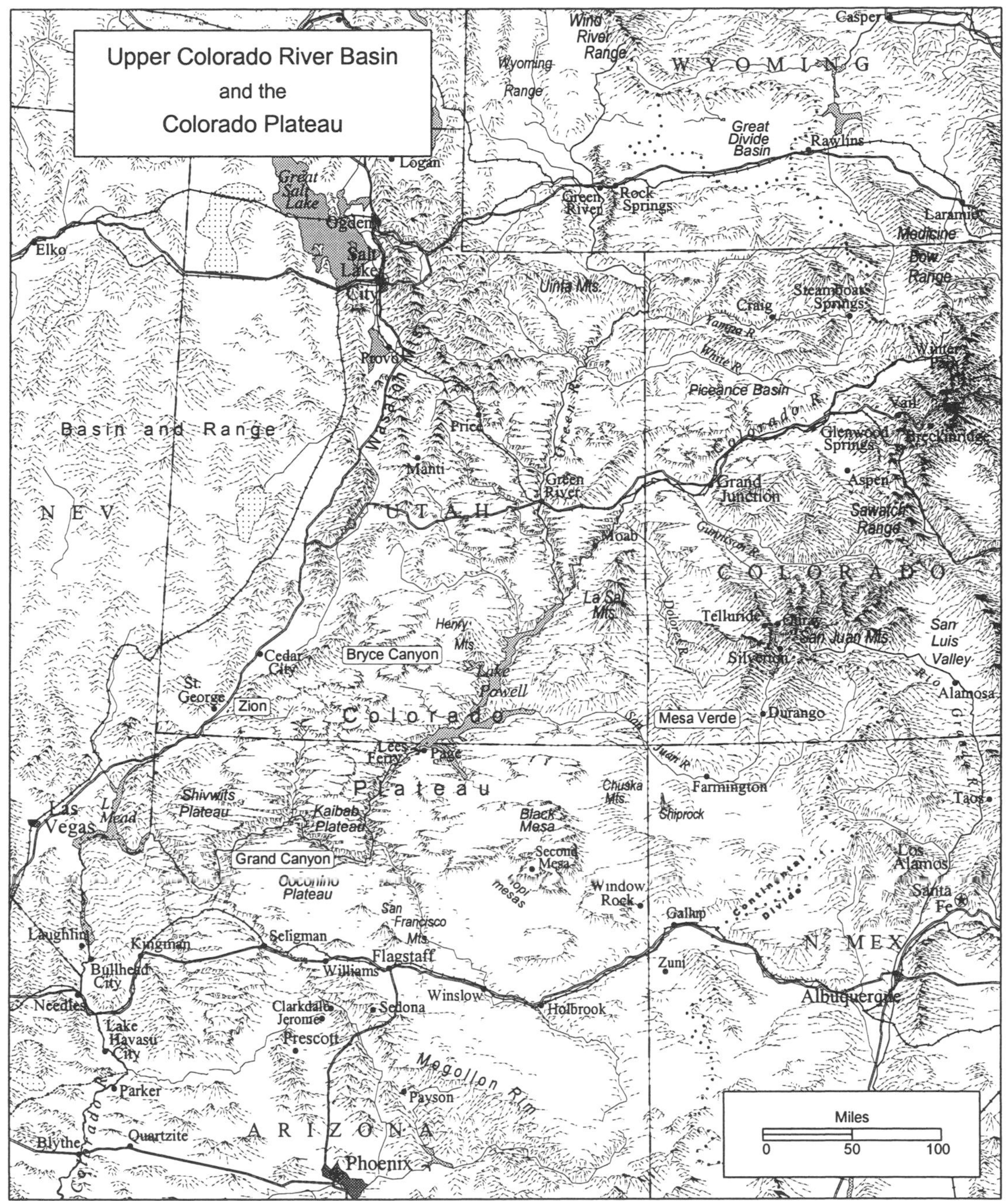


MAP 21.1

mining centers of the region, including Ouray, Telluride, and Silverton, were fabled boomtowns during Colorado's gold and silver mining era, although they have long since shifted to recreation and tourism for their economic support.

These three, as well as Aspen, Vail, Breckinridge, Winter Park, and other western slope communities, developed rapidly during the 1970s. Although alpine skiing was the main attraction to outsiders, the ski industry itself was secondary to real estate development in most of the western Rocky Mountains, and so it has remained. Some communities restricted the amount of new building, which drove up the price of existing condominiums and resort properties and thereby diverted investment into a broad range of endeavors. Problems of overcrowding, air pollution, and other ills more typically associated with cities appeared in Colorado's mountain valleys by the end of the 1970s. As elsewhere in the Rocky Mountains, the transition from "wilderness" to "overcrowded" was alarmingly swift once a real estate boom began.

## Colorado Plateau Resources

The geologic history of the Colorado Plateau is nearly identical with that of the Rocky Mountains up to the time of the Laramide orogeny (the mountain-building phase that created the Rockies, 65 million years ago). Newer sedimentary layers burying the future peaks of the Rocky Mountains were stripped away by erosion as the mass of the mountains was heaved upward. Because fault planes separated the Colorado Plateau from this uplift, the Plateau's upper layers of sedimentary rock remained intact and have been altered only slightly by underlying movements in the earth's crust. The Colorado Plateau is a region of multicolored sandstones, shales, and limestones and of high plateaus, deep canyons, and flat-topped mesas, stretching from the Rocky Mountains to the Wasatch Range.

The Plateau's roughly circular outline is centered near the four-corners meeting point of Utah, Colorado, New Mexico, and Arizona. The region extends southward to the Mogollon Rim of Arizona, reaches west to the Great Basin, and is bounded on the east by the Continental Divide, marking the drainage area of the Rio Grande in New Mexico. In all, the Colorado Plateau encompasses 150,000 square miles, about 90 percent of which is drained by the Colorado River and its tributaries.

Apart from small irrigated sections, such as the peach orchards near Grand Junction, Colorado, and the San Juan meadows of northern New Mexico, the region lacks suitable agricultural land, even though it does possess adequate irrigation water. Flatland in the Colorado Plateau lies primarily at high elevations, not low, which limits the reliable growing season; hence there are few opportunities for crop production. Recent attempts at irrigation, based on a diversion of waters from the San Juan River to plateau lands of the Navajo reservation, have been only modestly successful. Grazing is the most important agricultural land use. Especially at the cooler, high elevations, abundant grasses support thousands of sheep and cattle in year-round pastures.

Coal, oil shale, and natural gas are found in the Colorado Plateau's almost flat-lying sedimentary strata. The coal is a high-grade but low-sulfur bituminous type that has come into increased demand in recent years both for metallurgical and steaming (electricity generation) purposes. Mines in the Yampa Valley supply steel mills from Pueblo, Colorado, to the Chicago area; Utah's coal mines supply its local steel industry and provide steam coal for the western intermontane region. Large quantities of bituminous coal now are mined on the Black Mesa of northern Arizona to provide energy for local electric-power generation.

Oil shale was a resource thought ready for development during the oil-short 1970s, and plans for its exploitation helped stimulate an economic boom in Denver, but the industry collapsed when oil prices dropped. Natural gas produced in the Piceance Basin of Colorado supplies part of the energy needs of southern California. Producing oil wells are scattered around the major gas and oil-shale deposits, but the region's oil production is small.



The Colorado Plateau also is the major source of uranium in the United States. As was typical in the West, the discovery of one mineral often accidentally revealed the existence of another that was considered waste until a use for it developed. Colorado's 19th-century gold miners were troubled by a black, tar-like substance that adhered to their mining tools. It was pitchblende, and they discarded it as a nuisance. In 1898 Madame Marie Curie discovered that pitchblende was a source of the radioactive material radium, and her discovery caused an immediate interest in reworking the gold tailings of the Colorado Rockies. The most promising source of radium in the United States was carnotite, a yellowish, uranium-bearing rock abundant in the Colorado Plateau. The manufacture of radium from carnotite was complicated by its association with vanadium until vanadium was discovered to be valuable in the manufacture of high-tensile-strength steel. Each new mineral discovery stimulated further efforts at exploration for richer veins.

Given the significance of uranium in the Cold War era, the U.S. government was understandably eager to confirm the existence of domestic deposits. The first important uranium discovery, made near Moab, Utah, in 1952, stimulated a dramatic although short-lived mining boom. By the late 1950s the federal government (which purchased the entire output of Colorado Plateau uranium) believed its stockpile was sufficient and thereby ended the mining boom.

Interest in uranium was revived during the 1960s when the nuclear power industry appeared ready to supply much of the nation's future electricity needs. Even before the second uranium boom was cut short by growing public distrust of nuclear power, however, the mining of uranium was revealed to be a deadly business for those involved. Uranium was mined in poorly ventilated underground shafts and tunnels where miners breathed radon dust that, through the process of radioactive decay, eventually was transformed into lead in their bodies. Waste by-products of uranium refineries on the Colorado Plateau were shown to present a hazard to those who lived near them. The Plateau region remains as the major source of uranium in the United States, but the prospects for a revival of the industry are uncertain.

The Colorado Plateau's fascinating topography of plateaus, mesas, and canyons has made tourism the region's most important growth industry. The Grand Canyon of the Colorado River is only one of many on the Plateau produced by the downcutting action of stream flow against the rising bed of a river as the land mass experiences slow geologic uplift. Spectacular views of the Grand Canyon from its south rim stimulated interest in building a resort industry by the late decades of the 19th century, and the area was set aside as Grand Canyon National Park in 1919. Zion and Bryce Canyon National Parks in Utah, which also offer examples of differentially eroded sandstones, limestones, and shales into castellated rock formations, were established shortly thereafter. By the 1990s national parks and recreation areas enveloped most of the Colorado River as well as some of its tributaries.

## Native Populations in the Colorado Plateau

The high plateau of northern Arizona and New Mexico would seem to offer only a lean sustenance for people who live there, but its Native American occupants have enjoyed a measure of prosperity rarely encountered on reservations elsewhere in the United States (fig. 21.1). Puebloan people were the only occupants of the area before an in-migration of Athabaskan-speakers from northern Canada took place sometime between A.D. 800 and 1500. The northerners divided into two groups. The Apache established themselves south and east of the Colorado Plateau, and the Navajo took up residence on the Plateau itself. Not only are the Apache and Navajo close relatives, but both are also linked to other Athabaskan-speaking groups of the Northwest Territories. Their common usage of the name *diné* (dene), meaning "people," is a manifestation of the link.

The Navajo are by no means the aboriginal occupants of the Colorado Plateau. Their entrance into the Southwest may have been only slightly before the Spanish in the 16th century, although their date of arrival remains unknown. They acquired the use of the rifle and the horse from the Spanish and supplemented a life based on hunting and agriculture with herding sheep (which they also learned from the Spanish). Navajo pastoralism provided food as well as wool for clothing and was naturally suited to the high plateaus they occupied.

FIG. 21.1. *Navajo legislative assembly building, Window Rock, Arizona.*

Early Navajo settlements were concentrated in New Mexico and Arizona, somewhat east of the current Navajo reservation. The people prospered, and by the 1860s some fifteen thousand Navajo, owning perhaps half a million sheep, dominated the southern portion of the Colorado Plateau.



The area had not yet been fully integrated into American territory (following its acquisition from Mexico in 1848) when the Civil War began, adding a new dimension to the general lawlessness that prevailed. Knowing that they could not control the Navajos' possible role in the war, army generals dispatched Colonel Kit Carson and a force of soldiers to round up the entire Navajo population. In 1863 they were marched to an inhospitable stretch of bottomland along the Rio Grande River in New Mexico known as the Bosque Redondo. Although their forced residence at Bosque Redondo probably was the worst of times for the Navajo, they learned valuable arts, including iron working and silversmithing. They did not learn to become self-sufficient farmers as the government had hoped, however, and by 1868 the Navajo had become such a financial burden that they were allowed to leave Bosque Redondo to occupy a new reservation along the Arizona–New Mexico border.

The return to Arizona marked the beginning of an era of great prosperity for the Navajo. Their herds expanded as a part of the general growth of the range livestock industry in the Southwest once railroads entered the region in the 1880s. Their reservation was enlarged several times in area, and their population increased steadily. Navajo people also became expert in producing blankets and jewelry for sale to tourists who traveled aboard the trains that passed through Navajo country. Nonetheless, they maintained a largely isolated existence, retained most tribal practices intact, and lived in dispersed, kin-based settlements of a few houses each that also included one or more of the Navajo's traditional hogans (ceremonial structures).

The Navajo were experiencing growth and success at the same time that most other native American groups were suffering devastating losses of land and people. Overgrazing of Navajo lands, their practice of running herds outside the reservation, and the general advance of livestock numbers in both the Anglo and Hispano portions of the region led to a more active intervention by the federal government. Beginning in the 1930s, the Department of the Interior announced policies to reduce the size of Navajo sheep and goat herds in order to protect the range. Governmental institutions were introduced that were patterned more after standard American practices than Navajo experience.

As in other cases in which a deliberately isolated group has become increasingly drawn into a surrounding culture, the Navajo have experienced challenges to long-standing customs. Better than many native populations, however, they have adapted to change without losing their traditions, while keeping most outside influence to a minimum. Navajo tribal numbers expanded from tens of thousands to hundreds of thousands during the 20th century.

Puebloan groups on the Colorado Plateau, such as the Zuni and the Hopi, have not grown so rapidly. They occupy reservations that are small enclaves within or adjacent to Navajo lands, but their ancestral ties to the area extend much farther back in time than do those of the Navajo (see chapter 15, map 15.3). Several Hopi villages have been occupied continuously for six hundred to eight hundred years, and Hopi occupation of the Colorado Plateau in northern Arizona probably is at least twice as old as that. They are descendants of the Anasazi, whose culture traces back to at least 300 B.C. and among whose villages were the cliff dwellings at Mesa Verde, Colorado. Rumors of the Hopi and Zuni pueblos, which the Spanish imagined contained fortunes in gold, first attracted the Spaniards northward from Mexico in the 16th century.

The heart of the Hopi reservation is a series of small villages built at the steep, rock-faced precipices of First Mesa, Second Mesa, and Third Mesa, an irregular line of bluffs with an unobstructed view of the valley below. Hopi maize-based agriculture, itself of great antiquity, incorporates practices that allow corn production in an arid climate. Hopi cornfields lie at low elevations, where they are sustained by the invisible flow of water below ground level along streams, but most of their villages are on the high mesas, for protection. Hopi religious beliefs and social organizations, as well as their matrilocal residence pattern and their habits of egalitarianism and village autonomy, have attracted generations of cultural anthropologists to the mesa-edge villages.

Spanish colonizers who entered the Southwest in the 16th century were most successful in the Rio Grande Valley, less so to the west, and had only a slight impact on the Navajo. Attempts to convert the Hopi to Christianity met with only limited success. The "pueblo revolt" of 1680 caused the Spanish to retreat to El Paso for a time, and although the Spanish managed a successful reconquest in 1692, the Hopi remained outside the zone of strongest Spanish influence thereafter. The Hopi maintained their society and religion, despite their small numbers (only in the thousands) and in the face of strong pressure from the Spanish, even as they also endured nearly constant intrusions by the numerically superior Navajo who lived on all sides of them. Their strategy of avoiding military confrontation probably saved the Hopi from annihilation.

Latter-day conflicts between Navajo and Hopi were intensified by resource developments on the Colorado Plateau during the 1960s. The Hopi mesas are southward extensions of Black Mesa, a massive bituminous coal formation that lies within the Navajo and Hopi reservations. Natives had mined this coal well back into prehistoric times, but commercial production did not begin until electric-power plants were constructed around the fringes of the Colorado Plateau. The Navajo and the Hopi made land swaps to share the resource-rich lands and to grant more territory for the expansion of Navajo settlements. Generating plants burning Colorado Plateau coal, near Shiprock, New Mexico, Page, Arizona, and Laughlin, Nevada, stimulated the construction of new railroads and coal slurry pipelines across the Colorado Plateau. The coal-fired generating plants have been controversial for their alleged role in local climate alteration as well as for disrupting traditional resource-use patterns.

## The Mogollon Rim

Two zones defined on the basis of human occupancy and elevation separate the Native American lands of the Colorado Plateau from the more urbanized Arizona desert to the south. The main corridor of travel, through Gallup, New Mexico, and Holbrook, Flagstaff, Seligman, and Kingman, Arizona, to Needles, California, was the route of the land-grant railroad built along the 35th parallel west of Albuquerque in the 1880s (map 21.2). Today the Santa Fe/I-40 corridor is the busiest rail and highway route across the Southwest. East of Seligman it is an all-Plateau route, roughly fifty miles north of the Mogollon Rim, a major landscape divide that forms the Colorado Plateau's south-facing border across Arizona.



The 7,000-foot elevations of the Mogollon Rim represent the slightly upturned edges of Plateau strata on which a narrow belt of more rugged topography has formed. Eastward-moving air masses are forced to rise against the Plateau, which produces additional precipitation (fig. 21.2). The Mogollon Rim receives enough moisture to support the growth of an open, ponderosa pine forest, which forms the basis for a small lumber industry, badly needed in the arid Southwest. The Mogollon Rim's greater precipitation, combined with moderate-elevation temperatures, made this an attractive area to early settlers just as it is to tourists today. Prescott (Arizona's first capital) is in this zone, and so are a series of small communities founded in the 1870s by Mormons from Utah, who originally attempted agricultural settlements along the Little Colorado River on the open Plateau. Sedona, Payson, and the early copper towns of Clarkdale and Jerome also are in this midelevation band between high plateau and low desert.

## The Sonoran Desert

Arizona's ecosystems unfold downslope from the pine-clad Mogollon Rim, through shrubby canyon woodlands, to the true desert that blankets the southern portion of the state. As elsewhere in the West, vegetation and land use are determined largely by elevation and slope aspect, with the greatest moisture stress on lower, south-facing slopes and in basins of interior drainage. Southern Arizona is part of the Sonoran Desert, an ecological zone named for the Mexican state of Sonora, which is adjacent on the south. The Sonoran zone receives roughly the same amount of precipitation as the Colorado Plateau, but because it lies at a lower elevation and experiences a maximum of available sunshine, it has much hotter temperatures.

The Sonoran Desert has the largest values of potential evapotranspiration of any region of the United States, meaning that an open pan of water will evaporate more moisture back into the atmosphere in a given amount of time than it will anywhere else in the country. This gap between the sun's ability to evaporate water and the local supply of water makes irrigation an important activity in the desert environment.

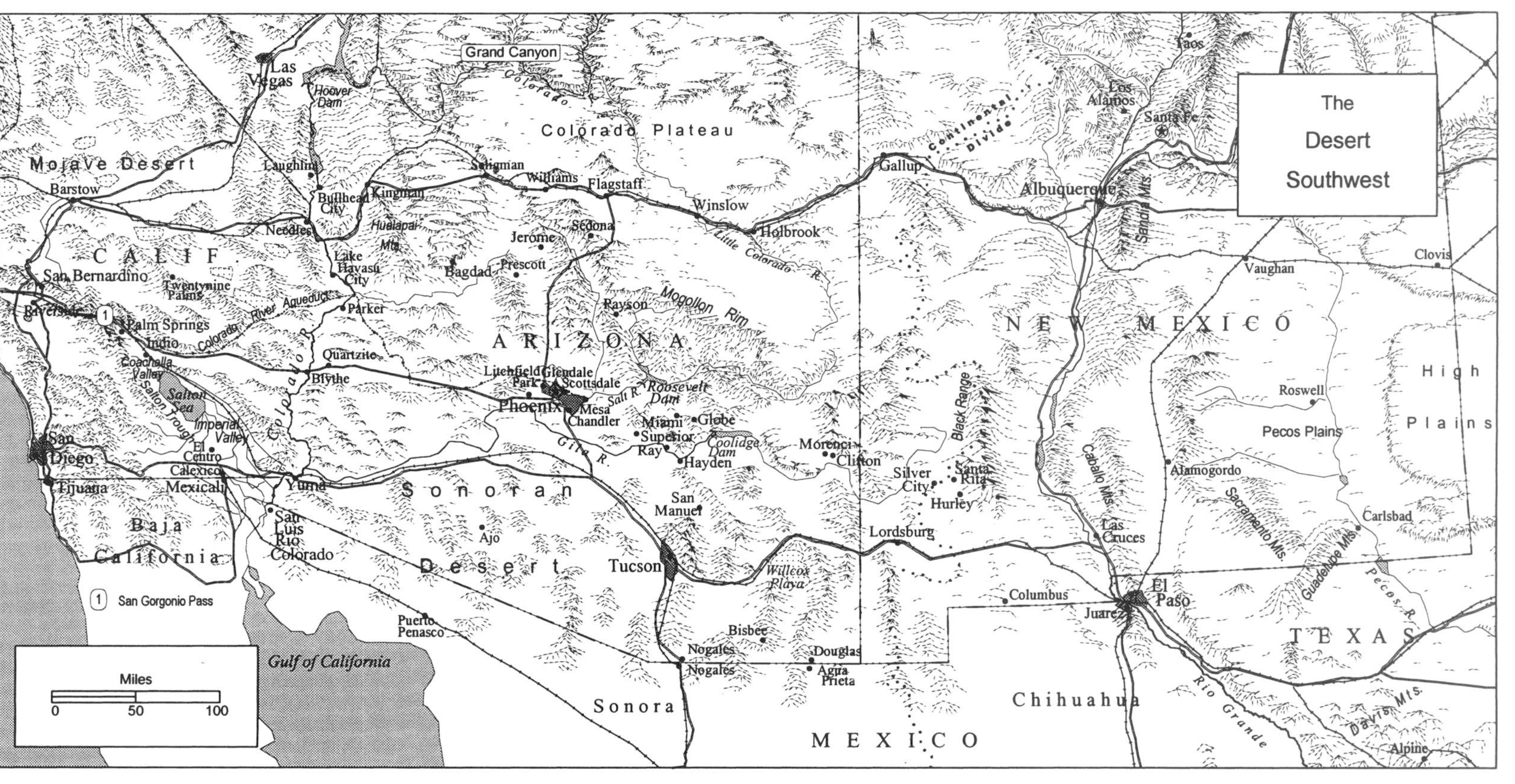
The Desert Southwest
Grand Canyon
Las Vegas
Hoover Dam
Colorado
Colorado Plateau
Mojave Desert
Laughlin
Barstow
Bullhead City
Kingman
Seligman
Williams
Flagstaff
Winslow
Holbrook
Little Colorado R.
Gallup
Continental Divide
Los Alamos
Santa Fe
Taos
Albuquerque
Sandia Mts
Needles
Hualapai Mts
Jerome
Sedona
Prescott
Bagdad
Lake Havasu City
CALIF.
San Bernardino
Twentynine Palms
Riverside
Palm Springs
Indio
Colorado River Aqueduct
Parker
Quartzite
Blythe
Coachella Valley
Salton Sea
Salton Trough
Imperial Valley
El Centro
Calexico
Mexicali
San Diego
Tijuana
Yuma
San Luis Rio Colorado
Baja California
Payson
Mogollon Rim
ARIZONA
Litchfield Park
Glendale
Scottsdale
Phoenix
Mesa
Chandler
Salt R.
Roosevelt Dam
Miami
Globe
Superior
Ray
Hayden
Coolidge Dam
Gila R.
Morenci
Clifton
Sonoran Desert
Ajo
San Manuel
Tucson
Willcox Playa
Bisbee
Douglas
Agua Prieta
Nogales
Nogales
Puerto Penasco
Gulf of California
Sonora
San Gorgonio Pass
Miles
0
50
100
Black Range
Silver City
Santa Rita
Hurley
Lordsburg
Columbus
NEW MEXICO
Caballo Mts.
Las Cruces
El Paso
Juarez
Chihuahua
MEXICO
Vaughan
Clovis
High Plains
Roswell
Pecos Plains
Alamogordo
Sacramento Mts.
Guadalupe Mts.
Carlsbad
Pecos R.
TEXAS
Rio Grande
Davis Mts.
Alpine

MAP 21.2

FIG. 21.2. *A pine forest on the Mogollon Rim near Payson, Arizona.*

Two vegetation types characteristic of the Sonoran zone reflect the region's extreme level of moisture stress to plants. One type is illustrated by the palo verde, which from a distance appears to be a clump of green poles, as its Spanish name suggests. The woody stems have no branches but are covered in very small green leaves. This type of plant-physiological adaptation, known as microphyllous habit, minimizes water loss through transpiration from leaf surfaces. A second type is the desert succulent plant, exemplified by the many species of cactus; cactus stores water in expandable chambers of the plant's trunk or leaves in order to survive the months without precipitation (fig. 21.3). Because succulent plants contain water, they are especially vulnerable to freezing and are often found only on the lowest desert slopes. The saguaro is perhaps the best known of the Sonoran desert cacti.

FIG. 21.3. *Cacti of the Sonoran zone include the tall saguaro (extreme left) and cholla (jumping cactus) (right foreground).*

Geologically speaking, the Sonoran Desert is part of the Basin and Range section, and thus it exhibits the same characteristic topography of low, widely separated mountain ridges flanked by pediments and alluvial fans. The Sonoran Desert has fewer basins of interior drainage than the Great Basin portion of the Basin and Range, however, and despite the existence of some large playas, it has a better-developed network of streams. Much of the area is tributary to the Gila River, which rises in the Black Range of New Mexico and flows west to enter the Colorado River at Yuma. (Because of upstream dam developments, the Gila now is normally a dry streambed west of Phoenix.)

## The Gadsden Purchase

At this latitude, desert environments are continuous from southern California's coastal mountains to the Pecos River in Texas. In addition to the unifying conditions of desert vegetation and a topography of broad basins separated by low mountains, these southwestern borderlands offer the closest approximation of a mountain-free crossing of the continent anywhere in the United States. In fact, the U.S.-Mexican border between El Paso and Yuma, which resulted from the Gadsden Purchase of 1853, was drawn to contain a low-elevation route for a transcontinental railroad entirely within American territory.

Although the primary objective of the United States in pursuing a war against Mexico in 1846 was the acquisition of California (Texas was annexed as a state the year before), there also was substantial American interest in acquiring various portions of Mexico's northern borderlands. The Treaty of Guadalupe Hidalgo, signed by the two countries in 1848, established the Rio Grande and Gila Rivers as the international boundary. Faulty maps used in the treaty negotiation misrepresented the location of El Paso, the acquisition of which mattered a great deal to Southerners, especially, who sought a route for a transcontinental railroad.

At El Paso del Norte the Rio Grande River cuts through the single mountain ridge separating Texas on the east from the smooth-floored basins of New Mexico on the west. The many mountain ranges of far western Texas are all of the basin-and-range variety. Some, such as the half-dozen ranges in the Big Bend region, have peaks over 6,000 feet, but they are isolated and pose no obstacle to travel. Others, such as the Davis Mountains near Alpine, Texas, can be crossed on relatively easy grades. These conditions, plus the availability of a lowland route through the Sonoran Desert, along the Gila River, meant that an all-weather transcontinental route could be constructed from San Antonio to California that was free of high mountain grades.

The most prominent advocate of such a route, roughly following the 32d parallel of latitude, was Jefferson Davis, then secretary of war. His great interest in a desert routeway from the Cotton South to the Pacific Coast even led Davis to experiment with using camels as pack animals, in the fashion of African or Asian long-distance trade routes, as a means of bridging the long gap before the railroad was built. Had this route been chosen for the first transcontinental line, as Davis urged, it would have given the southeastern states an obvious advantage for access to the Pacific Coast. The need to gain control of the Rio Grande crossing at El Paso plus the acquisition of enough additional land to construct the railroad line south of the Gila River led to a reopening of the border question with Mexico.



At Davis's urging, James Gadsden, a South Carolina politician and railroad builder, was appointed U.S. minister to Mexico in 1853, and he immediately began negotiating for the tract of land known ever since as the Gadsden Purchase. It eventually was reduced to only 30,000 square miles, but it added enough territory to secure the route to the mouth of the Gila River. The Gadsden Purchase cost the United States $10 million, whereas the entire Mexican cession of 1848—more than 525,000 square miles—had cost only $15 million. Davis, later president of the Confederate States of America, never was able to force adoption of the all-southern transcontinental route. The Southern states had seceded by the time the Pacific Railroad Act was passed in 1862, and the chosen route crossed Wyoming and Utah, not New Mexico and Arizona.

## The Copper Industry

The railroad across southern Arizona was built, but not until the 1880s. One of its first functions was to stimulate what became by far the most important industry of southern Arizona and New Mexico, the mining and smelting of copper. The first important mines at Bisbee and Morenci were followed by developments at Superior, Ajo, Jerome, and Globe-Miami, all in Arizona (fig. 21.4). New Mexico's copper mines are concentrated at Santa Rita and Hurley, near Silver City. Since 1945 new mines have opened near San Manuel, Bagdad, and in the Tucson area.

Arizona and New Mexico copper ores are low-grade, averaging only 4 percent pure copper even at the beginning of mining; today they average roughly

FIG. 21.4. *Copper-mining town of Miami, Arizona.*

0.6 percent purity. The substantial amounts of earth-moving and processing required to mine ores of this grade call for massive capital investments in mines and machinery. Most Arizona copper mines use the open-pit method, although some underground mines still operate. In recent years leaching methods have been introduced that remove dissolved copper by percolating a weak acid through the vein. Some mines operate with all three methods of extraction at once.

As in Montana and Utah, mining low-grade ores demands that the first stage beyond mining, the production of copper concentrates, be done at the mine site. Concentrates average 25 to 30 percent pure copper and are valuable enough that they can be transported over at least short distances to a smelter, where, in very high-temperature furnaces, the chemical bonds between the copper and various impurities are broken. Blister copper (97% pure) and anode copper (99% pure) from the smelters is valuable and bears transportation costs over a longer distance to be further refined into copper sheets, bars, wire, rods, and other products. There are copper refineries in El Paso, Texas, and several East and West Coast locations.

Arizona's copper workings demanded not only large amounts of capital but also, especially in the early years, a large labor force. No towns existed near the mines that would have provided the necessary infrastructure. Copper companies constructed company-owned towns near the mines and smelters to house the large numbers of workers that were needed. Many of them were totally owned by the copper companies and were supplied with company-owned stores where miners purchased necessary goods. Southern Arizona and New Mexico acquired a scatter of small cities as a result, pairing the mine/concentrator complexes with their usually nearby smelting centers. They include Ray-Hayden and Santa Rita–Hurley (Kennecott Copper) and Ajo, Clifton-Morenci, and Bisbee-Douglas (Phelps-Dodge).

A worldwide drop in copper prices in the 1970s threatened the future of some of the mines and smelters. The industry revitalized in the 1980s around joint ownership ventures (especially involving Japanese firms) and an increased tendency toward vertical integration. Several mines and smelters have been closed, whereas new methods of ore extraction and reduction have replaced the more labor-intensive operations in others. Taken together, Arizona's and New Mexico's copper mines and smelters account for roughly three-fourths of the copper industry in the United States at present, with the Bingham Canyon–Magna complex in Utah contributing much of the rest.

## Agriculture, Urbanization, and Water

Arizona's present crisis over water supply and water use involves a complex of legal, geographical, and economic factors. Central to all discussions of how the desert's scarce water supply should be allocated is the fact that most agriculture is a virtual impossibility in the Sonoran zone without irrigation water. Hohokam people were irrigating portions of the Gila River Valley as early as 300 B.C. Seventeenth-century Spanish missionaries introduced the *acequia* (community irrigation ditch), which was based on the use of diversion dams to channel water out of flowing streams and across fields. Anglo farmers began irrigating portions of the Salt River Valley at Phoenix in the 1860s. Irrigation thus has a long history in the region, and given the western legal doctrine of prior appropriation of water when determining rights of access, agricultural uses are prior to almost all others.

Urban water uses came early as well, but for many years Arizona's level of urbanization was low, especially in contrast with that of southern California. Arizona's first major city was Tucson, an early Spanish outpost, which began growing when it was reached by the first railroad built across the state. Tucson has no surface water to speak of, but groundwater supplies were abundant and adequate for urban and agricultural uses until recent decades. Some of Tucson's residential neighborhoods still favor the desert aesthetic of gravel-only landscaping and the use of desert plants for shrubbery rather than displaying the verdant greenery possible with unlimited irrigation water.

Phoenix was not founded until the late 1860s. Its early growth was stimulated by the development of irrigated agriculture over a tract of excellent land near the confluence of the Salt and Gila Rivers. The territorial capital was moved to Phoenix in 1889. By 1920 the city surpassed Tucson in size and had become a major transportation center. Phoenix is ringed by a series of smaller centers—Glendale, Mesa, Scottsdale, Chandler—that also emerged mostly from irrigation schemes, somewhat in the fashion of Los Angeles and its satellite cities. Another Phoenix-area community, Litchfield Park, was created in 1916 by the Goodyear Tire and Rubber Company, which converted thousands of acres of desert land to the production of irrigated, long-staple cotton (pima cotton), which at one time was considered essential in the manufacture of automobile tires.

Agriculture in central Arizona has relied on some pumped groundwater, but its major expansion was made possible by the construction of Roosevelt Dam on the Salt River in 1911 and Coolidge Dam on the Gila River in 1930 (fig. 21.5).

FIG. 21.5. *Coolidge Dam on the Gila River stores irrigation water for central Arizona.*

Arizona's first large cotton crops were from Goodyear's pima cotton fields at Litchfield Park, but those were surpassed when short-staple upland cotton was grown beginning in the 1920s. More than a quarter million acres of irrigated upland cotton have been produced annually in Arizona since the late 1940s. Cotton is Arizona's largest crop. Since more than 80 percent of Arizona's water consumption goes for agriculture, it is clear that cotton production is a major drain on the state's water supply.

Arizona's growing water needs eventually were interpreted in the context of utilizing the Colorado River. The state was a latecomer in laying claim to the Colorado's flow, after diversions had been made to irrigate the Colorado Piedmont and California's Imperial Valley and to supply the water needs of southern California generally. Although only a narrow fringe of western Arizona cropland is directly irrigable from the Colorado, the possibility of diverting water from the river and pumping it across central Arizona to Phoenix and Tucson was seen as a logical course to pursue.

An interstate agreement made in 1922 divided the drainage area of the Colorado River into upper- and lower-basin components, with the divide drawn at Lee's Ferry, just south of the Utah border in Arizona. The agreement, known as the Colorado River Compact, was ratified quickly by all basin states except Arizona, which waited until 1944 to declare its participation. Arizona opposed the large diversions made by California and fought the construction of Parker Dam in 1934, which stored water for the 240-mile Colorado River Aqueduct that supplied water to Los Angeles. Arizona's share of the water finally was confirmed by a U.S. Supreme Court decision in 1964, making possible the Central Arizona Water Project four years later.

The Central Arizona Project involves a series of ten aqueducts and pumping stations beginning at Lake Havasu to deliver water as far as Tucson. None of the water is to be used for agriculture. The project acquired notoriety when it was shelved (only temporarily, as it turned out) by President Jimmy Carter in 1977 on a variety of economic, social, and environmental grounds.

Urban growth in the lower Colorado Valley has boomed in the past three decades. Gambling casinos and retirement communities appeared in the all-new settlements of Laughlin, Nevada, and Bullhead City, Arizona, built on opposite banks of the river. Entrepreneurs purchased London Bridge, dismantled it, and moved it in pieces to Lake Havasu City, where it was reconstructed as a tourist attraction. The "trailer village" of Quartzite, Arizona, attracts thousands who winter there every year. New cotton fields have appeared on both sides of the river between Yuma and Needles (fig. 21.6).

In addition to the controversies between states, between economic sectors, and between interest groups, there are questions involving the broader impacts of water projects. The largest hydroelectric power installations on the Colorado are Hoover Dam, near Las Vegas, constructed in the late 1930s to supply power to southern California; and Glen Canyon Dam, built in the late 1950s, which mainly supplies power to central Arizona. Both have provided much-needed electricity that has stimulated further urban and industrial growth in the two

FIG. 21.6. *Despite the parched look of the soil, the Colorado River provides ample water for this cotton crop, growing in 110° heat near Blythe, California.*

states. The reservoirs impounded behind these and other dams act as massive evaporation pans in the intense desert heat that produce water losses to the atmosphere rivaling the entire contributions of the flow from some of the Colorado's upper-basin tributaries.

Debate over the future of Colorado River water and its uses will continue for the indefinite future. The total amount of water in the Colorado Basin is not adequate to serve all of the needs that have been promised, and there is no possibility for a transfer of water into this basin—short of a massive rearrangement of water flows at the continental scale—that would make up the difference. A curtailment of water use thus seems inevitable. That the river's flow is fully accounted for in existing diversions is revealed in the fact that essentially no water remains in the Colorado's channel as it meanders toward the Gulf of California south of the Mexican border. It is only an irrigation drain, carrying the runoff from fields through which its waters have already passed.

## The Imperial and Coachella Valleys

The ultimate sink of the Colorado River's water is not the Pacific Ocean (the Gulf of California), which the river enters, by definition, at sea level. Rather, it is the Imperial Valley of California and its sink, the Salton Sea, whose average surface level is 235 feet below sea level. The Salton Trough is a graben (a block-faulted valley, dropped down in elevation) that is part of the sequence of such depressions that are common close to the San Andreas fault in southern California. The Salton Trough's northern end, known as the Coachella Valley, contains the resort community of Palm Springs and a small area of irrigated agriculture at Indio, which remains one of the few areas of commercial date production in the United States, although now it is confined by urban sprawl.

Palm Springs and the Coachella Valley lie more than fifteen hundred feet below the level of San Gorgonio Pass, a few miles to the west. Air settling from the

Mojave Desert downslope through San Gorgonio Pass and into the Salton Trough creates a rush of wind sufficient to turn the blades of dozens of wind-powered turbines that have been placed on these slopes to generate electricity. Air passing through this "wind farm" is warming as it settles and reaches its maximum temperature on the valley floor, making Palm Springs one of the warmest winter resorts in the United States. The Salton Sea occupies the lowest portion of the Salton Trough, separating the Coachella Valley to the north from the larger Imperial Valley, which borders it on the south.

The Salton Sea owes its modern origin to an irrigation accident resulting from early efforts to channel water from the Colorado River into the Imperial Valley. A dam constructed just south of the Mexican border for irrigation diversion broke during a 1905 flood; for nearly two years, the river flowed unchecked across the Imperial Valley and into the Salton Sea, where, lacking an outlet, the waters accumulated and evaporated. The inflow of water was sufficient to establish the Salton Sea at approximately its current size, and it has been maintained since that time by agricultural runoff.

In 1942 the All-American Canal (referring to its point of diversion north of the international border) was completed to divert Colorado River water to new irrigation districts near El Centro, California. Most of the land here is more than 100 feet below sea level. Crops of irrigated sugar beets provide raw material for the Imperial Valley's sugar industry, and the waste products from beet-sugar production, along with irrigated crops of alfalfa, established the basis for a cattle-feeding industry in the Imperial Valley. Winter vegetables are another important Imperial Valley crop.

What little water remains in the Colorado beyond its diversion into the All-American Canal supports a final irrigation zone south of the Mexican border, the Mexicali Valley, which raises crops of winter vegetables sold primarily in markets north of the border. Given the northwestward slope of the land here, irrigation waters from the Mexican side enter the United States and drain toward the Salton Sea. Its last water having been diverted, the Colorado River approaches the Gulf of California as little more than a sandy streambed.

## *References*

Allen, James B. *The Company Town in the American West.* Norman: University of Oklahoma Press, 1966.

Bailey, Garrick, and Robert Glenn Bailey. *A History of the Navajos: The Reservation Years.* Santa Fe: School of American Research Press, 1986.

Comeaux, Malcolm E. *Arizona: A Geography.* Boulder: Westview Press, 1981.

Erickson, Kenneth A., and Albert W. Smith. *Atlas of Colorado.* Boulder: Colorado Associated University Press, 1985.

Fradkin, Philip L. *A River No More: The Colorado River and the West.* New York: Alfred A. Knopf, 1981.

Garber, Paul Neff. *The Gadsden Treaty.* Philadelphia: Press of the University of Pennsylvania, 1923.

Goodman, James M. *The Navajo Atlas: Environments, Resources, People, and History of the Diné Bikeyah.* Norman: University of Oklahoma Press, 1982.
Graf, William L. *Wilderness Preservation and the Sagebrush Rebellions.* Savage, Md.: Rowman & Littlefield, 1990.
Griffiths, Mel, and Lynnell Rubright. *Colorado: A Geography.* Boulder: Westview Press, 1983.
ÓhUallacháin, Breandan, and Richard A. Matthews. "Economic Restructuring in Primary Industries: Transaction Costs and Corporate Vertical Integration in the Arizona Copper Industry, 1980–1991." *Annals of the Association of American Geographers* 84 (1994): 399–417.
Powell, John Wesley. *The Exploration of the Colorado River.* Ed. Wallace Stegner. Chicago: University of Chicago Press, 1957.
Ringholz, Raye C. *Uranium Frenzy: Boom and Bust on the Colorado Plateau.* Albuquerque: University of New Mexico Press, 1989.
Whitely, Peter M. *Deliberate Acts: Changing Hopi Culture through the Oraibi Split.* Tucson: University of Arizona Press, 1988.
Wyckoff, William. *Creating Colorado: The Making of a Western American Landscape.* New Haven: Yale University Press, 1999.

*PART IX*

# The North

## CHAPTER 22

# The Upper Great Lakes

Less than one-tenth of the total area covered by boreal forest in North America lies within the lower forty-eight states. The northern Maine woods are part of the boreal forest, as are the rugged uplands of northern New England and the Adirondack Mountains of New York. The largest area, however, is the Upper Great Lakes region, adjacent to the main body of boreal forest that covers the interior of Canada. Nearly ninety thousand square miles of northern Michigan, Wisconsin, and Minnesota are included within the boreal zone (map 22.1). Although the Upper Great Lakes region is part of these three Middle Western states, its physical base, settlement history, and subsequent economic development more closely resemble central Canada than the U.S. Middle West.

### Furs and Forests

*Boreal* is a Latin word that refers to northern areas. The early French who sailed to North America did not come in search of boreal forests and fur-bearing animals, yet these were the most readily exploitable resources they found in the vast interior region drained by the Great Lakes and the St. Lawrence River. By 1670 French explorers had seen all of the Great Lakes and were about to discover the Mississippi River and its northern tributaries. The trade in the Upper Great Lakes region at that time was between native groups only. Gradually the French coureurs de bois took on the role of intermediaries, specializing in the barter of furs for European trade goods. Native people accepted arms, ammunition, utensils, cloth, and blankets in exchange for beaver peltry. All of the items were relatively easy to transport across the many canoe portages that formed successive links in the French supply network.

Recognizing the lucrative nature of the North American fur trade, British interests formed the Hudson's Bay Company in 1670. King Charles II granted the company exclusive trading privileges in the territory named Rupert's Land, which consisted of all of the area drained by rivers emptying into Hudson Bay. The nonoverlapping nature of drainage basins thus initially guaranteed that the Hudson Bay–based British trade and the Montreal-based French trade would not come into conflict. When Quebec fell to the British in 1760, Montreal's fur trade also came under British control. Organized as the Northwest Company, Montreal traders then extended their reach westward to Lake Manitoba and the Saskatchewan River. The Northwest and Hudson's Bay Companies competed

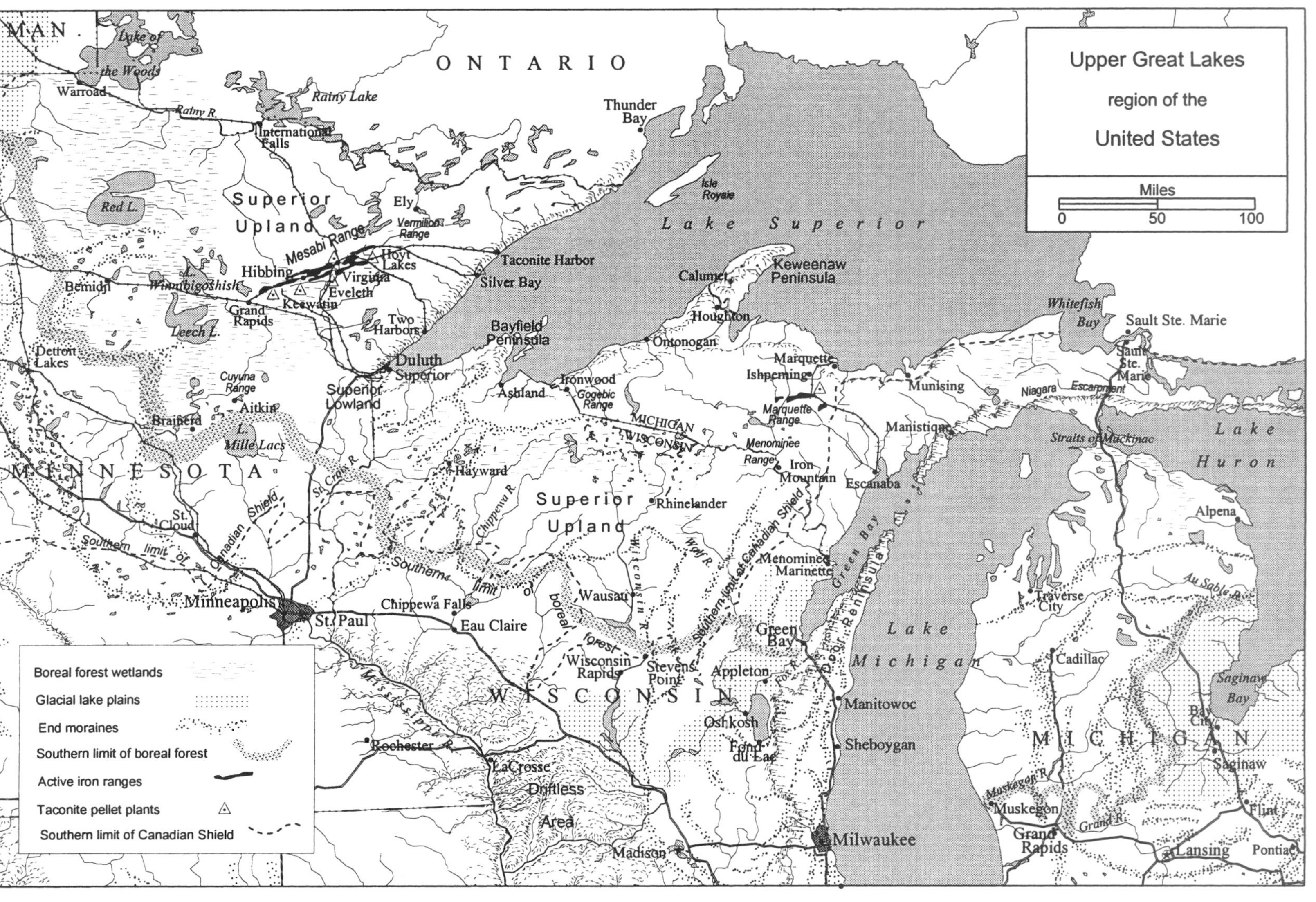

Upper Great Lakes
region of the
United States
Miles
0
50
100
Boreal forest wetlands
Glacial lake plains
End moraines
Southern limit of boreal forest
Active iron ranges
Taconite pellet plants
Southern limit of Canadian Shield
ONTARIO
MAN.
MINNESOTA
WISCONSIN
MICHIGAN
Lake of the Woods
Warroad
Rainy R.
Rainy Lake
International Falls
Thunder Bay
Isle Royale
Lake Superior
Lake Huron
Lake Michigan
Red L.
Superior Upland
Ely
Vermilion Range
Mesabi Range
Hoyt Lakes
Taconite Harbor
Silver Bay
Hibbing
Virginia
Eveleth
Keewatin
Bemidji
L. Winnibigoshish
Grand Rapids
Leech L.
Two Harbors
Duluth
Superior
Superior Lowland
Bayfield Peninsula
Detroit Lakes
Cuyuna Range
Aitkin
Brainerd
Mille Lacs L.
St. Cloud
Minneapolis
St. Paul
Southern limit of Canadian Shield
St. Croix R.
Hayward
Chippewa R.
Chippewa Falls
Eau Claire
Southern limit of boreal forest
Mississippi R.
Rochester
LaCrosse
Driftless Area
Madison
Calumet
Keweenaw Peninsula
Houghton
Ontonogan
Ashland
Ironwood
Gogebic Range
MICHIGAN
WISCONSIN
Rhinelander
Wisconsin R.
Wolf R.
Wausau
Wisconsin Rapids
Stevens Point
Appleton
Oshkosh
Fond du Lac
Fox R.
Green Bay
Manitowoc
Sheboygan
Milwaukee
Marquette
Ishpeming
Marquette Range
Menominee Range
Iron Mountain
Escanaba
Menominee
Marinette
Door Peninsula
Munising
Manistique
Whitefish Bay
Sault Ste. Marie
Niagara Escarpment
Straits of Mackinac
Alpena
Au Sable R.
Traverse City
Cadillac
Saginaw Bay
Bay City
Saginaw
Flint
Pontiac
Lansing
Grand R.
Grand Rapids
Muskegon R.
Muskegon


MAP 22.1

with one another for trading territory, especially in Manitoba and western Ontario, until the two groups merged in 1821.

The association of boreal forests with the trapping of live animals for their skins is not coincidental, of course. Very cold winters produce animals with the most desirable pelts. Beaver were the most eagerly sought furs in the boreal zone, with marten a somewhat distant second. Only late in the fur-trade era, when beaver populations had become depleted and attention turned to the slaughter of bison, was there a noticeable shift away from the boreal forest onto the plains of the West. The volume of the fur trade, which reached hundreds of thousands of skins annually in the mid–18th century, was made possible in part by the enormous area covered by boreal forest and the relatively uniform habitat its broad extent offered fur-bearing animals.



Thousands of Europeans and native North Americans were employed in the fur trade. One legacy of the contact between the two populations was the creation of a new ethnic stock. French or British traders with Indian wives typified many frontier outposts. Their mixed-blood offspring, known as métis, had French, English, or Scots surnames but followed native lifestyles. Métis people eventually occupied nearly all of the habitable portions of the boreal forest fringe, in both Canada and the United States. These were the men and women who greeted the first Euro-American settlers coming to the new frontier outposts, such as Milwaukee, in the 1830s.

Although the present U.S.-Canadian boundary dividing Lakes Superior, Huron, Erie, and Ontario was established by treaty with Great Britain in 1782, many Americans wished to push the border northward. Political miscalculations led the American government to invade Canada in 1812. British forces responded by capturing some key American military outposts, including Detroit and Mackinac. Michigan Territory, which had been created out of the original Northwest Territory in 1805, fell to the British. But the War of 1812 ended in stalemate, with the British returning the Upper Great Lakes region to the United States under the 1815 Treaty of Ghent. By terms of the Rush-Bagot Agreement of 1817, the United States and Great Britain reestablished the former boundary dividing the Great Lakes, and both countries promised to keep it unfortified thereafter. A second Michigan Territory, which included the present states of Wisconsin and Minnesota as well, was established in 1818.

Even with the international border question settled, however, a weak sense of nationalism persisted within the boreal forest zone. The economic capital of the Upper Great Lakes region was distant Montreal, regardless of whether one resided north or south of the border. John Jacob Astor's organization of the American Fur Company in 1807 cut the direct link to Montreal for traders who lived on the American side, but many individuals who remained from the fur-trade era did not move elsewhere when new Euro-American settlers arrived.

## The Pineries

In stages the Upper Great Lakes acquired a regional identity that differed from Yankeeland to the south. The Canadian-born outnumbered the Yankees among

FIG. 22.1. *Manitowish River, Wisconsin. The Upper Great Lakes region has many short streams connecting myriad lakes and ponds. The species composition of the boreal forest has changed markedly since pre-European times.*

the early settlers of the boreal forest zone in the United States for several reasons. Early Michigan had a substantial component of Loyalist descendants who moved west from Ontario to the United States. When large-scale lumbering operations began in the Upper Great Lakes region in the mid–19th century, many lumberjacks from the Ottawa and St. Lawrence Valleys moved west. Northern New England and the St. Lawrence Valley of New York, where lumbering and sawmilling also were major activities, similarly funneled labor to the Upper Great Lakes region.

In contrast to the wheat-farming Yankee frontier to the south, the boreal forest zone had little agriculture, because of its poor soils and short growing season. The agricultural mode of early settlement formation in the Middle West, characterized by a mosaic of farms and small, trade-center towns, was largely absent in the boreal zone. Northern Michigan, Wisconsin, and Minnesota had dense forest, essentially unpopulated save for the seasonal workers who cut timber and worked in the sawmills.

As a vegetation formation, boreal forests are defined primarily with respect to climate, specifically with severe winter temperatures that many common hardwood species native to more southern latitudes cannot survive (fig. 22.1). Generally speaking, conifers such as spruce and pine are more winter-hardy than broadleaf deciduous varieties, although trees such as the sugar maple survive winter temperatures of −40°C. Most pine species are not as tolerant of cold winters as are spruce and fir. The boreal forest is characterized by a series of loosely overlapping ecotones in which successively more cold-tolerant species gain dominance in the northward direction.

The most valuable trees in the Upper Great Lakes region were the white pine and the sugar maple, both characteristic of the southern limits of boreal forest. Maple is a desirable hardwood for furniture manufacture, whereas white pine is a sturdy, easily worked softwood useful for many purposes ranging from handicrafts to building construction. These two trees became the foundation of the late-19th-century Great Lakes lumber industry. White pine was so closely associated with lumbering that the whole complex of logging, river driving, log ponds, and sawmills was referred to as the pineries.

Unlike the widespread pioneer practice of cutting trees to clear land for settlement, the boreal forest was cut for its commercial timber alone, with virtually no thought given to subsequent land use. The first timber cessions were made along the Wisconsin River in 1836. Within three years nearly all of the timber rights in the upper Wisconsin Valley were owned by the makers and vendors of pine boards and shingles. Many were commercial lumbermen from upstate New York and northern New England who had relocated westward. Seasonal camps, rather than permanent towns, were the typical form of early settlement in the region.



Trees were cut in the winter and driven downriver in booms or rafts as soon as the spring breakup of ice permitted. Rivers were the only practical means of transporting logs and, at least initially, the most common source of power for operating sawmills. St. Anthony's Falls on the Mississippi River at Minneapolis was an important sawmill site years before Minneapolis gained importance in flour milling. Eau Claire and Chippewa Falls on the Chippewa River; Wausau, Wisconsin Rapids, and Stevens Point on the Wisconsin River; Oshkosh, at the mouth of the Wolf River; and Green Bay, on the Fox River, all became important sawmilling centers in Wisconsin. By 1879 Oshkosh was producing more than 100 million board feet of lumber each year and fabricating some two hundred thousand doors and six hundred thousand window sash, plus other wooden goods such as carriages and wagons. Because the rivers flow southward, the cutting of the boreal forest stimulated urban growth along the forest's southern margins, where the sawmills and furniture factories were located, establishing the nuclei for most of the cities of central Wisconsin. In Michigan, Grand Rapids, Muskegon, Bay City, and Saginaw commanded locations where logs, cut in the interior, could be floated out to the mills on or near Lakes Michigan and Huron.

Railroads built into the north woods during the last decades of the 19th century extended the lumber barons' reach for more logs and also enabled sawmilling to migrate northward toward the supply of timber. River-driving of logs disappeared. Eventually, so did the forest itself. So thorough was the exploitation of white pine that it ceased to be economically important by 1920. Lumber companies then relocated their operations to the northern Rocky Mountains and the Pacific Northwest, where the next lumber bonanza already was under way.

Just as the fur trade had ended as a result of the overharvesting of animals, by the early decades of the 20th century the Upper Great Lakes region had been stripped of a second valuable resource—its forests. Although a forest cover regenerated in time, the species composition of the 19th century never has been replicated under natural conditions of vegetation succession. White pines have been planted in many areas, but they have not regenerated spontaneously to form anything like the forest blanket encountered by the early lumbermen. One or more species of aspen became the most typical trees of the Upper Great Lakes forest once the white pine was gone.



## The Cutover

Another climatic restraint in the boreal forest arises from the interaction between soil-forming processes and the nature of the forest. Although the boreal zone typically has fairly warm summers, the growing season is short, and thus the amount of time for dead vegetation to decompose is compressed into only a few months. Needleleaf trees like spruce and pine leave an acid litter on the forest floor, one that often does not decompose entirely during a single growing season. Boreal forest soils formed under these conditions are termed Spodosols. They are mostly acid, low in nutrients needed by most crops, and what nutrients they possess are leached out by the downward movement of soil water. When Spodosols form on quartz sands weathered from granitic rocks, the fertility drops even further. There is a sharp reduction in agricultural land use north of the temperate forest margin and an almost total absence of agriculture on the granitic rocks of the Canadian Shield.

Despite agriculture's poor prospects, attempts were made to establish farms on land that had been cleared of all timber except the stumps. In the early 1900s, the north woods became known as the Cutover region, a land where hay fodder crops such as clover were grown among decaying tree stumps on land that had only recently been covered with heavy pine forests. "Stump farmers," as the unfortunate individuals who attempted agriculture in the north were known, faced the insurmountable odds of poor soils, a short growing season, and remoteness from markets for the dairy products that were to be their main source of income. By the 1930s the Cutover region was among the poorest in the United States; thousands of farm families there were forced to live on welfare. Agricultural abandonment has been steady within the region over the past fifty years (fig. 22.2).

FIG. 22.2. *An abandoned dairy barn near Bruce Crossing, Michigan. Farming was marginal at best in the Cutover region.*

FIG. 22.3. *Pine reforestation in Jackson County, Wisconsin. Planted for harvesting as pulpwood, the industrial forest is more a crop than a true forest.*

Agriculture never has been important to the Upper Great Lakes economy. What the land obviously produced in abundance was the tree crop. As forest-fringe farming in the Cutover declined, new efforts were made at reforestation to support a forest-product industry based on paper rather than sawn wood. Unlike lumber production, which is restricted to a few desirable species, wood pulp and wood chips can be produced from a variety of boreal forest trees. Paper manufacturing, which began in the Fox River Valley of Wisconsin in the 1870s, eventually became more important than lumber production ever had been. Former sawmill towns grew into paper-milling centers throughout the boreal forest zone. Planted industrial forests, designed to be thinned and eventually clear-cut on a regular basis, scarcely have the visual appeal of the big woods of a century ago, but they have returned the Upper Great Lakes region to a forested condition (fig. 22.3).

The nearly ubiquitous aspen, which was once considered a nuisance because it regenerated spontaneously after a forest fire, has become economically valuable as well. Aspen now is grown as a forest crop for the manufacture of oriented strand board, a less expensive alternative to plywood that is used in the home-building industry. In recent years restrictions on timber harvesting in the Pacific Northwest—imposed as part of an effort to halt the disappearance of old-growth forests—stimulated Great Lakes forest industries, especially those producing lesser-valued but nonetheless commercially salable building materials made from aspen chips. Printing papers, toilet tissue, and carton stock are the leading products of paper mills in the Fox and Wisconsin River Valleys and in the Upper Peninsula of Michigan. Northern Minnesota is the major producer of oriented strand board and other wood fiberboard products for the construction industry.

## The Canadian Shield in the United States

The boreal forest offers one defining characteristic of the Upper Great Lakes region. The Canadian Shield's southward extension into the United States is another. The Canadian Shield (sometimes called the Laurentian Shield or the Precambrian Shield) represents the greatest span of geologic time that can be seen in North America. Its surface exposes the ancient core of the North American continent, the Precambrian rocks, all of which are more than 600 million years old. In places, exposed Shield rock formations are more than a billion years old. Many parts of the Shield have been buried under sedimentary rock layers one or more times since the beginning of the Cambrian, but these newer rocks have been stripped away by erosion as the land surface has gradually risen.

The Upper Great Lakes portion of the Shield has two geological subdivisions. The larger one is the Superior Upland, a smooth, glacially sculpted surface south and west of Lake Superior. Near its center lies a broad, downwarped basin, the Superior Lowland, through which several major geological faults trend in a northeast-southwest direction. These structures are associated with the region's mineral wealth, the copper range of the Keweenaw Peninsula of Michigan and the several iron ranges of Minnesota, Wisconsin, and Upper Michigan.

The existence of copper and iron has been known since prehistoric times, when native people fashioned tools, jewelry, and weapons from nearly pure, glittering specimens of the metallic ores. Industrial quantities of iron and copper were first discovered in Michigan's Upper Peninsula during the 1840s. Between 1840 and 1887, when Minnesota's Mesabi Range was discovered, the Upper Great Lakes region boomed in anticipation of the wealth in mineral resources that would come from these discoveries. Thousands of European immigrants, including Finns, Slovenians, Croatians, and Italians, were recruited by the mines, where they worked alongside former lumberjacks who were attracted by the good wages the mining companies paid. A large industrial laboring class became established in the mining towns.

Unlike the other Great Lakes, Lake Superior occupies a deep basin that is flanked by rocky tablelands. From Lake Superior's shoreline north of Duluth, one climbs more than a thousand feet vertically in ten miles' travel to reach the crest of the Superior Upland. Rock formations dip under the lake, forming a basin whose eastern edge is a series of parallel, upturned ridges in Upper Michigan's Keweenaw Peninsula. Smaller ridges, such as the Bayfield Peninsula of Wisconsin and Isle Royale, an island in Lake Superior, are secondary features of the Lowland (fig. 22.4).

A large vein of native copper (a scarce but relatively pure copper ore), running more than one hundred miles through the center of the Keweenaw Peninsula, was mined intensively from the 1850s until recent times. Before the discovery of copper at Butte, Montana, Keweenaw's mines were the largest source of copper in North America. Michigan's Copper Range, centering on the cities of Houghton, Hancock, and Calumet, yielded hundreds of millions of dollars for the copper companies that developed it. Outside these cities the Copper Range was a bleak terrain of company towns carved out of the boreal forest. Most of the Upper Peninsula's copper veins have been depleted, although one or more mines operate sporadically depending on the world's demand for cop-

FIG. 22.4. *Rock layers dipping under Lake Superior near Grand Marais, Minnesota, emerge east of the lake as a ridge of hills forming Michigan's Keweenaw Peninsula.*

per. In recent years a new copper-mining industry has emerged at the fringe of the Shield in northern Wisconsin, but it is small in comparison with mining on the Copper Range in years gone by.

Keweenaw has the coolest summer temperatures and the heaviest winter snowfalls in the Middle West. The peninsula is sparsely populated today, bypassed by most of the tourism that now supports the Upper Great Lakes economy.

Iron mining remains an important economic activity within the United States portion of the Shield. The ore, which formed millions of years ago in cooled bodies of molten rock, consists mostly of hematite, a reddish-colored iron mineral. Millions of tons of hematite have been mined in six iron ranges around the western and southern margins of Lake Superior: the Vermilion, Mesabi, and Cuyuna Ranges in Minnesota; and the Marquette, Menominee, and Gogebic Ranges of Michigan (the latter two also extend into Wisconsin). Of these six, only the Mesabi and Marquette Ranges are of current importance.

By 1950 heavy mining to supply American and Canadian steel mills had removed all but the lower grades of iron ore, making yet a third resource exploited nearly to the vanishing point in the Upper Great Lakes region. Hematite ore was formed under the earth's surface from the melting of its parent rock, taconite, which contains iron in small flakes. Metallurgists at the University of Minnesota discovered that by grinding taconite rock to the consistency of flour, removing the iron particles, and mixing them with a fine clay, it was possible to form iron-rich pellets that are valuable enough to bear transportation costs to the steel mills.

Taconite pellets revitalized the Lake Superior iron-ore industry in the 1950s. Pellet manufacturing plants were constructed near the mines or adjacent to Lake Superior ore docks where the ore was transshipped. Employment in the pellet plants stimulated local economic growth in both the Mesabi and the Mar-

FIG. 22.5. *A taconite-processing plant at Hibbing, Minnesota.*

quette Ranges (fig. 22.5). At Silver Bay, Minnesota, on Lake Superior, tailings from pellet manufacture were dumped in the lake until the practice was forbidden by the courts. All of the other taconite plants have inland locations, from which the pellets are moved by rail to lakeside ore docks in the traditional pattern of the region's iron-ore industry.

Iron-range economies depend on the health of the steel industry in general and thus have revived in recent years as North American steel production has recovered from the low levels of the 1980s. Iron-range communities, such as Hibbing, Minnesota, and Ishpeming, Michigan, have more diversified economies than in former times when the steel industry's iron-ore demands were larger and steadier. But Lake Superior's iron ranges, like the copper range, never have captured more than a small fraction of the wealth created by the mines.

## Contemporary Trends

Cycles of resource discovery, exploitation, and disappearance do not explain all of the Upper Great Lakes' development. Tourism is the leading economic support today, although it, too, emerged out of the declining fortunes of the virgin-forest lumber industry in the early decades of the 20th century. In the 1920s lumber companies began selling land stripped of salable trees in small parcels that would interest resort operators and vacation-home builders. By the 1960s good roads had been extended into all parts of the region, bringing the north woods within a day's drive of the major metropolitan centers of the Middle West. The construction of a highway suspension bridge across the Straits of Mackinac in the late 1950s led to the expectation of a major tourist boom in Michigan's Upper Peninsula, but its effects have been modest.

Residents of Detroit, Chicago, Milwaukee, Minneapolis, and St. Paul all have their favorite resort hinterlands in the north, around which vacationers have congregated for generations. Outdoor recreation has an overwhelming presence. The outboard motor was invented in Wisconsin, Minnesota claims to be "the birthplace of water skiing," and snow skiing was popularized by the resort community surrounding Iron Mountain, Michigan. As in New England, some summer visitors discovered they could relocate their businesses to their favorite summer haunts. By the 1970s, nonrecreational employment was common in communities formerly dependent entirely on summer tourism.

The adjacent harbors of Duluth, Minnesota, and Superior, Wisconsin, remain the busiest freshwater port in the United States, although the anticipated growth in shipping following completion of the St. Lawrence Seaway has been far less than anticipated. The port is an economic focus, connecting the Upper Middle West and the Great Plains to the Great Lakes region. Quantities of wheat, durum, and barley grown in North Dakota are shipped from Duluth-Superior to foreign and domestic markets. Mesabi Range iron ore and low-sulfur coal from Wyoming and Montana are the other significant outbound shipments.

For many Canadians, the Upper Great Lakes region of the United States is also a transportation corridor, a shortcut in traveling across Canada that avoids the more circuitous route north of Lake Superior. The two-lane blacktop highway crossing Michigan's Upper Peninsula is peripheral to developments on the American side of the border. But the few railroads and highways north of Lake Superior are the only infrastructure on the Canadian side. They are slender connections between the two developed "halves" of Canada. The contrast reflects the very different course of development in the Canadian Shield within Canada compared with its small extension into the United States.

## *References*


Alanen, Arnold R. "Years of Change in the Iron Range." In *Minnesota in a Century of Change*, ed. Clifford E. Clark, Jr., 155–94. St. Paul: Minnesota Historical Society Press, 1989.

Borchert, John R. *America's Northern Heartland: An Economic and Historical Geography of the Upper Midwest.* Minneapolis: University of Minnesota Press, 1987.

Flader, Susan L., ed. *The Great Lakes Forest: An Environmental and Social History.* Minneapolis: University of Minnesota Press, 1983.

Fries, R. F. *Empire in Pine: The Story of Lumbering in Wisconsin, 1830–1900.* Madison: Wisconsin State Historical Society, 1951.

Hart, John Fraser. "Population Change in the Upper Lake States." *Annals of the Association of American Geographers* 74 (1984): 221–43.

———. "Resort Areas in Wisconsin." *Geographical Review* 74 (1984): 192–217.

Hudson, John C. "Cultural Geography and the Upper Great Lakes Region." *Journal of Cultural Geography* 5, no. 1 (1984): 19–32.

Jesness, Oscar B., and Reynolds I. Nowell. *A Program for Land Use in Northern Minnesota: A Type Study in Land Utilization.* Minneapolis: University of Minnesota Press, 1935.

Kaups, Matti E. "Cultural Landscape—Log Structures as Symbols of Ethnic Identity." *Material Culture* 27, no. 2 (1995): 1–20.

CHAPTER 23

# The Canadian Shield

Many regions of the United States have their Canadian counterparts, and vice versa. Both countries have rockbound Atlantic coasts and mountains fringing the Pacific. They share the Great Plains, the Rocky Mountains, and the Great Lakes. Apart from its cold northern fringes, what most distinguishes Canada's geography from the United States is Canada's lack of a populated heartland. In the center of the continent, where the United States has industrial cities and expanses of farmland, Canada has only boreal forest. Between Washington, D.C., and Omaha, a distance of approximately 1,300 miles, lies a substantial share of American industry and agriculture. To take a similar distance, of the 1,500 miles between Ottawa and Winnipeg, across the middle of Canada, all but a few miles on either end is part of the unpopulated Canadian Shield (map 23.1).

Before the Canadian Pacific Railway was completed in 1885, continuous land travel across Canada was not even possible. The Shield formed so effective a barrier between southern Ontario on the east and the prairie frontier of Manitoba on the west that all traffic across the Dominion of Canada had to detour through the United States. Travel across the rocky, swampy Shield terrain north of Lake Superior was so difficult that no wagon routes for trade or migration, similar to the Oregon or Santa Fe Trails in the United States, ever developed. When a railroad finally was constructed around the northern edge of Lake Superior, its route had to be blasted out of the granitic Shield rocks that form the lake's northern margin.

Today many populated sections of North America have no agriculture, and thus its presence is not absolutely associated with human settlement. But in the 19th century, land that was impossible to farm held little interest for European-derived settlers. Nonfarming areas, including almost all of the Shield within Canada, simply were bypassed. Substantial areas of the Shield north of Lake Superior consist of rocky, barren ground (fig. 23.1). Where cultivable soils exist, they are limited in productivity by their acidic nature, having developed on sands weathered from granitic rock and from a forest litter of boreal conifers, both of which produce infertility of the soil. (In the Canadian soil nomenclature, these soils are known as Podzols and are the equivalent of Spodosols of the United States.) Where even these limits can be overcome, the circumstance of a very short growing season, with late-spring and early autumn freezes, adds yet another impediment.

The general climatic model of Canada is that of a high-latitude continent with fairly warm currents along its west coast and a cool current along its east coast. The ecosystem boundaries in Canada's interior trend in a northwest-southeast direction, with warmer conditions prevailing on the west at a given latitude. The effect of continentality—being in the interior of a land mass—creates larger seasonal temperature differences in continental interiors than along oceans. This effect is less pronounced in North America than it is in Asia. Hudson Bay and its southward extension, James Bay, moderate climate through the marine temperature effect, making winters less severe but also reducing summer temperatures near the water. The Shield would be even more Siberia-like were it not for Hudson Bay.

Proof that urbanization is not necessarily tied to agriculturally productive land is offered in the growth of two of the Shield's largest cities, Sault Ste. Marie and Thunder Bay, at opposite ends of Lake Superior. Iron mines near Atikokan (west of Thunder Bay) and in the Michipicoten Range, near Wawa, Ontario, supplied a steel industry that grew to moderate size at Sault Ste. Marie, despite the distance from markets. National policy favored its growth, at an inland location, especially during World War II. In recent years mines in the Marquette Range of Michigan have supplied ore for the Sault Ste. Marie mill.

Port Arthur and Fort William (which merged to become the single city of Thunder Bay in 1970) are the Canadian counterparts of Duluth and Superior. Their primary role is to transship grain grown in the Prairie Provinces from railway to lake boat (fig. 23.2). Fort William was a headquarters city during the fur trade and subsequently became a railroad and lake transportation center. Port Arthur's complex of grain elevators grew into one of North America's largest grain ports by the 1930s, at a time when Canadian wheat exports were directed primarily toward the European market. After the St. Lawrence Seaway was completed, larger ocean ships could reach the grain elevators at Port Arthur and Fort William, but it is more economical for large ocean liners to load cargoes at the St. Lawrence ports, such as Port-Cartier and Baie-Comeau, Quebec, and let the smaller Great Lakes vessels shuttle back and forth to Lake Superior.

Paved highways were not continuous north of Lake Superior until the last link of the Trans-Canada Highway was completed in 1961. Even today, isolated railroad towns of northern Ontario, such as Chapleau and Sioux Lookout, are tied to the outside world through employment with the railroads they were built to serve. They have only small tourism-based economies and still rely on forwarding the endless shipments of freight across country for their existence.

Sawmills and paper mills are the major industries of Shield towns north of Lake Superior. Many towns depend on a single mill for their support. Spruce and balsam fir, which are the best sources of wood pulp for paper manufacturing, are so abundant that despite years of harvesting, the forests are far from depleted. Newsprint mills at Kapuskasing, near the northern limit of dominantly spruce-fir forest, supply newspapers in the United States. Like agriculture, wood-pulp production is limited by climate. North of the spruce-fir limit, the forest consists of smaller, widely spaced trees and sees little commercial cutting.

The exposed rocks of the Canadian Shield, whether in Canada or the United

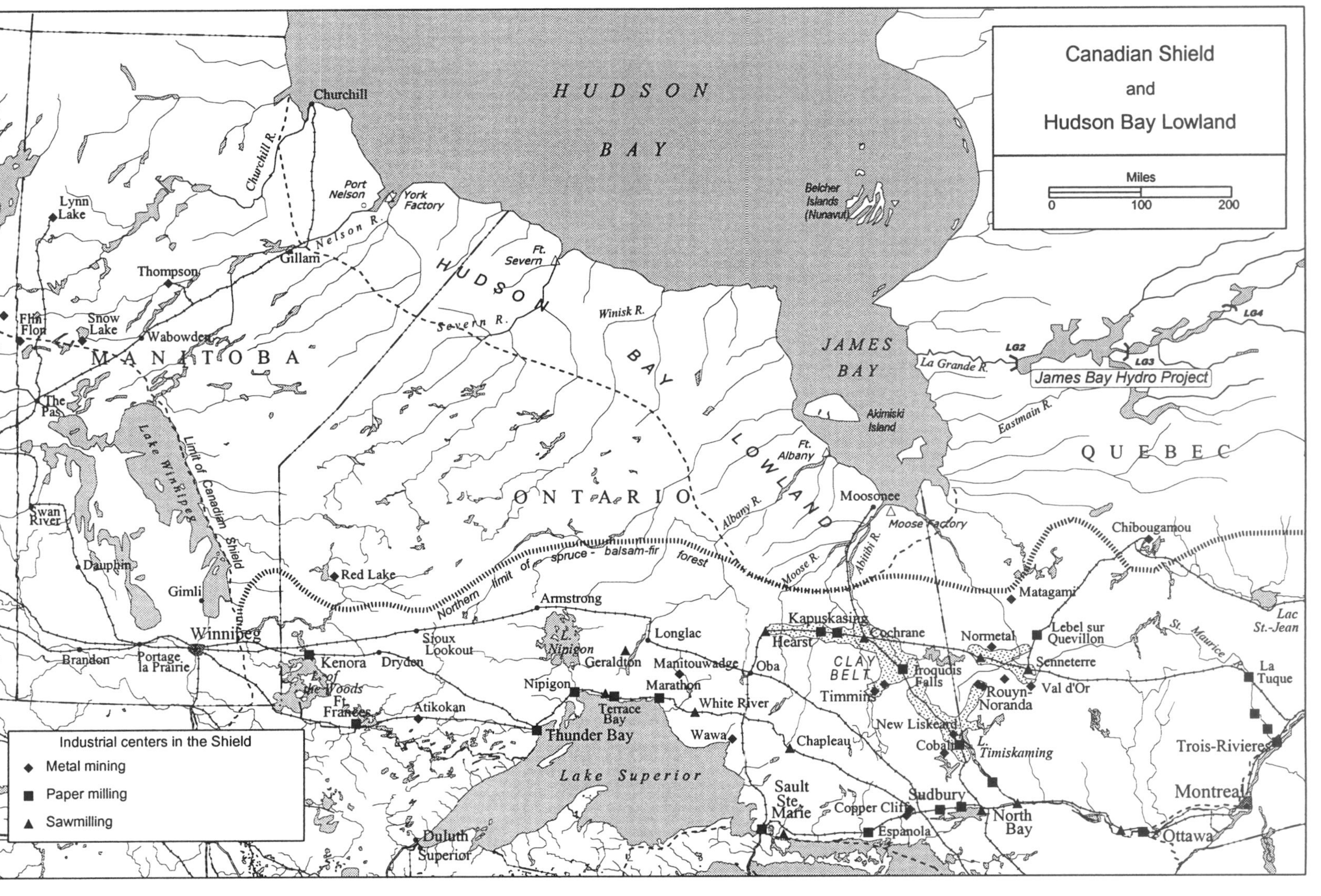
Canadian Shield
and
Hudson Bay Lowland
Miles
0
100
200
HUDSON
BAY
JAMES
BAY
Belcher Islands (Nunavut)
Akimiski Island
James Bay Hydro Project
LG2
LG3
LG4
La Grande R.
Eastmain R.
QUEBEC
MANITOBA
ONTARIO
HUDSON BAY LOWLAND
CLAY BELT
Northern limit of spruce - balsam-fir forest
Limit of Canadian Shield
Churchill
Churchill R.
Port Nelson
York Factory
Nelson R.
Gillam
Lynn Lake
Thompson
Flin Flon
Snow Lake
Wabowden
The Pas
Swan River
Dauphin
Gimli
Lake Winnipeg
Winnipeg
Brandon
Portage la Prairie
Kenora
L. of the Woods
Ft. Frances
Dryden
Sioux Lookout
Red Lake
Atikokan
Ft. Severn
Severn R.
Winisk R.
Ft. Albany
Albany R.
Moose R.
Abitibi R.
Moosonee
Moose Factory
Armstrong
L. Nipigon
Geraldton
Longlac
Nipigon
Terrace Bay
Thunder Bay
Manitouwadge
Marathon
White River
Wawa
Lake Superior
Duluth
Superior
Oba
Hearst
Kapuskasing
Cochrane
Timmins
Iroquois Falls
Chapleau
Sault Ste. Marie
Copper Cliff
Espanola
Sudbury
North Bay
New Liskeard
Cobalt
L. Timiskaming
Rouyn-Noranda
Normetal
Val d'Or
Senneterre
Lebel sur Quevillon
Matagami
Chibougamou
Lac St.-Jean
St. Maurice
La Tuque
Trois-Rivieres
Montreal
Ottawa
Industrial centers in the Shield
Metal mining
Paper milling
Sawmilling

MAP 23.1

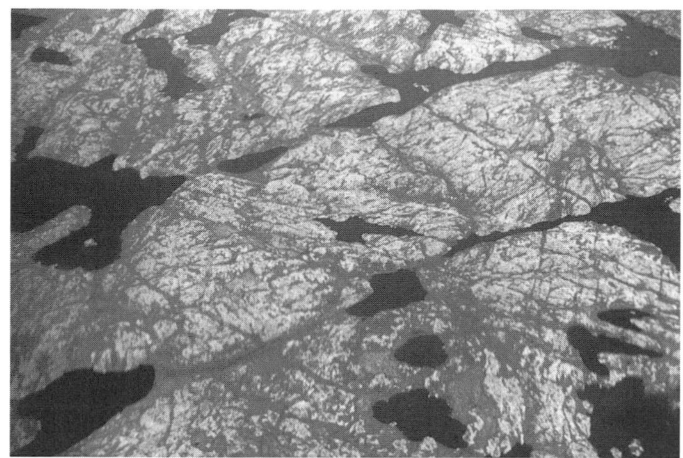

FIG. 23.1. *The irregular surface of the Canadian Shield is a maze of rock joints and fractures. Soils develop in the cracks and allow trees to take root. Larger crevasses and depressions contain lakes or short streams.*

States, are among the world's oldest surface features. Apart from the rocks themselves, however, very little that is "old" can be seen in the Shield. Most of the region's towns and cities are based on resource-extracting or processing industries, especially metal mining and wood-pulp and paper production, which emerged after 1900. Links to a past landscape also have been severed in many places by the effects of one of the Shield's worst natural hazards, forest fires. Devastating fires that raged out of control for weeks were common in the early decades of the 20th century. The residents of isolated communities had no choice but to abandon their homes and property, returning to rebuild after the fires had burned out.

Dozens of fur trading posts, constructed either by Hudson's Bay Company employees or by those engaged with the Montreal traders, might have formed

FIG. 23.2. *Thunder Bay, Ontario, where the transcontinental railroad meets the west shore of Lake Superior, is a major shipping port for grain raised in the Canadian Prairies.*

the nuclei for urban settlements in the Canadian Shield, as they did elsewhere to the south. Even though they occupied accessible sites along major water routes, the trading posts rarely acquired new functions in later years. If not abandoned outright, they remained only as focal points for small bands of native people and did not develop into cities. The Hudson Bay trade was organized around fewer (but larger) posts built at the mouths of the Churchill, Severn, Albany, Moose, York, Eastmain, and Rupert Rivers, but these, too, had little to sustain urbanization after the fur trade ended.

## The Role of the North

It is sometimes observed that the place of the North in the Canadian mind, and in the life and literature of the country generally, is comparable to the role played by the West in American history. Just as American imaginations were stimulated by the vision of what might be possible in the little-known terrain of the West, Canadians have dreamed that the future of their country might lie ultimately in developments that would take place in the North. In other countries with extensive fringes of pioneer settlement, such as Russia, Brazil, and Australia, it also is common for people to seek some sort of clue about their national identity in the unsettled territory.

The existence of a "North" has long been a favorite device of Canadian politicians, who have promised to tap its unused resources, to defend it from possible intruders, or to find within it new forms of expression for national ambitions. Over time, the definition of North itself has been rolled northward as frontiers have advanced (fig. 23.3). No line demarcates the North, of course, but rather there has been the sense that extending economic development and human institutions northward was a desirable thing for Canadians to do, perhaps as part of a manifest-destiny view of the natural sphere of expression for a greater Canadian nation.

FIG. 23.3. *Wawa, Ontario. Seaplanes are ideal for reaching the scattered lakes of the Canadian Shield, and they have drawn isolated settlements into closer contact with population centers.*

It is easy to dismiss such notions as being the product of greed for more resources or as emanating from nationalistic motives on the part of one people to dominate another's territory. More often, schemes to develop the North have arisen from a much simpler motivation, the Canadians' desire to occupy more than the fraction of total national territory to which their settlement has thus far been confined.



Musings about the possible future role of the North have not been restricted to the national scale. Among the provinces, Quebec has been most notable in developing its own strategies of northern development. Ontario has turned its attention northward on several occasions as well. Nowhere in Canada does the North come as far south as it does in the Shield of Ontario and Quebec. It lies at the back door of Montreal and Ottawa and comes within a hundred miles of Toronto. Given the proximity, it is understandable that the first attempts at colonizing the Shield would take place directly north of those cities.

## The Clay Belt

The construction of transcontinental railways created small northern outpost cities like North Bay, Ontario, by 1890. Settlers moving into the region extended the frontier northward in the Ottawa River Valley to Lac Timiskaming, along the Ontario-Quebec border. In the 1900s Ontario's provincial government began building a railway north from North Bay, first to connect the Timiskaming settlement and later to extend colonization still farther north. In 1903 construction workers accidentally discovered traces of silver in rocks near the present cities of Cobalt and New Liskeard, thus touching off the first mining bonanza in northern Ontario. The cycle was repeated several times in the years that followed. Each new mineral discovery associated with railway construction added to the popular demand for further northward extensions.

An immediate goal of the railroad was to reach the zone of arable land that lies in several swaths along the Ontario-Quebec border. Roughly 16 million acres of land are covered in a grayish-colored clay that formed in temporary glacial lakes during the late Pleistocene. The soils that developed on the glacial-lake clay have lent the region its popular name, the Clay Belt; and although they are not particularly fertile, they are better than those nearly anywhere else in the Canadian Shield. Beginning in 1905, a new transcontinental railroad was built from Quebec City to Winnipeg through nearly two thousand miles of unpopulated Shield on an alignment that would make the Clay Belt, then totally undeveloped, an agricultural heartland of sorts in a country that lacked one.

Ontario's interest in the Clay Belt was by no means independent of Quebec's. The provincial border cross-cuts the Clay Belt, and the two governments devised parallel strategies for its development. Northern colonization in Quebec had the added dimensions of religious and cultural expansion, to extend the territory that was both Roman Catholic and French-speaking. Early settlers in Ontario's portion of the Clay Belt were almost entirely Anglophone and Protestant. The two provinces were determined to extend their occupied territory northward along the provincial border to prevent colonization by the other.

The clearest expression of purely ideological reasons for settlement expansion came in Ontario during the 1920s, after more minerals had been discovered and thousands of settlers had taken up farms in the Clay Belt. By then the ultimate goal of northern expansion was no longer just the Clay Belt but the attainment of a tidewater port on James Bay, which would offer an all-Ontario link between Toronto and the world's ocean commerce. The United States had Atlantic and Pacific ports, like Canada, but it had a "third ocean," the Gulf of Mexico. Hudson Bay obviously was Canada's third ocean, and it provided a short water route to Great Britain as well.

No valuable resources ever were found along the railway's line to the would-be ocean port of Moosonee at the southern tip of James Bay. North of the Clay Belt the fringes of Hudson Bay consist of a cover of more recent sedimentary rocks that lie atop the Shield. The Hudson Bay Lowland, as the region of sedimentary rock cover is known, is poorly drained, and much of it is muskeg—a swampy mixture of stunted boreal forest trees and open tundra. Having a James Bay port—and having one before Quebec did—was so important that Ontario officials gave little thought to practicalities. No seagoing port ever was built at Moosonee, nor were more than the vaguest of plans laid for its development. The railway was completed during the 1930s, and it has been operated ever since, now primarily as a summer excursion route for tourists.

Agriculture has struggled to maintain itself in the Clay Belt. By the 1940s, Anglophone settlers had abandoned much of the Ontario side. The Quebec government continued to promote Clay Belt farm settlements into the 1950s. Some Francophones migrated from Quebec to the Ontario side in search of better land. Families were assisted financially to maintain themselves on their farms in Quebec but not in Ontario. Eventually, agriculture began to retreat from the Clay Belt's margins (fig. 23.4). The fringe of abandoned farms, now being reclaimed by boreal forest, like the semiabandoned agricultural villages in the lit-

FIG. 23.4. *The partly abandoned settlement of Despinassy, Quebec, on the retreating fringe of the Clay Belt.*

tle clearings, are reminders of the odds against agriculture in the North. The limitations of the Shield for settlement are more realistically understood today than when outsiders first contemplated the region's "possibilities" more than a century ago.



## Mining in the Shield

Mining employment and the expansion of wood-pulp production have accounted for most of the Clay Belt's growth since initial colonization. Ontario gold discoveries include those made in the Porcupine-Timmins district in 1909 and at Kirkland Lake in 1911. A complex of copper, gold, silver, and nickel mines was developed around Rouyn-Noranda and Val d'Or on the Quebec side during the 1920s. At present, mines in the Timmins and Rouyn-Noranda districts dominate mineral production in the Clay Belt. The chief products are copper and precious metals recovered from a variety of ores, all processed in local smelters.

The largest metal mining and smelting complex in the Shield is the oldest one—the nickel and copper mines of the Sudbury, Ontario, district, which came into production in the 1880s. Sudbury's nickel mines rank among the most valuable mineral workings in the world and include associated deposits of copper as well. Smelters at Sudbury and the nearby communities of Copper Cliff and Falconbridge have created an industrial complex based on the mines. Copper accounts for over half of the metallic mineral workings in the Canadian Shield, with gold mines ranking a close second, although many of the gold workings are small in scale.

The growth of copper and nickel mining has expanded Shield mining activities northwestward. Copper mines were developed near Flin Flon, Manitoba, during the 1920s, and nickel mines at Thompson and Lynn Lake came into production in the late 1950s. Continuing an already established pattern, the newer resource-based communities are company towns, with many aspects of housing and community infrastructure left to the mining companies to arrange. As the search for valuable metals has expanded into nearly all parts of the Shield, some of the most remote mines have developed a transient labor system. Workers fly in for specified periods of concentrated work, then fly out when they are due for a rest. Such a system avoids the costs of creating new resource-based towns.

The pattern of mining activities in the Shield reflects both the existence of deposits and the history of settlement and exploration. After decades of intensive testing, most of the Shield's mineral reserves are known. The knowledge of their existence does not mean that the ore bodies will be exploited soon, however, because the world's demand for precious metals is limited. Mining frontiers have expanded incrementally northward, occasionally taking long leaps in the direction of large, easily extracted, high-value minerals. Each expansion has provided a new base from which to search for more. Single-mine communities in the North have a limited life span, based on the quantity of extractable resources they offer as well as on the level of demand for the product. In the long run, the creation of new mining towns around richer deposits often accompanies the closure of older ones where the ore thereby becomes relatively less economical to extract.



## The James Bay Project

Perhaps the most ambitious scheme ever launched in the Canadian Shield is little known outside the region. It is the hydroelectric power project along the eastern edge of James Bay, constructed by Quebec's electric-power authority, Hydro Québec, beginning in the 1970s. The James Bay project is based on a unique combination of geographical circumstances.

Viewed in continental perspective, Canada's relief map resembles a saucer. The lowest elevations are in the center, in the large Hudson Bay depression, which is now mostly below sea level. Mountains fringe the rim of the country, including the western cordillera (the Rocky Mountains) and the mountains of Baffin and Ellesmere Islands to the north. On the east are the highlands of Quebec and Labrador and to the south the Superior Upland.

Eighteen thousand years ago, during Pleistocene full-glacial conditions, Ontario, Hudson Bay, and Quebec were buried under thousands of feet of ice. The world's oceans, which were the major source of water frozen as ice, were hundreds of feet below today's level. The land surface of the eastern James Bay region flexed downward more than four thousand feet, compensating for the weight of the ice load. When the ice sheets melted, sea levels rose. The weight of the ice was first removed around coastlines, then in the interior of large land areas like Quebec, which had been the center of one ice sheet. Interior northern Quebec thus continued to rise, in a process known as crustal rebound, even after sea levels had stabilized.

Hudson and James Bays are fed by numerous short streams and several major rivers. The heaviest seasonal runoff is on the Quebec side, in the same region where crustal rebound produced a substantial rise in the land surface with respect to sea level. Although no single major river drains this region, the La Grande (formerly Fort George) River, which drains areas 500 miles east of James Bay, has the largest flow. It drops a total of eighteen hundred feet, one-third of it in the last 150 miles before it reaches sea level.

The La Grande River has an excellent profile for hydroelectric power development, as do many others in Quebec. All of the province's electric-power utilities were nationalized under the existing Crown corporation, Hydro Québec, as part of the early 1960s restructuring in the province. The search for power-development sites in the North led to the selection of the La Grande River and to plans to build four power dams along it. Because the river's volume is seasonal, with a peak following spring snowmelt, it was necessary to create reservoirs where the seasonal runoff could be stored to even out the annual flow. Millions of acres were flooded for the reservoirs, a large share of which constituted hunting grounds traditionally used by the Cree and Inuit inhabitants of the region. Legal actions brought by native people led to the signing of an agreement that gave them $200 million compensation.

Three of the four La Grande hydro power dams were constructed, with the first, LG2, inaugurated in 1979; LG3 and LG4 came into operation in the early 1980s. Power from the project is transmitted southward, where much of it is sold on long-term contract in the United States, especially in New York and southern New England. In addition to James Bay, Hydro Québec also draws power from the Churchill Falls, Labrador, project, which it developed with the Newfoundland government, and from several smaller hydro dams on rivers tributary to the St. Lawrence. Summing all its capacities, Hydro Québec can sell power to Canada or to the United States in the equivalent energy range of 200 million barrels of oil per year. In the process of expanding its electricity empire, Hydro Québec came to have the largest assets of any corporation in Canada.



The most controversial aspect of the James Bay project was the disruption of native groups, who, although they received financial compensation for loss of hunting lands, ultimately were unable to stop the project. The flooding of large areas in northern latitudes may produce ecological changes that are detrimental in the long run. Current plans to build still more dams and reservoirs in the North have produced a reexamination of their impacts. But James Bay's hydroelectricity is arguably as "clean" as any power sold in the United States or Canada. Quebec has profited handsomely from it, and its stature as a political entity was greatly enhanced by its ability to complete successfully an undertaking of such magnitude.

## The Arctic Seaport

Hydroelectric power developments of a smaller scale have been built in northern Ontario and Manitoba, but there the rivers flow on a comparatively gentler gradient, over the easily eroded rocks of the flat Hudson Bay Lowland. The mouths of all the rivers entering Hudson Bay through the Lowland are swampy, and thus there is much less hydroelectric power potential on the Lowland than there is on the Shield. Although winter temperatures are slightly moderated by the bay, summer temperatures are cooled—enough to make the growth of woody plants marginal. Most of Hudson Bay is fringed with tundra as a result.

For all these limitations, the area lies no farther north than the latitude at which wheat is grown in Alberta. James Bay is at the same latitude as England, and the center of Hudson Bay is no farther north than the northern tip of Scotland. The usefulness of an ocean shipping route from central Canada to Great Britain, through the Hudson Strait and past Greenland, was known to the Hudson Bay traders. It saved hundreds of miles compared with the Great Lakes–St. Lawrence route to the Atlantic. Discussions of an Arctic seaport that would use the Hudson Bay–Liverpool shortcut for shipping western Canadian grain began in the 1870s.

No matter how sound the economic reasoning for such a shipping route, the Hudson Bay Lowland was even less attractive to settlement than the Shield. As railroads gradually were extended northward in Manitoba, however, the goal of reaching Hudson Bay came into sharper focus. Wheat farmers in the Prairie Provinces supported the project because it would give them a more direct access to European markets. As a national issue, the Arctic seaport was popular because it confirmed Canada's northern frontier ambitions. In the 1920s construction began north from The Pas, Manitoba, on a line that was to follow the Nelson River to Hudson Bay.

The construction of "Port Nelson" was begun, but the final portion of the route would have required building through muskeg that would not support a railroad. A quick decision was made to switch the line's terminus northward to the mouth of the Churchill River, where a seaport, to be known as Churchill, was built in the 1930s. A 5-million-bushel grain elevator was constructed, and a town was built to house workers at the port. Although it was possible to build a grain elevator on the tundra at Churchill, the sea ice surrounding it, which seldom disappears from Hudson Bay, was a powerful deterrent to the port's success (fig. 23.5). Because of the iceberg hazard, London underwriters refused to insure vessels passing through Hudson Strait before July 25 or leaving Churchill after October 10. With such a short season, and with the lack of commodities to ship other than Saskatchewan wheat and barley, the Arctic route eventually became a burden to maintain even though it was shorter.

There is little need for the port of Churchill at present, but the Canadian government continues to search for ways to make it more productive. Most shipping seasons in recent years have seen several vessels dock at Churchill to load export grain, some of which is shipped to Mexico and thereby avoids part of the North Atlantic iceberg hazard. Ships loaded with grain for southern destinations have left Churchill as late as November.

The technology on which Churchill is based is outmoded, yet abandoning the port appears to be out of the question. It would be a signal that northern development was retreating. Just as those who agitated for its construction

FIG. 23.5. *Ice floes surrounding the grain-shipping dock at Churchill, Manitoba, in July illustrate one of the port's major limitations.*

stressed the geopolitical themes of natural development and territorial expansion, Churchill's most important function continues to be symbolic rather than economic. Its very existence confirms the significance of Canada's northward reach.



## *References*

Bourassa, Robert. *Power from the North.* Scarborough, Ont.: Prentice-Hall Canada, 1985.

Fenge, Terry. "Ecological Change in the Hudson Bay Bioregion." *Northern Perspectives* 25, no. 1 (1997).

Fleming, Howard A. *Canada's Arctic Outlet: A History of the Hudson Bay Railway.* University of California, Publications in History, vol. 54. Berkeley: University of California Press, 1957.

Gill, Allison M. "Enhancing Social Interaction in New Resource Towns." *Tijdschrift voor Economische en Sociale Geografie* 81 (1990): 348–63.

Hogue, Clarence. *Québec: Un siècle d'électricité.* Quebec City: Libre Expression, 1979.

McDermott, George L. "Frontiers of Settlement in the Great Clay Belt, Ontario and Quebec." *Annals of the Association of American Geographers* 51 (1961): 261–73.

Osborne, Brian S., and Susan E. Wurtele. "The Other Railway: Canadian National's Department of Colonization and Agriculture." *Prairie Forum* 20 (1995): 231–53.

Randall, James E., and R. Geoff Ironside. "Communities on the Edge: An Economic Geography of Resource-Dependent Communities in Canada." *Canadian Geographer* 40 (1996): 17–35.

Richardson, Boyce. *Strangers Devour the Land.* New York: Alfred A. Knopf, 1976.

Robinson, Ira. *New Industrial Towns on Canada's Resource Frontier.* Department of Geography, University of Chicago, Research Paper no. 73. Chicago: University of Chicago Press, 1962.

Tucker, Albert. *Steam into Wilderness: Ontario Northland Railway, 1902–1962.* Toronto: Fitzhenry & Whiteside, 1978.

Wightman, Nancy M., and W. Robert Wightman. "Road and Highway Development in Northwestern Ontario, 1850 to 1990." *Canadian Geographer* 36 (1992): 366–80.

CHAPTER 24

# The Far North

In 1825 Russia and Great Britain established the eastern boundary of Alaska roughly where the Alaska-Canada boundary lies today. Between Alaska and the existing trading preserve of the Hudson's Bay Company known as Rupert's Land lay still another enormous but little-known region. The British designated it the North-Western Territory, and they attached it to Canada. When the Dominion of Canada was organized in 1867, Rupert's Land and the North-Western Territory were merged as a single entity, the Northwest Territories. It stretched across most of Canada, bordering Newfoundland on the east and British Columbia on the west.

Then, beginning with Manitoba in 1870, each of the Prairie Provinces was carved out of the Northwest Territories, and the lands east and south of Hudson Bay were attached to Ontario and Quebec. A separate Yukon Territory was created in 1898. Eventually all that remained of the initial Northwest Territories was the area north of 60° latitude and east of the crest of the Mackenzie Mountains. Although neither the Northwest Territories nor Yukon has achieved provincial status in the century since those borders were drawn, political evolution in the North continues. In 1999 a new territory, Nunavut, was created from the eastern portion of the Northwest Territories (map 24.1).

## Arctic Environments

The 60th parallel of latitude separates the northern territories from the western provinces. Although it is a political rather than environmental boundary, the 60° line and all other parallels are a proxy for the amount of incoming solar energy received at the top of the atmosphere. Temperatures decline gradually northward corresponding to a decreasing amount of solar radiation as latitude increases. At higher latitudes the earth radiates more energy to the atmosphere than it receives directly from the sun, which is another way of saying that temperatures are cooler at high latitudes.

Theoretically, all points north of the Arctic Circle (latitude 66°33′N) experience continuous darkness ("polar night") around the time of the winter solstice on December 22. Because of twilight conditions when the sun is slightly below the horizon, however, polar night is more like twilight as far north as 72°. Very few people actually live north of 72° and thus total winter darkness is not much of an impediment to human activity.

MAP 24.1

Continuous daylight ("midnight sun") around the summer solstice, June 22, can be experienced anywhere north of 60° because of the same twilight conditions produced when the sun is actually below the horizon. The longer days of summer do not compensate for the shorter winter days in terms of annual energy receipts, because at high latitudes the sun is at a low angle above the horizon even on the longest days. A given amount of solar energy is spread over a much larger area than is the case farther south. Long days—even twenty-four-hour days—do not produce much heating in the Arctic as a result.

FIG. 24.1. *Stunted boreal conifers alternate with patches of open tundra in the forest-tundra transition zone near Great Slave Lake.*

Also there is a west-to-east gradation of temperatures produced by continentality. The warm currents that increase temperatures along the Pacific Coast provide additional heat for air masses traveling over the mountains into the Yukon and occasionally beyond. January minimum temperatures in southern Yukon are similar to those in Winnipeg—very cold, but not polar. The lowest January temperatures are found west of Hudson Bay. Baker Lake has an average daily low temperature of −33°F in January. Temperatures decrease eastward because of a progressive diminution of Pacific air influence. In summer the Mackenzie Valley is as warm as the Maritimes. Such large annual swings in temperature are typical of high-latitude environments.

Both permafrost and tundra result from the severe Arctic climate (fig. 24.1). Most people who have experienced tundra have done so in mountainous environments at lower latitudes, where there is a tree line above which low, shrubby plants represent the largest growth. This alpine tundra also exists in the Arctic. But as one travels in a northward direction, the tree line moves downslope to progressively lower elevations until it merges with tundra at sea level.

Tundra environments have short, cool summers that are not warm enough for trees to produce the growth they need to survive. Tundra is dry, in part because cool air is not able to hold much moisture, resulting in "polar desert" conditions. Little soil moisture is available for plants during the brief summer season, and all plants are effectively freeze-dried during winter. True tundra vegetation dies back completely during winter. In most places tundra consists of a hummocky ground surface with grasses, sedges, mosses, and lichens as the principal plants. The warm season is too short, and temperatures are too cool, for total decomposition of dead vegetation each year. Organic matter accumulates in the soil and over time forms peat. Eventually a thick, spongy mat of plant material accumulates at the surface.

The mat of undecomposed vegetation helps insulate the ground underneath to some depth, acting as a blanket to prevent buried ice from melting even during summer. This condition is known as permafrost. Nearly all of the tundra has an underlying layer of permafrost, and so does the northern fringe of boreal forest. Like the tundra, permafrost extends farthest south in the eastern Arctic, where winter temperatures are coldest and summers are coolest. There is no precise southern limit to the distribution of permafrost; it occurs sporadically in a broad, semicircular band across Canada, extending as far south as James Bay.

The "permanently frozen" portion of the ground that gives permafrost its name is a layer of varying thickness that lies several feet below ground level. Beneath this layer the ground does not freeze because it is too far from the surface, although permafrost conditions may extend downward for hundreds of feet. The surface layer, in turn, freezes and thaws seasonally in response to air temperatures. Soil water occupies approximately 10 percent greater volume in a frozen state compared with a liquid state; hence the freeze-thaw layer expands upward in winter and contracts during summer. The freeze-thaw layer is quite active in terms of heaving up buried rocks, forming cracks and crevasses, and generally rearranging loose materials on the surface.

Wherever the permanently frozen layer gets too close to the surface, it melts and produces surface ponds. The Mackenzie and Yukon River deltas are dotted with thousands of thermokarst lakes—small depressions filled with water following subsidence of the ground layer above a zone of melting ice. The effects of freezing and thawing are much in evidence on the Arctic landscape, even during summer when no ice is visible on the ground surface. Patterned ground, pingoes (ice-cored mounds of earth), and stone polygons (what appear to be careful rearrangements of surface rocks in a looped pattern but are produced naturally) all result from the freeze-thaw cycle.

Construction crews building the Alaska Highway during World War II discovered that roads they had scraped out of the terrain became impassable quagmires after the spring thaw. Any removal or compaction of the surface layer produces melting underneath. During summer one sees faint trails of water across the Arctic landscape; they are tire tracks resulting from a trip made in an off-road vehicle. Buildings are best adapted to permafrost if they are "minimum footprint" structures. Raising buildings on short stilts to allow air to circulate under the structure is a common practice in the Arctic (fig. 24.2). Such a design simultaneously prevents melting of the permafrost underneath the structure during summer and raises it above snow depth to allow access in winter.

## Native Populations in the North

The total population of Nunavut, Northwest Territories, and Yukon is less than one hundred thousand. The sparsely settled North accounts for roughly 40 percent of Canada's total area but only 0.3 percent of its population. It is a multiethnic society in which people of European descent form a slight majority because they dominate the populations of the two largest cities, Yellowknife and Whitehorse. Native people constitute a majority outside of a few urban centers.

FIG. 24.2. *Hudson Bay Co. store at Tuktoyaktuk, Northwest Territories. Visible under the building are the timber joists resting on small concrete blocks that support the structure. Air is allowed to circulate under the building to prevent the permafrost from melting.*

The composition of the native majority is highly variable geographically, but the total numbers are divided about equally between Inuit and Dene. The term *Dene* is applied to native people who are part of the Athapaskan linguistic family; however, they are often grouped with the much smaller numbers of Dene-métis who live in the region. Apparently neither the Inuit nor the Dene are descended from the first people—those who migrated to North America across the land bridge of Beringia between twelve thousand and fourteen thousand years ago—but rather date from somewhat later migrations. The ancestors of today's Inuit emigrated from Siberia over the ice or in small boats. These Paleo-Eskimos are more closely related, racially and linguistically, to the people of northern Siberia than they are to the Dene.

A series of eastward migrations took place within the Arctic, beginning about four thousand years ago, in which progressively more adapted hunting strategies can be identified. More than a dozen relatively isolated groups established themselves from Alaska's North Slope to the fringes of Greenland and Labrador. Some were primarily sea oriented and lived on seals, walrus, beluga whales, and birds; others moved inland, hunted caribou, and fished. The climatic cooling that took place before the Inuit's encounters with European explorers had forced changes in their lifestyle. Inuit migrations and adaptive responses produced a dynamic pattern of human occupancy in the Arctic that continued even as the early European explorers arrived. Inuit ways were not as ancient and fixed as many once regarded them as being.

In historic times the Inuit lived in seasonal habitations or camps of several families each that were scattered widely around the mainland coast and the Arctic islands. They earned money from hunting and trapping, on land and on the sea, although their participation in the Canadian economy was slight. Most of their efforts went into subsistence. Dene natives occupied more inland locations, either in the boreal forest or in mixed forest-tundra zones, but they, too, evolved a lifestyle that focused on fur trading, the trading posts, and scattered missions. These patterns persisted until the economic depression of the 1930s caused the market for luxury furs to collapse. Native living standards then sank below the level of subsistence in many areas.

The occasion of fighting a war in the Pacific during the 1940s brought new kinds of people to the North, including Canadian and American doctors who soon observed that many natives were in poor health. Some were starving. Tuberculosis was common, especially among the Inuit. Efforts were begun to attract native people into large enough clusters to enable the government to provide medical care and social services. New programs of public housing were created to draw both Dene and Inuit out of their scattered habitations and into the designated health and service centers.

Both groups continued to depend on subsistence hunting and fishing ("country food"), but they had to travel farther to their old hunting grounds than before, and they began to rely on gasoline-powered boats and snowmobiles. They gained access to electricity, then to television. Inuit lifeways were radically transformed in the span of a few decades. Better education only prepared young people for jobs that do not exist in the North. Better health and nutrition reduced the death rate and increased the birthrate to the point where the Inuit are one of the fastest-growing population groups on the continent. Both Dene and Inuit rates of alcoholism, drug addiction, and teenage suicide are dangerously high.

## Nunavut

In April 1999 the new territory of Nunavut became an official part of Canada's political map. Although its appearance did not mark the end of political evolution in the North, the recognition of a dominantly Inuit polity within Canada is a step toward self-government for the Inuit. Of the world's total Inuit population of 125,000, roughly 25,000 live in Canada; of those, 17,500 reside in Nunavut. The Nunavut Land Claim Agreement of 1993, which called for capital transfers from the federal treasury in excess of $1.1 billion, established clear rules of ownership and control of land and resources in the territory. In a 1995 election, Iqaluit (formerly the historic whaling center of Frobisher Bay), a community of 3,600 people on Baffin Island, was chosen as the new territory's capital.

Nunavut is basically the eastern end of the old Northwest Territories and is largely confined, although not totally so, to tundra lands of the Canadian Shield. Resources played a large role in setting the value of the land claim, but Nunavut's creation also reflects changes in how resources have come to be perceived. A major stipulation of the agreement was the establishment of three new federally funded national parks in Nunavut and a framework for local participation in decisions involving sport and commercial development of renewable resources. Iqaluit is little more than a three-hour flight from Montreal or Ottawa. Tourism —especially ecotourism—will become a major component of the Nunavut economy.

Nunavut's western border coincides roughly with the tree line (the southern limit of tundra), and thus it approximates the traditional Dene-Inuit divide. The so-called Copper Eskimos living at Kugluktuk (Coppermine), on the Coronation Gulf, are thereby included, but the greater the stretch of Nunavut's borders, the greater the diversity that had to be accommodated. The language of the Baffin Island Inuit is the traditional, syllabic rendering of the Inuit's language, Inuktitut. In the central Arctic the Roman alphabet is used, and the language is known as Inuinnaqtun. Nunavut thus has three official languages, those two and English.

Two significant Inuit populations were not included in Nunavut. Because their lands lie outside the Northwest Territories, the Inuit in more than a dozen villages of the Ungava region of northern Quebec (known as Nunavik) were excluded. The western Arctic Inuit (Inuvialuit) people of the Mackenzie Delta region were left attached to the Northwest Territories because of their closer ties with populations in the west. The boundaries of Nunavik are only approximate, because the Quebec Inuit's land claim has not yet been settled, nor has the Quebec-Labrador boundary question been resolved to the satisfaction of all participating governments.

## The Northwest Passage and the Arctic Archipelago

Europeans of the 16th and 17th centuries who explored Atlantic coastal waters were seeking a quick route to the Orient, a Northwest Passage around the continent. Cabot, Frobisher, Hudson, Baffin, Davis, James, and others live on in the capes, straits, bays, and islands named for them, although they found no passages. Eventually it became clear that a number of such routes existed, but all of them involved navigating the icy waters of the Arctic Archipelago at latitudes north of the Arctic Circle. The routes have been little used apart from exploratory sailings designed to test their feasibility.

Fewer than four dozen complete transits of the Northwest Passage—between Baffin Bay and the Beaufort Sea—have been recorded. The first crossing, in 1905, was by the Norwegian Roald Amundsen in a herring boat. Two adventure crossings have been made, one each by a Dutch and a Japanese pleasure yacht. The gargantuan oil tanker *Manhattan*, owned by Atlantic-Richfield Oil Company, tested a possible route for shipping Alaskan oil in 1969, but the experiment was not repeated after the ice ripped a gaping hole in the *Manhattan*'s doubly reinforced hull. Most of the other crossings have been made under Canadian government auspices. The long-sought Northwest Passage, which was the costly preoccupation of Europeans for two centuries, proved of little interest once it was confirmed to exist.

The Arctic islands long remained unknown to outsiders, except to those who hunted whales, and Great Britain did not transfer their ownership to Canada until 1880. Apart from some questions raised by the Danish and Norwegian governments during the 1920s, the islands have remained an unchallenged part of Canadian territory ever since. During the 1950s the United States constructed forty radar stations along a cross-Canada line slightly north of the Arctic Circle. This was the DEW Line, the chain of distant early warning stations that comprised one element of the North American air defense system during the Cold War era.

Canada was concerned about possible American ambitions and began to take a greater interest in the region. Resolute (latitude 74°), on Cornwallis Island, and Grise Fjord (77°), on Ellesmere Island, were created in the 1950s after the removal of Inuit families from villages farther south. Establishing the new settlements and furnishing them with a native population was part of Canada's demonstration of its sovereignty in the high Arctic. Cornwallis Island and Grise Fjord are the northernmost outposts of the Inuit in North America. Government stations operate year-round at Eureka (latitude 80°) and Alert (83°), farther north on Ellesmere Island, but they are not native settlements. The North Pole lies 475 miles over water and sea ice beyond Alert.

The remoteness of places like Ellesmere Island is only relative, and it can change rapidly. Between two hundred and three hundred military personnel are stationed at the Canadian Forces base at Alert. New laboratories to measure greenhouse gases and atmospheric pollution have been installed as part of the worldwide effort to monitor global warming. The growth of ecotourism has created demands for recreation adventures in such places, especially because of their past inaccessibility. Parks Canada administers the Ellesmere Island National Park Reserve at the northern tip of the island. Park personnel cater to summer visitors wishing to climb mountains, watch whales, and marvel at the sight of thousands of sea birds nesting in the sheer rock faces bordering the ocean, literally at the top of the world. Today, anyone with the means to purchase an airline ticket can fly there.

## The Mackenzie Valley–Northwest Territories

The Nunavut–Northwest Territories division very roughly accompanies three westward transitions: from the barren Canadian Shield to the somewhat more hospitable Mackenzie Valley; from tundra to boreal forest; and from Inuit to Dene as the majority native population. When Alexander Mackenzie journeyed down the river that bears his name in 1789, the native guides who accompanied him regarded the Inuvialuit people who lived in the Mackenzie Delta region as hostile. The Inuvialuit moved into the Canadian Arctic later than the Inuit who lived farther east. They lived in permanent log and sod houses built near the Arctic Ocean and made a living by hunting and whaling.

Arctic beaches near the Mackenzie Delta are littered with the driftwood that rivers have floated out of the boreal forest zone. The delta region represents the northernmost extension of boreal forest on the continent, one that divides tundra into separate domains to the east and west. The climate is less severe in part because the fringing Mackenzie Mountains produce chinook winds and a rainshadow effect. The added warmth created an ice-free corridor between the mountains and the Mackenzie River during the Pleistocene, which thereby extended a transportation route for the first people crossing Beringia. They were able to continue walking southward into the heart of the North American continent.

The Mackenzie Valley is comparatively smooth, with a cover of glacial materials over sedimentary rocks. It is part of the Interior Plains of North America that stretches from Texas through Montana and Alberta to the Arctic Ocean. As elsewhere in this extensive region, underground structures in the Mackenzie Valley contain oil and gas. In 1944 the Canol Pipeline was completed to send oil from Norman Wells, on the Mackenzie River, across the mountains to the Alaska Highway at Whitehorse to supply military activities. Present-day oil and gas operations are concentrated in the Mackenzie Delta and the Beaufort Sea.

The Northwest Territories, like all of the Far North, has a poorly developed road network and yet a population that is widely scattered. Bringing in quantities of fuel oil for heating and electricity generation requires low-cost transportation. Supplying goods to stock the shelves of the numerous isolated stores and trading posts presents other challenges. The resupply problem—to maintain the flow of commerce on which the far-flung network of communities depends—has been solved by creating an intermodal transportation system that is unique to the North.

Barges are the favored means of transportation, and hence the resupply season is concentrated in the summer months when rivers are navigable. The barges are more like floating fuel tanks whose decks have tie-downs to secure containers for handling goods. The system begins at Edmonton, whence shipments move north by rail or truck to the Peace River District and then either by rail or the Mackenzie Highway on north to the port of Hay River on Great Slave Lake. At Hay River everything is loaded aboard barges for an eight- or nine-day trip through the lake and on north down the Mackenzie River to the port of Tuktoyaktuk on the Arctic Ocean (fig. 24.3). Here, cargoes are reshuffled once more and placed aboard ocean-going barges that are towed east as far as the community of Spence Bay. A haul road paralleling the Mackenzie River assumes

FIG. 24.3. *A Mackenzie River barge and tug being readied at the Hay River terminal for a trip downriver to the Arctic Ocean.*

FIG. 24.4. *Yellowknife is the capital of the Northwest Territories. The "old town" is in the foreground. A newer business district and government center are on the hill in the distance.*

part of the resupply function during winter months, when the ground is frozen and sleds loaded with goods can be pulled overland by tractors.

A smaller resupply system operates on the Athabasca and Slave Rivers north from Fort McMurray, Alberta, but rapids on the Slave River near Fort Smith prevent operating through to Great Slave Lake. Western Nunavut settlements are supplied from a barge terminal at Churchill, Manitoba, and the eastern Arctic is served by sealifts from the Atlantic coastal ports.

Nearly every Dene and Inuit community has a Hudson's Bay Company store, which often is the only retail outlet except in communities where local cooperative stores have been organized. Frozen foods and other perishables are transported in refrigerated containers that are tied down next to the lumber, prefabricated homes, bulldozers, piles of insulation bats, and anything else ordered by anyone living near one of the northern stores, all of it lashed to the decks of the fuel-tank barges.

One of the communities served by the barge-container system is Yellowknife, on Great Slave Lake, which has been the capital of the Northwest Territories since 1967, when the devolution of some government functions from Ottawa began. Yellowknife originated as a gold mining camp during the 1930s (fig. 24.4). It has grown large enough to have affluent neighborhoods, a downtown core of offices and commercial buildings, and city parks. Farther north, down the Mackenzie Valley, lies Inuvik, a totally planned community with curving streets and tract homes that resemble 1960s suburbia. Inuvik's "utilidors"—the above-ground network of water mains and other utilities that were elevated, not buried in the permafrost—reveal its Arctic location. Although one can drive to Inuvik from points south in Canada and Alaska, the road system ends there. Only trails meander the short distance across the tree-dotted tundra north to the Arctic Ocean.

Inuvik is the major community for more than three thousand Inuvialuit who inhabit the Mackenzie Delta region. The Inuvialuit, who negotiated independently with the Canadian government to settle their land claim in 1984, used part of the monetary payment to acquire ownership of the Mackenzie River barge line. They negotiated their settlement through COPE, the Committee for the Original People's Entitlement, an organization formed in 1969 that aimed to represent all native groups living in the Mackenzie Delta. Eventually, it became an Inuvialuit association because its other members joined their own ethnic-based groups, such as the Métis Association. The Indian Brotherhood changed its name to Dene Nation and proposed creation of a Dene state in the Mackenzie Valley, to be known as Denendeh.

Because the evolution of political movements in the Canadian North has involved issues of native land-claim settlement, political boundaries have tended to follow ethnic lines based on historical patterns of occupancy. Part of the stimulus for proclaiming Nunavut in the 1990s came from the fact that the Inuvialuit already had settled their land claim, which prevented any pan-Inuit approach to further negotiations with the federal government. The inclusion of so large a fraction of the Inuit population in Nunavut had the incidental effect of increasing Euro-Canadian dominance of the Northwest Territories, forcing the Dene into more of a minority-group status compared with former times when Inuit, Dene, and nonnatives each constituted a presence large enough that none was able to dominate the other two.

## The Yukon Basin

The Yukon and the Mackenzie are the two great rivers of the Far North. Both are navigable over most of their lengths, and together they drain nearly all of the northern Rocky Mountains. They, plus their bordering lowlands, may be the longest-used human transportation corridors on the continent. The Yukon rises in the coastal mountains of British Columbia only a few tens of miles from the Pacific Ocean, but then it flows north and west nearly two thousand miles to empty into the Bering Straits opposite Nome, Alaska (map 24.2). Nearly all of Yukon Territory and central Alaska between the Brooks and Alaska Ranges are tributary to the Yukon River. The Yukon and Kuskokwim Basins are home to widely but thinly scattered groups of Athapaskan speakers, part of the Dene domain, which extends across much of the Mackenzie Valley as well.

Human history in the Yukon Basin was shaped by two great migrations that took place more than ten thousand years apart. The first was "the entry" itself, the arrival in North America of the very first people to inhabit the Western Hemisphere. There was a rapid sequence of environmental changes toward the end of the Pleistocene. During that period, but before sea levels rose, the land known as Beringia was a dry tundra or steppe grassland that stretched unbroken from western Siberia to central Alaska. Though some archaeologists have sought to push back the date of the first migration to perhaps 40,000 years ago, others have suggested that the first people may have crossed less than 12,000 years ago.

Once they crossed to Alaska, people followed the valleys of either the Yukon

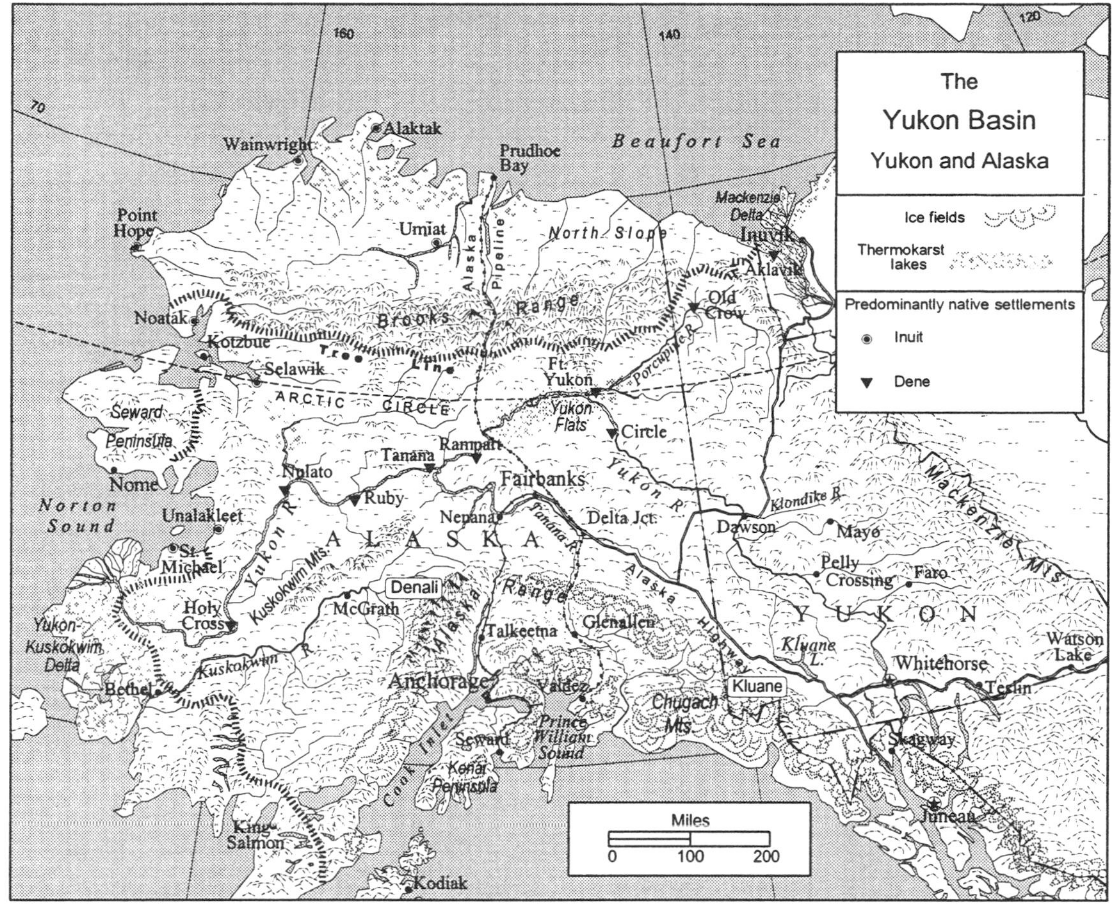


MAP 24.2

or the Kuskokwim River far into the interior. Archaeological sites near Fairbanks, Alaska, establish human occupation there by 11,000 years ago. Although one might expect the presence of massive ice sheets during the late Pleistocene, most of Alaska and the Yukon—including all of the Yukon Basin—were free of continental ice sheets. An ice-free corridor extended along the eastern base of the Rocky Mountains from the Yukon south into Alberta beginning 14,000 years ago.

The second migration in the Yukon Basin was in a direction opposite to the first. It was the great rush to the Klondike gold fields made in the fall of 1897. Thousands boarded ships at Seattle and Vancouver and made the trip north to one of two destinations: Skagway, Alaska, whence they crossed the mountains to the Yukon River, which took them downstream to Dawson; or Nome, where they boarded river steamers for the upstream trip to Dawson, at the mouth of the Klondike River. In 1898 Dawson was the largest Canadian city west of Winnipeg. Months later, Nome itself became the focus of another gold rush when flakes of gold were discovered washing up on its black sand beaches.

FIG. 24.5. *The headwaters of the Yukon River are in the coastal mountains where Alaska, British Columbia, and Yukon meet.*

Skagway was the staging point for the least difficult assault on the coastal mountains, through either the Chilkoot Pass or the White Pass, both of which lead to the headwaters of the Yukon River (fig. 24.5). The narrow-gauge White Pass & Yukon Railway was constructed from Skagway to Whitehorse by British interests, although not until the gold rush had ended. The White Horse Rapids is a major impediment to navigation on the Yukon River. The rapids required portaging, and a community developed at the site. The city of Whitehorse, which succeeded Dawson as the Yukon's largest city, grew from its transportation role, where the narrow-gauge railway met both the Alaska Highway and the head of steamboat navigation on the Yukon River (fig. 24.6).

Whitehorse, Dawson, Fairbanks, and Nome all were products of the gold rushes. Their populations have remained mainly white with a smaller native minority. The urban populations declined once the gold rushes ended, but boom conditions returned during World War II when the Alaska Highway was constructed from Edmonton, through the Peace River District, Fort Nelson, and Whitehorse, to Fairbanks. The purpose was to expedite the flow of war materiel that otherwise had to move via ocean vessels, then under possible attack in the Pacific. The highway was completed in 1942, just eighteen months after construction began, although it was an extremely primitive road that has been rebuilt almost continuously ever since.

Nearly everything purchased in Yukon Territory is brought from British Columbia or Puget Sound ports, handled in shipboard containers to Skagway, then trucked to the interior settlements. Dene villages along the Yukon and its tributaries are resupplied by a system of rail and barge transportation similar to that in the Northwest Territories. Fuel and other supplies are hauled four hundred miles north by rail from the ocean ports of Seward or Whittier, Alaska, to Nenana on the Tanana River, a Yukon tributary a few miles south of Fairbanks. There cargoes are offloaded to barges that serve more than two dozen Yukon River communities. Some outposts, such as Old Crow, the only community in the northern Yukon Territory, are most accessible to the outside world during winter when the rivers freeze to form smooth—although sometimes treacherous—routeways for vehicle travel.

Miles above the Yukon River today fly the intercontinental jet cargo and passenger aircraft plying routes between the United States and the Far East. Fairbanks is on the great circle (the shortest air distance) route between Tokyo and Chicago, whereas Anchorage lies close to the great circle connecting Los Angeles with Beijing. Air bases at both cities are important refueling stops for trans-Pacific traffic. For years Americans thought of Alaska as the "front line" between the United States and the Far East or the Soviet Union. More recently, the easing of restrictions on travel between Russia and Alaska has created new commercial linkages for Russians who fly east from Vladivostok or even Moscow to conduct business in Fairbanks and Anchorage.

## Alaska's Oil Boom

Alaska's North Slope—the portion of the state that drains north to the Beaufort Sea—is an inhospitable environment walled off by a chain of rugged mountains. The Brooks Range, which represents the northernmost extension of the Rocky Mountains, forms the northern edge of the Yukon Basin for more than five hundred miles across northern Alaska. Although the peaks in the Brooks Range are lower than those of the Alaska Range, the terrain is extremely difficult to cross, consisting of a maze of glaciated mountain valleys. Even the foothills of mountains at this latitude are devoid of trees. The North Slope's Inuit inhabitants, known as the Inupiat, are close relatives of the Inuvialuit who live in the adjacent Mackenzie Valley.

Major oil discoveries were made near Prudhoe Bay on the North Slope in 1968. An international consortium, Alyeska, was formed to develop the petro-

FIG. 24.6. *Whitehorse, on the Yukon River, is the largest city in Canada's northern territories.*

leum resources and forward the crude oil to market. There was no easy way to move the oil out of Prudhoe Bay, and many options were considered, including the use of icebreakers and oil tankers navigating the Arctic islands. The project was developed in the 1970s, soon after several major oil spills had heightened public sensitivity to the ecological damage caused by accidents involving oil tankers. The possibility of a pipeline across Canada was carefully considered, and so was the option of constructing a railway along the Mackenzie River. Gigantic airplane–oil tankers and even nuclear submarine–oil tankers were proposed and discussed. From both a political and an economic perspective, the best option was to construct an eight-hundred-mile pipeline across Alaska to a port on the Pacific where the crude oil could be transported by tanker to California oil refineries.

Built between 1974 and 1977, the Alaska Pipeline was among the first "environmentally sensitive" engineering projects in the world (fig. 24.7). Roughly equal portions were above ground (where the line had to cross permafrost) and buried. Above ground, the pipe was mounted on shoe assemblies that were not rigidly tied to the support structure. The wrapped and insulated pipe was mounted on supports and engineered to expand, contract, and withstand seismic disturbances. Because the crude oil had to be moved through a cold environment, it was necessary both to heat it and to speed its flow. Three pumping stations accelerate the flow in the pipe as it crosses the Brooks Range, four send it over the Alaska Range, and two more speed the flow of hot oil through the Chugach Mountains on the coast. Radiators, designed to dissipate heat and avoid melting the permafrost, were attached to the support structures.

The $12 billion pipeline feeds crude oil to an ocean terminal built at Valdez, Alaska, on Prince William Sound. Although the pipeline itself has not damaged its surroundings and has proved to be an excellent means for transporting crude oil, operations at Valdez were blighted in 1989 when the tanker *Exxon Valdez* ran aground and dumped a quarter million barrels of crude oil into Prince William

FIG. 24.7. *The trans-Alaska pipeline near Glennallen, Alaska.*

Sound, killing half a million sea birds and disrupting the Alaska fishing industry perhaps for decades. The eventual cost of cleaning up the spill and paying for damages exceeded the original cost of the pipeline.

Future oil developments on the North Slope and the Beaufort Sea will involve participation by the Inupiat, Inuvialuit, and other native groups who now hold clear title to certain oil-rich lands. The creation of new national parks and wildlife reserves in these same areas also will complicate any future negotiations involving resource extraction and pipeline construction. The oil reserves are large, and their eventual development seems almost certain; therefore, a long future for political negotiations involving resource policies is guaranteed.



## Land Claims and Regional Corporations

The current practice of land-claim negotiation in the Far North began in Alaska in the late 1960s as a direct result of the discovery of oil at Prudhoe Bay. Until that time native people had no assurance that they would benefit from the sale of natural resources from their lands. Nor was it clear what "their" lands actually meant, since a far-off government held title to everything. After the Prudhoe discovery, there was strong pressure to bring the oil to the market, which meant that the pipeline had to be completed and the natives compensated as soon as possible. What natives, where, and how much compensation might be due them had to be negotiated.

The agreement, which President Richard M. Nixon signed into law in December 1971, was negotiated through the Alaskan Federation of Natives. It provided the native people with title to 40 million acres of land for their own use and $962 million in compensation (half to be paid directly from the federal treasury, the other half from mineral revenue sharing). Although some tribal groups occupying land with more valuable resources might have expected to get more, others, with fewer resources, could have received less. The settlement was not based on actual resources, however, but treated all Alaskan resources together. Land and money would be distributed on the basis of the size and territory of the group. It was to be a settlement that would encompass all other claims, past or future.

Most native Alaskan communities had no financial institutions or administrative infrastructure to handle lands and manage large sums of money. Twelve regional corporations were established to take over the management role, acting as administrators for the settlement. Each person entitled to be compensated was given shares of ownership in a regional corporation. The corporations were established on a regional basis, but the actual regions kept ethnic divisions in mind so that Aleuts, Inuit, and Dene would work mainly with their own people. Individual communities, comprised mostly of natives, had the option of forming their own corporations. Dozens of community corporations soon dotted the map.

The regional corporations invested their cash payments in banks, entertainment complexes, hotels, fish processing plants, and a variety of other businesses and real estate ventures. Some corporations invested nearly all of their funds locally; others diversified and purchased properties outside the state. Some regional corporations went bankrupt and had to be reorganized, whereas others made a great deal of money for their members.

This system—which may be the best solution to Native American problems that the U.S. government has devised in nearly two centuries of trying—was copied by native groups settling their land claims with the Canadian government in the years that followed. With some exceptions, what worked in Alaska turned out to work equally well in Nunavut. Local councils, participation in land-use planning decisions, community involvement in resource management issues, and a long list of other, traditional native concerns are included in the procedures now followed by the locally elected and appointed officials who manage the corporations. Fears that some groups would be unable to manage their own affairs or would fall victim to unscrupulous white "advisors" have been borne out in some instances, but overall the record has been positive. Financial failure has been no more common than is observed generally in remote, resource-based frontier development schemes.

Among the problems left unsolved by the regional corporations are those involving future resource discoveries and allocation on lands owned by the corporations. Accepting title to some land and in principle forsaking the long-established native right to hunt and fish or gather food anywhere, on any land of their choice, clearly threatens the traditional basis of native subsistence patterns. The settlement of the land claims had the important effect of strengthening the economies and the governments of native groups, providing them with economic challenges and opportunities without sacrificing their ancestral ties to the places where they live, a benefit that would appear to outweigh any disadvantage that has thus far emerged.

## *References*

Bone, R. M. *The Geography of the Canadian North: Issues and Challenges.* Toronto: Oxford University Press, 1992.

Burn, C. R. "Thermokarst Lakes." *Canadian Geographer* 36 (1992): 81–85.

———. "Where Does the Polar Night Begin?" *Canadian Geographer* 39 (1995): 68–74.

DiFrancesco, Richard J. "A Diamond in the Rough? An Examination of the Issues Surrounding the Development of the Northwest Territories." *Canadian Geographer* 44 (2000): 114–34.

Gallant, Alisa L., Emily F. Binnian, James M. Omernik, and Mark B. Shasby. *Ecoregions of Alaska.* U.S. Geological Survey, Professional Paper 1567. Washington, D.C., 1995.

Hamilton, John David. *Arctic Revolution: Social Change in the Northwest Territories, 1935–1994.* Toronto: Dundurn Press, 1994.

Hamley, Will. "Problems and Challenges in Canada's Northwest Territories." *Geography* 78 (1993): 267–80.

Hoffecker, John F., W. Roger Powers, and Ted Goebel. "The Colonization of Beringia and the Peopling of the New World." *Science* 259 (1993): 46–53.

Purich, Donald J. *The Inuit and Their Land: The Story of Nunavut.* Toronto: J. Lorimer, 1992.

Rea, Kenneth J. *The Political Economy of the Canadian North: An Interpretation of the Course of Development in the Northern Territories of Canada to the Early 1960s.* Toronto: University of Toronto Press, 1968.

Smiley, Terah L., and James H. Zumberge. *Polar Deserts and Modern Man.* Tucson: University of Arizona Press, 1974.

Stager, Jack C., and Harry Swain. *Canada North: Journey to the High Arctic.* New Brunswick, N.J.: Rutgers University Press, 1992.

Whittington, Michael S., ed. *The North.* Royal Commission on the Economic Union and Development Prospects for Canada. Toronto: University of Toronto Press, 1985.

Wonders, William C., ed. *The North.* Studies in Canadian Geography. Toronto: University of Toronto Press, 1972.

*PART X*

# The Pacific Realm

## CHAPTER 25

# The Pacific Northwest

Alaska's breadth—from the tip of the Panhandle at 130°W longitude to the farthest Aleutian Island at 172°E—should require at least four time zones to span (map 25.1). The state's presence both east and west of 180° ought to require the western Aleutians to observe a different date (one day later) than the rest of Alaska. Both of these conditions would be true were it not for a zig-zag bend of the International Date Line and the definition of Alaska Standard Time, a single time zone that keeps the entire state on the same day and the same hour. The 58° of longitude between Alaska's eastern and western limits represents almost one-sixth of the distance around the world at this latitude, yet for all its breadth, end to end, Coastal Alaska is a fairly homogeneous region geologically, climatically, and even culturally. The region can be extended southward, along the coasts of British Columbia, Washington, and Oregon without much change in definition.

Coastal Alaska and British Columbia lend vivid meaning to the term *Pacific Rim.* The coastline resembles the joining of two broad arcs. One, which is concave to the north, is the chain of Aleutian Islands; the other, concave to the southwest, begins in the Gulf of Alaska and extends southward along the Panhandle and the coast of British Columbia. The entire distance along these two arcs, roughly three thousand miles, lies directly above the convergent boundary between the Pacific and North American lithospheric plates. The thinner Pacific Plate is plunging underneath the thicker and more buoyant North American Plate in the process of subduction, a condition that is accompanied by volcanic activity and earthquakes as one rock layer submerges beneath another and melts as it enters the earth's mantle.

## The Aleutian Islands

The Aleutian Islands are a chain of volcanic peaks that have emerged through the edge of the North American plate, forming an extensive island arc that has a deepwater trench along its southern margin in the zone of subduction. The Aleutians have a mild climate despite the northern latitude. Warm, moist air masses moving past Japan and driven northeastward by the Pacific high-pressure circulation pass over cold ocean currents flowing south from the Bering Sea. Moisture condenses to produce mists that envelop the islands much of the year, making the Aleutians one of the foggiest places on earth. The maritime cli-

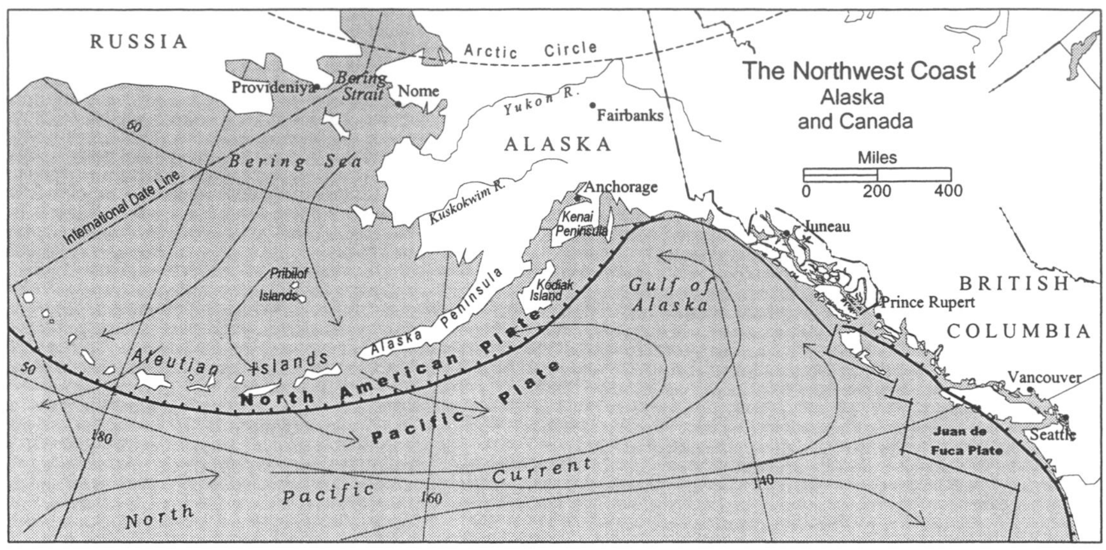


MAP 25.1

mate of the Aleutians, as well as most of coastal Alaska, means that temperatures vary only a few degrees between seasons.

Although the Aleutians are close to Asia, they were not occupied by humans as early as some parts of the Yukon Basin. The Aleut people appear to be more closely related to the Inuit than they are to the Dene natives of western Alaska and probably came to North America at the time of the early Inuit migrations. Aleuts traditionally garnered most of their subsistence from the ocean and were experts in hunting the sea otter, a sleek-skinned, fur-bearing mammal that thrives in the coastal waters of the island arc. As one of the least migratory groups of native people in North America, the Aleuts still inhabit the islands their ancestors settled thousands of years ago.

## The Russian Fur Trade

Russian interest in the North Pacific dates to the reign of Peter the Great (1682–1725), when Siberia was brought under the control of Moscow. The Danish navigator Vitus Bering was employed by the Russian government to explore the waters east of the Kamchatka Peninsula. After Peter died in 1725, his widow and successor, Catherine I, continued the project. In 1728 Bering sailed into the strait separating Alaska from Russia that bears his name and thus verified that the two continents were not linked by land. Subsequent voyages were more difficult, and Bering and several in his crew perished on a voyage made in 1741.

Bering's explorations confirmed that the area was rich in furs, especially sea otters, for which there was already a market in China. Numerous Russian fur-trading expeditions visited the Aleutian and Pribilof Islands in the half century that followed. One consequence was the near extinction of the sea otter in the more western islands. Russian traders pushed farther down the Alaska coast in search of more furs and forced the obliging Aleuts to serve as hunters on the expeditions. Russians constructed permanent habitations near the mouths of major rivers, on Kodiak Island, and along both the Alaska and Kenai Peninsulas by 1800.



The Russian success in the fur trade attracted the attention of Spanish authorities and drew British and American trading interests to the Northwest Coast as well. Spain claimed the North Pacific coast because it had claimed all the lands bordering the Pacific Ocean. New Spain extended its missions and presidios northward to Alta California (the San Francisco Bay area) and sent naval expeditions to the Northwest Coast to observe Russian activities. The first British entry was by Captain James Cook, who explored the Aleutians in 1776, searching for the Northwest Passage (in Cook's case, from the west). Although the Spanish had little impact on the Northwest Coast, the British viewed it as an extension of the fur trade from interior Canada and eventually dominated much of the coastal trade.

The Russians expanded their operations, even in the face of competition, and in 1812 they founded the outpost of Fort Ross just north of San Francisco Bay as a base for hunting sea otters and raising foodstuffs to provision the trade. The nearly simultaneous arrival of American and British fur-trading parties at the mouth of the Columbia River (Astoria) in 1811 eventually led to an agreement that the Pacific Northwest would be jointly occupied by the two countries, a condition that lasted until the Oregon boundary question was settled in 1846. The simultaneous presence of so many nations with trading ambitions did not lead to boundary disputes as often as it produced conflicts over trading rights, however, and the Pacific fur trade was conducted with a minimum of military conflict. Its growth was not crucial to any of the countries involved.

By the early decades of the 19th century, coastal Alaska was dotted with Russian trading outposts, where both British and Boston-based traders visited regularly to barter for goods and furs. Among the Russian settlements, Sitka (New Archangel) was the largest, with more than two hundred inhabitants (map 25.2). Orthodox priests established several successful missions among the Aleuts, and they also built churches southward in the Panhandle, which remain today as the most visible signs of Alaska's Russian heritage. Although their trading empire was vast, the Russians' grip on it was weak. The total Russian population of Alaska never exceeded one thousand. Attempts at diversification into agriculture, lumbering, and mining met with only limited success. Faced with debts from its European wars and distracted by attempts to colonize closer frontiers in Asia, the Russians eventually decided to withdraw.

## American Settlement of Coastal Alaska

The British, chiefly through the Hudson's Bay Company, were the most successful in pursuing the Northwest Coast fur trade, but it was the Americans who came to hold the greatest interest in acquiring the land. The purchase of Alaska

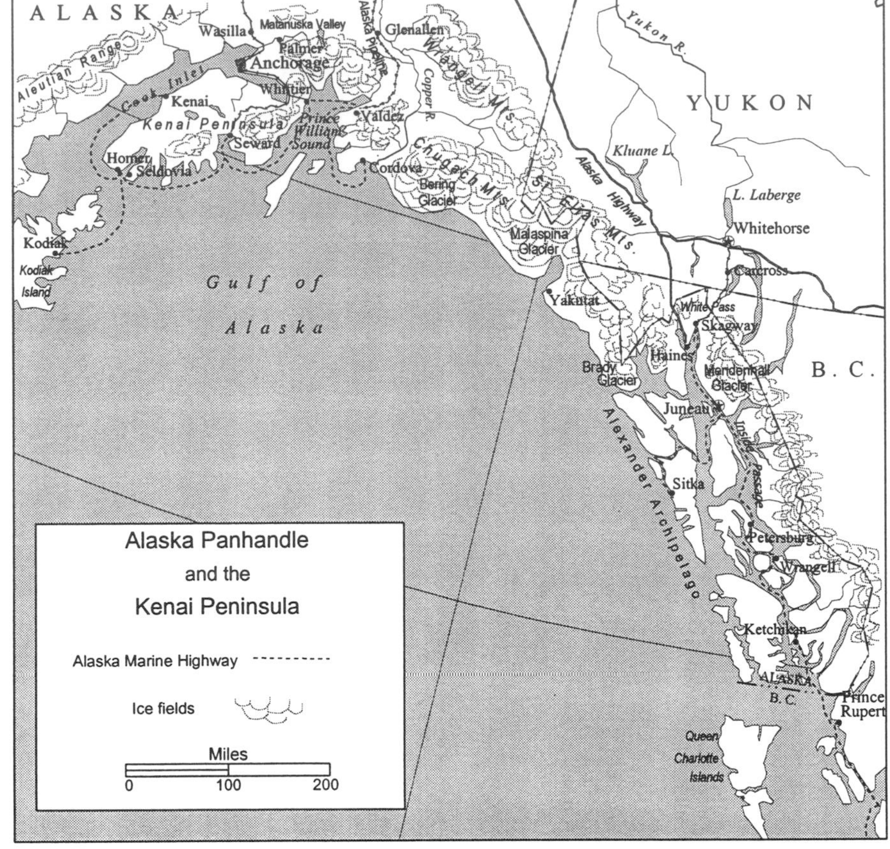


MAP 25.2

by the United States for $7.2 million was simultaneous with the passage of the British North America Act by the British Parliament, creating the Dominion of Canada, in March 1867. The United States purchased Alaska intact, and, apart from some minor adjustments to include more area in the Panhandle, Alaska's borders still follow the division of territory that the Russians and the British agreed upon in 1825.

The Russian interest in selling its North American territory was primarily financial, based on the poor prospects for the fur trade. The interest of the United States in acquiring Alaska was largely geopolitical—to curtail any possible westward expansion plans of Canada (or Great Britain). Apart from furs and salmon and the minor deposits of coal and oil the Russians had found, Americans knew practically nothing about Alaska's resources at the time of the purchase. But the search for precious metals commenced soon thereafter and led to gold strikes at Juneau (1881), Nome (1898), and Fairbanks (1902). A syndicate including J.P. Morgan and Company and the Guggenheim family of New York acquired copper mines along the Copper River inland from Cordova in 1898, invested in tracts of coal land, purchased steamship companies, and built railroads.

The U.S. government did a miserable job of administering Alaska in the first fifty years after the purchase. Abuse of the land-alienation laws led President Theodore Roosevelt to withdraw Alaska's coal lands from the market in 1906. Self-government was so slow in coming that Alaska's first territorial legislature did not even meet until 1913. Alaska's colonial status persisted until statehood was achieved in 1958, a development that many Alaskans had long desired, given the federal government's seeming inability to entrust it with the direction of its own affairs during the near-century that it endured territorial status.

The first permanent developments emerged in 1915 when the government-financed Alaska Railroad was begun from the port of Seward north to Fairbanks. Anchorage, platted by the railroad that year, grew to become Alaska's largest city. The railroad also passed near Mount McKinley, the highest point (20,320 feet) in North America, which was set aside as a national park (now called Denali National Park) in 1917. Substantial deposits of high-grade bituminous coal were discovered near the line as well. Alaskan coal, from mines near Denali National Park, is exported from Seward to industrial customers in the Pacific Rim, especially to steel mills in Korea.

Having acquired a "North" somewhat like Canada's, Americans became fascinated with Alaska, especially with agricultural settlement possibilities. Russian traders had raised potatoes successfully on Kodiak Island in the early 19th century but never had succeeded with grain crops. Coastal Alaska's climate is limited mainly by its cool summers, whereas the interior has the more obvious limitation of a very short growing season. Seven agricultural experiment stations were established by the U.S. government beginning in 1898 as part of an effort to explore the possibilities for future rural settlement.

The great need was to find a suitable crop. Although barley was (and still is) grown on an experimental basis as far north as Fairbanks, grain crops do not ripen well in Alaska, as the Russians already had discovered. Nonetheless, the notion that Alaska might support a larger population remained tied to efforts to stimulate agricultural settlement. The federal government's strongest entry was the Matanuska Colony at Palmer, Alaska, about forty miles north of Anchorage, founded in the mid-1930s as a New Deal Resettlement Administration project. A broad lowland, nestled in the coastal mountains along the Matanuska River, offered reasonably fertile soils. The Matanuska was close enough to the coast to have a less severe climate, and it was close enough to Anchorage to serve a local market.

Settlers—especially impoverished "stump farmers" from the Upper Great Lakes Cutover region—were recruited, assisted with housing, and given help clearing the land of its growth of spruce and aspen. Many Americans were drawn to the idea of farming this last American frontier in the Matanuska Valley, and for a time there was a steady inflow of new homesteaders. Various "back to the land" movements, especially during the late 1960s, saw a revival of interest, but the colony never grew beyond a modest size (fig. 25.1). Farms remaining in the Matanuska Valley specialize in producing milk and summer vegetables for the Anchorage market, but production costs are so high that imports from Seattle are cheaper.

FIG. 25.1. *A dairy farm in the Matanuska Valley of Alaska.*

Coastal Alaska includes nearly four-fifths of the state's population, with Anchorage's quarter of a million inhabitants making up more than 40 percent of the state's total. Anchorage is not a typical, limited-purpose northern settlement. Its economic base is diversified, and it has a large white-collar labor force employed by private- and public-sector agencies to administer the state's oil, coal, timber, and land resources. Alaska's regional corporations also have invested heavily in real estate and in business ventures in the Anchorage area.

Although it is possible to drive to Anchorage from Canada or the "lower 48," nearly everything Alaskans consume is brought in by barge and container ship, mostly from Seattle. One shipping system serves the Panhandle, where islands of the Alexander Archipelago shelter coastwise vessels from the high seas of the Pacific. Ships traveling this Inside Passage, which ends at Skagway, also call at Haines, Juneau, Petersburg, Ketchikan, and Sitka (fig. 25.2). In summer this

FIG. 25.2. *Skagway, Alaska, at the head of navigation on the Inside Passage.*

route sees the almost daily passage of cruise ships, which have become the most common way for tourists to reach Alaska. Much of the tourism industry is owned by multinational tour operators who also own the ships, hotels, and restaurants serving summer visitors.

The Gulf of Alaska has no such sheltered route among coastal islands, however, and larger ocean-going vessels are needed to reach the ports of Seward, Whittier, and Anchorage, all three of which serve the Anchorage metropolitan region and the rail-highway corridor north to Fairbanks. The Alaska Marine Highway resumes the coastwise traffic westward, beginning at Cordova and Valdez, in the sheltered waters of Prince William Sound, along the Alaska Peninsula, and on to Kodiak Island.



## Environments and Resources of the Northwest Coast

Native peoples living along the Northwest Coast spoke dozens of different languages when they were first encountered by Europeans in the late 18th century. The linguistic diversity makes it unlikely that they came from a single source region, but their exact origins remain unknown. Some of the coastal natives have close ties with Athapaskan speakers in the interior, although they share material culture and social rituals much more with other coastal groups. Sophisticated wood crafts, including the construction of totem poles, for example, were common among the coastal Alaskan Tlingit, the Haida and Tsimshian of British Columbia, the Salish of Washington, and many others.

Despite the discrete, localized groupings of natives along the coast, nearly all of them participated in trading furs with the Russians, the British, or both. They lived from the sea (salmon was the basic food) and supplemented their diet by hunting and gathering in the forests. It was an especially rich environment, where foodstuffs were abundant, building materials were at hand to fashion everything from houses to watercraft to utensils, furs and skins for clothing were readily obtainable, and the climate was mild most of the year, limited only by its dampness.

The resource base of the Northwest Coast is unique. It derives from a set of circumstances beginning with the coastal mountains themselves. The northeastward plunge of the Pacific Plate (as well as that of the much smaller Juan de Fuca plate along the Washington and Oregon coasts) acts like a very slow conveyor belt, delivering whatever islands, seamounts, or bits of would-be continents that once were isolated land areas within the Pacific Ocean up against the west coast of North America. From Alaska to Mexico, the coastal region consists of dozens of small accretions, known as terranes, that spent millions of years inching eastward, only to be scraped off the Pacific Plate and effectively welded to the North American continent along the subduction zone. This process is responsible for creating the nearly unbroken chain of mountains that fringe the Pacific Coast. The melting of subsurface rocks in the subduction zone created a series of volcanoes in a belt that extends from Alaska to Mexico. Each small segment of the mountain zone thus has a separate geologic history as well as different geographical origin, but together they form a nearly continuous highland margin for the Pacific Coast.

Because the general circulation of the atmosphere favors the presence of westerly winds in these latitudes, the coastal mountains intercept moisture-laden air masses that form over the Pacific. Although most of the ocean-surface currents along the Pacific Coast are cool (which means that they decrease surface air temperatures and thus inhibit rainfall at very low elevations), westerly winds forced aloft by the mountains bring rain or snow to the higher elevations over much of the year. In the colder temperature regime of Alaska, the precipitation falls as snow and produces massive glaciers, such as the Bering and Malaspina glaciers along the coast and the Mendenhall Glacier near Juneau.

The Coast Mountains of British Columbia also have glaciers, but there they are confined to higher elevations about fifty miles inland (map 25.3). Because temperatures continue to increase in the southward direction, glaciers retreat upslope and occupy only the highest elevations in the Cascade Mountains of Washington. During the Pleistocene epoch, these glaciers were much larger, and they moved downslope through existing river valleys, overdeepening the valley bottoms. When the climate warmed and much of the ice melted, rising sea levels drowned the glacially sculpted valleys to produce a highly indented coastline of fjords and small islands. The Northwest Coast has numerous glacier-fed streams that cascade down the mountain slopes and enter the ocean through these fjords.

The environment is ideal for the reproduction of anadromous fish (those that spawn in freshwater streams but spend most of their lives in the open ocean). Most prominent among them are the several species of Pacific salmon that have a preprogrammed life cycle that includes returning to the upstream pools where they were born before they reproduce. The annual Pacific salmon runs are intense periods of fishing activity to catch the salmon as they begin their upstream journey, jumping over rapids and low waterfalls, to reach the upstream environments where they spawn and die. Native peoples knew this cycle well and took full advantage of it, but because their fishing methods were not calculated to obtain the maximum catch, the overall stock of salmon was unaffected for centuries.

Moist air masses, releasing their precipitation as they rise against the mountain slopes, create optimal conditions for clothing the coastal mountains in a dense blanket of forest. Except for the Willamette Valley and portions of the Puget Sound Lowland, which are partial rain-shadow areas and have a different vegetation type, the entire Northwest Coast has a native cover of needleleaf evergreen forest (see chapter 19, map 19.2). The dominant species include Sitka spruce in the Alaska Panhandle and the British Columbia islands; cedar, spruce, and hemlock on the Olympic Peninsula and the Oregon Coast Ranges; and Douglas fir nearly everywhere else, including the west-facing slopes of the Cascade Mountains. The ecological dominance of the various species reflects the importance of microhabitats, including variations in temperature, moisture, and elevation.

The Pacific Northwest
Miles
0 50 100
Mountain passes
1 Stevens Pass
2 Stampede Pass
Masset
Prince Rupert
Terrace
Skeena R.
Kitimat
Queen Charlotte Islands
Vanderhoof
Nechako R.
Nechako Res.
Mitchell Pk.
Prince George
Fraser R.
Cariboo Mts.
Quesnel
Tete Jaune Cache
Fraser Plateau
Bella Coola
Williams Lake
BRITISH COLUMBIA
Mt. Waddington
Clinton
Thompson R.
Port Hardy
Lillooet
Kamloops
Campbell River
Powell River
Kelowna
Vancouver Island
Squamish
Strait of Georgia
Princeton
Penticton
Vancouver
Port Alberni
Nanaimo
Bellingham
Mt. Baker
Victoria
Mt. Logan
Strait of Juan de Fuca
Glacier Peak
Olympic
Mt. Olympus
Puget Sound
Everett
Seattle
Peninsula
Bremerton
Wenatchee
Auburn
Olympia
Tacoma
Ellensburg
Hoquiam
Aberdeen
Puget Sound Lowland
Mt. Rainier
WASHINGTON
Raymond
Yakima
Mt. St. Helens
Astoria
Longview
Mt. Adams
Columbia R.
Vancouver
The Dalles
Tillamook
Portland
Mt. Hood
Deschutes R.
Salem
Willamette R.
Albany
Madras
Newport
Corvallis
Willamette Valley
Mt. Jefferson
Springfield
Florence
Eugene
Sisters Mts.
Bend
Reedsport
OREGON
Coos Bay
Umpqua R.
Roseburg
Crater Lake
Rogue R.
Grants Pass
Siskiyou Mts.
Medford
Klamath Falls
Ashland

MAP 25.3

FIG. 25.3. *A temperate rain forest on the Olympic Peninsula of Washington.*

The Alaska Panhandle, the Coast Mountains of British Columbia, and the Olympic Peninsula of Washington are the wettest environments in North America (fig. 25.3). Evergreen conifers dominate a moist zone that extends more than twelve hundred miles along the coast and from ten to fifty miles inland. It is the largest remaining temperate rain forest in the world. The combination of mountains to provide lift and nearly constant onshore movement of moist air results in rainfall or snowfall more than two hundred days of the year. Rarely is the luxuriant rain-forest vegetation dry at ground level.

## Forestry and Fishing

Not only is this environment unique; most of it was in a natural condition, not much affected by the people who used it, until the early decades of the 20th century. Lumbering on the Northwest Coast began on a small scale in the 1850s, but it was severely limited by the lack of markets. Ocean shipping consumed most of the profits that lumber suppliers were able to earn. The industry was stimulated by the arrival of transcontinental railroads at Vancouver, Seattle, Portland, and Tacoma in the 1880s. Not until 1910, when the Upper Great Lakes forests had been cut, did the Pacific Northwest come into its own as an important supplier of lumber to the U.S. national market. Transportation costs to reach markets were high, and closer alternative sources continued to supply the national demand before that time.

The lag preserved Pacific Northwest forests until after the major lumber companies had acquired the timberlands they would eventually cut. A major source of timber (and mineral) lands was the land grants that had been made to railroads such as the Northern Pacific. Under its grant, the railroad was entitled to alternate sections of the public domain, in a checkerboard of uniformly square, 640-acre (one-square-mile) parcels, which it was to sell in order to finance the construction of its line. The railroad was chronically short of funds and desired to sell its lands quickly, in large blocks and at low prices, rather than to withhold them from the market until their value had appreciated. Frederick Weyerhaeuser and others from the Great Lakes domain purchased railroad timberlands and held them until logging was profitable.



Unlike the Upper Great Lakes region, there was no river-driving or rafting of logs in the Pacific Northwest. The streams are too swift and unpredictable to be used for transportation. Railways were extended in various directions from lowland centers like Seattle, Tacoma, and Aberdeen, up the tributary valleys, and were built farther into the mountains as the cut proceeded upslope. Logging technology was changing rapidly at the same time. The progression from the double-bitted axe to the cross-cut saw and then to the chain saw was compressed into only a few decades. By the 1950s heavy trucks and bulldozers were making the old logging railroads obsolete, and soon thereafter large helicopters were employed to conduct logging operations in areas where surface trails could not be attempted.

Much of western Washington had been logged over by the 1940s, after which the Oregon lumber industry expanded in scope and intensity. British Columbia's logging activities were concentrated in Vancouver Island and the coastal fjords, where booms could be assembled for towing logs to sawmilling centers at the mouth of the Fraser River. Lumber—for building construction and later for manufacturing plywood veneer—always has been the primary product of the Pacific Northwest forest industry. The shift to wood-veneer products in the 1960s was in part an economy move to make better use of the increasingly expensive wood, but it also increased the volume of by-products, especially wood chips, which were ideal for manufacturing paper. The production of pulp and paper by the Northwest Coast forest industry has increased in response.

Because its enormous zone of boreal forest stretches from the Atlantic to the Pacific, statistics about the "remaining" forests of Canada have little significance compared with similar figures for the United States. Before nationwide efforts at reforestation began in the United States during the 1930s, the near disappearance of commercially exploitable timber stands elsewhere kept increasing the Pacific Northwest's share of the nation's remaining forests even as its own rate of cutting increased. It became the principal supplier of lumber for the entire nation, a position that could only grow stronger in the absence of replanting efforts elsewhere.

Pacific Northwest forests are mountain forests, and they typically grow on steep slopes that are almost totally unfit for farming. They also produce soils that are infertile for agriculture even when the land is level enough to cultivate. No serious attempts were made to farm the "cutover" of the Pacific Northwest similar to those undertaken in the Upper Great Lakes region beginning in the second decade of the 20th century. Widespread attempts at reforestation began during the conservation-conscious 1930s, a time when, not coincidentally, a large unemployed labor force was available to undertake what seemed like the very uneconomical task of replanting forests.

Clear-cutting—the practice of harvesting all the trees on a parcel, then burning the slash to provide a new surface for regeneration or replanting—was not invented in the Pacific Northwest. It had been the common practice for some time before the shift to the Northwest began and, in fact, was regarded as an optimal practice because it produced even-aged stands of trees. It became a natural adjunct to the practice of reforestation—treating the forest as a crop rather than a once-only resource—because it provided a clear surface on which small seedlings could be planted. Reforestation efforts were much slower than the rate at which the forest was being cut on privately owned lands of the Northwest, however, and the prospect of a substantial decrease in forest yield became a threat by the 1960s.

The rest of the forest—the publicly owned sections of the land-grant checkerboard—had seen little harvesting of timber before that time. In most parts of the Pacific Northwest, they had been set aside as national forests and were administered for a variety of uses, including recreation, grazing, watershed management, soil conservation, and perhaps eventual logging for commercial purposes. Whereas many deplored clear-cutting of the privately owned alternate sections, public reaction against logging national forest land became a far more explosive issue. Given the existence of clear-cuts on most of the privately owned land, logging the national forests meant that the entire landscape, for miles in all directions, would have its tree cover removed. The land would be opened to erosion, and nearly all wildlife habitat would either be lost or so severely altered that the ecosystem itself could not be maintained.

By the 1970s privately owned forest lands were becoming ecologically distinct from the tracts of national forest with which they alternated. The private lands were more like tree farms than natural ecosystems, with even-aged stands of timber that produced a maximum number of board feet of lumber per acre when they were cut prior to reaching the "old growth" stage. Forests on public land also had gone beyond their natural state. They had been protected from fire and left undisturbed for so long that they included substantial amounts of dead and fallen wood, not useful for sawing into lumber but essential as habitat for creatures such as the spotted owl, a bird whose diet consisted of beetles and other decomposer organisms found in areas where wood is rotting.

Several endangered-species laws enacted by the U.S. government during the 1970s provided the basis for halting the spread of logging in the national forests. A series of federal court tests of the laws in the 1990s produced rulings that protected old-growth forests, not so much to save the trees as to preserve the old-growth habitat. "Saving the spotted owl" thus became equivalent to protecting old-growth forests from further cutting.

These issues impacted forests and forest industries throughout the Northwest Coast region. Canada enacted logging bans in parts of Vancouver Island as a result of a similar series of arguments, although more through the involvement of native groups (First Nations, in Canadian terms) than in the United States. Appeals to halt logging on the grounds of endangering old-growth habitats reduced timber cuts in the Pacific Northwest (where most of the old-growth habitat remains) but increased cutting elsewhere, especially in the Southeast and the Great Lakes, where most forests already are in the regrowth category.

Northwest Coast fisheries have undergone an equally drastic series of adjustments during the past several decades. Fishing fleets and the salmon canneries that processed the catch once were common in Northwest Coast ports from the mouth of the Columbia River to the Bering Strait. Halibut, cod, tuna, crab, and oyster fisheries were of lesser importance but still contributed to the large Pacific Coast catch. The shift from canned fish products to fresh and frozen fish fillets changed the industry. As a result many salmon canneries closed, although a greater impact has come from the drastic reduction of salmon stocks.



The fast-moving streams where salmon spawned also offered ideal sites for the construction of hydroelectric power dams. Fish weirs, allowing the salmon to pass around the lower dams, were impossible at the large dams, such as Grand Coulee, and thus dam construction closed the upper reaches of many important salmon streams. Attempts have been made to mitigate these circumstances by intercepting salmon on their upstream journeys and placing them in new areas where they can reproduce.

Despite these changes, it has been the overharvesting of salmon that has most severely reduced the annual catch. Dwindling salmon stocks have led to almost continuous conflict between Canada and the United States and between native peoples and commercial fishing organizations, over fishing rights and allowable procedures for harvesting fish. From a peak in the 1960s, the salmon catch was reduced year by year until, in the mid-1990s, the annual salmon runs had dwindled to insignificance in rivers like the Fraser. Salmon fishing is already overregulated and is not likely to revive through changes in those regulations. Its future remains in doubt.

## Urbanization in Coastal British Columbia

The earliest important city of the Pacific Northwest was Victoria, on Vancouver Island, which was a headquarters city and the most important center of settlement formation during the fur trade (fig. 25.4). There were two British colonies in the early years, Vancouver's Island (named for Captain George Vancouver, an early maritime explorer) and New Caledonia, which was coextensive with the southern portion of the mainland. New Caledonia was renamed British Columbia at the suggestion of Queen Victoria, and the colonies were united as British Columbia in 1866, a prelude to the formation of the Dominion of Canada the next year.

The urban development of the Northwest Coast was delayed in part because the region's political geography remained uncertain for years after European settlement began. The rallying cry "Fifty-four-forty or fight" of the 1844 presidential election in the United States referred to the strong desire of many Americans to extend the Oregon boundary northward to that latitude, which was then, as now, the southernmost limit of Alaska. The international boundary was adjusted for the last time in 1846 by continuing the 49th parallel line (already established as the border as far west as the Rocky Mountains) to the Pacific.

FIG. 25.4. *British Columbia parliament buildings, Victoria.*

Even with the border in place, it was by no means certain that the separate colony of British Columbia would become part of Canada. A gold rush to the lower valley of the Fraser River in 1858 established a heavily American-born population in the future province. Its loyalties were directed more south than east. The United States was poised to acquire Alaska and gladly would have purchased British Columbia as well, but the British government needed its naval base at Esquimault, on Vancouver Island, and was unwilling to withdraw. British Columbia lay thousands of miles west of the settled portion of Canada at that time. Even the creation of the new province of Manitoba in 1870 did little to bind the East to the Pacific Coast. Determined to entice British Columbia to join Canada, the Canadian government promised to build a transcontinental railroad, like the line that had just then been completed across the western United States. Although the Canadian Pacific Railway was not completed until 1887, the promise of a rail link helped secure British Columbia's loyalties, and in 1871 it became the sixth province to join the Dominion of Canada.

Victoria's somewhat drier and very mild climate at the southern tip of Vancouver Island made it a favorite tourist destination even in the early years of settlement, a role that it has continued to play up to the present. But Victoria was on an island and could not possibly be the recipient of a transcontinental railroad built across Canada. By the mid-1880s a collection of sawmilling settlements and trading posts had appeared around the mouth of the Fraser River on the mainland, but no city was there until the Canadian Pacific platted a substantial new town named Vancouver in 1886. Vancouver was thus deliberately created to be the necessary seaport at the end of the transcontinental railroad corridor across Canada. Docks, shipyards, and warehouses appeared around Burrard Inlet, where the railroad terminated. The city later expanded to the south, around the numerous bays and inlets, and also spread inland up the Fraser River Valley.

For the next thirty years Vancouver was the only seaport on the west coast of Canada. Because of the great difficulties of land access through the mountains fringing the Strait of Georgia, it was easier to transport logs by water to Vancouver-area sawmills than to attempt new sawmilling centers to the north. Vancouver remained the leading center of the mainland coastal lumber industry. It also grew from its seaport and manufacturing roles and developed a diversified economy, becoming the third largest city in Canada. The commodities shipped from Vancouver's port facilities today reflect the resources of all of western Canada: wheat, barley, and canola from the Prairie Provinces; potash from Saskatchewan; coal from the Rocky Mountains; and lumber from the Pacific Coast region. Although a relative latecomer to the ranks of North American metropolises, Vancouver has developed into one of the continent's most attractive cities.

Since at least the 1840s, those who schemed about trade with the Orient had understood that the best choice of a Pacific Coast seaport would be the northernmost possible site. The argument in favor of a northern port is the minimum-distance, great-circle route—exactly the same logic that places Fairbanks or Anchorage on the optimal air navigation route. In 1915 the Grand Trunk Pacific (later, the Canadian National) Railway constructed a line to another new seaport, Prince Rupert, at the mouth of the Skeena River and only a few dozen miles south of the Alaska border—the northernmost Pacific seaport possible in Canada. Although the trade through Prince Rupert was slow to develop, the port became much more active beginning in the 1970s, when Canada's foreign trade around the Pacific Rim responded to the marked increase in Asian demand for grain, coal, and forest products.

The only other significant urban center in coastal British Columbia developed in the 1950s when the Aluminum Company of Canada constructed a smelter at Kitimat, about one hundred miles southeast of Prince Rupert. Kitimat's attraction was the hydroelectricity made possible by damming the Nechako River, which offered a cheap source of power for aluminum manufacturing. Ocean vessels carrying bauxite from the western Pacific Rim (or anywhere else in the world) could unload the raw material at Kitimat's deep-water harbor, a counterpart of the conditions that led to creating an aluminum manufacturing industry on the lower Saguenay River in Quebec years before.

## The Puget Sound Lowland

Although the institutional and political contexts were different, Americans were projecting the development of new seaports in the Pacific Northwest at roughly the same time that the Canadians established Vancouver, and for largely the same reasons. No cities existed in the Northwest when the first transcontinental railroads began building across the northern Great Plains. When they reached the coast, the railroad builders would exercise the advantage of priority in creating new towns at locations they thought advantageous for conducting business.

Two geologically distinct mountain zones bracket the Puget Sound Lowland. The Coast Ranges, including the Olympic Peninsula, extend from California to Washington. They originated as marine sediments that were compressed into a maze of folded and faulted mountains. Their summits are comparatively low, although Mount Olympus (7,965 feet) is high enough to be the center of an ice field. Midelevation west-facing slopes of the Coast Ranges have a dense forest cover and belong in the temperate rain-forest category. The Cascade Mountains, bordering the Puget Sound Lowland on the east, have higher peaks of volcanic origin. Mount Baker (10,775 feet), Mount Rainier (14,411 feet), Mount Adams (12,276 feet), Mount Hood (11,245 feet), and a dozen lower summits (including 8,366-foot Mount St. Helens) are volcanic features that have erupted during geologically recent times. Their north-south alignment marks the subduction zone where rocks of the oceanic lithosphere are plunging under the North American continent.

The Puget Sound Lowland is part of a continuous, north-south-trending structural depression that also encompasses the Willamette Valley, the Georgia Strait, and the Inside Passage of Alaska. Sea levels have fluctuated repeatedly, and the northern part of the depression is now largely submerged, whereas the southern end is dry land. Various portions of the lowland have been the sites of river deltas where streams from the east have deposited sediments in the shallow waters. Puget Sound's numerous bays and inlets—some of which are called canals, although they are natural features—are the product of glaciation and sea-level change.

Puget Sound was the obvious choice for the terminus of the northern transcontinental line across the United States, not only because of its northern latitude but also because the coastline of Washington offered few other sites for a seaport. The best harbor on Puget Sound was Elliott Bay, where the small community of Seattle had become a sawmilling center by the 1860s. Rather than to extend their tracks to an existing port, where the investments made by others would limit the possibilities for profits, the builders of the Northern Pacific Railroad chose to create their own seaport on Puget Sound. They chose Commencement Bay, forty miles south of Seattle, and in 1872 the Northern Pacific platted its new town there, which they named Tacoma. The plans to make Tacoma a major seaport were postponed by the bankruptcy of the Northern Pacific and its subsequent redirection by a group of investors who believed that Portland was a better terminus for the line. Tacoma's growth was thus delayed for two decades.

These Pacific Coast ports needed routes through the mountains so that the railroads could reach them. Access was accomplished easily in the cases of Vancouver, at the mouth of the Fraser River, and Portland, on the Columbia. No other rivers cross the Cascades; for both Seattle and Tacoma, it was necessary to discover a nearby mountain pass to carry the railroad's line down to tidewater. With the opening of its line through Stampede Pass in 1887, the Northern Pacific finally reached Tacoma directly, the same year that the Canadian Pacific reached Vancouver. Although Seattle received a railroad branch from Tacoma, it did not acquire transcontinental status until 1893, when James J. Hill constructed his Great Northern Railway through Stevens Pass, east of Everett, Washington, and thereby made the small city on Elliott Bay a transcontinental terminal.

These developments, which were many years in the planning, led to the creation of all the major ports of the Pacific Northwest in a single decade. The subsequent growth of the cities depended largely on the rate at which traffic developed through their ports and the extent to which each city diversified its economy. Seattle eclipsed Tacoma as a seaport as soon as its eastward connection was in place, and it eventually became the busiest port in the Pacific Northwest. All of the cities remained as logging and lumbering centers for years after their founding, and it was not until World War II that local economies diversified.



War with Japan increased traffic volumes through the ports. William E. Boeing began manufacturing airplanes in Seattle in 1917, but Boeing did not become an important local industry until World War II. The company expanded into new manufacturing facilities in Seattle, Renton, and Vancouver. The U.S. Naval Yard at Bremerton, Washington, brought the Puget Sound region wartime growth that later continued in defense-contracting industries. The growth of the aircraft industry was assisted by the Pacific Northwest's aluminum industry; the eventual triumph of jet aircraft brought further growth to the area. Seattle became a major center for computer software innovation in the 1990s, continuing its role as a leader in fostering new industries. Like Vancouver, it typically ranks as one of the continent's most livable cities (fig. 25.5).

The Puget Sound Lowland is now an almost continuously urbanized strip that merges with the Willamette Valley on the south and the Vancouver metropolitan region on the north to form a densely settled region of more than 6 million people. It is still home to some of the largest wood-product and paper-making industries in North America even after decades of economic diversification. Intermodal cargo terminals at Seattle, Tacoma, and Roberts Bank (Vancouver)

FIG. 25.5. *Warehouses and piers on downtown Seattle's waterfront have given way to restaurants and condominiums.*

vie with one another for prominence in handling time-sensitive shipments in the transpacific trade, continuing the very role for which these cities were created more than a century ago.



## The Willamette Valley

The mouth of the Columbia—the one great river that cuts through the Cascade Mountains and the Coast Ranges to reach the Pacific—might have been the site of an important seaport. Astoria, Oregon, the city that grew there, certainly had the advantage of priority. It was reached by Lewis and Clark in 1805 and in 1811 was selected by John Jacob Astor's Pacific Fur Company for a trading post that would coordinate all the company's trade in the Northwest. Astor lost interest in the venture, and the post was taken over by the British. The most significant limitation of Astoria as a seaport was the nature of the Columbia's estuary. The river's broad mouth is windswept, and navigation is treacherous because of sand bars. Astoria's shortcomings meant that the Columbia's port would have to be somewhere upriver.

Even the fur traders who lived on the damp, windy coast preferred to locate their operations farther inland. Vancouver, Washington, began as a trading post of the Hudson's Bay Company, one hundred miles up the Columbia from Astoria. Traders sought to raise food to supply the post in the broad valley at the south bank where the Willamette River enters the Columbia. Stretching more than one hundred miles south from this point is a lowland averaging about twenty-five miles in width where the Willamette River meanders through a fertile plain covered with grass and fringed with oak woodlands. The lowland is the most favorable tract of potential farmland in the entire Pacific Northwest. Although it was not yet a part of the United States in the early 1840s, popular opinion held that British territory would not extend south of the Columbia River. In 1843 the first great migration arrived down the Oregon Trail, and by 1850 nearly ten thousand people had made the overland trek to Oregon. The great majority of them chose to take lands in the Willamette Valley.

In many respects, Oregon was a dry run for the settlement of California a few years later. In fact, the discovery of gold in California in 1848 ended the Oregon boom just a few years after it began. More significantly, the Willamette Valley was perhaps the most distant frontier imaginable when it was settled. In the early 1840s the zone of continuous settlement in the United States had just reached southern Wisconsin and had not yet crossed the Missouri River. The faint outlines of the Corn Belt were just emerging in the Middle West when the Oregon migration began. The new Oregonians were much like those who took lands on the new Middle Western frontiers of that period—a blend of people from the Ohio Valley, Illinois, and the Upper South, most of whom were interested in establishing farms on new lands. The blending of population origins in the Middle West was further mixed in the new households that formed on the Oregon frontier.

The Oregon Trail began at Kansas City and St. Joseph, because there were no Euro-American settlements to speak of west of there at the time. Although it meant a journey of more than eighteen hundred miles, the Willamette Valley had attractions that the Middle West totally lacked. One was its mild climate, which permitted the growth of fruit orchards and vineyards on a scale that Americans rarely had seen before that time (fig. 25.6). Another was the nearly unbroken forest of Douglas fir, cedar, and hemlock that reached down into the Willamette Valley from the surrounding mountain slopes. A lumber industry grew at water-power sites along the navigable, lower reaches of the Willamette River, and the industry provided the nuclei for such cities as Portland, Oregon City, and Milwaukie. Portland was as far up the Willamette as ocean-going vessels could safely reach, and thus it became the most important transportation center of the region.



What Oregon lacked was a market. Wheat yielded well in the Willamette Valley, and exports were sent on ocean ships to California, especially during the gold rush era when farming had not become well established to the south. Like the slow growth of lumber exports from Puget Sound, distances to the market were excessive, which limited the possibilities for specializing in staple exports. Portland's first important link to the rest of the United States came in 1883, when a railway was completed along the Columbia River, connecting Portland to transcontinental routes to St. Paul and Omaha.

Portland has been the most accessible city in the Pacific Northwest since that time, with direct connections to California, the trans–Rocky Mountain region, and Puget Sound. Its port function was not as important as Seattle's, however. For many years Portland lacked the innovation-based industries that emerged in the Seattle-Tacoma area. Portland also differed in that the rich agricultural region along the Willamette dispersed urban growth through the scatter of small cities down the Willamette Valley. By the 1980s Portland had evolved a mix of local industries that included the high-tech, computer-based firms that also characterize Seattle and Silicon Valley.

FIG. 25.6. *In addition to its fruit, nut, and vegetable crops, the Willamette Valley of Oregon is a major producer of nursery stock.*

FIG. 25.7. *Sisters, Oregon. Some highways through the Cascade Mountains remain closed by winter snows even though warm spring weather prevails in the lowlands.*

The Willamette Valley continues as a structural lowland south of Eugene, although it is not a single drainage basin beyond there. Rivers flowing to the Pacific, including the Umpqua and the Rogue, drain the interrupted lowland as it continues to the vicinity of Ashland, where the Siskiyou Mountains form the valley's southern limit. Roseburg, Grants Pass, Medford, and Ashland have economies based on cutting timber in the Cascade Mountains and the Coast Ranges. Despite the closure of many mills during the 1980s, the Oregon forest-products industry revived during the 1990s. The state's largest industrial complex is a collection of lumber and wood-product mills near White City, on the Rogue River, a few miles north of Medford.

Mountain ranges divide Oregon into distinct regions (fig. 25.7). The well-settled Willamette Valley is bordered by sparsely settled ranching country in the rain shadow east of the Cascades. On the south, the Siskiyous and the Klamath Mountains separate the Willamette Valley from the Central Valley of California. All traffic between Oregon and California must either cross the Cascade Mountains and enter the Great Basin or follow the narrow mountain valleys through the Siskiyous. The coastal mountains, the Cascades, and the Sierra Nevada all merge into the single Klamath mountain system, which blocks the lowland from stretching continuously from southern California to the Inside Passage of Alaska.

## References

Abbott, Carl. *Portland: Planning, Politics, and Growth in a Twentieth-Century City.* Lincoln: University of Nebraska Press, 1983.

Barman, Jean. *The West beyond the West: A History of British Columbia.* Rev. ed. Toronto: University of Toronto Press, 1996.

Bowen, William A. *The Willamette Valley: Migration and Settlement on the Oregon Frontier.* Seattle: University of Washington Press, 1978.

Freeman, Otis W., and Howard H. Martin. *The Pacific Northwest: An Overall Appreciation.* New York: John Wiley, 1954.

Gibson, James R. *Imperial Russia in Frontier America: The Changing Geography of Supply of Russian America, 1784–1867.* New York: Oxford University Press, 1976.

MacDonald, Norbert. *Distant Neighbors: A Comparative History of Seattle and Vancouver.* Lincoln: University of Nebraska Press, 1987.

MacKay, Donald. *The Asian Dream: The Pacific Rim and Canada's National Railway.* Vancouver: Douglas & McIntyre, 1986.

Miller, Orlando. *The Frontier in Alaska and the Matanuska Colony.* New Haven: Yale University Press, 1975.

Naske, Claus-M., and Herman E. Slotnick. *Alaska: A History of the 49th State.* 2d ed. Norman: University of Oklahoma Press, 1987.

Robinson, J. Lewis. *British Columbia.* Studies in Canadian Geography. Toronto: University of Toronto Press, 1972.

Schwantes, Carlos A. *Railroad Signatures across the Pacific Northwest.* Seattle: University of Washington Press, 1993.

Williams, Michael. *Americans and Their Forests: A Historical Geography.* Cambridge: Cambridge University Press, 1989.

CHAPTER 26

# California

Beyond the crest of the Sierra Nevada, California's geography is divided into three parallel regions: the Sierra foothills, the Central Valley, and the Coast Ranges (map 26.1). The Coast Ranges and the Sierra also are connected by the Klamath Mountains at the northern end of the Central Valley and by the Tehachapi Mountains at the southern end, creating an almost continuous perimeter of mountains. The mountain ring is broken only by the Carquinez Strait, a sea-level passage through the Coast Ranges leading to San Francisco Bay.

Underlying these topographic divisions is the state's major geologic divide, the San Andreas Fault, which is the surface trace of contact between the Pacific and North American lithospheric plates. It runs from the Salton Trough through the Los Angeles Lowland and the Coast Ranges and strikes out to sea north of San Francisco. Everything west of the San Andreas fault is moving very slowly in a northwesterly direction relative to everything east of it. Five million years ago the Central Valley's outlet to the sea was near present-day Santa Barbara; 5 million years from now San Francisco Bay will be an arm of the ocean hundreds of miles north of its present location.

## The North Coast

The topography of the Coast Ranges is controlled by a series of small northwest-southeast fault lines that are associated with the San Andreas system. The pattern of low mountain ridges and valleys produced by the faults is cross-cut by the Pacific coastline, which trends in a more nearly north-south direction. As a result, the coastline is a sequence of truncated headlands separated by inlets where the coastal rivers enter the ocean. Periodic episodes of uplift of the coastal zone have produced a series of marine terraces that give a stair-step appearance to the landscape.

The beauty of northern California's rugged coastline once was matched by a forest of massive redwood trees that occupied the low- to midelevation slopes up to fifty miles back from the coast and extended south, sporadically, to the vicinity of Monterey. Lumbering began even during the Spanish period in California and grew after San Francisco became a seaport during the mid–19th century. Eureka and Fort Bragg were the important centers of the lumber industry on the North Coast. The inland redwood forests were difficult to reach, and they

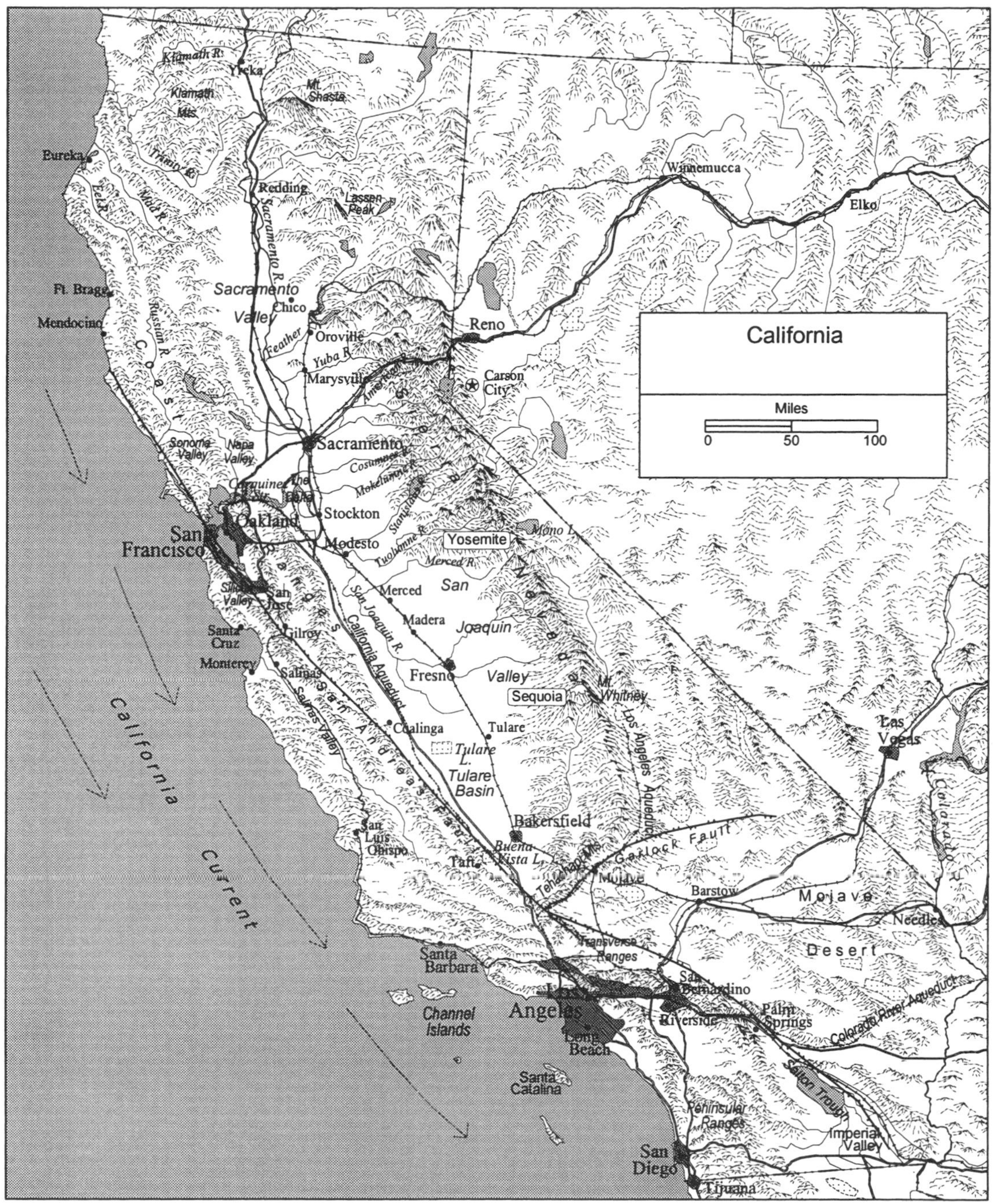


MAP 26.1

were not exploited significantly until railroads reached into the coastal valleys of the Klamath, Eel, and Russian Rivers in the late 19th century.

Lumber companies owned or leased almost all of the "redwood empire." So little was protected from cutting that its final disappearance became an issue of concern by the 1960s. Associated stands of cedar, hemlock, and Douglas fir still form the basis for northern California's lumber industry, but the magnificent stands of redwoods today are confined to small areas where they have been preserved.

The North Coast remains a sparsely populated section of California despite the metropolitan overspill of recent decades, which brought a new lifestyle to coastal communities like Mendocino. Although official production figures obviously cannot exist, because the crop is illegal, the North Coast is alleged to be one of the most important areas of marijuana production in the United States.

## The Mediterranean Climate

The redwood forest, like nearly all of California's vegetation, is controlled more by elevation and slope aspect than by latitude. Southern California is warmer and drier than the north, to be sure, but climate and vegetation zones are nonetheless demarcated primarily as a series of elevation-controlled bands, parallel with the coast. California is North America's principal zone of Mediterranean climate, a type that occupies analogous west-coast locations on all of the world's continents, including southwestern Europe's Mediterranean Sea, from which the climate takes its name. Mediterranean climates have a winter maximum of precipitation and a dry season at high sun. Though their temperature regimes vary, no Mediterranean climates have cold winters, and their winter and summer temperatures contrast only moderately.

The cause of California's Mediterranean climate is found in the seasonal movement of high-pressure circulation systems (clockwise, in the northern hemisphere) over the adjacent ocean. In January the Pacific high is centered at latitude 30°N, and California is downwind for moist air masses flowing eastward off the Pacific. In July this same system is centered at about 35°N, and the onshore movement of moist, oceanic air masses is confined to latitudes north of California. During summer, the atmospheric circulation along the California coast favors a north-to-south movement of air. In early autumn the circulation shifts to a northeast-to-southwest pattern, bringing hot air from the inland valleys to the coast.

California's climate is a product of all three of these conditions. The precipitation maximum is in December and January, when the Pacific high is farthest south; the precipitation minimum is in July and August, when it is farthest north and both wind and water currents flow southward along the coast; and the hottest temperatures are in August and September, when the circulation is directed from land to sea.

California has a temperate climate year round especially because of the cooling effects of the north-to-south flow during summer. The cool California Current produces an upwelling of cold water from the depths, further reducing surface temperatures near the coast. The Sierra Nevada shelter the state from cold air masses originating in the continent's interior during the winter. The effect is well illustrated at San Francisco, where July and January temperatures average 62.1°F and 50.9°F, respectively. On the Atlantic Coast one would have to go north to St. John's to find summer temperatures this cool and south to Charleston to find winter temperatures this warm.

## The San Francisco Bay Region



San Francisco was California's first important city. The Spanish extended their network of missions, presidios (military forts), and pueblos (agricultural villages) northward through the Coast Ranges to "Alta California" during the last decades of the 18th century, in part from a concern about British and Russian interest in the region. San Francisco Bay received both a mission and a presidio, and half a dozen other Spanish outposts were planted in the coastal valleys between San Francisco and Monterey. These early Spanish beginnings influenced the modern city of San Francisco, but its phenomenal growth during the 1850s had little to do with its early heritage.

Under the terms of the Treaty of Guadalupe Hidalgo, the United States purchased Alta California (and the rest of the southwestern United States) from Mexico at the conclusion of the 1846–48 war between the two countries. Almost simultaneously, gold was discovered on a tributary of the Sacramento River. News of the California gold strike was transmitted as rapidly as communication in that era allowed, and eventually word spread around the world. This was the first major gold rush in the American West and the most important in terms of stimulating "gold fever." The migrations lasted nearly a decade as the new discoveries shifted from the American River (the first strike) to the Yuba, the Cosumnes, and the Mokelumne and then south of there to the Tuolumne and Mariposa Rivers. The Sierra foothills region became known as the Mother Lode, although no single source of its gold-bearing stream deposits ever was found.

In 1848 fewer than 15,000 Euro-Americans lived in California, but by 1857 that number had grown to 500,000. Nearly half of the new arrivals had been born in New England, New York, and Pennsylvania. Getting to California was not easy for them. The Mormon and Oregon Trails provided a route as far as the Great Basin, but no satisfactory land routes existed across Nevada—let alone across the Sierra Nevada—when the gold rush began. Despite the difficulties, between 20,000 and 40,000 new arrivals crossed the Sierra Nevada (or the Colorado River) into California every year as the gold fever spread. An even larger number sailed from East Coast ports to Mexico or Nicaragua, crossed Central America as best they could, then resumed their journey northward on small coastwise vessels. Some sailed the entire distance around Cape Horn and landed at San Francisco.

San Francisco was the major point of entry despite the fact that it lay west of the gold fields and required yet another journey—up the Sacramento River on a steamboat. San Francisco boomed, growing from a city of perhaps 1,000 people in 1848 to 35,000 in 1850. San Francisco ranks with Chicago in the national chronology of urban settlement, even though it is twenty-seven hundred miles farther west and was not connected to the rest of the continent by railroad for another twenty years. Building the transcontinental railroad created another population boom in San Francisco, with the importation of Chinese laborers as construction workers, thus creating the first sizable Asian population in the United States.



Like Chicago, San Francisco was platted as a gridiron of city blocks—a real estate promoter's dream, where tracts of land were bought, sold, and traded in boom-bust cycles of land prices—even though the grid was totally inappropriate to San Francisco's site, more hilly than any other major American city (fig. 26.1). The city's cable cars were a way of adapting the inappropriate street system to the terrain. San Francisco became the major banking and financial center for the Pacific Coast and the western headquarters for many national corporations. When it was rebuilt following the 1906 earthquake, San Francisco was the major urban center of western North America.

By 1906 it was no longer the most important seaport, even in California. San Francisco's peninsular location forced travelers crossing the country to detrain at Oakland and complete the last few miles of their journey by ferry boat, a practice that continued until the Oakland Bay bridge was completed in the 1950s. Oakland was the real terminus of the transcontinental railroads, and it emerged as the major seaport. Manufacturing, including automobile assembly plants operated by U.S. and Japanese car builders, were concentrated in the East Bay. Heavy industry is represented by steel mills at Pittsburg and oil refineries at Richmond.

Oakland's growth as an international port and a manufacturing center reflected a cluster of blue-collar economic specializations that were counterpart to San Francisco's white-collar financial and business communities. Oakland also attracted waves of new immigrant groups to the city, including African Americans, who began migrating from Texas to California in large numbers following World War II. The two cities thus became specialized portions of a single metropolitan economy.

FIG. 26.1. *Traditional San Francisco, with its steep hills and cable cars, lacked the tall buildings common in other large American cities.*

Urbanization also spread north and south. Marin County, north of San Francisco, grew rapidly after the Golden Gate Bridge was completed in 1937. Coastal valleys farther to the north have remained more rural, although they have developed a unique industrial base as well. Italian, German, and Swiss immigrants to the Napa and Sonoma Valleys established the foundations of California's wine industry in the 1880s. Wine grapes flourished in the Mediterranean climate. The industry expanded in the 1960s, when California wines started to become popular nationally, and has grown steadily since that time. The Napa Valley is North America's largest wine region, with 240 wineries and thirty-five thousand acres of producing vineyards crowded into a thirty-mile-long valley between Napa and Calistoga. A single fault-block ridge separates the Napa from the parallel Sonoma Valley on the west, where another 190 wineries and an equal acreage of grapevines are in production. The two valleys together market wine valued in the billions of dollars each year.

The Santa Clara Valley, which opens into the San Francisco Bay lowland on the southern end at San Jose, is topographically analogous to the Napa and Sonoma Valleys. The southern Santa Clara Valley is also a significant wine-producing area, and until the 1950s the entire valley was famous for its apricots and other fruit orchards. It is far better known today as Silicon Valley, the label it acquired after the orchards were bulldozed to make room for computer factories and tract homes. Much of the theoretical and practical research for the semiconductor and computer industries has taken place here, in a twenty-mile corridor between Palo Alto (Stanford University) and San Jose.

Electronics researchers at Stanford helped David Hewlett and William Packard commercialize the audio-oscillator in the 1930s. The Hewlett-Packard firm then joined Stanford in developing a cluster of research companies in the electronics and computer industries. The invention of the transistor by William Shockley, also on the Stanford faculty, in the 1950s was another step toward creating a practical digital computer, because it enabled electronic images to be stored in a very small space and then magnified, eliminating the inefficient vacuum tubes that had been a standard feature of all electronic instruments.

Major military contractors, such as California's Lockheed Corporation, settled in the San Jose area during the 1950s. Federal expenditures for defense-related research and development led to further growth. Semiconductors procured by the government accounted for nearly 40 percent of the industry's output into the 1960s. As the number of firms grew, Silicon Valley's manufacturing and research core expanded through the freeway-linked cluster of communities between Palo Alto and San Jose. Eighty-five manufacturing companies, producing computers, semiconductors, disk drives, printers, workstations, circuit boards, and imaging equipment, were in Silicon Valley by the end of the 1990s. Dozens more companies specialized in the production of computer software.

San Francisco's outward growth was limited by the geographical circumstances of its location. For many years it was also prevented from growing "up" because of the earthquake hazard, which limited the height of buildings. Advanced engineering techniques permitting taller buildings that would also withstand earthquake shocks were introduced in the 1960s, and since that time San Francisco's skyline has grown taller until it resembles that of most other major cities. The city itself accounts for only 15 percent of the San Francisco–Oakland–San Jose metropolitan population of 6.4 million, but it is the most important workplace in the Bay area and is connected by rapid transit to the East Bay and the southern peninsula.



## The Sierra Nevada

Although the geologic structures are not continuous, the Sierra Nevada continues the chain of high mountain peaks of Washington and Oregon's Cascade Range southward into California. Lava plateaus near Mount Shasta in northern California break the mountains, which then resume with the Lassen volcanic peaks. The Sierra Nevada is a massive fault-block mountain range, upraised along its eastern margin and tilted toward the west. The Sierra's steepest slopes are on the Nevada side (see chapter 20, fig. 20.1). On the gentler, California slope, elevations decrease from 2,700 feet at Donner Pass to a few feet above sea level at Sacramento. The distance requires nearly one hundred miles of travel, making a remarkably gentle mountain grade.

California's highest elevation is Mount Whitney (14,494 feet) at the edge of Sequoia National Park. Other high peaks are clustered in the southern portion of the Sierra, although the best-known part of the range is Yosemite National Park, near the middle. Yosemite was set aside as a national park in 1890. It is a "preservationists' park," in contrast to many other national parks, which were established by interests that sought to develop the parks as tourist attractions. Yosemite's glacially sculpted peaks and valleys were a favorite of naturalist John Muir, who, with two dozen others, founded the Sierra Club in 1892. The wilderness-preservation aesthetic was tied to Yosemite in particular and to the Sierra Nevada and northern California more generally.

The Sierra's high peaks and gentle western slopes perform another, more utilitarian function. Between 40 and 75 inches of precipitation fall on the Sierra's high, west-facing slopes each year, most of it during the winter months and mostly as snow, making an annual snowpack of hundreds of inches in a typical winter. As the snow melts gradually during the following summer, the water flows down dozens of streams that descend to the Central Valley. The valley floor receives less than 20 inches of rainfall per year, which is insufficient for producing crops, but almost any crop can be raised in its mild-winter, hot-summer climate if enough water is applied (fig. 26.2). All Californians needed to do to turn this natural system into an irrigation project was to construct storage reservoirs in the Sierra foothills to hold the water until it was needed.

## Irrigation and Agriculture in the Central Valley

Warm temperatures, adequate irrigation water, flatland, and fertile soils are combined in the Central Valley to create the largest area of intensive agriculture in the United States. The valley stretches more than four hundred miles north

FIG. 26.2. *Irrigated orange groves often occupy foothill locations for better air drainage.*

to south and averages fifty miles in width. It is a structural depression, divided into several drainage areas, that has been infilled for millions of years by the deposition of sediments from the Sierra Nevada and Coast Ranges.

The northern one-third is the Sacramento Valley, which is drained by the south-flowing Sacramento River and its tributaries. The north-flowing San Joaquin River and its tributaries, most of which rise in the central Sierra Nevada, drain the valley's largest segment. The San Joaquin's drainage basin, from Fresno to Stockton, has the greatest agricultural diversity. South of Fresno alluvial fans built by streams descending the Sierra's western slope block the northward drainage of the San Joaquin River and divert the stream flow into several basins of interior drainage. The largest is the Tulare Basin, between Fresno and Bakersfield, where surface waters drain into Tulare Lake and evaporate. A final drainage area, south of Bakersfield, is tributary to Buena Vista Lake, which also is an evaporation basin.

The confluence of the San Joaquin and Sacramento Rivers takes place in "The Delta," west of Stockton, an area known for its marsh vegetation of tule reeds and bullrushes. The marshy lowlands have been reclaimed by diking and drainage to produce additional cropland. The lower Sacramento Valley has undergone several improvements for navigation, first to Stockton, later to Sacramento, to allow ocean vessels to dock at the inland ports. The delta's complex of canals and levees has been reworked over the years to allow the transfer of water from the Sacramento side to the San Joaquin side of the valley.

Nature has provided most of the engineering required to deliver water to the Central Valley, including a continuous backdrop of high mountains to intercept eastward-moving air masses off the Pacific, but there is a geographical imbalance of water supply and demand. Water supply is most abundant in the Sacramento Valley, whereas water demand is concentrated in the San Joaquin and the basins of interior drainage to the south. The northern Sierra Nevada has fifteen to twenty inches more precipitation per year than the southern portion.

Until the 1930s, irrigation in the San Joaquin Valley diverted water from

Sierra-side streams into canals that took water not only to fields at the foot of the mountains but also to the drier western side of the valley and to the Tulare and Buena Vista Basins. The Sacramento Valley had ample irrigation water, but by the 1930s the San Joaquin was beginning to feel the limits of its smaller supply, and plans were made to begin diverting water from the Sacramento to the San Joaquin to even out the imbalance.

The main concern in the Sacramento Valley, in contrast, was protection from the seasonal floods that inundated large areas of low-lying floodplain and damaged croplands and urban settlements. Floods on the Feather River (a Sacramento tributary) in 1955 caused the water plan to be expanded as a flood-control system by constructing a large reservoir north of Oroville for water storage. The overall water-control and -supply effort became identified as the California Water Project, which has created a comprehensive system for moving water from the "surplus" area in northern California southward through the Central Valley and beyond to the Los Angeles area.

Canals take the lower Sacramento's flow across the delta, where it is pumped upgrade (to the south). The largest components of the project are the pumping stations and canals of the California Aqueduct, which transfer water upslope the entire length of the San Joaquin Valley, carry it across the Tulare and Buena Vista Basins, and then pump what remains over the low ridges of the Tehachapi Mountains and the Transverse Ranges into the Los Angeles Lowland. Irrigation waters released from the aqueduct in the San Joaquin Valley flow through a series of laterals and ditches, through orchards, vineyards, and croplands, and then drain into the San Joaquin River, which carries the agricultural wastewater back north and discharges it into the delta. The arrangement has many advantages and also some disadvantages, including the sometimes lethal concentrations of pesticides and other agricultural chemicals that accumulate in low-lying areas downstream on the San Joaquin River.

Irrigated agriculture in the Sacramento Valley begins near Redding and broadens southward to span the entire Central Valley, from the Coast Ranges to the Sierra foothills. Prunes, walnuts, almonds, and rice are produced in the northern counties. Rice performs well under flood irrigation of the extremely flat lands in the middle Sacramento Valley. California's rice crop is well situated to supply export markets in Asia through the port of Sacramento, although it has been prevented from reaching markets in countries such as Japan that seek self-sufficiency in rice production.

Thousands of new acres of nut trees, including almonds, pistachios, and English walnuts, have appeared in the Sacramento Valley in the past several decades in response to a growing domestic market for these crops. They have replaced some peach, pear, and apricot orchards, although soft fruits remain an important Sacramento Valley crop. Farther south in the Sacramento Valley, as well as in the northern San Joaquin, tomatoes are the most important crop; they are grown both for fresh consumption and for processing.

Grapes are the San Joaquin Valley's leading crop. They are produced for table use, for wine-making, and for drying as raisins. Wineries near Stockton, Mo-

FIG. 26.3. *A California dairy farm.*

desto, and Fresno produce most of the California wine that does not come from the Coast Range valleys. Southern San Joaquin Valley counties produce onions, winter potatoes, durum (for manufacturing pasta), and asparagus. Orange groves are concentrated on the Sierra slope, east of Fresno. Chickens (broilers), eggs, and turkeys are produced near Merced and Modesto.

The dairy industry of the San Joaquin Valley (plus that of suburban Los Angeles) makes California the top milk-producing state in the United States (fig. 26.3). California's dairy farms have comparatively low production costs because of the mild climate. Crops of alfalfa hay are rotated with fruits and vegetables, providing support for dairying in the Central Valley. Dairies have appeared in the Sierra foothills of the Tulare Basin as well. By the end of the 1990s California's milk production was roughly one-fourth larger than Wisconsin's.

The southern San Joaquin Valley and Tulare Basin are the principal areas of cotton production in California. The state is the nation's largest producer of long-staple (pima) cotton, which is used in making higher-quality garments. California ranks second to Texas in the production of the more general-purpose upland cotton. Both crops thrive in the hundred-degree summer temperatures of the Fresno-Bakersfield portion of the valley; they are irrigated by a combination of ditch and sprinkler methods. The displaced cotton farmers of Oklahoma and Texas brought cotton culture to the Central Valley during the Dust Bowl years of the 1930s. Although cotton production declined during the 1970s, California increased its share of the nation's crop when U.S. cotton acreage doubled between the early 1980s and late 1990s.

The Central Valley has long depended on a hired farm workforce based on immigration from Mexico. Although many Mexican-Americans still are employed as agricultural laborers, others have become farm owners and operators, especially in the San Joaquin Valley. Efforts to organize California farm workers for higher wages and improved living conditions have produced relatively high wages (at least when compared with other farm-labor regions of the United

States), although the continued mechanization of all agricultural industries has reduced the demand for labor.

Issues in Central Valley agriculture today include pricing irrigation water as well as rights of access to it. Because the previous winter's snowfall in the Sierra must meet each year's water demand, farmers can forecast water availability well in advance of the crop season. Water marketing—in which urban and agricultural users of irrigation water bid competitively—has been slow to emerge in California compared with other western states largely because of the opposition from the state's agricultural users.

California's Central Valley is an agribusiness complex that includes farms, canneries, wineries, packing plants, cold-storage warehouses, and the transportation infrastructure necessary to send its products to market. Food industries are the most important economic activity in most of the valley's cities. Nearly all of the land is used to produce crops, and much of it produces more than one type of crop in a single year. The aroma of agricultural chemicals often hangs heavy in the air above its fields. Scores of semitrailer trucks travel the rural roads, along the roadside irrigation ditches, shuttling between fields where laborers harvest the produce and processing plants where it is packaged for sale. Trainloads of fresh, bottled, canned, and frozen food products are loaded up and down the Central Valley every day of the year. Without this region the American supermarket would be an impossibility.

## The Coast Ranges

In addition to its wealth of agricultural resources, California is the third-ranked oil-producing state (behind Texas and Alaska). Major oil fields lie offshore in the Santa Barbara Channel, as well as on land. The largest oil fields are in the southern Coast Ranges (at Taft, west of Bakersfield) and in the Los Angeles Lowland. The oil-refinery complex at Richmond, on San Francisco Bay, is supplied with some crude oil from California's fields, although much of its output depends on the arrival of tanker ships from Valdez, Alaska. Despite substantial production within the state, California is a net importer of crude oil.

Northwest-southeast-trending fault lines associated with the San Andreas Fault continue the pattern of parallel mountains south from San Francisco Bay to the vicinity of Santa Barbara. Low ridges separating broad, shallow valleys support a Mediterranean oak woodland vegetation with both evergreen and deciduous species. Oak woodlands are a California vegetation type that occupy drier, midelevation slopes in a broad band encircling the Central Valley and the Coast Ranges (fig. 26.4). The open vegetation canopy allows grasses to thrive, making an excellent range that has been used for cattle ever since the Spanish arrived.

Franciscan padres founded a dozen missions in the southern Coast Range valleys between 1770 and 1804. From there they launched expeditions across the coastal mountains to explore the interior. Mexican authorities made dozens of land grants, especially in the Coast Ranges, which became the foundation

FIG. 26.4. *A California oak woodland, Maricopa County.*

of large cattle ranches that persisted into the period of Anglo settlement. The mission-based settlements were only marginally successful, however, and some of them lay in ruins by the time settlers from the eastern United States came to California in the late 1840s. Unlike New Mexico, where the early Hispano villages persisted and became the main foci of settlement thereafter, California's "mission period" was an isolated episode in the state's history.

The southern Coast Range valleys account for California's second-largest agricultural region. California leads the nation in the production of strawberries, cauliflower, avocados, artichokes, broccoli, brussels sprouts, celery, garlic, lemons, spinach, and lettuce. These crops are specialties of the southern Coast Ranges, beginning on the north with garlic and artichokes near Gilroy and Watsonville. The eighty-mile-long Salinas Valley, which opens to the Pacific Coast north of the city of Salinas, is the most intensive area of lettuce production in the United States, a distinction it has retained for many years. Broccoli, strawberries, lemons, and avocados are produced in the coastal valleys near Santa Barbara and on the Ventura-Oxnard Lowland, a level plain bordering the Pacific south of Santa Barbara (map 26.2).

Coastal mountains and valleys in southern California have one of the most salubrious climates in the United States. Much of the year has warm, sunny days and cool nights, and the climate is drier than San Francisco's. The south-flowing California Current cools summer temperatures and also helps to make the summer dry by creating low-level temperature inversions. The Mediterranean climate zone of California has most midlatitude woodland and grassland soil types. Here they carry the prefix *xer-* (Xerolls, Xeralfs, Xerults), indicating a period of soil-moisture deficiency at high sun. Natural vegetation has evolved in California's Mediterranean zone to match the annual soil-water regime. Many species of evergreen oaks and coastal chapparal have the sclerophyllous traits of hard, thick, shiny leaves, which minimize transpiration water loss during the dry, high-sun season.

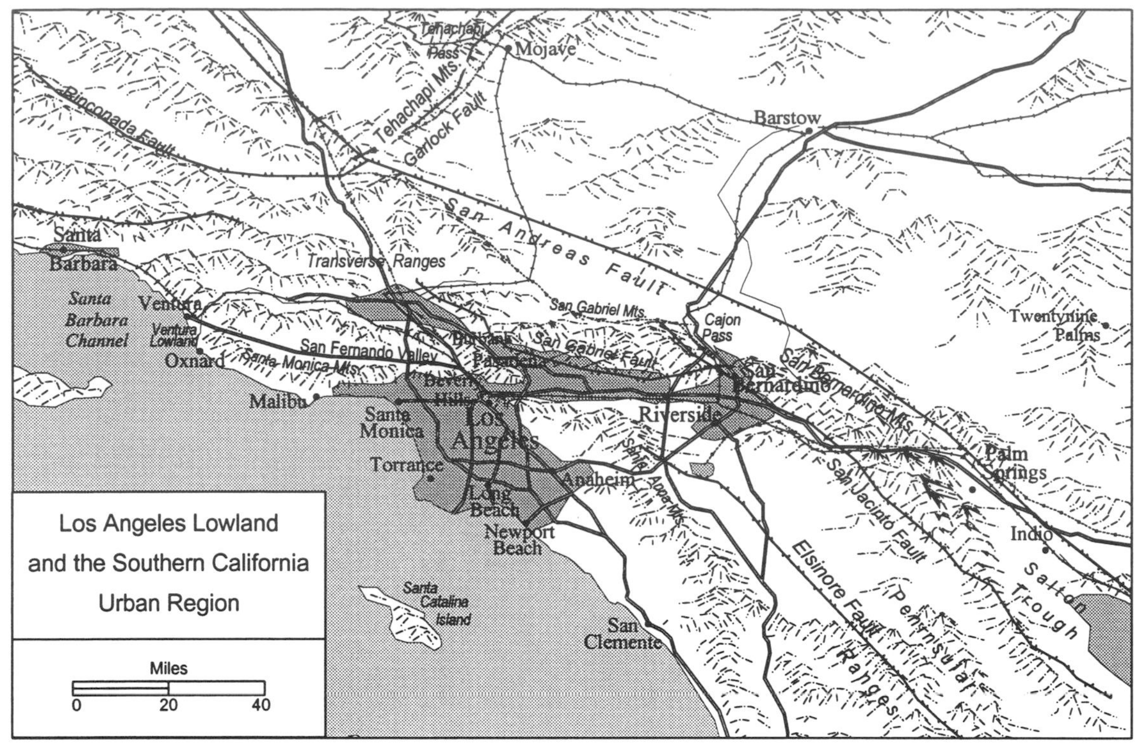


MAP 26.2

## Southern California

The Tehachapi Mountains are the divide between northern and southern California. Along their southern edge runs the 150-mile-long Garlock Fault, which separates the Sierra Nevada from the Basin and Range portion of California. The San Andreas and Garlock Faults intersect about 40 miles south of Bakersfield, where major fault zones radiate toward the northwest and the southeast. Tehachapi Pass, just east of this intersection, is a major routeway connecting the relatively high elevations of the Mojave Desert with the much lower elevations of the Central Valley at Bakersfield. The entire area is seismically active. Mountain passes here—which carry the routes of both of the interstate highways and the two major rail lines connecting northern and southern California—are especially vulnerable to earthquakes.

Southern California's mountains have varied orientations. Some parallel the San Andreas Fault, and others follow the east-west trend of the Garlock Fault. The Los Angeles Lowland is bordered by the San Gabriel and San Bernardino Ranges, which form a continuous, high-elevation fringe lying roughly thirty miles back from the coastline. Half a dozen fault planes segment the lowland into blocks subject to movement in various directions. Lower summits within the Los Angeles Lowland include the Santa Monica Mountains, which separate downtown Los Angeles from the San Fernando Valley, and the Santa Ana Mountains, between the coastal communities of Orange County and the Riverside area at the eastern end of the lowland.

South of Los Angeles the coastal mountains are called the Peninsular Ranges. They continue south beyond the Mexican border at Tijuana–San Ysidro. Although agricultural land is at a premium in this mountainous section, San Diego County is a major producer of avocados, nursery stock, and indoor decorative plants. The peaks of the Peninsular Ranges are not especially high, but the mountains are a tight barrier limiting access to the Salton Trough/Imperial Valley, which lies immediately to the east.

San Diego's mission and presidio were founded in 1769, the first of the chain of outposts established in the area that was to become part of the United States. The Mexican government designated San Diego as a pueblo (agricultural settlement) in 1834. The city was gradually redesigned around the perimeter of San Diego Bay for its anticipated role as a port, once the international boundary was confirmed and it became obvious that this would be the southernmost seaport on the Pacific Coast of the United States. This possibility stimulated the interests of politicians in the southeastern states toward the Gadsden Purchase (1857) and the construction of a transcontinental railroad that would have given the Cotton South a direct line of access to the Pacific.

No mountain pass capable of carrying a railroad line ever was found in the Peninsular Ranges north of the border, and thus San Diego was denied status as a West Coast port in the same class with Seattle or Oakland or Los Angeles. San Diego and its environs were chosen as sites of major military installations in the 20th century, and the city also prospered from its border role in trade with Mexico. The San Diego metropolitan area now contains approximately 2.5 million inhabitants and is the third largest in California.

Southern California's coastal mountains form a barrier that traps the inland movement of ocean breezes. Atmospheric stagnation is reinforced by the frequent presence of upper-atmosphere temperature inversions that block the vertical movement of air. Hazy atmospheric conditions are common, especially in the Los Angeles Lowland, which extends eastward more than eighty miles to San Bernardino. A health hazard is created when aerosol pollutants are added to the entrapped air mass, making smog. California led the nation in the 1970s by requiring that vehicles be equipped with antipollution devices. Los Angeles's notorious smog still exists, but it has abated to a significant extent compared with the 1960s.

## Los Angeles

The total population of greater Los Angeles now exceeds 15 million. It ranks a close second among U.S. metropolitan areas, with a population more than four-fifths that of New York and nearly twice Chicago's. Los Angeles was founded as a civic pueblo as part of the general effort to extend Spanish settlements northward in 1781. For the better part of the next century, it remained little more than a village. After 1869, when San Francisco was linked directly overland by the transcontinental railroad, Los Angeles became an outpost of San Francisco. Los Angeles's only railroad extended south, not east, to the docks at San Pedro, where the connection was made by coastwise shipping to San Francisco. Los Angeles remained smaller than San Francisco until the 1920s.

The first major growth of Los Angeles took place in the 1880s when it acquired direct railroad access overland to Chicago and to New Orleans. California's railroads vigorously promoted migration to the state, migration that, from the 1880s onward, was directed more toward southern California than to the San Francisco Bay area. Los Angeles, roughly in the middle of the lowland at the foot of the San Bernardino and San Gabriel Mountains, was the largest community, but by the end of the 19th century a scatter of other towns had appeared as well. Many of Los Angeles's suburbs were not the product of urban overspill but originated as separate communities, often with an agricultural base. German immigrants established vineyards and a wine industry at Anaheim beginning in the late 1850s. Pasadena began as a communal settlement. Mormons from Utah who sought to expand their domain into the Southwest founded San Bernardino.

The many independent starts of communities helps explain the extraordinarily dispersed nature of urbanization in the Los Angeles Lowland. The attractions that lured immigrants were typically more rural and agricultural than urban or industrial. California did not become a major manufacturing state until the 1940s, and given the timing and the nature of its labor pool, it developed mainly the types of clean industries that employed an educated, white-collar workforce.

Comparatively few European immigrants came to southern California, but Asians did—at least during the periods when American immigration policy allowed it. Chinese, Japanese, and Filipinos were drawn more to California than to any other state. San Francisco was the major center for the Chinese, whereas the Los Angeles area attracted the largest numbers of Japanese. Between the 1890s and 1924, when the door closed, thousands of Japanese took up residence in southern California. Even though they were forbidden to own land, many Japanese small-scale farmers settled among the scatter of villages in the Los Angeles Lowland. They introduced many cultural practices, including techniques for fruit, vegetable, and nursery-stock production and generally grew crops suited to small amounts of land in high-density environments. They were the issei (first-generation) Japanese, and most of them were raising families by the time they were forcibly relocated to internment camps during World War II.

Hispanics were relatively unimportant before the 20th century. Despite the early Spanish settlements in the state, there was little continuity with later times. Even Canadians outnumbered Mexicans in the population of Los Angeles as late as 1910. California's Hispanic "heritage" does not bear close historical scrutiny. It is more apparent than real, and it derives from two unrelated developments in the early 20th century. The transformation of the Central Valley into an irrigation empire required thousands of laborers. Mexico was the closest source of

FIG. 26.5. *Spanish colonial architecture became popular in California long after the Spanish colonial period ended. The buildings and landscaping for San Diego's Panama-California Exposition of 1915 illustrate both the architecture and the habit of importing exotic vegetation for beautification.*

cheap labor, and migration from Mexico to California was a direct result. Los Angeles acquired its Mexican-born population partly from immigration but received many thousands more people of Hispanic ancestry from migration within the state. Spanish heritage has little to do with California's large Hispanic population today, which, at more than 7.5 million, is the largest of any state's.

Nor does the widespread popularity of "mission-style" architecture in California relate directly to the distant Spanish past (fig. 26.5). It is more a product of the international expositions at San Francisco and San Diego in 1915, in which Spanish, Moorish, and Italian architectural themes were blended in a style that became almost instantly popular for houses, public buildings, and even commercial structures. Urban views of Los Angeles before that time reveal a city that was basically Middle Western in its architecture and design. It was not until the second decade of the 20th century that the mission style became dominant.

Southern California landscapes became familiar to many people because of the motion picture industry's concentration in Los Angeles. The creation of illusion, so necessary to the movie business, seems to have been accomplished easily in southern California. In few places did the distinction between "real" and "make-believe" matter so little. The stucco houses and bungalows that became a hallmark of California were imports or inventions, but they became models for the rest of the nation. The orange groves, lemon orchards, palm trees, and rose bushes that beautified California—and came to typify its landscape, even though they required copious amounts of irrigation water—also were imported from around the world. "California" became a style as well as a place, one that seems to have been held in exceptionally high regard by many Americans.

The idea of moving to California has persisted as one of the most commonly entertained notions in modern American history. Between 1870 and 1950, migration to California accounted for more than one-fourth of all net redistribution of population in the United States. It was this migration, more than any

FIG. 26.6. *Los Angeles's skyline is less famous than its freeway system. The expansion of automobile ownership enhanced the southern California growth booms that began in the 1920s. Freeways further dispersed urban growth across the Los Angeles Lowland.*

other, that tugged the center of the U.S. population southwestward at a steady pace for more than a century. In the peak year of 1923, net migration to California reached 90 per 1,000 residents of the state. By that time automobiles had become the favored means of reaching California, and they had become almost a necessity in the dispersed metropolis around Los Angeles.

Whereas the San Francisco region grew primarily from in-migration within the state, Los Angeles's growth was based primarily on the Middle West and the Upper South. The "Arkies" and "Okies" who came in the 1930s were but one component. Illinois, Ohio, Indiana, Iowa, and Missouri sent many thousands. Those who came were a mixture of city-born and small-town dwellers, farmers and laborers, who were attracted by the promise of a better life, in a more benign environment than the one they had known. In 1965 California passed New York to become the nation's most populous state. Almost simultaneously California's image began to change, away from natural attractions to overcrowding, congested freeways, smog, and high taxes.

Although Los Angeles has a central business district, it was one of the first American cities to grow in peripheral areas well away from the city center (fig. 26.6). Long Beach, its largest satellite, became Los Angeles's ocean port. Heavy industry appeared there and in the eastern end of the Los Angeles Lowland, where steel mills were constructed at Fontana during World War II. Residential properties were valued in terms of elevation and nearness to the ocean. Land favorable for residential growth occupied a broad band, from Santa Monica through Beverly Hills and Pasadena on the north and through a series of oceanfront communities that eventually extended unbroken south to the San Diego metropolitan area.

As Los Angeles and its environs grew, land once used for orchards and small farms became too valuable not to convert into residential uses. The San Fernando Valley, which was once a small-scale farming area, was transformed into residential suburbs during the 1950s. The dispersed metropolitan area stimulated the construction of the extensive freeway system with which Los Angeles became identified. Los Angeles's aerospace industry had grown up in the open spaces where airfields could be built during World War II. It was one of several industries that depended heavily on federal defense expenditures for its research and development activities. The growth of aerospace drew thousands of more migrants to the Los Angeles region during the 1950s.



The economic recession during the mid-1960s caused stagnation in the aerospace industry and led to retrenchments in the local economy. California grew slowly, although the trend was reversed in the following decade. Concerns about taxation led to the passage of popular initiatives limiting public expenditures. These and other proposals aimed at curtailing growth were enacted by a population that wished to preserve the amenities that had led to the great migrations. By the late 1990s nearly half of California's population was in the Los Angeles Lowland, which was solidly urbanized. California was the largest state, and southern California alone had as many inhabitants as New York, the third largest. Los Angeles is becoming the largest metropolitan area in the United States.

## *References*

Dasmann, Raymond F. *The Destruction of California.* New York: Macmillan, 1965.

Gordon, Margaret T. *Employment Expansion and Population Growth: The California Experience, 1900–1950.* Berkeley: University of California Press, 1954.

Lantis, David W., Rodney Steiner, and Arthur E. Karinen. *California, Land of Contrast.* 2d ed. Belmont, Calif.: Wadsworth, 1970.

McWilliams, Carey. *Southern California Country.* New York: Duell, Sloan & Pearce, 1946.

Palm, Risa. *Earthquake Insurance. A Longitudinal Study of California Homeowners.* Boulder: Westview Press, 1995.

Parsons, James J. "The Uniqueness of California." *Nation* 7 (1955): 45–55.

———. "A Geographer Looks at the San Joaquin Valley." *Geographical Review* 76 (1986): 371–89.

Rand, Christopher. *Los Angeles: The Ultimate City.* New York: Oxford University Press, 1967.

Rubin, Barbara. "A Chronology of Architecture in Los Angeles." *Annals of the Association of American Geographers* 67 (1977): 521–37.

Saxenian, A. L. *Regional Advantage: Culture and Competition in Silicon Valley and Route 128.* Cambridge: Harvard University Press, 1994.

Soja, Edward W., and Allen J. Scott, eds. *The City: Los Angeles and Urban Theory at the End of the Twentieth Century.* Berkeley: University of California Press, 1996.

Vance, James E., Jr. "California and the Search for the Ideal." *Annals of the Association of American Geographers* 62 (1972): 185–210.

## CHAPTER 27

# Hawaii

Hawaii is the most youthful American state. It was admitted to statehood only in 1959 after six decades of territorial status. The principal Hawaiian Islands are relatively youthful on the geologic time scale. They began as bumps on the sea floor and erupted to become a chain of formidable volcanic islands entirely within the past 5 million years. Hawaii was not occupied by people until approximately A.D. 500, giving the state one of the shortest records of human settlement of any place on earth.

## The Native Hawaiians

Despite these indicators of recency, Hawaii is an old land when viewed through the cultural traditions of its native inhabitants. Native Hawaiians are descended from Polynesian islanders who took long sea voyages into the vastness of the South Pacific beginning more than two thousand years ago. From a base in Tonga and Samoa, which was the origin of Polynesian languages and material-culture traits, they sailed in double-hull canoes to discover group after group of reef-fringed islands in the open ocean. From Samoa they sailed east and discovered the Society Islands (Tahiti). From there they reached the Marquesas and eventually came upon remote Easter Island in the South Pacific. Although the Polynesians' discoveries of new islands were accidental, their understanding of ocean currents and celestial navigation allowed them to repeat the voyages, linking the islands they discovered. Between 300 B.C. and A.D. 700, voyagers from the Marquesas Islands sailed north of the equator and discovered Hawaii.

No written records are available, but an understanding of the process of discovery and certain aspects of its dating is possible from interpretations of native Hawaiian oral traditions. The period of long-distance voyaging to Hawaii lasted for more than a thousand years. Migrants from Tahiti later outnumbered the early arrivals from the Marquesas. The population grew slowly, and the voyages appear to have ceased during the 15th century. By 1565 the Spanish were sending one or more galleons from the Philippines to Mexico every year as part of their colonial system of worldwide trade. For nearly two hundred years the Manila galleons must have passed near Hawaii, although Spanish accounts do not mention the islands, nor is there evidence that Spanish traders ever landed there.

Europe's recorded discovery of Hawaii came in 1778 when Captain James

Cook, aboard HMS *Resolution* and accompanied by HMS *Discovery*, happened upon the southern shores of Kauai while on a voyage seeking a northern sea route from the Pacific to the Atlantic. Cook's account leaves little doubt that the inhabitants of Kauai had never before encountered Europeans—or their weapons, clothing, or ships. Cook named his discovery the Sandwich Islands after his patron, the earl of Sandwich, lord of the British Admiralty. Native fascination with Cook and his party in 1778 turned to hostility when he returned the next year, this time landing on the "big island," Hawaii. A brief quarrel with natives on the beach led to Cook's murder by those who had turned out to greet him.

The ugly mood that Cook found upon his return may have reflected native awareness that his men had brought a terrible disease to the islands. Native Hawaiian population numbers, estimated between 200,000 and 300,000 in 1778, declined steadily after European contact. One problem was the spread of venereal disease, which Cook tried to prevent by ordering his men to have no contact with native women. His orders, as well as those of hundreds of ship captains who followed, were ignored. Prenatal infections, birth defects, and sterility resulting from venereal disease prevented normal population replacement through natural increase. Other diseases, including leprosy (apparently introduced from China) and smallpox (European), took their toll as well. By 1870 the islands contained fewer than 100,000 inhabitants.

Considered as a population descended entirely from the Polynesian voyagers, native Hawaiians were almost extinct by the late 20th century; but through intermarriage with haoles (outsiders), their numbers increased. After the minimum was reached in 1870, the population grew steadily, mainly from the immigration of Asians and Americans. Hawaii's current population of 1.2 million is composed of roughly equal numbers of people of white (Caucasian), Japanese, part-Hawaiian native, and mixed ancestry.

Native Hawaiians traditionally settled in favorable valleys near the coast. They raised taro (a root crop, used for making poi, the traditional staple food) in small, irrigated plots (fig. 27.1). They derived a rich subsistence from the sea

FIG. 27.1. *Taro fields in the Hanalei Valley, Kauai.*

as well. After 1900, when American institutions began to be introduced to the islands, various attempts were made to improve native living conditions. In 1920 the U.S. government passed a special version of the homestead law, the Hawaiian Home Lands program, to turn aboriginal Hawaiians into commercial farmers. The law applied only to natives, but small farms were unable to compete with the large plantations that had been established by that time and the attempt at land reform produced few lasting results.

## The Hawaiian Environment

Contemporary theories about plate tectonics were inspired in part by observations about the Hawaiian Islands. The source of magma for Hawaii's volcanoes is believed to be deep below the solid, lithospheric portion of the earth. The origin of the molten rock is fixed relative to the mantle on which the lithosphere rests. As the Pacific Plate moved past the magma-producing "hot spot," volcanoes erupted on the surface, but they were carried away, to the north and west, as volcanic activity continued at its fixed position (map 27.1). The theory is confirmed in the fact that the westernmost Hawaiian Islands are the oldest and most

MAP 27.1

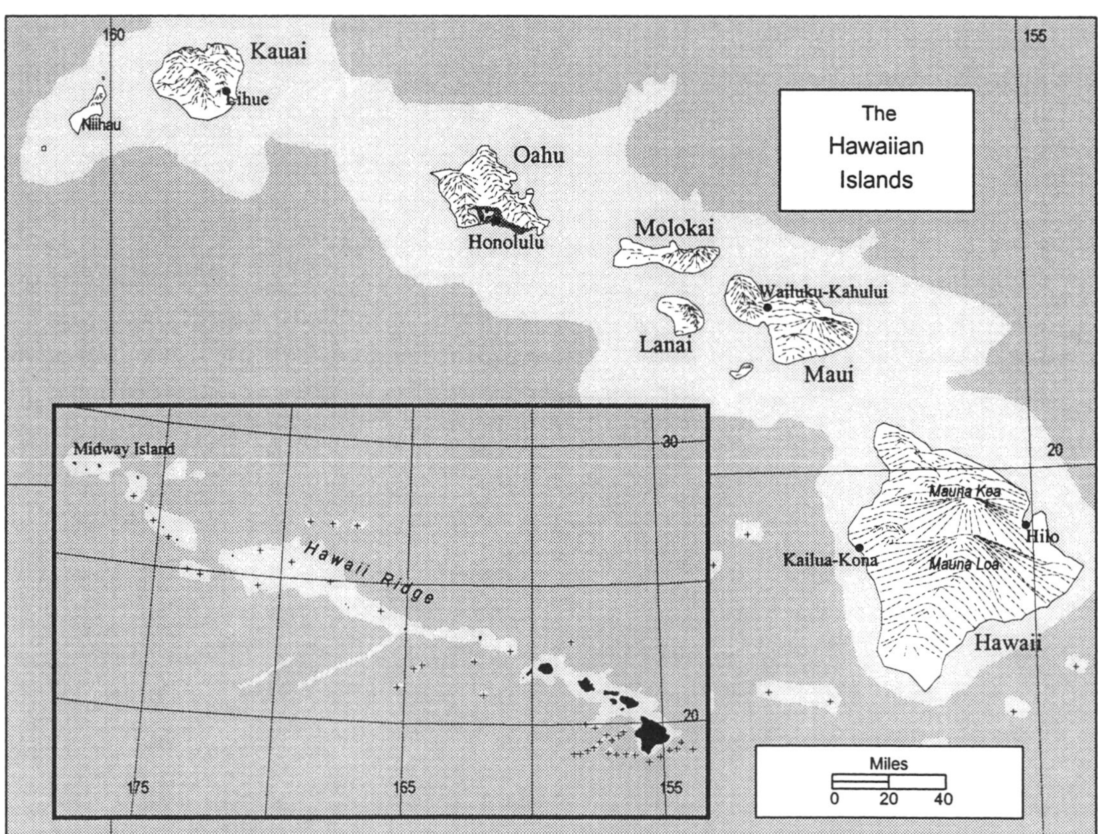

eroded, whereas those to the east are the youngest and smoothest. In millions of years of age, the islands range from Kauai (5.1), through Oahu (3.7), Molokai (1.8), and Maui (0.8) to Hawaii (0.4). A new seamount, about twenty miles southwest of Hawaii, is the next island-in-the-making, although it has yet to emerge above sea level as a volcanic peak. West of Kauai, in contrast, small islands of the Hawaiian Ridge increase in age (and decrease in size) with origin dates exceeding 30 million years near Midway Island. A single theory thus explains the linear pattern of the islands as well as their relative age and size.



All of Hawaii's rocks, except the fringing reefs of marine origin, are volcanic. Hawaii thus has no metallic minerals, no coal, and no oil or natural gas. Its topography is developed on the sloping sides of volcanic cones that would be the largest mountains on earth if their great mass below sea level were included in reckoning their sizes. Mauna Kea and Mauna Loa, on the island of Hawaii, have summit elevations of nearly 14,000 feet, although the mountains extend an even greater distance below sea level, making them taller than Mount Everest and many times more massive.

Hawaii lies entirely within the tropical northeasterly trade-wind zone, which produces wet conditions on the islands' windward (northeast-facing) slopes and leeward rain shadows on the southwest. The high peaks on Kauai are some of the wettest environments on earth. Large portions of the island of Hawaii and sections of central Oahu and Maui have a grassy or shrubby vegetation rather than dense forest, although the extent of human influence on the plant cover is not known.

Hawaii is home to many thousands of species of plants, birds, insects, and small mammals, some eighty-eight hundred of which are found nowhere else on earth. Included within this number are half of the species classified as endangered in the United States at present. As a small catchment in the streams of oceanic currents, drifts, and countercurrents, the Hawaiian Islands intercepted the seeds of plants and became a haven for migrating birds and sea creatures. The *hala* (pandanus) tree is but one species that found a niche on Hawaii's beaches after its seeds floated in from some other island.

Biogeographic theory predicts that the number of species on an island will reach an equilibrium, in balance between the rate of chance arrival of new species and the rate at which existing species go extinct. Given Hawaii's remoteness, its large number of species suggests a very long period of accumulation (and speciation) accompanied by fairly low rates of extinction, a scenario that is consistent with the fact that people have lived there for a relatively short period of time. Extinction rates have increased rapidly during the past several decades of widespread habitat alteration on the islands.

Before the arrival of Europeans, Hawaii had no large mammals that might have been hunted, and thus native Hawaiians had little incentive to manipulate their surroundings to enhance the habitats of such animals. In contrast to nearly all inhabited areas of North America, Hawaiian environments seem to have been only slightly enhanced by their aboriginal occupants (fig. 27.2). Apart from the infrequent typhoons and tsunamis that reshaped coastal zones by provid-

FIG. 27.2. *Huleia Stream, Kauai. Hawaii's coastal valleys were among the first areas to receive human settlement. The lagoon at the lower right, known as the Menehune Fish Pond, is regarded by some as having been the work of Hawaii's earliest Polynesian inhabitants.*

ing fresh surfaces for colonizing species, environmental change was slow until recent times.

## Nation, State, and Annexation

One figure, King Kamehameha I, who ruled Hawaii from 1795 until his death in 1819, looms larger than any other in Hawaiian history. As a young chief from the island of Hawaii, Kamehameha had witnessed Cook's expedition. He developed a fascination with—and a desire for—the weapons that the British carried, which were totally unknown in the islands at that time. Victorious in a series of power struggles, Kamehameha first acquired control of the island of Hawaii. He skillfully bargained local trade goods for British arms and equipped both an army and a navy to extend his domain leeward. His forces overwhelmed those of the local chiefs on Maui, Molokai, and Oahu, and he controlled all of the major islands except Kauai by 1795. The king's authority gradually absorbed the westernmost islands in succeeding years.

Kamehameha turned a tribal society of islands into a nation-state. Once the warfare he unleashed had subsided, Kamehameha presided over a tranquil Hawaiian society. Lesser royalty (*alii*), both male and female, maintained authority on the individual islands. Europeans understood that Hawaii had changed dramatically in the few years following Cook's visit: they would have to recognize that the islands were a monarchy, ruled by a powerful king.

After Kamehameha's death in 1819, each successor to the throne in the Kamehameha dynasty seems to have had slightly less success in dealing with outsiders than his or her predecessor. The first American missionaries came to the Sandwich Islands the year after Kamehameha I died (fig. 27.3). They were New England Congregationalists, pious descendants of the Puritans, who imposed their views about marriage, redemption, adultery, dancing, nudity, monogamy, the Sabbath, alcohol, and a host of other issues on a Hawaiian population that, before that time, was relaxed about such matters. Hawaii became an almost stereotypical case of missionary-versus-native confrontation.

The missionaries' attempts to proselytize the natives had limited success, but their very presence created an entering wedge for American influence. Although they frowned on interracial marriages, some missionaries, and their sons and daughters, married native Hawaiians, especially the local royalty. Because Hawaii was a monarchy, where the Crown owned all lands not specifically assigned, the involvement of Americans with those who held title to vast acreages of Hawaiian land produced a small group of very wealthy landowners whose control was out of all proportion to their numbers.



A succession of outsiders, mostly British and American, assumed duties as principal government functionaries under the monarchy. They became the agents for Europeans or Americans to contact when doing business with the Hawaiian government. Members of the Hawaiian royalty were more inclined to seek recognition from the king or queen of Great Britain than from the United States. Although Great Britain had concluded several treaties that acknowledged its rights in the Sandwich Islands by virtue of Cook's discovery, British influence declined steadily as trade with the United States grew.

The Hawaiian sugar industry first began in opposition to the interests of California sugar refiners, who desired a high tariff to prevent foreign sugar from flooding the American market. In 1876 the refiners reversed their position and concluded a reciprocity treaty allowing unmanufactured Hawaiian sugar to enter the United States duty free; the tariff simultaneously erased all duties on American goods entering Hawaii. When sugar planters began to show an interest in Hawaii during the 1870s, ownership of large blocks of land became crucial. The rush to assemble large acreages for sugar plantations on Hawaii, Maui,

FIG. 27.3. *A New England–style church steeple is visible over the waterfront at Kailua Kona on the island of Hawaii. The first Christian missionaries from New England landed here in 1820.*

Oahu, and Kauai began at once, even though this was a rather late date in world history to begin a new plantation industry.

Labor had to be obtained, because there were too few native Hawaiians for an adequate labor force. Slavery was no longer an option, and the importation of Africans (which would have been the strategy one or two generations earlier) was not feasible. The solution was to turn to the well-populated countries of Asia, especially Japan, China, and the Philippines, where labor could be recruited easily. It was at this time, and to secure a labor force on the sugar plantations, that Asian migration to Hawaii began on a large scale.

By the late 1880s the Hawaiian economy depended almost entirely on the sugar industry plus a few other crops, such as pineapples, associated with tropical plantation agriculture. Sanford B. Dole, a missionary's son and a sugar planter, whose family name would be known for its connection to the pineapple industry, was a prominent politician, one of a group of Americans who had grown increasingly vocal in their criticism of the monarchy by the late 1880s. Dole also was a justice of the Supreme Court of Hawaii, the body charged with maintaining the country's constitution. Changes in the constitution in 1887 had limited the powers of the monarch and had imposed property or nationality restrictions on voting that were designed to dilute the rights of Hawaii's dark-skinned majority (which grew larger each year through the importation of Asian laborers) and to enhance those of the white-skinned, property-owning minority.

In 1891 the sugar planters suffered two setbacks. The passage of the McKinley Act, removing the tariff on all raw sugar entering the United States, meant that Hawaii's crop would no longer enjoy special access to the American market. Also in that year Queen Liliuokalani succeeded her brother to the throne. Although she was the last monarch of the Kamehameha dynasty, she did more than any other ruler since Kamehameha I to assert Hawaiian nationalism. In 1893 she proclaimed a new constitution that would have restored the monarchy's powers and simultaneously returned the franchise to her native Hawaiian subjects.

The planters decided to seek protection under the laws of the United States, simultaneously erasing the tariff problems and the unpredictability of Hawaiian national politics. U.S. president Benjamin Harrison sent word to those plotting Liliuokalani's overthrow that his administration would be sympathetic to the idea of annexing Hawaii. A Committee of Safety, with Dole as its president, seized control of the government in January 1893 with little bloodshed. Dole's group then approached the U.S. government. The Harrison administration had played no direct role in the coup, although John L. Stevens, the U.S. minister to Hawaii, had advised the revolutionaries in their actions and American troops from the USS *Boston* were on hand to observe activities.

The timing proved to be embarrassing. Harrison was still in office, but Grover Cleveland had defeated him in the preceding November's election. Cleveland, who opposed the action Dole's group had taken and appears to have favored returning Liliuokalani to power, sent a special envoy to Honolulu to depose all parties. The envoy was shocked by the queen's response to questions concerning what she would do to the insurgents if returned to power: she is alleged to have replied that she would have Dole and the others beheaded. President Cleveland's support for Liliuokalani vanished after he received the envoy's report, and he took no further action.



Without the welcome acceptance they had planned to receive from the United States—and given Liliuokalani's views as to their future—Dole's group had little choice but to proclaim an independent Hawaiian republic. Although the much more sympathetic William McKinley succeeded Cleveland as president in 1897, the gap still was not closed. The new Hawaiian government's insistence on entering the United States with its de jure system of race-based voter qualifications made it unacceptable to the Congress, because it violated the U.S. Constitution (even though a similar, but de facto, disfranchisement operated in the American South). Others argued that Hawaii was too unlike the United States ever to be absorbed as a part of it.

The matter might have lingered in Congress for years had it not been for Admiral George Dewey's decisive victory over the Spanish naval fleet at Manila in 1898. The subsequent American purchase of the Philippines and Guam suddenly made Hawaii a strategic way station on the transpacific route. The United States assumed control over Hawaii that year, and it became a territory in 1900.

## Economic Change

Outsiders, primarily Americans, gained control of large blocks of land through intermarriage with members of the Hawaiian royalty or through their business dealings with the government during the latter years of the monarchy. Some land trusts included entire islands—such as Lanai, which operated for many years as a single pineapple plantation. Large grazing properties, including the Parker Ranch, which still controls the largest tract of land on the island of Hawaii, also came out of the land trusts. Perhaps the greatest beneficiary was the sugar industry, which was organized around the substantial acreages needed to sustain commercial levels of production.

The largest two dozen private landholders controlled almost half of the land in Hawaii until recent times, including nearly all of the good farmland in the state. Ownership of the compresses (sugar mills), workers' villages, docks, and steamship companies serving the sugar industry was linked to the land trusts. The sugar camps had an Asian labor force that was largely Japanese but included smaller numbers of Chinese. The federal government passed exclusion laws, designed to prevent Asian immigration to the United States, during the 1920s. Agricultural migrants from the Philippines were employed in greater numbers after that time.

The first generation of Japanese (issei) were forbidden to become naturalized citizens under U.S. law, but their children (nisei) automatically became citizens because they were born in the United States. To some extent, the legalized racism of the late monarchy-republic and early territorial years was self-erasing. The nisei generation came into adulthood in the World War II era, and by the 1960s they had become one of the most influential groups in Hawaii, both politically and economically.

The high points of the sugar and pineapple industries came before World War II. The plantation labor force dropped from 50,000 in 1930 to approximately 15,000 in 1960. The decline was due in part to mechanization but also to lower production costs for pineapples in the Philippines and Indonesia. The result was a drastic reduction in Hawaii's pineapple acreage. Population growth leading to urban sprawl drove up land prices in the plantation areas, making the land too valuable to use for agriculture. Central Oahu has a few remaining pineapple fields, although much of the former acreage has disappeared into the expanding zones of suburban tract housing (fig. 27.4).

Sugar cane remains Hawaii's major crop, and it is produced on all of the larger islands, although its acreage has been greatly reduced by the same factors that have caused pineapples to decline. As a crop, sugar cane has a prodigious thirst, requiring eight feet (ninety-six inches) of water per year in Hawaii's temperature regime. On most of Hawaii's cane fields, only a little more than half of that total is produced by precipitation; the rest must come from irrigation. Land trusts controlling the sugar crop typically have long-established rights of access to irrigation water, but they are now in conflict over water rights with the demands of a growing, dispersed, nonagricultural population.

Between 1950 and 1970 Hawaii was transformed from an agricultural base to a diversified economy focused on construction, the service industries, and tourism. Construction booms lasted for years and were punctuated by only short periods of slow growth. As late as 1950, only 40,000 visitors came to Hawaii each year by ship and airplane. By the early 1970s that number had increased to 1.5 million. Jet air travel made Hawaii accessible as a resort for Japanese and Koreans as much as for Americans and Canadians. By the late 1990s, more than 7 million visitors were coming to Hawaii every year, although many of the "visi-

FIG. 27.4. *Pineapple fields near Wahiawa in the Central Oahu Plain.*

FIG. 27.5. *Honolulu's pattern of high-rise apartments and office towers is a hybrid of Asian and North American urban styles.*

tors" were property owners as well, who came to the islands to spend time in the thousands of resort condominiums that were constructed during the 1970s and 1980s booms.

Honolulu, whose urban area now contains more than three-quarters of a million people, has been Hawaii's capital since 1850. It has been both the leading port and the largest city of the islands for more than a century. The city has grown as the coalescence of three formerly separate urban nodes. Farthest west is Pearl Harbor, which the United States began leasing from the Hawaiian government as a naval base and coaling station in 1887. The Japanese attack on Pearl Harbor on December 7, 1941, destroyed Pearl Harbor and also began a period of martial law that was to remain in force in Hawaii for the duration of the war.

The traditional urban core of Honolulu lies to the east of Pearl Harbor and includes the old Chinatown neighborhood and other ethnic enclaves as well as the government district, which still is centered around the palace grounds of Hawaii's monarchy. Flanking downtown Honolulu on the east is Waikiki Beach, itself now a high-density neighborhood (fig. 27.5). The once-swampy zone behind Waikiki's beachfront boomed in the 1960s from hotel construction. Since then it has developed into a specialty retail district catering to the world's wealthiest shoppers.

Honolulu is a Western city by design and construction, although it is more Asian in its population composition and pace of activity. It is by far the largest city on any of the Pacific islands, the most important banking center specializing in East-West transactions, and a leading center for all types of trade between Southeast Asia and the United States. Honolulu's sprawling growth has long since encircled the island of Oahu and spread through its central lowland. Only Oahu's two diagonal mountain spines, on the east and on the west, remain largely unaffected by Honolulu's expansion.

To many people Hawaii is simply a tropical, mid-Pacific complex of hotels, beaches, golf courses, and shopping centers, tied through frequent direct air service to all of the world's major cities. Most who visit Hawaii see little else. The "old" Hawaiian landscape—the sugar plantation, the isolated farm, the rural church, the beachfront village—are found in a few places, but little thought was given to preserving the islands' history when the rush to construct resorts and condominiums began three or four decades ago. Apart from the palaces of kings, little of the past can be seen. An active undercurrent of political activity in Hawaii seeks to regain access to this past, in part so that the native Hawaiian people's true role can be understood and appreciated.

During the past several decades, the renewed appreciation of native Hawaiian culture, including the rediscovery of folk practices and religious beliefs, has accompanied the resurgence of Hawaiian native issues, which now play an important role in the state's political economy. Such efforts may restrain the pace of development before all of the islands have been transformed in a way that forever obliterates their history.

## *References*

Daws, Gavan. *Shoal of Time: A History of the Hawaiian Islands.* New York: Macmillan, 1968.

Department of Geography, University of Hawaii. *Atlas of Hawaii.* Honolulu: University of Hawaii Press, 1973.

Horwitz, Robert H., and Norman Meller. *Land and Politics in Hawaii.* 3d ed. Honolulu: University of Hawaii Press, 1973.

Kane, Herb Kawainui. *Voyagers.* Bellevue, Wash.: WhaleSong, 1991.

Nordyke, Eleanor C. *The Peopling of Hawai'i.* 2d ed. Honolulu: University of Hawaii Press, 1989.

Wright, Theon. *The Disenchanted Isles: The Story of the Second Revolution in Hawaii.* New York: Dial Press, 1972.

# Index

Page numbers in italics refer to maps.

## About the Author

John C. Hudson was born in Milton, Wisconsin, in 1941. He was educated at the University of Wisconsin–Madison (B.S., geography) and the University of Iowa (M.A. and Ph.D., geography). He has been a John Simon Guggenheim Memorial Foundation Fellow, and he served as editor of the *Annals of the Association of American Geographers* from 1976 through 1981. He is the author of *Geographical Diffusion Theory* (1972), *Plains Country Towns* (1985), which was awarded the John Brinckerhoff Jackson Prize, and *Making the Corn Belt* (1994). He is also editor of *Goode's World Atlas* (2000). He is a professor of geography at Northwestern University in Evanston, Illinois.